CHEMICAL ANALYSIS

(*continued on back*)

The Analytical Chemistry
of Silicones

CHEMICAL ANALYSIS

A SERIES OF MONOGRAPHS ON
ANALYTICAL CHEMISTRY AND ITS APPLICATIONS

VOLUME 112

A WILEY-INTERSCIENCE PUBLICATION

JOHN WILEY & SONS, INC.

New York / **Chichester** / **Brisbane** / **Toronto** / **Singapore**

The Analytical Chemistry of Silicones

Edited by

A. Lee Smith

Analytical Research Department
Dow Corning Corp.
Midland, Michigan

A WILEY-INTERSCIENCE PUBLICATION

JOHN WILEY & SONS, INC.

New York / Chichester / Brisbane / Toronto / Singapore

In recognition of the importance of preserving what has been written, it is a policy of John Wiley & Sons, Inc., to have books of enduring value published in the United States printed on acid-free paper, and we exert our best efforts to that end.

Library of Congress Cataloging in Publication Data:

The Analytical chemistry of silicones / edited by A. Lee Smith.
    p. cm.—(Chemical analysis, ISSN 0069-2883; v. 112)
    "A Wiley-Interscience publication."
    Includes bibliographical references and index.
    ISBN 0-471-51624-4
    1. Organosilicon compounds—Analysis. I. Smith, A. Lee (Albert Lee), 1924– . II. Series.
    QD412.S6A474 1991
    547'.08046—dc20   90-39514
                                                                   CIP

Printed in the United States of America

10 9 8 7 6 5 4

CONTRIBUTORS

N. C. Angelotti, Dow Corning Corp., Midland, Michigan

M. A. Becker, Dow Corning Corp., Midland, Michigan

R. L. Durall, Dow Corning Corp., Midland, Michigan

D. K. Fillmore, Dow Corning Corp., Midland, Michigan

Ora L. Flaningam, Dow Corning Corp., Midland, Michigan

H. A. Freeman, Dow Corning Corp., Midland, Michigan

M. D. Gaul, Dow Corning Corp., Midland, Michigan

Helen M. Klimisch, Dow Corning Corp., Midland, Michigan

Neal R. Langley, Dow Corning Corp., Midland, Michigan

E. D. Lipp, Dow Corning Corp., Midland, Michigan

Nelson W. Lytle, Dow Corning Corp., Midland, Michigan

John A. Moore, Dow Corning Corp., Midland, Michigan

Michael J. Owen, Dow Corning Corp., Midland, Michigan

B. Parbhoo, Dow Corning Corp., Midland, Michigan

R. D. Parker, Dow Corning Corp., Barry, Glamorgan, South Wales, UK

D. R. Peterson, Dow Corning Corp., Midland, Michigan

A. Lee Smith, Scientist Emeritus, Analytical Research Department, Dow Corning Corp., Midland, Michigan

R. D. Steinmeyer, Dow Corning Corp., Carrollton, Kentucky

R. B. Taylor, Dow Corning Corp., Midland, Michigan

PREFACE

Since the publication of our book "Analysis of Silicones"* in 1974, analytical chemistry has developed at a pace that could not even have been imagined at that time. Innovations such as Fourier transform spectroscopy, near-infrared spectroscopy, a variety of surface analysis techniques, chemometrics, and the pervasive use of computers have largely been developed during this period. The literature has likewise mushroomed. For example, a literature search on infrared spectra of silicones turned up over 4000 references for the 15-year period; a search for NMR spectra of silicones gave over twice that many references. Clearly much of the technology described in the 1974 book is outdated, and it seemed an appropriate time to attempt to capture the essence of the new analytical technologies.

Also during this period, uses for organosilicon materials have expanded dramatically, so that they have become almost ubiquitous in application, although often invisibly. Their importance as interface modifiers (release agents, coupling agents, antifoams, and the like) has grown; organofunctional silicones are used in new applications that rely on their unique physical and chemical properties. Even the older well-established polymers are finding new uses in new product concepts. Familiarity with organosilicon materials will become more and more important for chemists and engineers in the years to come. Quality has also become an important issue; clearly, excellent quality requires excellence in analytical technology.

This book is directed to technical personnel who have at least a nodding acquaintance with analytical chemistry. The current trend is for technologists in whatever field to do most of their own analytical chemistry, with only the most complex and sophisticated techniques left to the analytical specialists. We anticipate our readership will include technical personnel such as chemical engineers or research chemists who may have only limited experience with either analytical chemistry or silicones; analytical generalists in small laboratories who have broad experience but no detailed knowledge of silicone chemistry; and analytical specialists who are experts in their fields but

---

* A. L. Smith, *Analysis of Silicones*, Wiley, New York, 1974.

again have little knowledge of silicone chemistry. With this audience in mind, we have included some basic material on silicone chemistry and good analytical practices. Also included are references to basic texts and papers that can provide additional background, if needed, in the more specialized analytical techniques and in organosilicon chemistry and technology.

We have attempted to stress principles and general approaches to problems, and cite examples of some more or less typical approaches to problems of current interest. Although this book is not a literature review, we have tried to cite key references that the reader may find useful. Inevitably, however, we will have missed some important ones; time constraints limit any attempts at comprehensive coverage of the literature, even through abstracts. We have included titles as part of the references, as the title seems to us to be an essential part of the citation.

In order to control the length of this volume, we touch only lightly on the inorganic aspects of silicon chemistry such as semiconductor and ceramics technologies.

We are indebted to the Dow Corning Technical Information Service, who cheerfully provided numerous literature searches as well as abstracts and copies of papers. Without their help the effort involved would have been overwhelming. We also thank the Dow Corning Corporation and C. A. Roth, Manager of Analytical Research, for support and encouragement.

A. LEE SMITH

Midland, Michigan
January 1991

CONTENTS

ix

**CHAPTER 3. ANALYSIS OF POLYMERS, MIXTURES, AND
COMPOSITIONS**   47

N. C. Angelotti

CHAPTER 9.   MICROSCOPICAL CHARACTERIZATION . . .   219
H. A. Freeman and R. L. Durall

CHAPTER 10.   CHROMATOGRAPHIC METHODS   255
R. D. Steinmeyer and M. A. Becker

The Analytical Chemistry
of Silicones

PART

1

INTRODUCTION

CHAPTER

1

INTRODUCTION TO SILICONES

A. LEE SMITH

Analytical Research Department
Dow Corning Corporation
Midland, Michigan

When you can measure what you are speaking about and express it in numbers, you know something about it; but when you cannot express it in numbers, your knowledge is of a meagre and unsatisfactory kind; it may be the beginning of knowledge, but you have scarcely in your thoughts advanced to the state of science, whatever the matter may be.

WILLIAM THOMPSON, LORD KELVIN

1. INTRODUCTION

The importance of sound analytical technology in research, development, and manufacturing cannot be overemphasized. Analysis may be as simple as taking a temperature or density reading or as complex as a multinuclear nuclear magnetic resonance (NMR) study, but without it, the chemist has no control over the product or process. Analytical chemistry is truly the foundation on which all chemical research and manufacturing rests.

Organosilicon chemistry lies between organic and inorganic chemistry and draws liberally from both fields. The analytical chemistry of organosilicon compounds is also interdisciplinary and rapidly evolving. Thus, it is important that the essence of the best current knowledge from other fields be constantly distilled into a viable organosilicon analytical technology. The purpose of this book is to provide a source of reliable procedures for the analysis of organosilicon materials.

In this chapter, we provide some introductory material to those workers who may not be familiar with the chemistry and nomenclature of organosil-

The Analytical Chemistry of Silicones, Edited by A. Lee Smith.
ISBN 0-471-51624-4 © 1991 John Wiley & Sons, Inc.

icon compounds. Subsequent chapters (2–6) provide a perspective on analytical approaches to problems specific to materials and products. These discussions are followed by chapters (7–15) on specific techniques, with special emphasis on the unique aspects of their application to organosilicon materials.

2. NOMENCLATURE

The nomenclature of organosilicon monomers is basically simple. Naming the more complex heteroatom structures systematically is likely to be an exercise in ingenuity for both the writer and the reader. In such cases trivial names often serve well for all uses except indexing, and structural diagrams are indispensable; the structure is often simpler than the name. The term "silicone" was coined by Wöhler in 1857 to describe compounds having the empirical formula R_2SiO, by analogy to organic ketones of formula R_2CO. Although the anticipated ketone structure $R_2Si=O$ did not in fact exist, the name "silicone" persisted and is applied generally to organosilicon materials.

The existing nomenclature rules for organosilicon compounds conflict and are not entirely satisfactory. The following system, used by Eaborn and Bott (1), is suggested as being the least confusing and easiest to use in practice.

The following order of precedence is used for naming substituents on silicon:

1. Functional organic substituents.
2. Organic substituents.
3. Functional substituents.
Within each grouping alphabetical order applies.

The root compound in organosilicon chemistry is *silane*, SiH_4. Silicon-containing monomers are named as silane derivatives, for example: dichlorosilane (H_2SiCl_2), tetrabromosilane ($SiBr_4$), and ethylmethylphenylsilane (EtMePhSiH). Note that the silanic hydrogen is not named explicitly, but it is assumed that any bonds left over after substituent groups are named are hydrogenated. Accordingly, $Me(EtO)_2SiH$ is methyldiethoxysilane, $PhSiHCl_2$ is phenyldichlorosilane, and $MeSiPh_3$ is methyltriphenylsilane. Other species are named similarly: $ClCH_2SiHMe_2$ is (chloromethyl)dimethylsilane, $ClCH_2MeHSiCl$ is chloromethylmethylchlorosilane, and $Cl(CH_2)_3SiEtMe(OEt)$ is 3-chloropropylethylmethylethoxysilane. Polysilanes are named *disilane* (H_3SiSiH_3), *trisilane* ($H_3SiSiH_2SiH_3$), and so on.

Chains are numbered as in organic chemistry, starting as usual from a key atom and considering the longest chain as the parent compound.

Polymers consisting of alternating silicon and oxygen atoms are called siloxanes. The prefix designates the number of silicon atoms in the molecules:

$H_3SiOSiH_3$

1
Disiloxane

$H_3SiOSiOSiH_3$ (with H above and below central Si)

2
Trisiloxane

Cyclotrisiloxane structure with H_2Si, O, SiH_2, O, SiH_2, O

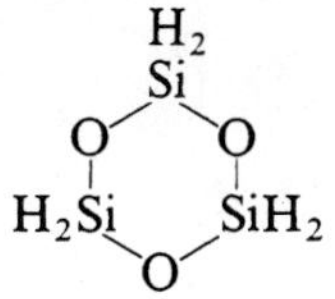

3
Cyclotrisiloxane

Substituted siloxanes are named in the same manner as silanes:

$Me_3SiOSiMe_3$

4
Hexamethyldisiloxane

$HCl_2SiOSiCl_2H$

5
1,1,3,3-Tetrachlorodisiloxane

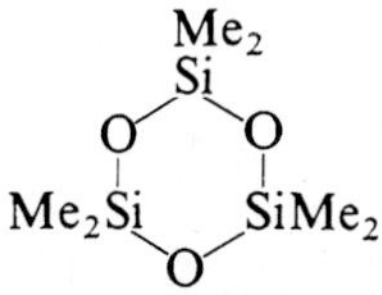

Cyclic structure with Me_2Si, O, $SiMe_2$, O, $SiMe_2$, O

6
Hexamethylcyclotrisiloxane

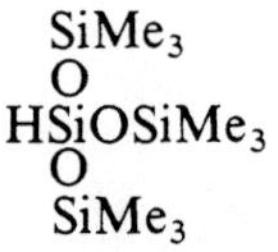

$SiMe_3$
O
$HSiOSiMe_3$
O
$SiMe_3$

7
Tris(trimethylsiloxy)silane

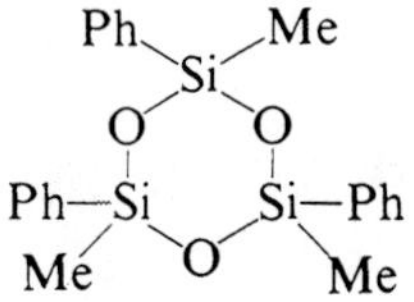

Cyclic structure with Ph, Me groups on Si atoms

8
2,4,6-Trimethyl-2,4,6-triphenyl-cyclotrisiloxane

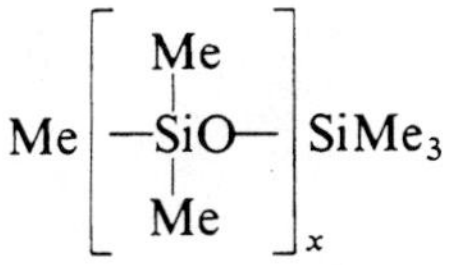

$Me\left[\begin{array}{c} Me \\ | \\ -SiO- \\ | \\ Me \end{array}\right]_x SiMe_3$

9
Poly(dimethylsiloxane)

Monosubstituted silicon–oxygen polymers are generally called silsesquioxanes; trivial names may be more convenient in everyday usage than International Union of Pure and Applied Chemistry (IUPAC) names.

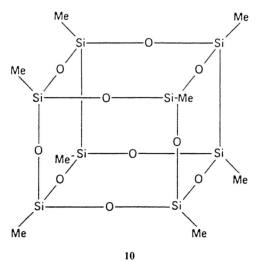

10

Octamethyl-*octaprismo*-octasilasesquioxane (IUPAC)
Methylsilsesquioxane, cubical octamer (trivial)

In a manner analogous to the siloxanes, polymers of alternating silicon and nitrogen are called silazanes.

$$H_3SiNSiH_3$$
(with H above N)

11
Disilazane

$$MeSi—N—SiMe$$
(with H, Me, H above and H, H below)

12
1,2,3-Trimethyldisilazane

$$Me_2Si \cdots SiMe_2$$
(cyclotrisilazane ring: N-H at top position 1, Si positions 6 and 2, HN position 5 and NH position 3, Si Me_2 position 4)

13
2,2,4,4,6,6,-Hexamethylcyclotrisilazane

The so-called silicates $(RO)_4Si$, where R = Me, Et, *i*-Pr and so on, or silicate esters of silicic acid are not, strictly speaking, organosilicon compounds. Nevertheless, such species are important commercially and the general chemistry of the functional OR group does not depend on whether or not one or more Si–C bonds also happen to be present in the molecule. Thus we shall make no further distinction in this book between organosilicon compounds and organic silicates. Both types of structures are correctly designated organic compounds of silicon.

The silicon analogs to organic alcohols are called silanols.

Me_3SiOH

14

Trimethylsilanol

$Ph_2Si(OH)_2$

15

Diphenylsilanediol

Metal salts of the silanols are termed silanolates.

$NaOSiMe_3$

16

Sodium trimethylsilanolate

Sulfur derivatives are designated silthianes

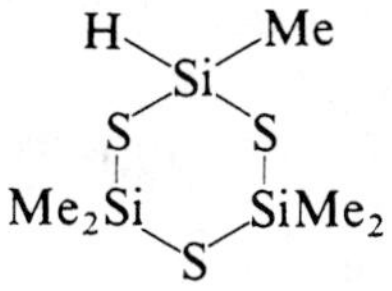

17

Pentamethylcyclotrisilthiane

and silanethiols

Me_3SiSH

18

Trimethylsilanethiol

$Me_3SiSC(CH_3)_3$

19

Trimethyl(*tert*-butylthio)silane

Complex structures containing two or more heteroatoms are named (as a last resort) using the oxa-aza-thia-sila convention.

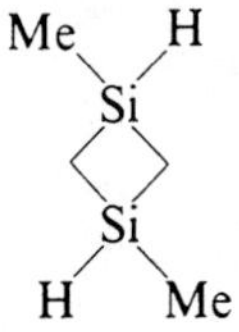

20

Trans-1,3-dimethyl-
1,3-disilacyclobutane

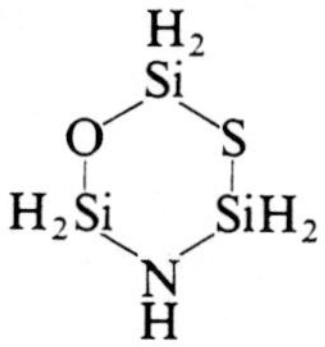

21

1-Aza-2,4,6-trisila-
3-thia-5-oxa-cyclohexane

If the structure contains an amine or suffix designated functional group, the functional moiety is sometimes treated as the parent compound and the silicon-containing part as a radical.

$$Me_3SiCH_2CH_2OH \qquad\qquad H_3SiNHMe$$

2-Trimethylsilylethanol N-Methylsilylamine

Other substituents are named similarly: H_3SiNH- (silylamino), Me_3SiS- (trimethylsilylthio), and H_3SiO- (siloxy).

The nomenclature systems used today are based on recommendations by Sauer (2); those recommendations were subsequently developed and adopted by the American Chemical Society (ACS) (3), the IUPAC (4) and the Chemical Society (5). An updating by the IUPAC (4) resulted in only minor changes in the earlier recommendations. In practical usage, it seems reasonable that the primary rule should be that the name is unambiguous and descriptive. Beyond those requirements, reasonable flexibility is evident in the application of nomenclature rules, and rigid formalism is often subjugated to the cause of smooth reading or brevity.

A convenient shorthand notation is sometimes used to designate the degree of functionality of the silicon atom as shown in Figure 1.1. Thus, structure **6** can be written D_3; structure **4** as M_2, structure **9** as MD_nM, and structure **10** as T_8. The system is sometimes modified by the use of superscript letters designating other substituents, for example, $D^H = (MeHSiO)$ and $M^{Ph} = (Me_2PhSiO_{1/2})$.

Me	Me	Me	$O_{1/2}$
$MeSiO_{1/2}$	$O_{1/2}SiO_{1/2}$	$O_{1/2}SiO_{1/2}$	$O_{1/2}SiO_{1/2}$
Me	Me	$O_{1/2}$	$O_{1/2}$
M (Monofunctional)	D (Difunctional)	T (Trifunctional)	Q (Quadrifunctional)

Figure 1.1. Shorthand notation for siloxane polymer units.

3. PROPERTIES

The most common organosilicon materials are those composed mainly of dimethyl-, phenyl-, and methyl-substituted polysiloxanes, perhaps combined with organofunctional substituents such as vinyl, aminopropyl, or hexenyl groups. These polymers appear in many forms (fluids, resins, and elastomers), but all have the same basic silicon–oxygen–silicon backbone. Silicones are uniquely man-made materials—no case of a naturally occurring organosilicon compound has ever been convincingly demonstrated. The properties of these hybrid molecules are also unique, and it is their unusual properties that account for the utilization of these materials in today's technology.

Compared with conventional organic polymers, many silicones show superior thermal properties. Some silicone elastomers, for example, remain flexible at $-100\,°C$ and retain their properties for long periods at $200\,°C$.

Most of the common silicone polymers are quite hydrophobic but are permeable to water vapor and to other gases. Their resistance to ultraviolet (UV) and other radiation means that they weather well—paints based on silicone resins have extremely long life. They are good electrical insulators, and they are frequently used for wire coatings, encapsulants, and potting compounds for electronic components and electrical applications.

The low surface tension of poly(dimethylsiloxane) (PDMS) fluids (~ 20 dyne/cm; cf. benzene, 28.9 and water, 72) permits them to spread easily over irregular surfaces. They are often used as additives in polishes and waxes to promote easier application and to give additional water repellency to the surface. Poly(dimethylsiloxane) fluids are not generally good lubricants for steel surfaces, although this property can be improved by additives and by changing from methyl to other alkyl substituents on the silicon.

Higher molecular weight PDMS exhibits minimal biological activity; it is essentially inert chemically in the body. For this reason, silicone polymers are successfully used in medicine for implants and therapeutic agents (silicones in medicine are discussed in Chapter 4).

Some silicones are excellent release agents; even the most aggressive organic adhesives do not adhere to them. Silicone mold release agents are commonly used for rubber and plastic items, and silicone-treated papers are used for self-adhering label backing. On the other hand, another type of silicone polymer is used as a pressure-sensitive adhesive.

Most silicones are not particularly solvent resistant; however, elastomers made from siloxanes substituted with 3,3,3-trifluoropropyl groups show good resistance to jet fuel and hydrocarbon solvents, but not to more polar solvents such as ketones and esters.

The use of organosilicon monomers as primers for the glass fibers used in composites and laminates provides an excellent illustration of the dual chemical nature of these materials. In such applications, monomers such as $(MeO)_3Si(CH_2)_3NH_2$ are applied to the glass. The methoxy groups react with the surface water and silanol groups on the glass to form SiOH and SiOSi bonds that link the silicon to the glass surface. The organofunctional end of the molecule reacts with the organic resin, and the silicon compound effectively "glues" the resin to the glass fiber.

Just a few of the many applications in which silicones are used are listed in Table 1.1, along with references to reviews that provide further information.

Table 1.1. Some Uses of Silicones

Use	References
Adhesives	6
Caulking material	6–8
Coatings	9
Cosmetics	10
Coupling agents	11
Drugs and pharmaceuticals	12
Drug delivery, transdermal	13
Electrical insulation	8
Gas chromatography (GC) column substrates	14, 15
Heat transfer fluids	16
Lubricants	16
Organic synthesis aids	17
Medical prostheses and devices	18
Pharmaceutical manufacturing	19
Polishes	20
Pressure sensitive adhesives	21
Printing inks	9
Silylation of organics for GC analysis	22
Surfactants and foaming agents	23, 24
Textile finishes	9, 25

4. CHEMISTRY OF THE SILICONES

In most applications, organosilicon compounds are used in the form of polymers; the properties of the polymer depend on the molecular size and shape, on the amount of cross-linking, and on the nature of the substituents on the silicon. The great diversity of products obtained from a relatively few starting monomers may best be understood by considering the chemistry of the monomers and their many possible combinations. The industrial preparation of silicones begins with elemental silicon at a purity of 98–99%. The silicon is ground to a fine powder and induced to react with methyl chloride (with a catalyst of copper powder) at 250–300 °C in what is called a "direct process" to give several important chlorosilane intermediates.

$$Si + MeCl \longrightarrow Me_2SiCl_2 + Me_3SiCl + SiCl_4 + HSiCl_3$$
$$+ MeHSiCl_2 + \cdots \tag{1.1}$$

The monomers are separated by careful fractional distillation. To preclude cross-linking sites in the polymers, it is essential in some cases that no more than a few parts per million (ppm) of higher functionality monomers be present in the difunctional monomers. The fractionation must be very efficient, for the boiling point spread is not large.

The purified chlorosilane monomers are then hydrolyzed, individually or after being recombined in the proper proportions, to yield hydrogen chloride and siloxane polymers.

$$\begin{array}{cc} \text{Me} & \text{Me} \\ \text{Me}_2\text{SiCl}_2 + \text{H}_2\text{O} \longrightarrow \text{HO(SiO)}_x\text{H} + \text{(SiO)}_{3,4,5} + \text{HCl} \\ \text{Me} & \text{Me} \end{array} \qquad (1.2)$$

Although the most common silicon substituent is the methyl group, phenyl substitution is also important; other organic substituents are also used. Vinyl, 3,3,3-trifluoropropyl, and 3-chloropropyl substituents as well as hydrogen, are found in many commercial products, and the list of specialized functionalities is constantly growing. (Most of the monomers containing these groups are not generated in the direct process, but the substituents are attached to silicon by other means.)

The polymer resulting from the hydrolysis may be a fluid, crystalline solid, gel, gum, or resin, depending on the average functionality of the monomers.

$$R_n\text{SiCl}_{4-n} + \text{H}_2\text{O} \longrightarrow \{R_n\text{SiO}_{(4-n)/2}\}_x \qquad (1.3)$$

If $n > 2$, the product is a fluid; if $n = 2$, the product is a cyclic or linear siloxane (a gum, if a high molecular weight siloxane); if $n < 2$, the product is a resin.

It may be instructive at this point to discuss briefly some of the chemical properties of silicon-functional groups, noting in particular those that differ from their carbon analogs.

Si–Halogen. The halogen bonds in SiCl, SiBr, and SiI compounds hydrolyze extremely rapidly with liberation of the halogen acid, as indicated in Eq. (1.2). The reaction is reversible, but the equilbrium lies far to the right; the equilibrium constant for the reaction 1.2 is $K = 10^{14}$ (26). The equilbrium for hydrolysis of the SiF bond favors the halogenated silane.

Silanols. Examples of silanols, $R_3\text{SiOH}$, silanediols, $R_2\text{Si(OH)}_2$, and silanetriols, $R\text{Si(OH)}_3$, are all known. The SiOH group presents a most interesting example of a single moiety showing a wide range of properties depending on the molecule in which it is found. In dimethylsilanediol, $\text{Me}_2\text{Si(OH)}_2$, the SiOH group is so sensitive that traces of atmospheric acids or residual alkali on glass surfaces promote its condensation, whereas $(t\text{-Bu})_2\text{Si(OH)}_2$ can be heated with concentrated H_2SO_4 without change. This fact makes the analysis for the SiOH group extremely difficult; it is not easy to cope with such a wide range of reactivities by any single analytical method. The stabilities of silanols lie in the order $R_3SiOH > R_2Si(OH)_2 > RSi(OH)_3$ and, within any homologous series of R groups, bulkier ones give more stability to the SiOH function.

SiH. The SiH group is a rather good reducing agent. For example, it fogs photographic film by reducing the silver halide to silver. The SiH group reacts readily with alkali to liberate hydrogen. Although silane itself (SiH_4) reacts explosively with air, and a few substituted silanes ($RSiH_3$ and R_2SiH_2) are unstable, most substituted silanes are air stable under ordinary conditions. In the presence of Lewis acids, however, disproportionation may occur.

$$2R_2SiH_2 \xrightarrow{\text{AlCl}_3} R_4Si + H_4Si \qquad (1.4)$$

The possibility of such a reaction occurring should be kept in mind when work with silanes is undertaken.

An extremely useful reaction is synthetic organosilicon chemistry is the platinum-catalyzed addition of SiH to an unsaturated organic group to give the corresponding alkyl derivative. The reaction is catalyzed by minute amounts of platinum; a great variety of organofunctional silicon derivatives can be prepared by this means.

SiOC. The SiOC compounds superficially resemble their organic analogs, but in the presence of water the bond slowly hydrolyzes to give the alcohol, SiOH and SiOSi polymers. If acids or alkalies are present, the hydrolysis rate may be much accelerated.

SiN. Silylamines, silylamides, and silazanes can all react with water, especially if acids present, to liberate the corresponding organic nitrogen compound (or ammonia). In the absence of moisture, however, the SiN bond has good thermal stability.

Organofunctional Silicon Compounds. Although the silicon atom still has some electronic impact on the third or fourth carbon distant, the analytical chemistry of an organofunctional silicon compound may not be much different from that of the corresponding silicon-free structure. The "organic" chemistry may be somewhat different, however; for example, alkyl groups substituted in the β position by electronegative groups are unstable; they are quickly eliminated to form unsaturated aliphatic compounds.

$$\text{>SiCH}_2\text{CH}_2\text{Cl} \longrightarrow \text{>SiCl} + CH_2{=}CH_2 \qquad (1.5)$$

Siloxanes. Although it is remarkably stable to thermal and oxidative degradation, the SiOSi bond is subject to chemical attack and subsequent rearrangement to form siloxane structures of varying molecular size. Eventually, a thermodynamic equilibrium of polymer species is established. At temperatures up to 200 °C the equilibrium mixture for PDMS contains ~ 88% linear and 12% cyclic siloxanes. If this equilibration is carried out under conditions such that the cyclics are allowed to volatilize, the polymer will eventually disappear. Because of the amphoteric nature of silicon, both strong acids and strong alkalies catalyze the rearrangement.

SiC. Redistribution or disproportionation of SiC bonds can occur in the presence of Lewis acids, Lewis basis, or free radicals, for example,

$$R_2SiCl_2 \underset{}{\overset{AlCl_3}{\rightleftharpoons}} RSiCl_3 + R_3SiCl \tag{1.6}$$

This reaction is useful for synthesis; it also may be a source of frustration for the chemist who attempts to distil a mixture of chlorosilanes containing appreciable quantities of $AlCl_3$.

For the reader with an interest in specific aspects of organosilicon chemistry such as structural or stereochemistry, synthetic applications, reaction mechanisms, photochemistry, polysilanes, siloxane polymers, and thermochemistry, the review volume edited by Patai and Rappoport (27) provides much useful information. The book by Noll (28) covers both the chemistry and technology of the silicones. Ranney (8, 9) discusses technology as disclosed in the patent literature. Other more focused reviews are available that cover both technology and chemistry, and some of these are cited in the chapter references of this book.

5. THE PHYSICAL CHEMISTRY OF ORGANOSILICON COMPOUNDS

In order to better understand the chemical and spectroscopic properties of organosilicon compounds, we review briefly some of the fundamentals relating to their molecular architecture.

Silicon is not carbon. This simple fact is obvious, yet the relationship of these two Group IV elements in the periodic table stimulates comparisons (29); one intuitively expects similarities and finds both similarities and profound differences. We have already noted how the chemical properties of SiH–, SiCl–, and SiN-containing compounds differ from those of their organic analogs.

One similarity for which many chemists have searched in vain is the double bond with silicon. Not until 1976 was a stable silene demonstrated.

Since that time, a number of silicon structures containing double bonds have been prepared and characterized. Some of these are stable at room temperature (in the absence of oxygen). They are named as follows:

Silene	Disilene	Silanimine	Silanone	Silanephos-phimine	Silaaromatic
$\underset{/}{\overset{\backslash}{\text{Si}}}=\overset{/}{\underset{\backslash}{\text{C}}}$	$\underset{/}{\overset{\backslash}{\text{Si}}}=\overset{/}{\underset{\backslash}{\text{Si}}}$	$\underset{/}{\overset{\backslash}{\text{Si}}}=\text{N}-$	$\underset{/}{\overset{\backslash}{\text{Si}}}=\text{O}$	$\underset{/}{\overset{\backslash}{\text{Si}}}=\text{P}-$	(aromatic ring with Si)

The preparation, chemistry, and analysis of these materials has been reviewed by Raabe and Michl (30, 31).

Some of the relevant fundamental properties of the Group IV elements are shown in Table 1.2. Of these properties, values for the atomic radius, electronegativity, and relative electron density clearly do not fit a smooth trend line. Bond lengths between silicon and certain elements (especially oxygen and fluorine) are abnormally short, even when corrected for the electronegativity difference of the atoms (32). The SiOSi bond angle in siloxanes (130–150°) is much larger than the COC bond angle in aliphatic ethers (105–115°), although both elements use sp^3 bonding. Trimethylamine, $(CH_3)_3N$, like ammonia, is pyramidal; the corresponding silicon compound, $(H_3Si)_3N$, is planar (33, 34). Also, the latter does not form a hydrochloride derivative, which can only mean that the nitrogen lone pair is not available for bonding to the halogen acid. Silanols should be weaker acids than carbinols, because silicon (being more electropositive than carbon) should release electrons to oxygen more freely. The reverse is in fact observed (35). A comparison of the bonding of carbon and silicon is shown in Figure 1.2. Silicon, in contrast to carbon, easily forms five- and six-coordinate bonds to electronegative elements.

Table 1.2. Fundamental Properties of the Group IV Elements

Property	Carbon	Silicon	Germanium	Tin	Lead
Atomic number	6	14	32	50	82
Atomic weight	12.01	28.09	72.59	118.69	207.19
Atomic radius	0.772	1.176	1.225	1.405	1.53
Electronegativity	2.5	1.8	1.8	1.8	1.8
ΔH atomization, kcal	171	108	90	72	47
Relative electron density: atomic number/ (atomic radius)^3	13.0	8.6	17.4	18.0	22.9

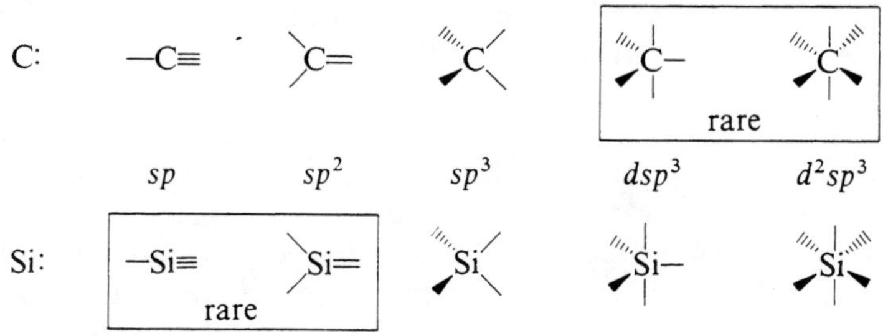

C: $-C\equiv$ $\diagdown C=$ $\diagup C\diagdown$ $\diagup C-$ $\diagup C\diagdown$

rare

sp sp^2 sp^3 dsp^3 d^2sp^3

Si: $-Si\equiv$ $\diagdown Si=$ $\diagup Si\diagdown$ $\diagup Si-$ $\diagup Si\diagdown$

rare

When carbon is replaced by silicon

C–C	1.54 Å
$\overset{\delta^-}{C}-\overset{\delta^-}{Si}$	1.90 Å
Si–Si	2.34 Å

Figure 1.2. Comparison of bonding in carbon and silicon (29). From *The Chemistry of Organic Silicon Compounds*, S. Patai and Z. Rappoport, Eds., © 1989, John Wiley & Sons, Ltd.

In 1954, Craig et al. (36) published a theoretical paper postulating that second-row elements could use their empty d-orbitals to form $(d-p)\pi$ dative bonds. Stone and Seyferth (37) in 1955 utilized this concept to explain bond shortening in siloxanes, the stereochemistry of certain bonds, and the anomalous acidity and basicity of silanols. The argument involved back-donation of lone-pair electrons from the oxygen to form bonds of the $d-\pi$ type, which exist concurrently with normal σ bonds. If the lone-pair electrons were thus engaged, they would not fully occupy the nonbonding oxygen orbitals, and the SiOSi bond angle would tend toward 180° (sp hybrid bond); the oxygen would be less basic than expected, and the SiO bond distance would be shorter than predicted (38). A similar explanation was given for $(H_3Si)_3N$; the nitrogen lone pair is presumably delocalized to give three partial double bonds of the $(d-p)\pi$ type between silicon and nitrogen.

Conclusions from subsequent work, however, including improved molecular orbital (MO) calculations (39–41) and experimental photoelectron spectra (42) have tended to minimize the importance of $(d-p)\pi$ bonding in silicon compounds. The effects noted can be explained by formation of ionic bonds of the type $R3Si^+X^-$, with d orbitals playing only a minor role. This viewpoint is not universally shared, however, as of this writing. More extended discussions of this topic may be found elsewhere (29, 43). Use of d orbitals for the formation of five- and six-coordinate structures is, however, very probable (44).

Polysilanes show an unexpected pseudoaromaticity, with structures of the type $R_3Si(R_2Si)_nR$ giving strong UV absorption above 200 nm (45). These structures are photochemically active (45, 46). Their unusual behavior is attributed to delocalization of the σ electrons in the Si–Si bond. The chemistry and properties of polysilanes have been reviewed (47–49).

The silicon atom is rather sensitive to inductive effects, that is, changes in bond hybridization induced by electronegative substituents. The chemistry of attached groups may therefore change with the nature of the other substituents on silicon, and correlations of reactivity with inductive parameters have been developed (50, 51).

Steric effects are also important, and the stability of $(t\text{-Bu})_2Si(OH)_2$ as contrasted with the reactivity of $Me_2Si(OH)_2$ is attributed largely to steric factors. Bulky mesityl substituents are used to stabilize some disilenes (31, 52).

Using the concepts mentioned above, we can rationalize many of the physical and chemical properties of the silicones. Bonds to oxygen and nitrogen have a large ionic component, and thus are less rigidly directional than sp^3 hybrid bonds to carbon, a fact that is consistent with the flexibility of the siloxane chain. Because of the less rigid bonding structure and the larger size of the silicon atom, silicon compounds occupy a larger molecular volume than do the corresponding organic compounds. The intermolecular (van der Waals) forces are lower. These facts offer a partial explanation for the low viscosity–temperature coefficient of the PDMS fluids and the desirable low-temperature properties of fluids and elastomers. The low surface tension contributes to the water repellent and spreading properties of silicone fluids, as well as to their release properties and antifoaming characteristics. The thermal and oxidative stability is related to the lack of vulnerable double bonds and the inherent stability of the SiO linkage.

REFERENCES

1. C. Eaborn and R. W. Bott, in *Organometallic Compounds of the Group IV Elements*, Vol. 1, Part 2, A. G. MacDiarmid, Ed., Marcel Dekker, New York, 1968.

2. R. O. Sauer, "Nomenclature of organosilicon compounds," *J. Chem. Educ.*, **21**, 303 (1944).

3. American Chemical Society Nomenclature, Spelling and Pronunciation Committee, *Chem. Eng. News*, **30**, 4517 (1952).

4. IUPAC, *Comp. Rend. 15th Conference*, 127–132 (1949); *Information Bulletin No. 31*, 1973, p. 87.

5. Editorial Report on Nomenclature, *J. Chem. Soc.*, **1952**, 5064.

6. W. C. Wake, "Silicone adhesives, sealants, and coupling agents," *Crit. Rep. Appl. Chem.*, **16**, 89 (1987). *Chem. Abstr.*, **108**: 7276b (1988).

7. B. Knop, "Chemistry of caulking compositions," *Kunststoffe*, **76**, 783 (1986). Through *Chem. Abstr.*, **105**: 228532x (1986).

8. M. W. Ranney, *Chemical Technology Review No. 91. Silicones, Vol. 1. Rubber, Electrical Molding Resins and Functional Fluids.* Noyes Data Corp., Park Ridge, NJ, 1977. *Chem. Abstr.*, **89**: 180776b (1978).

9. M. W. Ranney, *Chemical Technology Review No. 92. Silicones, Vol. 2. Coatings, Printing inks, Cellular plastics, Textiles and Consumer products.* Noyes Data Corp., Park Ridge, NJ, 1977. *Chem. Abstr.*, **89**: 181303g (1978).

10. S. R. Wendel, "Use of silicones in cosmetics and toiletries," *Parfums, Cosmet. Aromes*, **59**, 67 (1984). *Chem. Abstr.*, **102**: 84237t (1985).

11. E. P. Plueddemann, *Silane Coupling Agents*, Plenum Press, New York, 1982.

12. R. Tacke and B. Becker, "Sila substitution of drugs and biotransformation of organosilicon compounds," *Main Group Met. Chem.*, **10**, 169 (1987). *Chem. Abstr.*, **108**: 48551f (1988).

13. F. S. Rankin, "The use of silicones for the controlled release of drugs," *Chim. Oggi*, **1986** (10), 37.

14. J. K. Haken, "Developments in polysiloxane stationary phases in gas chromatography," *J. Chromatogr.*, **300**, 1 (1984).

15. L. Blomberg, J. Buijten, K. Markides, and T. Wannman, "Some aspects of current techniques for the preparation of capillary columns for gas chromatography," *J. Chromatogr.*, **279**, 9 (1983).

16. P. Lonsky, "Silicones as lubricants and heat transfer fluids," *Tribol. Schmierungstech.*, **34**, 87 (1987). Through *Chem. Abstr.*, **107**: 42657c (1987).

17. E. W. Colvin, *Silicon in Organic Synthesis*, Butterworths, London, 1981.

18. M. B. Habal, "The biologic basis for the clinical application of the silicones," *Arch. Surg. (Chicago)*, **119**, 843 (1984).

19. C. E. Creamer, "Organosilicon chemistry and its application in the manufacture of pharmaceuticals," *Pharm. Technol.*, **6**, 79 (1982).

20. R. J. Thimineur, "The chemistry of silicones in polishes," *Chem. Times Trends*, **4**, 40 (1981).

21. D. F. Merrill, "Silicone pressure-sensitive adhesives," *Natl. SAMPE Symp. Exhib. (Proc.)*, **30**, 340 (1985).

22. C. F. Poole, "Recent advances in the silylation of organic compounds for gas chromatography," in *Handbook Deriv. Chromatogr.*, K. Blau and G. S. King, Eds., Heyden, London, 1978, pp. 152–200. *Chem. Abstr.*, **89**: 156937j (1978).

23. G. Sonnek, "Organosilicon sufactants," *Naturwissenschaften, Tech*, **1986**, p. 153. *Chem. Abstr.*, **107**: 136346c (1987).

24. B. Gruening and G. Koerner, "Silicone surfactants," *Tenside, Surfactants, Deterg.*, **26**, 312 (1989). *Chem. Abstr.*, **111**: 235617f (1989).

25. V. A. Shenai, "Chemistry of silicones," *Text. Dyer Printer*, **19**, 21 (1986).

26. J. F. Hyde, P. L. Brown, and A. L. Smith, "Inductive effects in the chlorosilane hydrolysis equilibrium," *J. Am. Chem. Soc.*, **82**, 5854 (1960).

27. S. Patai and Z. Rappoport, Eds., *The Chemistry of Organic Silicon Compounds*, Wiley, New York, 1989.

28. W. Noll, *Chemistry and Technology of Silicones*, Academic, New York, 1968.

29. J. Y. Corey, "Historical overview and comparison of silicon with carbon," in *The Chemistry of Organic Silicon Compounds*, S. Patai and Z. Rappoport, Eds., Chapter 1, Wiley, New York, 1989.

30. G. Raabe and J. Michl, "Multiple bonds to silicon," in *The Chemistry of Organic Silicon Compounds*, S. Patai and Z. Rappoport, Eds., Chapter 17, Wiley, New York, 1989.

31. G. Raabe and J. Michl, "Multiple bonding to silicon," *Chem. Rev.*, **85**, 419 (1985).

32. V. Schomaker and D. P. Stevenson, "Some revisions of the covalent radii and the additivity rule for the lengths of partially ionic single covalent bonds," *J. Am. Chem. Soc.*, **63**, 37 (1941).

33. H. Buerger, U. Goetze, and W. Sawodny, "Vibrational spectra and force constants of silyl and trimethylsilyl compounds of Group V elements," *Spectrochem. Acta*, **26A**, 671 (1970).

34. K. Hedberg, "The molecular structure of trisilylamine $(SiH_3)_3N$," *J. Am. Chem. Soc.*, **77**, 6491 (1955).

35. L. Allred, E. G. Rochow, and F. G. A. Stone, "The nature of the silicon–oxygen bond," *J. Inorg. Nucl. Chem.*, **2**, 416 (1956).

36. D. P. Craig, A. Maccoll, R. S. Nyholm, L. E. Orgel, and L. E. Sutton, "Chemical bonds involving *d*-orbitals. Part I," *J. Chem. Soc.*, **1954**, 332.

37. F. G. A. Stone and D. Seyferth, "The chemistry of silicon involving probable use of *d*-type orbitals," *J. Inorg. Nucl. Chem.*, **1**, 112 (1955).

38. D. W. J. Cruickshank, "1077. The role of 3*d*-orbitals in π-bonds between (a) silicon, phosphorus, sulphur, or chlorine and (b) oxygen or nitrogen," *J. Chem. Soc.*, **1961**, 5486.

39. B. T. Luke, J. A. Pople, M.-B. Krogh-Jespersen, Y. Apeloig, J. Chandrasekhar, and P. v. R. Schleyer, "A theoretical survey of singly bonded silicon compounds. Comparison of the structures and bond energies of silyl and methyl derivatives," *J. Am. Chem. Soc.*, **108**, 260 (1986).

40. D. Schomburg, A. Blaschette, and E. Wieland, "Polysulfonylamines. VIII. A molecule with an extremely long silicon–nitrogen bond. Solid state structures of *N*-(trimethylsilyl)- and *N*-methyldimesylamine," *Z. Naturforsch., B: Anorg. Chem.*, **41B**, 1112 (1986).

41. B. Rempfer, H. Oberhammer, and N. Auner, "The $(p–d)\pi$ bonding in fluorosilanes? Gas-phase structures of $(CH_3)_{4-n}SiF_n$ with $n = 1$–3 and of $(tert\text{-}Bu)_2SiF_2$," *J. Am. Chem. Soc.*, **108**, 3893 (1986).

42. H. Bock and B. Solouki, "Photoelectron spectra of silicon compounds," in *The Chemistry of Organic Silicon Compounds*, S. Patai and Z. Rappoport, Eds., Chapter 9, Wiley, New York, 1989.

43. W. S. Sheldrick, "Structural chemistry of organic silicon compounds," in *The Chemistry of Organic Silicon Compounds*, S. Patai and Z. Rappoport, Eds., Chapter 3, Wiley, New York, 1989.

44. S. N. Tandura, M. G. Voronkov, and N. V. Alekseev, "Molecular and electronic structure of penta- and hexacoordinate silicon compounds," *Top. Curr. Chem.*, **131** (Struct. Chem. Boron Silicon), 99 (1986). *Chem. Abstr.*, **105**: 85312n (1986).

45. M. Ishikawa and M. Kumada, "Photochemistry of organopolysilanes," *Adv. Organomet. Chem.*, **19**, 51 (1981).

46. A. G. Brook, "The photochemistry of organosilicon compounds," in *The Chemistry of Organic Silicon Compounds*, S. Patai and Z. Rappoport, Eds., Chapter 15, Wiley, New York, 1989.

47. R. West, "Polysilanes," in *The Chemistry of Organic Silicon Compounds*, S. Patai and Z. Rappoport, Eds., Chapter 19, Wiley, New York, 1989.

48. E. F. Hengge, "Preparation and properties of chlorooligosilanes," *Rev. Inorg. Chem.*, **2**, 139 (1980).

49. R. West, Chapter 9.4 in *Comprehensive Organometallic Chemistry*, E. Abel, Ed., Pergamon Press, Oxford, England, 1982.

50. O. W. Steward and O. R. Pierce, "The effect of polar substituents on the alkali-catalyzed hydrolysis of triorganosilanes," *J. Am. Chem. Soc.*, **83**, 1916 (1961).

51. V. P. Mileshkevich and N. F. Novikova, "Use of Taft dependences for correlations of properties of organosilicon compounds. Role of effects of substituents," *Usp. Khem. (Progress in Chemistry)*, **50**, 85 (1981). Through *Chem. Abstr.*, **94**: 120339e (1981).

52. R. West, M. J. Fink, and J. Michl, "Tetramesityldisilene, a stable compound containing a silicon–silicon double bond," *Science*, **214**, 1343 (1981).

PART

2

THE PROBLEM-ORIENTED APPROACH

CHAPTER

2

THE ANALYTICAL CHEMISTRY OF ORGANOSILICON MATERIALS

A. LEE SMITH

Analytical Research Department
Dow Corning Corporation
Midland, Michigan

1. DEFINING THE PROBLEM

It is axiomatic that the first step in solving a problem is correctly defining it, and yet this step may be the most difficult part of the solution. Almost every analytical chemist has had the experience of being asked to perform a particular analysis, or to obtain a specific piece of information, which upon further inquiry proved to be totally irrelevant to the real problem. The analytical chemist must therefore assume nothing and constantly test the questions that are asked of him. In fact, asking the right questions is an art that is to be encouraged on the part of all chemists.

Among the problems the analyst is often asked to solve are those of identification, purity (contamination), structure, structure–property relationships, and composition. Several approaches are usually possible; it is important to choose the one that gives the best information with the least expenditure of time and expense.

Questions often asked about a material having unexpected or unusual properties are, Is it a silicone? Does it contain silicone? Is it contaminated by silicone? or Why does this material not perform as expected? Each of these questions may require a different line of attack. The amount of sample available can also have an important influence on the approach.

In the case of commercial products, the vendor often can suggest suitable methods for checking properties relative to product specifications, and may be able to offer assistance in identifying the cause of any abnormal behavior.

In some applications, silicones are used at very low levels of concentration, for example, a few parts per million, where the physical effect is significant but

The Analytical Chemistry of Silicones, Edited by A. Lee Smith.
ISBN 0-471-51624-4 © 1991 John Wiley & Sons, Inc.

analysis is difficult. Tests for silicon are not definitive, because inorganic silicon compounds are frequently present as additives or contaminants. The analytical chemistry of silicones therefore relies heavily on separations and instrumental characterization of specific molecular species.

The desire is frequently expressed for a simple spot test for silicones. However, the most commonly used silicone polymers, because of their inert nature, are not amenable to the usual kinds of chemical spot tests, but require relatively severe decomposition procedures to prepare them for chemical analysis. Such tests are discussed further in Chapter 8.

The literature on silicone analysis is scattered through many journals and some books, with only a few general reviews and summaries. Books on this subject include those by Kreshkov (1), Smith (2), and Crompton (3). Chapters in books have been authored by Noll (4), Urbanski (5), and Crompton (6).

The "decision tree" shown in Figure 2.1 may aid in defining the problem and selecting the most appropriate analytical technique(s).

Some general observations may be helpful to those involved with analysis of materials containing organosilicon compounds. First, infrared (IR) spectroscopy is a very powerful resource, and is usually the first choice for examination of an unknown sample. The information obtained from the IR scan can be used to plan the best subsequent analytical scheme. Second, it is important to be aware of the potential for contamination by silicones as discussed in Chapter 4, Section 1.2. Samples can also change during analysis, particularly if a siloxane comes in contact with acids or bases at elevated temperature (e.g., in a contaminated injection port of a gas chromatograph). Third, many of the functional groups, especially SiOH, SiCl, and SiN, are either hydrolytically or thermally sensitive, and the analytical approach must take this instability into account.

The approach one takes to an analytical problem depends on the resources available. In many cases satisfactory answers can be obtained with very little specialized equipment. Problems of molecular structure, trace analysis, or precise quantitative analysis understandably may require more sophisticated methods and instrumentation. Some simple screening tests are suggested in Chapter 3, Section 1.1.

Surfaces treated with PDMS or similar silicone polymers may often be recognized by the Scotch tape test (the tape will not stick), or by placing a drop of water on the surface. A high contact angle (the water froms a bead) indicates the possibility of a silicone. These tests are nonspecific; a fluorocarbon or oily surface shows the same behavior. More specific tests are discussed in Chapter 5.

A simple qualitative scheme for identifying polymers (including silicones) using only elementary laboratory equipment has been given by Saunders (7). However, IR spectroscopy is preferred if it is available.

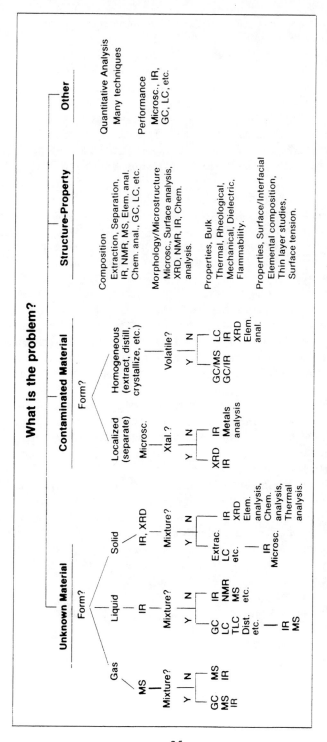

Figure 2.1. Analytical problem solving decision tree. © 1989 Dow Corning Corp. Used by permission.

25

Physical properties (refractive index, density, boiling point, and viscosity) are useful for characterization of monomers and some polymers (Chapter 7). Values for these properties are tabulated for various monomers and oligomers in the Appendix.

Among the physical methods, the value of optical microscopy in solving technical problems is often overlooked. This technique should be routinely employed for screening most solid and some liquid samples. Microscopy is frequently useful in solving problems involving abnormal behavior of materials, for example, failure of coatings, lubricants, elastomers, or emulsions. Some examples are given in Chapter 9.

Atomic spectroscopy, which includes atomic absorption, emission, plasma-induced emission, and X-ray fluorescence (XRF), gives elemental information over a concentration range of parts per billion (ppb) to 100%. The many uses of these techniques in organosilicon analysis are covered in Chapters 14 and 15.

Gas chromatography provides a superlative means of fractionating extremely small quantities of materials with great selectivity and sensitivity. Although it gives neither qualitative identification nor true quantitative analysis (unless response factors and an internal standard are used), the elegant simplicity of both the technique and the resulting chromatograms have insured this method a place in almost every laboratory. Unfortunately, many users tend to equate a single peak on the chromatogram with a pure sample. Nevertheless, if properly used, GC can be of great utility for the characterization of organosilicon monomers and oligomers (Chapter 10).

Supercritical fluid chromatography extends the range of separation for polymeric materials to higher molecular weights that can be accommodated by GC. Furthermore, the less severe treatment of the sample means that some materials containing moieties such as SiOH can be chromatographed directly without damage to the sample. Chapter 10 gives further details.

Nuclear magnetic resonance is an indispensible tool in organosilicon chemistry because of the valuable insights obtained from ^{29}Si NMR spectra. Problems of long relaxation times and poor sensitivity for the Si nucleus have been largely overcome (Chapter 12). Multinuclear NMR studies are essential for characterization of new organosilicon molecules.

Mass spectroscopy (Chapter 13) complements IR and NMR for molecular identification and structure determination. Mass spectroscopy combined with GC provides a particularly powerful combination of selectivity and sensitivity.

Many specialized techniques are available for studying surfaces. Some are sophisticated and expensive, and require skilled technologists to obtain and interpret the data. Applications to silicones are discussed in Chapters 5, 9, and 15.

The so-called hyphenated techniques (GC–IR, TGA–MS, and the like) are often useful, especially when a separation–identification procedure is indicated. The use of such techniques for solving industrial problems has been discussed by Crummett et al. (8).

The reader should not infer from our emphasis on instrumental techniques that the analyses are performed solely by the instruments. The competence of the technologist is the key to successful analyses. It is critical that an analytical approach be used for analytical problems. The analyst must be skilled in use of his tool kit (instrumentation), and familiar with the capabilities of other tools as well.

2. SAMPLING

Often the weakest part of an analytical procedure, sampling deserves the same attention that is given to insuring the reliability of the other parts of the analysis. *The analysis can never be better than the sample.* If the sample does not reflect the true composition of the material in question, the most accurate analysis is meaningless. For reactive organosilicon compounds (e.g., chlorosilanes, silazanes, and silanols), the possibility of inadvertent reaction or contamination during sampling must be considered and preventative measures taken. For materials in bulk storage, for example, rust from iron sampling valves can form iron chlorides in the presence of chlorosilanes and catalyze a series of confounding reactions (Chapter 10, Section 2.1). Hydrogen chloride in chlorosilanes can also catalyze rearrangement and other reactions. Even PDMS fluid rapidly absorbs atmospheric moisture until it reaches an equilibrium concentration of 200–500 ppm.

The simplest and probably most common means of sampling liquid and solid organosilicon materials involves drawing the material into glass bottles of various descriptions. This procedure is completely adequate in many cases, but it does expose the sample to potential reaction with atmospheric moisture and to silanol and other active sites on the surface of the glass. Such exposure may cause enough changes in the sample composition to invalidate the analytical results. Volatile components are easily lost even from tightly capped containers.

For moisture-sensitive materials, sampling with stainless steel bombs may be a practical alternative. These devices are inserted in-line, flushed and filled with sample, then sealed, removed, and taken to the laboratory for analysis. Rigid bombs are not recommended for liquids unless it can be assured that sufficient headspace is retained for sample expansion. Bombs constructed from flexible stainless steel bellows have been used for sampling both liquids and gases, but care must be taken not to exceed their rather low operating pressure ($\sim$ 25 psig).

Gases likewise may present problems in sampling. The simplest procedure for taking gaseous samples is by means of either stainless steel sampling bombs or gas sampling bags. Sampling bags constructed of Tedlar® appear to be most generally suitable because of their low reactivity and impermeability. Chlorosilanes and other reactive monomers, however, may be lost to the walls of even these containers, especially if traces of moisture are present, as they usually are.

Low concentrations of airborne siloxane vapors and aerosols may be sampled by either of two similar procedures, depending on the analyte. In the first, a measured volume of air is drawn through an impinger containing a suitable trapping solvent. In the second, the impinger is replaced with tubes containing activated carbon from which the siloxane is later displaced with solvent. In both cases, the concentration detected in the solvent is related to the concentration in the air. As some carbon adsorbents are catalytically active toward some silicones, prescreening, including recovery studies, is suggested to identify materials that are suitable for use.

Airborne chlorosilanes may be sampled using impingers containing an alcohol and its corresponding triorthoformate, which acts as an acid scavenger. The chlorosilanes are simultaneously trapped and converted to their higher boiling and more stable alkoxy derivatives, which are quantitated in the alcohol solution. It should be noted that silanic hydrogen is partially converted to alkoxy also, and chlorosilanes containing this group end up as mixed species. Examples of trace analysis for airborne organosilicon compounds are given in Chapter 4, Section 3.1.3.

Finally, the transfer and manipulation of samples in the laboratory should be carefully thought through so that sample integrity is not lost at this point.

3. ANALYSIS OF MONOMERS AND OLIGOMERS

As discussed in Chapter 1, silicone polymers are usually prepared by hydrolysis of chlorosilanes; this class of monomers is thus of great industrial and synthetic importance. Other types of monomers may also be encountered. Alkoxy and aryloxy silanes have specialized uses, as do acetoxy-substituted silanes. Silicon compounds with active nitrogen substituents are often used for derivatizing organic hydroxyl compounds. Low molecular weight silanols, because of their chemical fragility, have found very limited use per se, but are intermediates in every hydrolysis reaction involving silicon and may be encountered as unintentional products of incomplete condensation. Monomer silanes containing reactive organic functions, usually in combination with active silicon functions, are used extensively in many industries.

For our purposes, oligomers may be treated in the same way as monomers and, indeed, they are often isolated as intermediates in the synthesis of polymers.

3.1. Identification

If a monomer silane has been previously prepared and its properties documented, it can often be identified from such easily determined properties as boiling point, specific gravity, and refractive index (see Chapter 7 and the Appendix), or from its IR, NMR, or mass spectrum. The physical properties of many organosilicon compounds have been compiled by Bazant et al. (9).

If the material has not been previously characterized, an appropriate combination of elemental and molecular spectroscopic analyses usually provides unequivocable proof of structure (provided a pure specimen is available for the analysis). A tailored approach may be necessary if the sample is impure or transient (see Section 6).

3.2. Analysis of Pure and Mixed Monomers

Chlorosilanes are an important class of monomers, and analysis of pure specimens for parts per million levels of other chlorosilanes (e.g., $MeSiCl_3$ and $SiCl_4$ in Me_2SiCl_2) is critical in controlling branching in the finished polymer. Such analyses are usually done by GC, as discussed in Chapter 10.

Analyses for metals can be rather messy, usually involving hydrolysis of the monomer and elimination of the silicon component by ashing with HF and H_2SO_4. Direct analysis by XRF can be carried out in a sealed cell, but sensitivity varies depending on the element sought from ~ 1 ppm to several hundred parts per million.

Because chlorosilanes react so vigorously with moisture, it is important that they be handled carefully, both from the standpoint of safety and of maintaining sample purity. A drybox may be needed for manipulation of ultrapure or small ($< 100\,mg$) samples, but is usually not necessary for handling larger quantities if some hydrolysis can be tolerated. Use of a well-ventilated fume hood is essential, however. Containers for chlorosilanes should be carefully dried before use. If chlorosilanes are to be stored for more than a few days, they should be sealed in glass ampules. Storage in ordinary bottles results in a white deposit of hydrolyzed material on and around the bottle and, inevitably, corrosion of any metal in the vicinity from the HCl fumes. Even glass carries a layer of SiOH and adsorbed water on its surface, which will react with SiCl compounds. This water can be reduced by silanizing the surface (pretreating it with a chlorosilane or other reactive silane monomer), but it is virtually impossible to eliminate all traces of moisture without taking extreme precautions. Some small amount of hydrolysis is almost inevitable. This hydrolysis is scarcely noticed with large samples, but may become quite serious for small amounts of chlorosilanes.

Mixtures of reactive monomers can be fractionally distilled, provided moisture is excluded from the distillation apparatus. This technique is useful for concentrating extremely dilute impurities. The individual cuts can then be analyzed by GC–MS or GC–IR. Fractional distillation also serves as a preparative procedure for pure chlorosilanes used as reference materials and standards. Discussions of chlorosilane fractionation have been given (10, 11).

For some chemical procedures, it may be desirable to hydrolyze the chlorosilane prior to analysis to reduce its volatility or reactivity. Hydrolysis can be accomplished by dropwise addition of the sample to wet alcohol or cold aqueous ammonia.

Other common types of hydrolyzable silane monomers—alkoxy and aryloxy silanes, carboxy silanes, silylamines, and their dimers and trimers—are usually less reactive than chlorosilanes. They can be handled and characterized in a similar manner. Dilute acid or alkali may be needed to catalyze the hydrolysis of some alkoxy and aryloxy silanes. The alcohol liberated in the reaction can be separated and identified by standard techniques, and the polysiloxane residue characterized by methods described in Chapter 3.

The SiH function is easily identified by IR or NMR spectroscopy, or by a chemical test as described in Chapter 8.

Silanols deserve special comment. Although they are intermediates in every hydrolysis reaction of a silicon-functional group, they are not often isolated. Traces of acid or base catalyze their condensation to siloxanes and water.

$$2SiOH \xrightarrow{\text{H}^+ \text{ or OH}^-} SiOSi + H_2O \qquad (2.1)$$

Consequently, even when they are crystallized in a pure form, silanols may react with alkali in the glass container to form siloxanes and water. If pure, however, they may be stored safely in plastic bottles at low temperatures for extended periods.

Silanols pose special problems in analysis. The melting point of a crystalline silanol may not be repeatable because the material often decomposes even as it melts. It is difficult to distinguish chemically between silanol and water, so the analysis of a silanol and its condensation product by determination of –OH is not meaningful. Instrumental methods are therefore preferred. Nuclear magnetic resonance, IR, and near-IR can give unambiguous identification and at least semiquantitative analysis for silanols, but sensitivity is somewhat limited (on the order of 30–100 ppm). Derivatizing the silanol with $-Me_3Si$ groups (Chapter 10, Section 4.2) permits analysis by GC or liquid chromatography (LC) but care must be taken that the derivatization procedure does not further condense the silanol. Proper choice of a derivatizing group (e.g., $EtMe_2Si-$) permits detection of $HO(SiMe_2O)_nH$ in the presence of $Me(SiMe_2O)_{n+1}SiMe_3$. The silanol can be derivatized with a UV-absorbing group for analysis by LC with a UV detector.

Organofunctional monomers can often be analyzed by GC, although sometimes the organofunctional group needs to be blocked, for example, with the $-Me_3Si$ group (12). Extremely small amounts (low and sub ppm) of halogenated contaminants in nonhalogenated silane monomers may be detected by the use of GC with an electron capture detector.

3.3. Analysis of Oligomers

Oligomeric species such as low molecular weight linear and cyclic polysiloxanes can be identified, if pure, by their physical properties (Chapter 7) or spectroscopically. Gas chromatography and supercritical fluid chromatography (SFC) (Chapter 10) are useful for separation; the combination GC–IR and GC–MS has provided identification of trace amounts of unusual siloxane structures (13) by use of group frequencies together with retention times and molecular weight data. The chromatogram of the product of an alkali catalyzed rearrangement of PDMS is shown in Figure 2.2, and some of the structures that were identified are given in Table 2.1.

The lower members (up to Si_8) of the oligomeric siloxanes have distinguishable IR spectra (14, 15). Higher members, however, show absorption patterns that differ only slightly.

Physical data including melting point, boiling point, density, and refractive index, as well as references to the original literature, are given for an extensive list of cyclic siloxanes, cyclic silazanes, and cyclosiloxazane structures by Johannson and Lee (16). Some IR, NMR, and X-ray data for some selected cyclic compounds are also provided. We list physical property data for a few of the more common oligomers in the Appendix.

4. APPROACHES TO POLYMER ANALYSIS

Polymer analysis can involve questions of identity, composition, polymer size and size distribution, chemical properties, and physical properties. Standard methods of polymer analysis are applicable to silicone polymers, although

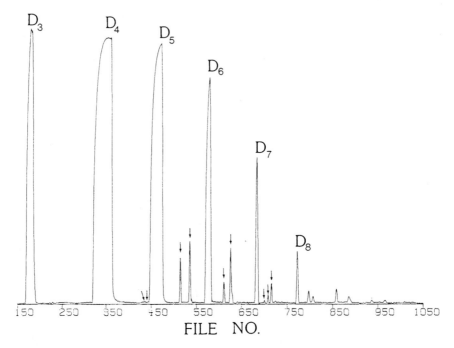

FILE   NO.

Figure 2.2. Gram–Schmidt reconstruction of the chromatogram of the reaction products resulting from a base-catalyzed rearrangement of PDMS. File numbers identify peaks whose structures are shown in Table 2.1. Minor peaks are marked with arrows (13). From *Chemical, Biological, and Industrial Applications of infrared Spectroscopy*, J. R. Durig, Ed., © 1985, John Wiley & Sons, Inc.

Table 2.1. Structures in the Si_3 to Si_8 Range from the Base-Catalyzed Rearrangement of PDMS, as Identified by GC–IR and GC–MS[a]

File number	Identification	Structure
182	D_3	△
361	D_4	□
433	T_2D_3	◇

Table 2.1. (*Contd.*)

File number	Identification	Structure
443	T_2D_3	
472	D_5	
515	T_2D_4	
535	T_2D_4	
580	D_6	
611	T_2D_5	
626	T_2D_5	
642	T_2D_5	
657	T_2D_5	
669	MD_5T	
683	D_7	
700	T_2D_6	
708	T_2D_6	
716	T_2D_6	

[a] The corresponding chromatogram is shown in Figure 2.2 (13). From *Chemical, Biological, and Industrial Applications of Infrared Spectroscopy*, J. R. Durig, Ed., © 1985, John Wiley & Sons, Inc.

some modifications may be necessary. The techniques outlined in Chapters 3 and 7 are usually adequate to identify and characterize silicone polymers.

A discussion of the chemistry, structure, and physical properties of siloxane polymers and copolymers is given by Kendrick et al. (17). The structure and chemistry of all types of silicon-containing polymers are reviewed in a book edited by Zeigler and Fearon (18).

5. DETERMINATION OF MOLECULAR STRUCTURE

A wide choice of approaches, including many of the techniques described in this book, is available for molecular structure determination. In addition, other sophisticated methods such as X-ray structure analysis, EXAFS (extended X-ray absorption fine structure), or neutron diffraction can aid in unraveling more complex structures. Fortunately, most of the problems involving organosilicon monomers can be solved by simpler chemical or physical methods, or a combination thereof.

Although NMR can cope well with impure specimens, most techniques require pure material for molecular structure analysis. A melting or boiling range can be helpful in evaluating purity. A single GC peak on a high resolution column is suggestive of purity but does not assure it. Materials that have been fractionally distilled or recrystallized are not of guaranteed purity; azeotropic mixtures and mixed crystals (or crystals containing solvent of crystallization) are not uncommon with organosilicon compounds. Using two or more different approaches to checking purity before undertaking extensive work can save much analytical time.

The approach to structure determination is dictated by the nature of the sample. Spectroscopic techniques are extremely useful. Chemical analyses, especially for carbon and silicon, give valuable information.

6. ANALYSIS FOR TRANSIENT SPECIES

Although the presence of transient silicon-containing intermediate species has long been postulated, only since 1976 has the combination of chemical insights and analytical tools permitted their isolation and study. Analyses for transient molecules generally utilize chemical trapping, physical (low temperature) trapping, or use of a method having rapid response such as photo-electron spectroscopy.

The long search for double bonds to silicon bore fruit only when the stereochemistry and fragility of such compounds was understood. Reports of multiply bonded silicon in 1979–1981 sparked intense activity in academia. Concurrent theoretical studies (19, 20) have given us a much clearer picture of bonding to silicon. Applications of these newly demonstrated structures to analytical chemistry, however, have been sparse as of this writing.

Structures containing multiply bonded silicon that are stable at room temperature (if O_2 is excluded) have been prepared using bulky substituents such as mesityl or t-butyl groups. Such materials have been studied by NMR, IR, Raman, and UV spectroscopies. Radical anions of several disilenes have been prepared, and have been studied using ESR. Comprehensive reviews covering the structure, chemistry, and analysis of multiply bonded silicon compounds have been given by Raabe and Michl (21, 22).

7. STRUCTURE–PROPERTY RELATIONSHIPS

Optimization of properties for formulated materials is a goal of almost every applications chemist. Instead of using a strictly empirical approach, however, it is highly desirable that the chemist establish meaningful relationships between the structure of the material(s) in the formulation and its properties. This process implies a quantitative measurement of the desired properties as well as a knowledge of the structure of the components. Usually the structure cannot be assumed—it must be determined by reliable methods. Purity of the material may also be of critical importance. Frequently, minor impurities have significant influence on the properties being studied. In all cases, analytical chemistry and physical measurements play key roles.

Because of the tremendous variety of materials and properties of interest, we cite only a few examples to illustrate some past approaches to the subject.

1. Flexibility, toughness, adhesion, and transparency of polyimide–siloxane segmented copolymers (23).

The properties of these copolymers were determined as a function of silicone content. Studies by X-ray photoelectron spectrocopy (XPS) showed that the top surface (~ 10 Å) was dominated by the siloxane.

2. Elastomeric properties and moisture sensitivity of PDMS–PTMO–PU copolymers (24).

The properties of elastomers based on urethane with polytetramethylene oxide (PTMO) and PDMS segments were investigated using differential scanning calorimetry (DSC), dynamic mechanical spectroscopy, tensile testing, and small-angle X-ray scattering.

3. Calculation of physical properties from molecular structure (25).

Boiling temperatures; critical temperatures, pressures, and volumes; molar volumes, and other properties are calculated from molecular data consisting of total molecular surface area, electrostatic molecular surface interactions, and hydrogen-bonding interactions. The method can be used to predict physical properties of compounds having flexible or rigid, symmetric or asymmetric, polar or nonpolar molecular structures. The key to this work was the availability of accurate and precise data for some well-characterized compounds. Further examples of structure–property correlations are given in Chapter 7, Section 3.

8. QUALITY IN THE ANALYTICAL LABORATORY

8.1. Safety Hazards and Practices

Commercial organosilicon polymers such as PDMS are essentially nontoxic as regards skin contact and ingestion, but if spilled can be very slippery. Their slipperiness on pavement surfaces is equivalent to that of hydrocarbon oils (26). Certain silicon-containing monomers, for example, $HSi(OMe)_3$ and $Si(OMe)_4$, are hazardous from the standpoint of vapor inhalation and, to a lesser extent, skin contact. Other monomers and oligomers show biological activity; as with any material whose toxicological properties have not been fully explored, new organosilicon compounds should be treated with respect.

Chloro- and other halosilanes react rapidly with water to give siloxanes and the corresponding halogen acid; thus skin and especially eye contact with these materials should be avoided. Accidental exposure should be treated by immediate flushing with running water for 15 min. Medical attention should be sought after washing in the case of eye contact or extensive skin exposure.

Silane (SiH_4) and some silanes and siloxanes that have high SiH content are easily ignited or spontaneously explosive when contacted by air (cf. the Appendix). Any compound containing SiH can slowly decompose with even mild catalysts to give hydrogen gas. Such materials should be treated with caution and their containers checked periodically for pressure buildup.

8.2. Good Laboratory Practices

Analytical methods and data are generated to support research, new products and processes, and problem-solving activities. Analytical data are also used to satisfy a variety of regulatory requirements including product registration, industrial hygiene, and environmental controls, as well as to support and defend litigation.

The manufacture of high quality products is totally dependent on having reliable ways of measuring quality. Such measurements usually involve physical or chemical (analytical) measurements at some stage. The quality of the analytical or test procedure must therefore be assured in order to certify the quality of the product.

It is essential that data generated through the use of analytical technology be both technically reliable and legally defensible. The former requirement is met by laboratory quality verification procedures to insure that any data generated is good enough for its intended purpose including valid scientific inference, quality products, and intelligent actions based on the data. The latter requirement is met by adequate documentation to prove that sound work was done.

8.2.1.   Sampling

It is impossible to define sampling procedures that are applicable to all cases, since the method of sampling depends on both the purpose and the method of analysis. Therefore reliable sampling procedures should be developed that properly relate to the information desired (27, 28). Some of the special problems encountered with organosilicon compounds are discussed in Section 2. Factors to consider during sampling include how well the sample represents the material under study, whether the sampling can be done safely and reliably (including proper labeling), and whether the material sampled can react with the atmosphere or with the sample container (soft glass bottles can be quite reactive towards some silicon functional groups), or degenerate with time (the latter can be a particular problem with nonhomogeneous materials such as emulsions that tend to separate or "cream").

8.2.2.   Quality Verification of Laboratory Operations

An experienced analytical chemist is fully aware of the necessity for frequent standardization of equipment and methods; nevertheless, these vital matters are sometimes neglected, and it seems worthwhile to reemphasize their importance.

Analytical instrumentation is subject to slow deterioration in performance, wavelength drift, and similar problems. Monitoring performance by checking key parameters (Table 2.2) is recommended. Chemical methods are also subject to many random and systematic errors, and it is important to make judicious use of standard reference materials, blanks, and synthetic samples. Even a simple analytical balance must be checked periodically for accuracy—the consequences of weighing errors are likely to be serious, a fact that has been appreciated for a long time (38). Monitoring of methods is

Table 2.2. Suggested Standardization Schedule for Instrumental Techniques

Instrument	Variables to Check	Standard Materials	Suggested Frequency	References
Gas chromatograph	Detector performance	n-butane	Weekly	29
	Column performance	Grob mixture[a]	Weekly	30
Liquid chromatograph	Detector performance	Methanol	Monthly	31
	Calibration	Polystyrene	Monthly	32
Infrared spectrometer, dispersive	Wavelength, response, resolution	Indene	Daily[b]	33, 34
NMR spectrometer	Shift calibration	Calibration mixture	Daily[b]	35
Mass spectrometer	Inlet system	Cholesterol	Weekly	36
	Peak intensities, resolution	Perfluorotributylamine	Weekly	36
Ultraviolet–visible spectrometer	Wavelength	Benzene, mercury arc	Monthly	37
	Sensitivity	Holmium oxide glass	Monthly	37
Near IR spectrometer	Wavelength	1,2,4-Trichlorobenzene	Weekly	37
	Photometric accuracy; S/N	Screens	Weekly	37
X-ray fluorescence spectrometer	Intensity, S/N	Glass disk (for Si)	Daily	(b, c)
Atomic absorption spectrometer	Intensities	Synthetic samples	With each sample set	
ICP spectrometer	S/N	Synthetic samples	With each sample set	
Electron microscope	Magnification calibration	Grating repica	Monthly	

[a] Except for columns used with chlorosilanes.
[b] Depends on the particular instrument and its conditions of use.
[c] The manufacturer's manual should be consulted.

simplified by the use of statistical process control methods (39, 40), which are capable of detecting reproducibility problems long before they become noticeable to the casual observer.

In many cases, nonsilicone standard materials are used for checking methods, for example, benzoic acid for carbon analysis and NaCl for hydrolyzable chloride. At times, however, it may be desirable to use an organosilicon material for standardization, as in SiOH, SiH, or Si determinations. Selection of suitable materials for standards can be a bit of a problem in some cases. Ideally such standards should be easily purified, indefinitely stable, insensitive to air or moisture, and have a reproducible composition that can be accurately calculated. Problems are likely to be encountered with stability of materials containing SiOH, SiCl, SiBr, SiOR, or SiN bonds. However, if these materials are properly stored as discussed below, minimal decomposition should result. Standards for trace analyses (e.g., vinyl groups in PDMS gums) may be prepared by diluting a selected standard material in a suitable matrix. Thorough blending is necessary to assure a homogeneous mixture. Some organosilicon materials that can be used for standards are listed in Table 2.3. Some vendors for organosilicon materials are given in Table 2.4.

Satisfactory storage of chlorosilanes, silazanes, alkoxysilanes, and similarly reactive materials for more than a few days can be accomplished by using sealed glass ampules. When well sealed, such containers permit indefinite storage of these materials. The ampule should be thoroughly dried before it is filled.

Silanols may decompose in soft glass bottles from the action of the alkali in the glass. The use of borosilicate glass containers is preferred for any material sensitive to traces of alkali.

8.3. Cleaning Glassware and Other Materials

Silicones have a reputation, not altogether undeserved, for being difficult to remove from glassware. The removal treatment chosen depends on the type of polymer present, its history, and the degree of cleanliness needed. For example, cleaning a stopcock for regreasing is less of a problem than cleaning silicone-treated glassware to achieve a hydrophilic condition.

The first step in cleaning glassware is to remove as much silicone as possible using a disposable tissue or cloth. For some situations, a solvent rinse will suffice. Chlorinated and aromatic solvents are suitable. One procedure calls for an acetone rinse, air-drying, and three or four quick washes with trichloroethylene (41). Some proprietary aqueous surfactant formulations are also claimed to be effective. More stubborn coatings require a more drastic approach (42). A bath consisting of 10 parts 50% KOH

Table 2.3. Organosilicon Chemicals that Can Be Used as Standards[a]

Analyte	Material	Analysis	
	Elemental Analysis Standards		
Si	Si (semiconductor grade)	100.00%	
	PDMS, > 300 cS	37.9	
	$\{F_3CCH_2CH_2(Me)SiO\}_x$	18.0	
	$Ph_2MeSi(OSiPhMe)OSiPh_2Me$	15.4	DC-705® Fluid
	PDMS–EOPO copolymer	Varies	
	$(Ph_2SiO)_4$	14:1	
Cl (hydrolyzable)	Me_2SiCl_2	54.8	
Cl (total)	$Cl(CH_2)_3Si(OMe)_3$	17.9	
H(Si)	Ph_3SiH	0.387	
	$HMe_2SiOSiMe_2H$	1.50	
F	$\{F_3CCH_2CH_2(Me)SiO\}_x$	36.5	
	Functional Group Standards		
Acetoxy(Si)	$ViSi(OAc)_3$	76.3	
Alkoxy(Si)	$Me_2Si(OMe)_2$	51.6	
	$Ph_2Si(OMe)_2$	25.4	
	$Ph_2Si(OEt)_2$	33.1	
	$(BuO)_4Si$		
Hydroxyl	$HOMe_2SiOSiMe_2OH^b$	20.5	Not commercially available
	$HO(Me_2SiO)_xOH$	Varies	
Vinyl(Vi)	$ViMe_2SiOSiMe_2Vi$	29.0	
	$ViSi(OAc)_3$	11.6	
	Physical Property Standards		
GC elution time	$Me_3Si(OMe_2Si)_nOSiMe_3,$		
	$(Me_2SiO)_n$(cyclic)		
Viscosity	PDMS		
NMR shifts	Me_4Si		
	3-(Trimethylsilyl)-1-propanesulfonic acid, Na salt (DSS)		
	2,2,3,3,-d4,3-(Trimethylsilyl)propionic acid, Na salt (deuterated TSP)		

[a] Other materials may be equally suitable for reference standards. The purity of any material should be verified before it is adopted as a standard.
[b] Should be refrigerated and stored in plastic.

Table 2.4. Suppliers of Organosilicon Chemicals[a]

Sources (Broad Selection)

Aldrich Chemical Co., 940 West St. Paul Ave., Milwaukee, WI 53201
Atomergic Chemetals Corp., 91 Carolyn Blvd., Farmingdale, NY 11735
Fluka, 980 South Second Street, Ronkonkoma, NY 11779
Huls America (Petrarch Systems), 2570 Pearl Buck Rd., Briston, PA 19007
Lancaster Synthesis, PO Box 1000, Windham, NH 03087
PCR, Inc., PO Box 1466, Gainesville, FL 32602

Sources (Limited Selection)

Alpha Products, 152 Andover St., PO Box 299, Danuero, MA 01923
Ashland Chemical Co., 2011 Turner St., Lansing, MI 48906
CPS Chemical Co., PO Box 162, Old Bridge, NJ 08857
Eastman Kodak Co., Rochester, NY 14650
Kay-Fries, Rockleigh, NJ 07647
K & K Laboratories, Plainview, NY 11803
Pfaltz and Bauer, Inc., 172 E. Aurora, Waterbury, CT 06708
Pierce Chemical Co., PO Box 117, Rockford, IL 61105
Regis Chemical, 8210 Austin Ave., Morton Grove, IL 60053
Sigma Chemical Co., PO Box 14508, St. Louis, MO 63178-9916
Silar Laboratories, 10 Alplaus Rd., Scotia, NY 12302
SWS Silicones Corp., Adrian, MI 49221

Sources (Specialty Materials)

NIST, Office of Standard Reference Materials, Gaithersburg, MD 20899
Ohio Valley Specialty Chemical Co., 115 Industry Road, Marietta, OH 45750
Supelco, Bellfonte, PA 16823-0048

[a] This list of suppliers may not be complete; other vendors may also supply these chemicals. Inclusion of a vendor on this list does not necessarily constitute a recommendation.

solution to 100 parts denatured EtOH is warmed (65 °C) in a stainless steel container. The glassware to be cleaned is immersed for a period of up to 10 min, then successively rinsed with mineral spirits, warm chromic acid solution, and finally water. Ultrasonic agitation may speed the cleaning. The solution etches glass, so exposure should be no longer than necessary. Volumetric glassware should be recalibrated frequently if cleaned with this solution. When a scum forms on the surface the solution should be discarded. This solution can produce severe and painful burns if allowed to contact the skin or the eye; therefore use of splash goggles and rubber gloves is mandatory. Eye or skin exposure is treated by flooding with water immediately and washing for 10–15 min. Medical attention should also be sought.

A procedure for cleaning low-scatter mirrors after exposure to silicone oil has been published (43). The solvent sequence is (1) 50% mineral spirits plus 50% ethylene glycol butyl ether (butyl cellosolve or Dowanol EB); (2) acetone; and (3) isopropanol.

Silicone rubbers and encapsulants can be stripped from electronic devices by tetramethylguanidine (44). This was the only solvent found to be effective for the purpose.

Silicone treatment on fibers, yarns, and textiles is reportedly removed by treatment with an alkaline surfactant solution (45).

Tin, aluminum, and zinc surfaces are cleaned by immersing them for 5–10 min in a warm (60 °C) solution made up of 50 parts ethylene glycol butyl ether, 50 parts mineral spirits, 1 part NaOH, and 1 part cresylic acid. The object is then rinsed, first with butyl cellosolve and then with water (42).

Steel, copper, brass, iron, and other metals may be cleaned in a warm solution consisting of 1 part NaOH, 100 parts butyl cellosolve, and 100 parts mineral spirits. Butyl cellosolve and water are used for rinsing (42).

REFERENCES

1. A. P. Kreshkov, Ed., *Practical Handbook of Analysis of Monomers and Polymers of Organosilicon Compounds*, State Publishing House for Chemical Literature, Moscow, 1962.

2. A. L. Smith, Ed., *Analysis of Silicones*, Wiley, New York, 1974.

3. T. R. Crompton, *The Analysis of Organic Materials. No. 4*: Chemical Analysis of Organometallic Compounds. Vol. 2: Elements of Group IVA–B, Academic, New York, 1974.

4. W. Noll, *Chemistry and Technology of Silicones*, Academic, New York, 1968.

5. J. Urbanski, "Silicones," in *Analysis of Synthetic Polymers and Plastics*, Chapter 20, G. G. Cameron, Ed., Ellis Horwood (Halsted Press), London, 1977.

6. T. R. Crompton, in *The Chemistry of Organic Silicon Compounds*, Chapter 6, S. Patai and Z. Rappoport, Eds., Wiley, New York, 1989.

7. K. J. Saunders, *The Identification of Plastics and Rubbers*, Chapman & Hall, London, 1966.

8. W. B. Crummett, H. J. Cortes, T. G. Fawcett, G. J. Kallos, S. J. Martin, C. L. Putzig, J. C. Tou, V. T. Turkelson, L. Yurga, and D. Zakett, "Some industrial developments and applications of multidimensional techniques," *Talanta*, **36**, 63 (1989).

9. V. Bazant and J. Hetflejs, *Handbook of Organosilicon Compounds. Advances since 1961*. Vols. 1–4. Marcel Dekker, New York, 1973. V. Chvalovsky and J. Rathousky, Eds., *Organosilicon Compounds*, Vols. 5–10. Institute of Chemical Process Fundamentals of the Czech. Acad. of Sci., Prague, Czechoslovakia, 1983.

10. Yu. K. Molokanov, T. P. Korablina, M. A. Kleinovskaya, and M. A. Shchelkun-ova, *Separation of Mixtures of Organosilicon Compounds*, Khimiya, Moscow, USSR, 1974.

11. J. A. McHard, in *Analytical Chemistry of Polymers*, Vol. XII, Part 1, G. M. Kline, Ed., Interscience, New York, 1959, p. 361.

12. C. F. Poole, "Recent advances in the silylation of organic compounds for gas chromatography," in *Handbook of Derivatives for Chromatography*, K. Blau and G. S. King, Eds., Heyden, London, 1978, p. 152.

13. A. L. Smith, "Infrared group frequencies for structure determination in organosili-con compounds," in *Chemical, Biological and Industrial Applications of Infrared Spectroscopy*, J. R. Durig, Ed., Wiley, New York, 1985.

14. N. Wright and M. J. Hunter, "Organosilicon polymers. III. Infrared spectra of the methylpolysiloxanes," *J. Am. Chem. Soc.*, **69**, 803 (1947).

15. C. W. Young, P. C. Servais, C. C. Currie, and M. J. Hunter, "Organosilicon polymers. IV. Infrared studies on cyclic disubstituted siloxanes," *J. Am. Chem. Soc.*, **70**, 3758 (1948).

16. O. K. Johannson and C-L. Lee, in *High Polymers*, K. C. Frisch, Ed., **26**, Wiley–Interscience, New York, 1972, pp. 459–686.

17. T. C. Kendrick, B. Parbhoo, and J. W. White, "Siloxane polymers and copoly-mers," in *The Chemistry of Organic Silicon Compounds*, S. Patai and Z. Rappoport, Eds., Wiley, New York, 1989.

18. J. M. Zeigler and F. W. Fearon, Eds., Silicon-Based Polymer Science, *Advances in Chemistry*, Vol. **224**, American Chemical Society, Washington, DC, 1990.

19. Y. Apeloig, "Theoretical aspects of organosilicon compounds," in *The Chemistry of Organic Silicon Compounds*, S. Patai and Z. Rappoport, Eds., Chapter 2, Wiley, New York, 1989.

20. W. S. Sheldrick, "Structural chemistry of organic silicon compounds," *The Chemistry of Organic Silicon Compounds*, S. Patai and Z. Rappoport, Eds., Chapter 3, Wiley, New York, 1989.

21. G. Raabe and J. Michl, "Multiple bonding to silicon," *Chem. Rev.*, **85**, 419 (1985).

22. G. Raabe and J. Michl, "Multiple bonds to silicon," in *The Chemistry of Organic Silicon Compounds*, S. Patai and Z. Rappoport, Eds., Chapter 17, Wiley, New York, 1989.

23. C. A. Arnold, J. D. Summers, R. H. Bott, L. T. Taylor, T. C. Ward, and J. E. McGrath, "Structure-property behavior of polyimide siloxane segmented copoly-mers," *Int. SAMPE Symp. Exhib.*, 32nd, 586, 1987.

24. R. A. Phillips, J. C. Stevenson, M. R. Nagarajan, and S. L. Cooper, "Structure-property relationships and moisture sensitivity of PDMS/PTMO mixed soft segment urethane elastomers," *J. Macromol. Sci., Phys.*, **B27**, 245 (1988).

25. S. Grigoras, "A structural approach to calculate physical properties of pure organic substances: the critical temperature, critical volume and related proper-ties," *J. Comput. Chem.*, **11**, 493 (1990).

26. T. S. Cox and D. N. Ingebrigtson, "Coping with spills of silicone fluids," *Env. Sci. Technol.*, **10**, 598 (1976).

27. American Society for Testing and Materials, "Probability sampling of materials," Method E105-58 (Reapproved 1975), *Annual Book of Standards*, Vol. 14.02, ASTM, Philadelphia, 1989.

28. B. Kratochvil and J. K. Taylor, "Sampling for chemical analysis," *Anal. Chem.*, **53**, 924A (1981).

29. American Society for Testing and Materials, "Testing thermal conductivity detectors . . . ," Method E-516-74 (Reapproved 1981); "Testing flame ionization detectors . . . ," Method E-594-77; "Testing electron-capture detectors . . . ," Method E-697-79; *Annual Book of Standards*, Vol. 14.01, ASTM, Philadelphia, 1989.

30. K. Grob and G. Grob, "Acidity test as part of the quality tests of a capillary column," *Chromatographia*, **4**, 422 (1971).

31. American Society for Testing and Materials, "Testing fixed-wavelength photometric detectors used in liquid chromatography," Method E-685-79, *Annual Book of Standards*, Vol. 14.01, ASTM, Philadelphia, 1989.

32. H. Coll, in *Gel Permeation Chromatography*, K. H. Altgelt and L. Segal, Eds., Marcel Dekker, New York, 1971, pp. 135–144.

33. American Society for Testing and Materials, "Describing and measuring performance of dispersive infrared spectometers," Method E-932-83, *Annual Book of Standards*, Vol. 14.01, ASTM, Philadelphia, 1989.

34. Coblentz Society Board of Managers, "Specifications for evaluation of infrared reference spectra," *Anal. Chem.*, **38**(9), 27A (1966).

35. J. L. Jungnickel, "Calibration of nuclear magnetic resonance chemical shift scale", *Anal. Chem.*, **35**, 1985 (1963).

36. J. G. Dillard, S. R. Heller, F. W. McLafferty, G. W. A. Milne, and R. Venkataraghavan, "Critical evaluation of Class II and Class III electron impact mass spectra. Operating parameters and reporting mass spectra," *Org. Mass Spectrom.*, **16**, 48 (1981).

37. American Society for Testing and Materials, "Describing and measuring performance of ultraviolet, visible, and near infrared spectrophotometers," Method E 275-83 (Reapproved 1989), Vol. 14.01, *Annual Book of Standards*, ASTM, Philadelphia, 1989.

38. Holy Bible, Proverbs 11, 1.

39. J. K. Taylor, *Quality Assurance of Chemical Measurements*, Lewis Publishers, Chelsea, MI, 1987.

40. D. J. Wheeler and D. S. Chambers, *Understanding Statistical Process Control*, Statistical Process Controls, Inc., Knoxville, TN, 1986.

41. C. J. Conner, "Rapid removal of silicone lubricant from glassware," *Chemist Analyst*, **50**, 84 (1961).

42. Dow Corning Corp., Bulletin 22-081A-72 (January 1972).

43. V. L. Williams and G. J. Fleig, "A method for cleaning low scatter mirrors after exposure to silicone oil," *Proc. Soc. Photo-Opt. Instrum. Eng.*, **107**, 170 (1977).

44. F. X. Ventrice, "Evaluation of chemical strippers for silicones," *Soc. Plast. Eng. Tech. Pap.*, **24**, 608 (1978).

45. F. Nickel, "Removing silicones from fibers, yarns, and textiles," German Patent DE 3,515,077, 19 June 1985.

CHAPTER

3

ANALYSIS OF POLYMERS, MIXTURES, AND COMPOSITIONS

N. C. ANGELOTTI

Dow Corning Corporation
Midland, Michigan

1. INTRODUCTION

Analysis of mixtures and compositions presents a challenge to the analytical chemist. It is here that breadth and depth of training and experience are invaluable. It is important that the chemist be able to recognize clues about the characteristics of a composition and possible approaches to its analysis. Some preliminary analytical tests may be necessary in order to make judgments on how to proceed to obtain a successful analysis.

Questions often asked in silicone analysis are, what is the polymer, and how much of it is present in the sample? Analysis of compositions is done for quality control of the products, to determine the migration of silicones into another material, to monitor competitive products, or to find the environmental fate of the material (1, 2). Before starting a complex analysis, it is important to know what use will be made of the information obtained, for the answer to this question determines the analytical approach.

As explained in Chapter 1, silicones (i.e., polyorganosiloxanes) are prepared from the chlorosilanes by hydrolysis followed by polymerization of the hydrolysate material. Siloxane polymers can be classified into three categories, fluids, gums, and resins. These siloxanes form the basis for hundreds of silicone products.

1.1. Screening Tests

A simple procedure to check for the presence of silicone in an unknown material consists of carefully burning a small quantity of the material on the

The Analytical Chemistry of Silicones, Edited by A. Lee Smith.
ISBN 0-471-51624-4 © 1991 John Wiley & Sons, Inc.

tip of a spatula in a Bunsen flame. The burning characteristics of the sample give indications of its chemical makeup. Silicones, in general, burn with a "sparkly" flame much like the old fashioned Fourth of July sparklers. Also, with a silicone, a white soot will be formed along with a white silica residue on the spatula. If the sample burns with a black sooty flame, and no solvents are present, the material may contain aromatic substituents as part of the polymer system. A further test for Si is to heat a small amount of the ash with a drop or two of hydrofluoric acid. If upon heating the residue is volatilized, it most likely is a silica residue from combustion of a silicone.

Other useful measurements for pure materials that are relatively simple and inexpensive are refractive index and density. From these data, the molar refraction (3) can be calculated and compared to listed values for silicone materials to assist in the determination of the structure (see Chapter 7).

Fluids that are components of the mixture may need to be separated before the silicone fraction can be identified. Separation schemes are outlined later in this chapter.

2. FLUIDS AND GUMS

2.1. Pure Fluids

2.1.1. General

Fluids are prepared from monomeric siloxanes that have a degree of substitution of two or greater. Fluids range from the simplest siloxane with a viscosity of 0.65 cS (hexamethyldisiloxane) to several million centistokes at 25 °C. The solubility of silicone fluids depends on their molecular weight and structure. The best solvents for PDMS fluids are aromatic solvents such as toluene and xylene, aliphatic hydrocarbons such as hexane and higher homologs, and chlorinated solvents such as carbon tetrachloride and trichloromethane. The solubility of fluids in a variety of solvents is shown in Table 3.1 (4).

If the structure is altered by substituting alkyl groups such as hexyl or octadecyl for one or both methyl groups, the polymer properties are significantly changed. The slope of the viscosity–temperature curve increases (5); the oxidative stability of the fluid is lowered; the fluid becomes more compatible with other organic compounds and its lubricating properties are enhanced.

2.1.2. Analysis

Infrared spectroscopy is undoubtedly the best way to identify a silicone polymer. The spectrum provides a fingerprint that identifies the material as a

Table 3.1. Solubility of Silicone Fluids in Various Solvents[a]

Solvents miscible in all proportions with poly(dimethylsiloxane) fluids in all viscosity grades:

Amyl acetate	Methyl ethyl ketone
Benzene	Methyl isobutyl ketone
Carbon tetrachloride	Methylene chloride
Chloroform	Mineral spirits
Diethyl ether	Perchloroethylene
Ethylene dichloride	Stoddard solvent
2-Ethyl hexanol	Toluene
Gasoline	Trichloroethylene
Hexane	Xylene
Hexyl ether	
Kerosene	

Nonsolvents for poly(dimethylsiloxane) fluids in viscosity grades above 5 cS:

Carbitol	Ethylene glycol
Cellosolve	Methanol
Cyclohexanol	Paraffin oil (liquid)
Dimethyl phthalate	Water

Partial solvents, miscible in all proportions with 0.65–10 cS poly(dimethylsiloxane) fluids, and generally immiscible with viscosity grade above 50 cS:

Acetone	Ethanol
n-Butanol	Heptadecanol
Dioxane	Isopropanol

[a] From J. A. McHard, in *Analytical Chemistry of Polymers*, Vol. XII, Part I, G. M. Kline, Ed., © Interscience publishers, 1959 (Ref. 4).

silicone and gives a great deal of information about its substituents and whether or not other materials are present. Its applicability to almost any type of sample, combined with the extraordinarily reproducible group frequencies found with silicones, make it one of the most useful laboratory tools for analysis or organosilicon materials (Chapter 11). Identification by IR is straightforward, provided reference spectra are available. The strong SiOSi absorption at 1000–1100 cm^{-1} is always present, and the 1262 cm^{-1} MeSi band is characteristic. Copolymers, however, cannot usually be distinguished from mixtures by IR. In this case, liquid chromatography (LC) can often provide discrimination.

Analysis by ^{29}Si NMR (Chapter 12) is extremely useful for polymers. It is possible to distinguish between monofunctional, difunctional, and trifunctional silicon species with this technique. Proton NMR is not as useful in characterizing silicones, particularly the methyl siloxanes, since their protons all resonate at nearly the same frequency; nevertheless, a large amount of data can be gleaned from a proton spectrum.

A pyrolysis–GC method has been described for characterizing poly-siloxanes (6) in which a directly coupled mass spectrometer was used for identification. Products of the pyrolysis were mainly cyclic oligomers with some recombination compounds formed from the pendant substituents.

Liquid chromatography, including gel permeation and high-performance LC (HPLC), is invaluable for the characterization of higher molecular weight polymers. Specialized detectors (IR, UV, and the like) extend the selectivity and usefulness of the technique (Chapter 10). Because molecular weights range from < 200 to several million, determination of molecular size may require several complementary techniques as outlined in Chapter 7. Low levels of branching in silicone polymers are difficult to quantify; some practical approaches are given in Chapter 7, Section 3.2.

If IR, NMR, and gel permeation chromatography (GPC) are not available, decomposition or depolymerization techniques may be used to obtain in-formation about the silicone substituent groups. Chapter 8, Section 3.7.3 describes a depolymerization procedure that uses gas chromatographic (GC) analysis for identification and quantitation (7). Specific gravity and refractive index values are helpful in determining the phenyl content of pure fluids. Some typical values for some phenyl and methyl containing polysiloxanes are shown in Table 3.2.

Table 3.2. Variation of Physical Properties with the Phenyl/Methyl Ratio of Silicone Fluids

Ratio of phenyl to methyl	0	1:20	1:3	1:1
Specific gravity at 25 °C	0.970	1.000	1.07	1.10
Refractive index n_D^{20}	1.430	1.42–1.43	1.48–1.49	1.52–1.54

Chemical analysis is capable of giving much valuable information about the structure, purity, and functional groups present in a material. Some recommended procedures are given in Chapter 8. An accurate value for silicon content is particularly useful for polymer analysis. From this figure an average polymer unit weight (R_2SiO) can be calculated.

$$\text{Weight per polymer unit} = (28.09 \times 100)/(\%Si) \qquad (3.1)$$

For polymers of 100 cS viscosity or higher, the end-group contribution is negligible. Thus 37.8% Si indicates a Me_2SiO polymer (unit molecular weight 74) a value of 20.6% Si indicates a PhMeSiO polymer (unit weight 136).

Carbon analysis, combined with the silicon figure, can be used to calculate a carbon/silicon molar ratio. Analyses for other elements, as well as for functional groups such as SiOH and $SiCH=CH_2$, often give useful information about minor but important constituents that sometimes are not detected by instrumental methods.

2.2. Fluids in Formulations

2.2.1. *In Situ Analysis*

Silicone formulations can be as simple as solvent solutions of a single siloxane polymer, or as complex as several siloxane polymers and copolymers in conjunction with solvents, emulsifiers, thickeners, fillers, stabilizers, and dyes. The analytical approach must be tailored to the nature of the formulation and the definition of the problem. Often IR spectroscopy can be used to determine the general nature of the material. Features such as methyl, ethyl, phenyl, vinyl, silane hydrogen, or alkoxy-silicone species may be discerned, provided they are present in greater than trace quantities. Additional information can be obtained from ^1H and ^{29}Si NMR studies. Low boiling materials can be detected by GC, and GPC will assist in determining if the polymer is a single entity or a mixture of two or more polymer species.

2.2.2. *Separation Techniques*

The separation technique that is chosen depends on what is learned from the screening test. A simple pH measurement on an aqueous dispersion of the sample indicates if it has any acidic or basic characteristics. An acidic reaction may indicate a silicon halide, or more likely, a hydrolyzed acetoxy group. If the material is basic the polymer may have amine functional substituents.

2.2.2.1. Polishes. Polishes are generally a complex mixture of many ingredients, each of which imparts certain properties to the product. Polishes differ in their composition depending on their intended use. They can be categorized into three groups: solvent based (water free), paste-type (also water free), and oil-in-water emulsions (8).

For a solvent based polish, the total solids are determined by evaporation of the solvent until a constant weight is attained. It is best to allow the solvent to evaporate in a well-ventilated fume hood first, and then in a 135 °C forced draft oven for 1 h. The residue is weighed and percent solids calculated. The abrasives and fillers can be determined by extracting the residue with a solvent such as toluene, heating if necessary to dissolve the organic materials, and either filtering or centrifuging the mixture to separate the solids. After drying, the solids are weighed and then examined by X-ray diffraction to speciate the filler. The toluene soluble materials are examined by IR after evaporation of the solvent. Identification and quantitation of the silicone can be made by IR, along with the identification of the wax used in the polish. Paste polishes can be handled in a similar manner.

For emulsion polishes, a ^{1}H NMR spectrum of the sample is useful. Based on this information, separations as described later in this chapter can be devised.

2.2.2.2.   Cosmetics and Toilet Goods. A broad range of silicone materials is used in such items, generically described as personal care products. The generally low concentration of silicone in these products sometimes presents analytical difficulties, but with good techniques, successful analyses can be preformed. A more complete description of these products and their analysis is given in Chapter 6.

2.3.   Hybrid Fluids

The hybrid silicone fluids are copolymers or organic polymers and polysiloxanes. One common type is comprised of a class of surfactants that have found widespread use in the polyurethane foam manufacturing industry for control of cell size. These materials are copolymers of PDMS and polyethylene and/or polypropylene glycols. They are also used as wetting agents for lithographic plates, antifog agents for glass and plastics used as optical parts, fiber and plastic lubricants, and as antifoamers in water-based coating materials.

The organic entity may be linked to the siloxane through a Si–C bond or through a Si–O–C bond (9). Polyether siloxanes with an SiOC linkage are prepared by the reaction of the hydroxyl end group from the polyether chain with an aminosilane ($SiNH_2$), silicon hydride (SiH), or a silicon alkoxy (SiOR) function. The Si–O–C bond is readily hydrolyzed by acid. A relatively easy procedure to quantitate the constituents is to dissolve the sample in CCl_4 and acidify it with a small quantity of concentrated hydrochloric acid. It is mixed thoroughly, water is added, mixed, and the sample centifuged. The sample phase separates into aqueous and CCl_4 layers, with the glycol fraction being in the water layer and the polysiloxane in the CCl_4. Infrared and NMR may be used to analyze the CCl_4 phase, while the aqueous phase can be examined with ^{1}H or ^{13}C NMR spectroscopy. From the data, it should be possible to determine the structures of the polyglycol and the silicone, and the ratio of the two in the sample.

Silicone-polyglycols that contain the silicon–carbon linkage are manufactured by the catalytic addition of a silicon hydride-containing siloxane to a polyglycol that is terminated with an olefinic substituent (most commonly, allyl) (10, 11). This structure is not cleaved by hydrochloric acid. In these cases, analysis most often is done using IR or NMR spectroscopy. It is also useful to obtain the molecular weight of the fluid using GPC.

2.4. Emulsions

An emulsion is a stable mixture of two or more immiscible liquids in which one of the liquids is in the form of fine droplets held in suspension by small amounts of substances known as emulsifiers (12). The function of the emulsifier is to reduce the surface tension at the interface of the suspended particles. The molecular structure of the emulsifier is such that one part of the molecule is soluble in one liquid, and another part in the other liquid. Such a wide variety of emulsifiers is available that emulsions can be tailored to have very specific properties (12–15). The stability of an emulsion depends on the emulsifier, the size and size distribution of the emulsified particles, and the presences of impurities.

Emulsifiers are categorized as anionic, cationic, or nonionic. The nonionic emulsifiers are among the most widely used. Typically they are poly-oxyethylene alcohols, along with their various derivatives such as the poly-glycol esters, alkylamine derivatives, and polyglycol–polysiloxanes (13).

Silicone emulsions can exist as ether oil in water or water in oil emulsions. Both types have found uses as antifoaming agents (16). They are used in processing foods such as fruit juices, canned vegetables, milk, instant coffee, and wine. Emulsions are also used as mold release agents in the rubber, plastic, and glass industries. Silicone emulsion formulations are used for applying silicone waterproofing and release treatments to such materials as paper, textiles, and leather. They have extensive use in polishes, cleaning agents, cosmetics, and pharmaceutical products.

2.4.1. Separation Techniques

For best results with emulsion analysis, it is important first to obtain as much information about the system as possible. Any insight that can be gained about the emulsion is helpful in deciding the method of attack for the analysis. One useful tool for this purpose is ^{1}H NMR, which pinpoints the major constituents and gives a rough idea of the amounts of each. Once that has been determined, the next step is to separate the emulsion into its components.

A procedure that we have employed successfully is to dilute approximately 2 g of emulsion with about 10 mL of methanol in a 1-oz vial with a polyseal screw cap. This mixture is shaken thoroughly for 1 min, then 10 mL of spectroscopic grade hexane is added. After the vial is sealed, it is placed on a shaker for 30 min. It is then centrifuged for 10 min, and in most cases, the emulsion breaks cleanly. The upper hexane layer contains the silicone (oil phase), and the lower methanol–water layer contains the emulsifiers and any other additives. The hexane layer is decanted into a tared evaporating dish, and the methanol–water is extracted a second time with a fresh 10 mL of hexane. This is again shaken and centrifuged, and the hexane added to the first evaporating dish. The hexane is then evaporated in a forced draft hood, under low heat such that the hexane does not boil. The dish is then further dried for $\frac{1}{2}$ h in an oven at 105 °C. After cooling, the dish is reweighed, and based on the original sample weight, the percent of silicone is determined.

2.4.2.   *Analysis*

An IR spectrum is obtained on the material for structure verification and purity. Usually, a GPC analysis is run on the silicone for molecular weight and distribution (Figures 3.1 and 3.2).

The methanol–water phase is transferred to a second tared evaporating dish, the vial is thoroughly washed, and the washings are added to the

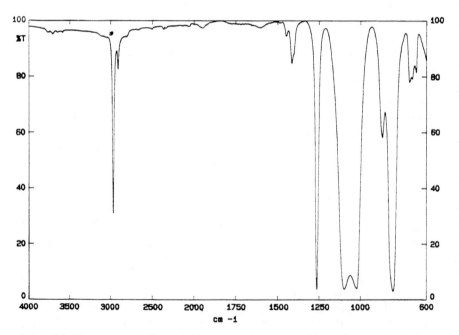

Figure 3.1. IR spectrum of silicone fluid separated from an emulsion. Film on a NaCl plate.

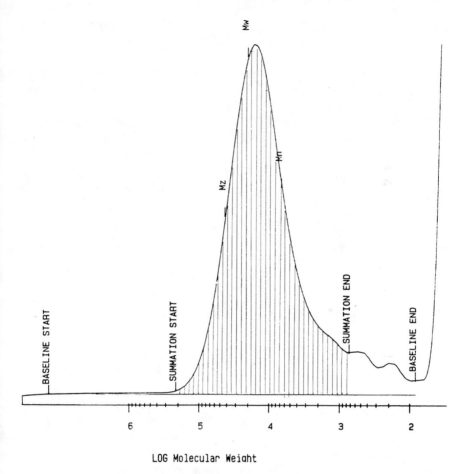

Figure 3.2. GPC curve of silicone fluid from Figure 3.1 ($\bar{M}_w = 20,000$).

evaporating dish. The dish is dried as before, again finishing by drying in the oven. The percent of emulsifiers is calculated from the weight of this residue. An IR spectrum is obtained on this fraction to determine the structure of the emulsifier. A GPC of the emulsifier is also obtained, generally using a polyglycol calibration curve. From the GPC data, one can often determine if a single emulsifier is present, or if the emulsifier is a mixture (Figures 3.3 and 3.4).

With antifoam materials, it is likely that a silica filler is present (13). During the breaking of the original emulsion and the centrifuging steps, the silica partitions to the bottom of the extraction vial. It can then be removed,

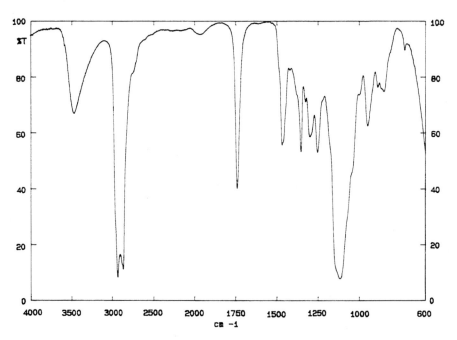

Figure 3.3. IR spectrum of separated emulsifier, identified as a polyglycol ester of a fatty acid. Film on a NaCl plate.

dried, and weighed to determine its concentration. Infrared spectra or X-ray diffraction analysis can be used to verify the composition of the filler.

2.4.3. *Emulsifier Characterization*

The ionic properties of the emulsifier are determined through the use of the methylene blue dye test (14, 15). This test is performed by taking 2 mL of the dye solution (in water), and 2 mL of chloroform. An anionic surfactant solution is added dropwise to this mixture and shaken after the addition of each drop until the color of the water phase and chloroform phase are the same. The unknown surfactant, as a methanol solution ($\sim \frac{1}{2}$% concentration), is added to the previously prepared chloroform–water–methylene blue solution. It is then thoroughly shaken and the organic and water phases are allowed to separate. If the aqueous phase (upper layer) becomes deeper blue in color, the surfactant is cationic. If the chloroform layer becomes deeper blue in color, the surfactant is anionic. If both the layers remain unchanged, the surfactant is nonionic.

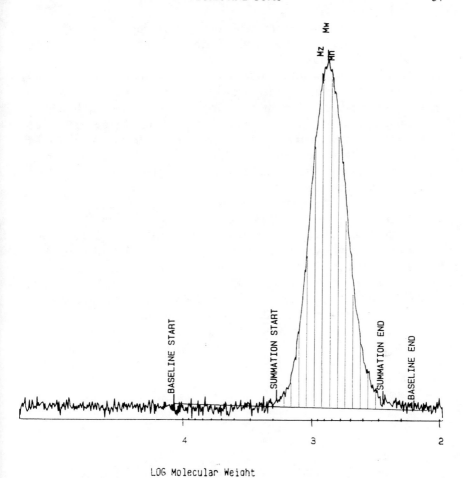

Figure 3.4. GPC curve of emulsifier from Figure 3.3 ($\bar{M}_w = 780$).

A second spot check on the classification of emulsifiers can be done by taking 5–10 mg of the separated emulsifiers and dissolving them in ~ 5 mL of water. To this is added 5 mL of a mixed indicator solution made from disulphine blue and dimidium bromide, and 5 mL of chloroform. This is shaken thoroughly and the phases allowed to separate. If the chloroform layer turns pink, the surfactant is anionic. If it is blue, the surfactant is cationic. If the mixture forms an emulsion that is difficult to separate into the aqueous and organic layers, the surfactant most likely is nonionic (15, 16).

Mixed emulsifiers can be separated on an ion exchange column (14). A mixed bed of anion and cation exchange resins is prepared. We have found Dowex 1 × 2-200 anion exchange resin and Dowex 50 × 8-100 cation exchange resin useful. The resin is washed and placed into a 10-mL buret. After the column is thoroughly washed free from any impurities, a solution of the emulsifiers in methanol, ~80-mL total volume, is slowly passed through the column at ~1 mL/min. This is followed with 80-mL of pure methanol to complete the elution of nonionic components. The two fractions are collected, the methanol allowed to evaporate, and the amount of the nonionic emulsifier is calculated from the weight of residue. The column is then eluted with $3N$ ammonium hydroxide to isolate the anionic surfactants. After a methanol rinse of the column, the cationic surfactants are eluted using a solution of HCl in methanol made by diluting 37% acid to give a $1N$ solution. The fractions are collected, the solvents evaporated, and the percentage of each type of emulsifier calculated from these weights (Figure 3.5). The identity of the various fractions can be ascertained from their IR spectra. In some cases, NMR is used to confirm the IR results. Also it is useful to obtain a molecular weight by GPC. A method such as X-ray fluorescence (XRF) spectroscopy can detect such elements as sulfur, chlorine, bromine, and phosphorous. It can also detect metals such as sodium and potassium in the emulsifiers.

We have tested these procedures by analyzing production samples of emulsions and antifoams along with some laboratory preparations of these types of materials with good success.

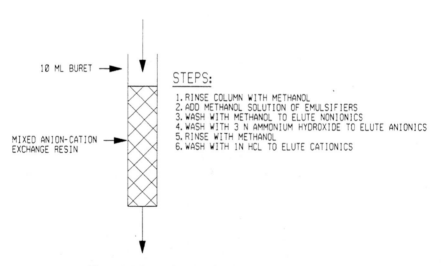

STEPS:

1. RINSE COLUMN WITH METHANOL
2. ADD METHANOL SOLUTION OF EMULSIFIERS
3. WASH WITH METHANOL TO ELUTE NONIONICS
4. WASH WITH 3 N AMMONIUM HYDROXIDE TO ELUTE ANIONICS
5. RINSE WITH METHANOL
6. WASH WITH 1N HCL TO ELUTE CATIONICS

10 ML BURET →

MIXED ANION-CATION EXCHANGE RESIN →

Figure 3.5. Ion exchange column for emulsifier separation.

2.5. Greases

Greases are formulated by compounding fluids with thickening agents, stabilizers, lubricity enchancers, oxidation inhibitors, and other additives. Typical thickeners are silica, or metallic soaps such as lithium stearate or lithium hydroxy stearate. Graphite or MoS_2 may be added to improve lubricity under extreme loads. Greases containing silica thickening agents are usually used in special applications such as stopcock lubricants, as release agents, and as dielectric coatings. The soap thickened greases have better load carrying characteristics and are used where high or low temperature stability of the grease is a necessity.

A preliminary IR spectrum of the grease reveals the characteristics of the silicone portion and in many cases give a good idea of the thickening agents and other additives. Based upon the preliminary data, the thickener, if it is a silica, can be isolated by using a thermal decomposition procedure as outlined in Section 6.2. The volatiles are collected in a cold trap, weighed, and a positive identification made based on their IR spectrum. The silica that remains in the combustion boat is quantified by weighing. It may be further

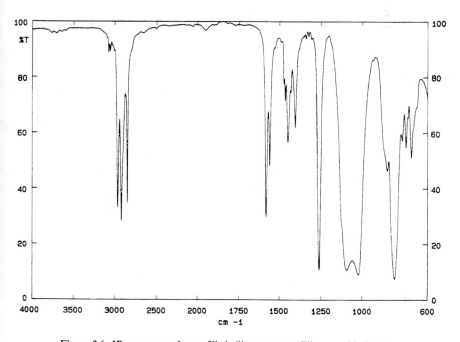

Figure 3.6. IR spectrum of soap-filled silicone grease. Film on a NaCl plate.

characterized by IR and X-ray diffraction. From the sample and residue weights, the percent of fluid and filler can be calculated.

Thickeners or lubricity additives are also revealed in the IR spectrum and in most cases can be identified (Figures 3.6 and 3.7). It is also useful to run an XRF scan on the original grease sample to ascertain what metals may be present; for example, barium or calcium salts are commonly used in greases.

Solvent extraction techniques using hexane or toluene can be employed to separate the fluid from additives. The extraction generally has to be repeated a second, and occasionally a third time to effect complete separation. Because of this difficulty, the pyrolysis procedure is the preferred method for best quantitation of the components.

Greases that contain organic additives are more difficult to separate cleanly. Again, the IR spectrum of the material provides much useful information about the overall composition and assists one in the right procedure to use for the analysis. Separation can sometimes be made by GPC with collection of the fractions as they elute. Identification of the fractions is done by IR spectroscopy. Figures 3.6 and 3.7 show examples of the IR spectra of typical grease formulations.

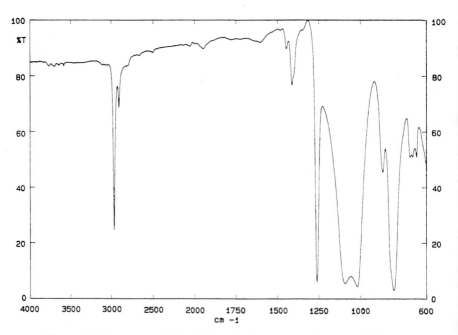

Figure 3.7. IR spectrum of silica-thickened silicone grease. Film on a NaCl plate.

2.6. Paper, Textile, and Leather Treatments

For paper, the preferred analytical procedure for % Si is XRF analysis (Chapter 15), although IR or chemical analysis for Si are also possible (17).

Textile treatments involve a cross-linking mechanism that ties the siloxane to the textile fiber to ensure durability through many washes. Paper and leather treatments are also cross-linked after cure. In general, it is not possible to achieve quantitative solvent extraction of silicone from a fabric, paper, or leather sample. Thus, a direct total silicon analysis is the preferred procedure. X-ray fluorescence on a ground sample is usually satisfactory. Infrared or ethylorthosilicate derivatization (Chapter 8, Section, 3.7.3) are also possible. The sample can also be decomposed in a Parr bomb and the residue analyzed using colorimetric silicon analysis or atomic spectroscopy (Chapter 14).

3. RESINS

3.1. General

Silicone resins are cross-linked structures having a polysiloxane network. They are used in high temperature paints and release coatings, laminates, molding compounds, and waterproofing compounds. Lightly cross-linked structures form a more rubbery or gellike matrix. Such materials are used for protection of circuit boards and electronic devices, as water-repellant fabric and leather treatment, and as release coatings for paper. The architecture of the silicone resins is developed from mono-, di-, tri-, and tetrafunctional silanes through a series of processing steps to form partially condensed siloxanes having a silanol content in the range of 3–6%. The resin then undergoes a bodying step whereby the molecular weight is increased through further condensation of the silanol functionality with a condensation catalyst. The finished resin has a silanol content in the range 0.5–1.5%. The user of the resin may add pigments, fillers, other resins, and a catalyst for the final cure (18).

The most common substituents in silicone resins are $MeSiO_{3/2}$, $PhSiO_{3/2}$, Me_2SiO, Ph_2SiO, and $PhMeSiO$ groups in various combinations. A high methyl content imparts flexibility, water repellency, fast cure rate, chemical resistance, gloss retention, and stability to UV and IR radiation (18). Resins with higher phenyl content display heat stability and oxidation resistance, toughness of the finished film, and retention of flexibility upon heat aging. Larger alkyl substituents cause slower cure and give a softer resin.

The degree of substitution on the silicon atom is an important factor that helps to predict the flexibility, cure rate, and chemical resistance of the final resin. It is simply defined as the average number of substituent groups per silicon atom.

3.2. Resin Characterization

The most useful instrumental tools for characterizing the composition of silicone resins are IR and ^{29}Si NMR spectroscopy, and GPC (19). Chemical analysis to quantitate specific functional groups is generally done after the basic structure of the resin has been determined by one or more of the above techniques (Figure 3.8).

If the sample is in solution, or soluble, a preliminary IR spectrum can be obtained as a solution spectrum. From this spectrum it is possible to identify the groups attached to the silicon atom, as well as determine the composition of the solvent.

A total solids content is obtained by drying a sample in a 105 °C oven (suitable for solvent use) for 2 h and weighing the residue. Generally, because the silanol functionality condenses under these conditions, the sample becomes insoluble. The fillers and/or pigments present in the sample of uncured

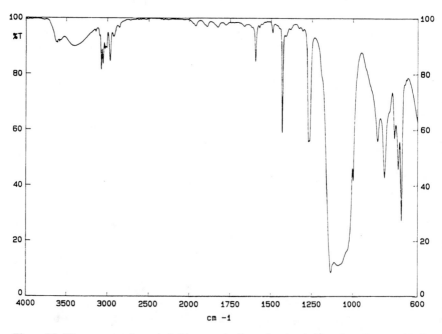

Figure 3.8. IR spectrum of a typical silicone resin (Dow Corning 840® resin). Film on a NaCl plate, air dried.

resin can best be separated by diluting a known weight of the resin solution approximately fourfold with either xylene or toluene. It is then centrifuged for $\frac{1}{2}$ h to separate the solvent-insoluble materials. The solvent is carefully decanted, and the residue is washed with a fresh portion of solvent, centrifuged, and the solvent combined with the first aliquot. The solids are carefully dried in a 105 °C oven for 1 h, cooled to room temperature, and weighed to obtain the percentage of filler in the resin sample. Identification of the filler is made by X-ray diffraction analysis.

The xylene or toluene solution of the resin is diluted to a known volume, and an aliquot of this is carefully oven dried to obtain the percent resin in the sample. Another aliquot can be taken and subjected to a silanol analysis as described in Chapter 8. If catalysts are present, another aliquot of this solution can be examined by emission spectroscopy or by XRF to determine the metal constituents and their concentration. An aliquot can also be examined by ^{29}Si NMR to determine the type and quantity of the groups attached to silicon. If NMR is not available, the functionality test (7) as described in Chapter 8, Section 3.7.3 can be used. This information should adequately describe the composition of the resin.

In the cases where the resin is not a pure silicone, but is a copolymer containing, for example, alkyd or epoxy moieties, the same general schemes of analysis can be applied. The organic portions can be identified by IR or ^{13}C NMR.

By proper choice of techniques, it is possible to determine the composition of any resin. If the sample is insoluble, it may still be possible to perform a direct IR analysis either by transmission techniques on the ground sample or by ATR (attenuated total reflectance) as described in Chapter 11. In a worst case scenario, the sample can be pyrolyzed and the volatiles examined by IR (20) or MS (6).

If the sample is in the form of a resin-treated fiberglass cloth laminate, it is often possible simply to extract the resin with a solvent such as acetone. The solvent is filtered to remove the glass fibers and analysis performed on the solvent solution of the resin as described earlier. If simple solvents do not remove the resin, a more stringent treatment may have to be applied, such as Soxhlet extraction of the resin using piperidine as the solvent. This process requires a 4-h extraction time, and is usually avoided if possible. In some cases, it is possible to obtain an ATR spectrum of the resin-treated glass cloth. Pyrolysis–IR can also be used.

4. COUPLING AGENTS

Although coupling agents are not strictly resins, they do form cross-linked networks during cure. Initially they are monomeric species $RSi(OR')_3$ that generally have two types of reactive sites: silicon-functional groups, usually alkoxy groups; and an organic tail that contains a functional organic group such as aminoalkyl, chloroalkyl, glycidoxyalkyl, or methacryloxyalkyl moieties. The former reacts with an inorganic substrate, and the latter with an organic coating. The coupling agent permits bonding of an organic resin to the surface of a material with which it would normally not be compatible. Theoretically, the silane coupling agent forms a monomolecular film but in actuality, the film is usually multimolecular. Coupling agents are used with a wide variety of organic polymers, including thermosets, thermoplastics, elastomers, and sealants. A variety of silanes is used to ensure compatibility with different substrates.

Analysis for the silane treatment on fillers and glass cloth is most commonly performed by IR spectroscopy, (Chapter 11, Section 5.3). Secondary ion mass spectrometry (SIMS) and electron spectroscopy for chemical analysis (ESCA) have been used to a limited extent. Most reports in the literature to date have been studies of model systems.

5. SEALANTS

5.1. General

Sealants are polymeric materials that are fluid before cure and elastomeric after cure. They are used for caulking, formed-in-place gaskets, adhesives, mold making materials, and encapsulants. They have excellent electrical properties, but mechanical properties are not as good as those of most silicone elastomers. Sealants are formulated from α,ω-functional PDMS, cross-linkers, plasticizers, catalysts, and fillers. Both one-part and two-part systems are produced. The former type cross-links by reaction of a hydrolyzable silane or siloxane with moisture, whereas the two-part systems do not require moisture for cure, but cure upon mixing.

5.2. Analysis

The cross-linking agent can often be identified by odor. An odor of acetic acid (from the uncured sealant) indicates an acetoxy silane or siloxane. The amount of the acetoxy in the sealant can be easily measured by a standard acid–base titration using a 50–50 solvent mixture of toluene–n-butanol with phenolphthalein as the indicator. Typically, the concentration of acetoxy groups is in the range 3–4%. A horseradish odor suggests an oxime cross-linker, and an alcohol odor indicates an alkoxy silane.

Also, with an uncured sample, it is possible to isolate the cross-linker from the total sample by vacuum stripping and collecting the volatile fraction in a cold trap. Infrared or NMR spectroscopy may then be used for the identification. It is worthwhile to obtain an XRF scan of the sample to check for any catalytic metals such as platinum that may be present in trace amounts. Catalyic metal compounds containing tin or titanium may be detected, along with metal oxides or salts present as fillers or extenders.

If the sample to be analyzed is in a cured state, little or no free cross-linker remains. In some cases, head space GC–MS analysis may permit detection of residual cross-linker. The polymer can be identified in several ways. Usually an IR spectrum of the material can be obtained by ATR for polymer identification. If sufficient sample is available, it can be subjected to the functionality test (Chapter 8, Section 3.7.3) to determine the moieties on silicon. A pyrolysis technique can be applied to intractable samples (6, 20). The volatiles (largely cyclic siloxanes) are collected and scanned by IR or GC–MS.

The filler content of uncured sealants may be determined using solvent extraction of the polymer with an aliphatic solvent such as hexane. The solid residue is dried, and the percent of polymer and filler is calculated. The filler may be identified by the use of X-ray diffraction. The polymer and cross-linking species in the hexane extract can best be identified using IR spectroscopy. The degree of polymerization (DP) of the polymer can be measured by GPC (Chapter 10, Section 9).

With the two part systems, analysis is simplified if samples of each part can be obtained prior to cure. If the elastomer has been cured, decomposition and depolymerization techniques may have to be used. Direct XRF analysis of the cured elastomeric material aids in the determination of the type of cure system. If platinum is detected, an SiH–vinyl hydrosilation cure system is likely. If a metal such as titanium is found, it may be the catalyst for an alkoxy–silanol reaction with the liberation of an alcohol. A metal such as tin may indicate an acetoxy or oxime-type cross-linking system. With cured systems, of course, it is not possible to determine the molecular weight of the starting polymer.

Dubiel et al. (21) give an example of an analysis for the reactive components of a silicone RTV foam. The SiOH and SiH were determined by IR spectroscopy, using an extremely dilute solution for the former to minimize hydrogen bonding. The cross-linking agent, tetrapropoxysilane, and its hydrolysis product, n-PrOH, were extracted with CS_2 and quantitated using GC with a surface coated open tubular capillary column. Diphenylmethylsilanol in the formulation was determined by GPC analysis. The polymer was not analyzed, but presumably could have been isolated by GPC.

6. ELASTOMERS

6.1. General

Silicone rubber stocks are prepared by milling polymer, fillers, plasticizers, vulcanizing agents, and any special additives necessary to give the product the desired properties. The stock is then molded, extruded or shaped into the final form, and heat cured to form the elastomeric product.

Silicone rubber is used in many applications that require good high and low temperature stability or excellent electrical properties. The most common type of silicone polymer used in the fabrication of elastomeric products is high molecular weight PDMS. It may contain other polysiloxane substituents such as vinyl, phenyl, or fluoroalkyl structures.

Simple cross-linking of PDMS does not give it optimal elastomeric properties. Physical properties are much improved by the incorporation of reinforcing fillers such as fumed silica, silica aerogel, or a precipitated silica made from sodium silicate solution.

Some of the fillers may be treated with functional silanes to incorporate organic groups on the surface of the filler particles. Other inorganic materials are used as extenders. These have a much larger particle size than does the reinforcing silica, and they provide bulk to the polymer. Common extenders are diatomaceous earth, calcium carbonate, finely powdered quartz, zirconium silicate and in some cases carbon black. Certain metal oxides may be added to the silicone rubber stock to enhance thermal stability, compression set, or to give a particular color.

Silicone elastomers are cured with the aid of vulcanizing agents, usually organic peroxides such as benzoyl peroxide or 2,4-dichlorobenzoyl peroxide. These free radical producing compounds induce cross-linking between the methyl groups. If the cross-linking is done through vinyl structures in the polymer, a less active free radical generator such as di-t-butyl peroxide, dicumyl peroxide, or di-t-butyl peracetate is used. These materials also induce hydrocarbon bridges between the polymer chains through the unsaturated linkages. Gamma radiation is employed infrequently for the cure of the polysiloxane elastomers. As with any analytical problem, background information about the material being analyzed is invaluable to the analyst.

6.2. Analysis

The previously described procedures for sealant analysis may, in general, be applied to elastomers. The uncured polymer may be separated from the filler by solvent extraction, and characterized by IR, NMR, and GPC.

Elastomers frequently contain additives or plasticizers that complicate solvent separation of the unvulcanized polymer from the filler. A reasonably clean separation can be achieved with a 1:1 mixture of solvent, such as toluene, and aqueous NH_3 (22). The silicone rubber stock or compound is cut into small pieces and placed in a centrifuge tube or glass vial. The toluene–aqueous NH_3 mixture is added, and the container is capped. It is then shaken for 10–15 min on a wrist-action shaker. The mixture is centrifuged and the solvent phase is separated and saved. The procedure is repeated three or four times. The combined decantates are evaporated in a forced-air oven or fume hood to give pure polymer. The extracted filler is dried in a forced-air oven and set aside for characterization.

The separation and identification of the peroxide catalyst in an uncured elastomer is not difficult. Quantitative measurements of the peroxide can be made by performing a Soxhlet extraction of the elastomer using ethanol or isopropanol. An aliquot of the extract is titrated for its active oxygen content using the iodimetric procedure (23), while a second aliquot of the extract is carefully dried and the residue examined by IR spectroscopy for the identification of the type of peroxide. From this information, the peroxide content of the original sample is calculated.

If the sample is already cured, it is more difficult to determine the curing mechanism. In most instances, cure is initiated by a free radical generating material, and with careful technique, it is sometimes possible to qualitatively identify that material. If the fragmented sample is carefully extracted with an alcohol such as ethanol or isopropanol using a Soxhlet extraction apparatus, traces of peroxide can be extracted in the alcohol. Qualitatively, by using iodometric techniques, the presence of an oxidant (peroxide residues) can be verified. In some cases, a peroxide free radical catalyst or its decomposition product manifests itself as "bloom" on the surface of the cured material. This powder can be very carefully scraped from the surface and examined by IR to identify it.

For cured material, the pyrolysis method can be used to measure the polymer-to-filler ratio. A sample of the elastomer is cut into thin sections and diced into pieces 2–3 mm on a side. This sample is weighed into a porcelain or a platinum boat and placed into a quartz or vycor tube in a tube furnace. The furnace is slowly heated to a temperature of 600 °C and held for 1 h, while a constant purge of dry nitrogen flows through the tube. Thermal depolymerization occurs and the volatiles are either vented, or trapped in a cold trap for identification and quantitation. The residue in the boat, after cooling, is carefully weighed to determine the amount of the filler, and an X-ray diffraction pattern obtained for identification. If it is suspected that the polymer may contain some phenyl structures, a modification of this procedure may be used. A bleed of anhydrous ammonia added to the nitrogen stream during the depolymerization process aids in giving good separation of phenyl siloxanes from the filler. From the sample, filler, and volatile weights, it is possible to calculate the filler content. Other additives, such as coloring agents, heat stabilizers, or other special-purpose materials may be present in an elastomer, but these are generally of secondary importance to the characterization of the polymer, filler, catalyst, and cross-linkers.

REFERENCES

1. L. G. Mahone, P. J. Garner, R. R. Buch, T. H. Lane, J. R. Tatera, R. C. Smith, and C. L. Frye, "A method for the qualitative and quantitative characterization of waterborne organosilicon substances," *Environ. Tox. Chem.*, **2**, 307 (1983).

2. R. R. Buch and D. N. Ingebrigtson, "Rearrangement of poly(dimethylsiloxane) fluids on soil," *Environ. Sci. Technol.*, **13**, 676–79 (1979).

3. E. L. Warrick, "The application of bond refractions to organo-silicon chemistry," *J. Am. Chem. Soc.*, **68**, 2455 (1946).

4. J. A. McHard, Chap. XIV, Silicones, in *Analytical Chemistry of Polymers*, Vol. XII, Part I, G. M. Kline, Ed., Interscience, New York, 1959.

5. A. J. Barry and H. N. Beck, "Silicone polymers," in *Inorganic Polymers*, F. G. A. Stone and W. A. G. Graham, Eds., Academic, New York, 1962.

6. S. Fujimoto, H. Ohtani, and S. Tsuge, "Characterization of polysiloxanes by high-resolution pyrolysis-gas chromatography-mass spectrometry," *Fresenius' Z. Anal. Chem.*, **331**, 342 (1988).

7. P. J. Garner and R. C. Smith, "Characterization of siloxane polymers by tetraethoxysilane derivatization and subsequent gas chromatography," Presented at the 1985 Pittsburgh Conference and Exposition, New Orleans, Louisiana.

8. L. Ivanovszky, "On the analysis of silicone-containing and other polishes," *Soap Perfum. Cosmet.*, **28**, 187 (1955).

9. W. Noll, *Chemistry and Technology of Silicones*, Academic, New York, 1968.

10. L. A. Haluska, US Patent 2,868,824 (1959).

11. L. A. Haluska, Fr. Patent 1,179,743 (1958).

12. P. Becher, *Emulsions—Theory and Practice*, 2nd ed., Rheinhold, New York, 1965.

13. C. C. Currie and M. C. Hommel, "Antifoam emulsion," US Patent, 2,595,928 (1952).

14. M. J. Rosen and H. A. Goldsmith, *Systematic analysis of surface-active agents*, 2nd ed., Wiley-Interscience, New York, 1972.

15. B. M. Milwidsky and D. M. Gabriel, *Detergent Analysis*, Wiley, New York, 1982.

16. B. R. Bluestein and C. L. Hilton, *Amphoteric Surfactants*, Surfactant Science Series, Volume 12, Marcel Dekker, New York, 1982.

17. E. D. Lipp, P. S. Rzyrkowski, and R. F. Geiger, "Comparison of X-ray fluoresence with other techniques for the quantitative analysis of silicone paper coatings," *Tappi J.*, **70**, 95 (1987).

18. L. H. Brown, "Silicones in protective coatings," in *Film Forming Compositions*, Vol. I, Part III, R. R. Meyers and J. S. Long, Eds., Marcel Dekker, New York, 1972.

19. P. J. Brandt, R. Subramanian, P. M. Sormani, T. C. Ward, and J. E. McGrath, "29-Si NMR of functional polysiloxane oligomers," *Polymer Preprints*, Division of Polymer Chemistry, American Chemical Society, 1985.

20. D. L. Harms, "Identification of complex organic materials by infrared spectra of their pyrolysis products," *Anal. Chem.*, **25**, 1140 (1953).

21. S. V. Dubiel, G. W. Griffith, C. L. Long, G. K. Baker, and R. E. Smith, "Determination of reactive components in silicone foams," *Anal. Chem.*, **54**, 1533–1537 (1983).

22. J. F. Hyde, Dow Corning Corp., private communication.

23. S. Siggia, *Quantitative Organic Analysis via Functional Groups*, 3rd ed., Chapter 6, Wiley, New York, 1954.

CHAPTER

4

TRACE ANALYSIS INVOLVING SILICONES

A. LEE SMITH

Analytical Research Department
Dow Corning Corporation
Midland, Michigan

and

R. D. PARKER

Dow Corning Ltd.
Barry, Glamorgan, South Wales,
United Kingdom

1. INTRODUCTION

The uniqueness of the analytical chemistry of organosilicon materials becomes strikingly evident in the practice of trace analysis. On one hand, silicones seem ubiquitous, and adventitous contamination is common. On the other hand, quantitation of small amounts of silicones may be extremely frustrating because of seemingly unexplainable losses from the sample, or their complete disappearance. In this chapter, we define some of the possibilities for both contamination and loss of the material to be measured. By being aware of both possibilities, and by carefully thinking through both the sampling and the analytical procedures, the analyst should be able to skirt many of the pitfalls that have trapped so many chemists. The literature is full of false reports (1), which upon reflection, could have been avoided had the authors been aware of the unique chemistry, physical properties, and unsuspected uses of silicones.

1.1. The Analytical Procedure

In order to obtain a reasonable result, one must know the chemical and physical properties of the species for which one is analyzing. For example, the

The Analytical Chemistry of Silicones, Edited by A. Lee Smith.
ISBN 0-471-51624-4 © 1991 John Wiley & Sons, Inc.

properties of PDMS, MW 10,000, are quite different from those of PDMS, MW 252 ($Me_3SiOSiMe_2OSiMe_3$). The former is viscous and nonvolatile (although it may contain traces of volatile oligomers), whereas the latter is volatile (bp 153 °C). The more viscous fluid has a C/Si ratio close to 2, whereas C/Si for the volatile species is 2.7. Some organosilicon polymers contain substituents that are reactive, for example, SiOH, SiOR, and SiN. A procedure appropriate for one "silicone" species may be wholly inappropriate for another.

Qualitative analysis may be approached with the usual tools: IR, NMR, MS, and GC–IR or GC–MS. The possibility that the species detected is a rearrangement or reaction product (i.e., an artifact of the analysis) must always be kept in mind. For example, detection of $(Me_2SiO)_3$ and $(Me_2SiO)_4$ could indicate thermal or catalytic rearrangement of PDMS. The structure $RSiO_{3/2}$ may result from hydrolysis of a $RSi(OMe)_3$ silanizing agent.

Once the target material is identified, an analytical procedure can be developed. It is important to rigorously exclude all possible sources of contamination or artifacts, and to carry through several "blank" samples that are treated in exactly the same manner as the actual samples. Recovery studies are also informative and are a necessary part of any serious investigation.

Studies of migratory behavior can be carried out with the help of labeled compounds, such as deuteromethyl substituted siloxanes. Or, even higher sensitivity can be obtained with radioactive substituents.

Sampling procedures are especially critical because in many cases, samples are gathered by nontechnical personnel, or by personnel not familiar with the special properties of silicones. It is therefore wise for the analyst to be involved in the entire sampling procedure as well as the actual analysis, because the latter will not be meaningful if the former is not done properly. General guidelines for trace analysis, including sampling, are given by the American Chemical Society Committee on Environmental Analysis (2).

1.2. Contamination

Contamination of samples by unsuspected sources of silicones is common, because these materials are commonly found in laboratory and urban or industrial environments. Silicone stopcock grease is used routinely in many laboratories, not only on stopcocks, but to seal desiccators and to lubricate glassware and rubber tubing or stoppers. Because of its low surface tension, PDMS (the fluid ingredient of stopcock lubricant) tends to "creep" along surfaces and may often be found some distance from the point of its initial application. The rate of creep depends on the temperature, the nature of the surface, and the viscosity of the fluid (3). An illustration of the viscosity

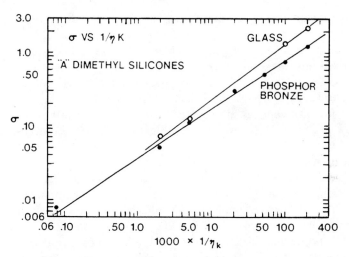

Figure 4.1. Spreading tendency of PDMS fluid on a smooth surface as a function of reciprocal viscosity (cS). Sigma (σ) is the spreading rate coefficient, in units of $cm^2 \cdot h^{-1/2}$. $\Delta A = \sigma \sqrt{t}$, where ΔA is the area increase and t is time in h. From N. M. Kitchen and C. A. Russell, *IEEE Trans. Parts, Hybrids, Packag.*, **12**, 24, © IEEE 1976.

dependence of creep is shown in Figure 4.1. Glass wool may be silanized and thus provide a source for organosilicon contamination. Poly(dimethylsiloxane) fluid is often used as a lubricant for rubber or plastic, as in plastic syringes. Or, it can be used as a mold release agent in the manufacture of these items. Organosilicon polymers are common ingredients of furniture and car polishes. Silicone vacuum pump fluids are short chain siloxanes substituted with methyl and phenyl groups, and while their vapor pressure is extremely low, they may backstream, as does any fluid in a high vacuum system. Glassware may be deliberately or inadvertently treated with silicone fluid (or with methyl chlorosilanes, which hydrolyze to form siloxanes). Such treatment can be extremely persistent and difficult to remove (cf. Chapter 2, Section 8.3), but may be a source of contamination in trace analysis.

Silicone rubber is also commonly used in laboratories in the form of tubing, septa, O-rings, and gaskets. Varying amounts of PDMS may be extracted if the rubber is contacted by organic solvents. Room temperature vulcanizing elastomers also can be sources of PDMS through solvent extraction, volatility, or fluid creep. Pressure sensitive adhesives used for self-adhering labels often use silicone-coated release backings. While such release coatings are essentially nonmigratory, they are not immune to solvent extraction or to depolymerization if heated.

Many personal care products such as hand lotions, antiperspirants, shampoos, hair conditioners, and sun screens contain cyclosiloxanes or other siloxane polymers. Photocopying machines use PDMS fluid, some of which is vaporized into the atmosphere.

Most GC columns use silicone polymers as liquid phase. Although these liquids are usually bonded to the substrate, they may bleed if overheated, or if alkaline or acidic samples are injected. They are particularly vulnerable to HF, and injection of samples containing this material or decomposable fluorine compounds will almost certainly give Me_2SiF_2 and $(Me_2SiO)_3$ reaction products.

Commercial grade PDMS fluids may be blended to reach a viscosity specification, and thus can contain low molecular weight components. In fact, most such fluids contain small amounts of cyclic and low molecular weight linear polymers as a result of the manufacturing process. Such components can creep and/or volatilize, even though the bulk polymer does not show such behavior. Similarly sealants, foams, and elastomers, if not fully cured, can release low molecular weight polymers,

Nonspecific analyses for Si, such as XRF, inductively coupled plasma spectrometry (ICP), and atomic absorption spectrometry (AA), are subject to interference from any source of Si, such as dust. Felby (4) noted that during graphite furnace AA analysis of hexane extracts of tissue, a series of high values was initially obtained for both sample and standards. The extreme values disappeared when precautions against dust contamination were taken. Storing tubes and micropipette tips in a dust-free container, wiping the working area with a moist cloth, and centrifuging the extracts substantially reduced the contamination problem.

1.3. Losses of Silicones

It is not good science to hope that losses and contamination will balance each other. In spite of the outstanding thermal and chemical stability of bulk silicones, traces of the same materials can be very vulnerable to losses by physical and chemical means.

Silicone polymers such as PDMS have strong affinity for glass, and are often adsorbed on the walls of glass bottles or on other siliceous materials. Water samples captured in glass bottles should be extracted several times from their original containers. Glass wool must be used with caution, as it can be both a source and a sink for PDMS polymers. We have already discussed the propensity of PDMS to creep over glass and other surfaces.

The rather high boiling point of many oligomers belies their actual volatility; their low heats of vaporization makes materials such as the cyclic PDMS oligomers subject to rapid volatilization.

Silicone polymers rearrange readily in the presence of strong acids or bases to give oligomers that may be volatile. Dust particles that contain clay fragments can catalyze such rearrangements. Traces of PDMS and related polymers are best handled in solution (keeping in mind their tendency to adhere to glass). Even the alkaline surface of soft glass (such as used for sample bottles) may cause siloxane rearrangement or silanol condensation.

Thermal rearrangement may also present a problem. Trace quantities of silicone fluids can be lost completely upon moderate heating, especially if traces of acids or bases are present.

2. ANALYSIS FOR SILICONES IN FOODS

Silicones may contact foods in many different ways. Small amounts (ppm levels) of silicone antifoams are often added to beverages, fruit and vegetable juices, and canned or frozen vegetables as process aids. Antifoams are also incorporated in some frying oils or fats to improve oxidative stability of the oils and to suppress foaming during frying. Other foods may contact silicone surfaces during processing or storage, for example, coated pans, baking papers, coated greaseproof papers, labels, coated cardboard storage containers, freeze drying containers, silicone tubing, and O-rings and seals. In most cases the residual level of silicone is very low, substantially < 10 ppm.

The interest in determining trace silicones in foods does not relate to toxicity, since PDMS is essentially inert physiologically (5–7). Rather, the interest is in knowing and controlling the level of any food additives.

2.1. The Analytical Method

The analytical approach chosen depends on the silicone polymer sought and the matrix. Silicones in aqueous matrices can be extracted and concentrated by organic solvents for a spectroscopic finish. For oleophilic materials such as fats and oils, the problem is more difficult, as the solubility characteristics of the silicone and of the matrix in the extracting solvent are similar. Nonspecific analysis for total silicon can sometimes be used, but many foods contain variable amounts of inorganic silicates or silica (8). Thus, nonspecific methods are limited to foods that contain low levels of these materials. Some values for naturally occurring silicon in foods are shown in Table 4.1.

When levels of inorganic silicon are high or variable, it is better either to use a specific method such as IR spectroscopy, or to physically separate the silicone from the matrix before quantification. Because matrices vary so widely, it is difficult to give a universal routine method. Problems of

Table 4.1. Concentrations of Naturally Occurring Silicon in Various Foods

Food	Silicon (ppm)	Reference
Milk	1	9
Ground beef	1	10
Cookies	< 1	10
Dinner rolls	< 1	10
Catsup	10	10
Bread	< 1	10
Canned lima beans	3	10
Canned corn	2–7	10
Spinach	2500	11
Oyster tissue	1300	11
Rice flour	150	11

incomplete extraction of silicone or of interfering coextractants are commonly encountered. A few examples of the approach to various types of foods are cited here.

2.2. Total Silicon Methods

A chemical method to determine total silicon in foods with low natural inorganic silicon has been reported by Horner et al. (12). The sample is decomposed with H_2SO_4 and HNO_3, and the residue is fused with Na_2CO_3 to convert the silica to silicate. After dissolution in HCl solution, the silicate is converted to heteropoly blue silicomolybdate, which is determined calorimetrically at 800 nm (Chapter 8, Section 2.1.3). The method is sensitive to about 1 ppm and recoveries of 86–98% are reported on ground beef over the range 2.8–22 ppm. Other applications included catsup, precooked meat loaf, canned ham, sausage, lard, wine, and whiskey. The chemical method, while sensitive, is somewhat tedious and may be subject to serious interference from phosphorus and fluoride compounds. These interferences can be masked if their concentrations are not too high, however.

Analysis of fats and oils by direct aspiration of the sample into an atomic absorption flame was reported by Doeden et al. (13). Standardization can be difficult, however, because of the differing response of silicon in different matrices. Bocca et al. (14) compared analyses for PDMS antifoam in olive oil using molybdenum blue photometry, atomic absorption spectrometry, and ICP spectrometry. Best results were obtained for ICP spectrometry. McCamey et al. (15) have developed an electrothermal atomic absorption method for fats and oils that used graphite furnace air oxidation to eliminate matrix interferences. The furnace program included drying (110 °C), oxidation (250 °C), charring (800 °C), and atomization (2600 °C) steps. The sensitivity of detection was 0.3 ppm. Recovery experiments averaged 95.2% with a relative standard deviation of 6.2% at the 3 ppm level. It is, however, very easy to lose part or all of the silicone polymer during the drying and oxidation steps, so recovery studies are very important when graphite furnace AA methods are used. Contamination can also be a problem (cf. Chapter 14, Section 4.2.3).

The foregoing method was used to study losses of silicone from commercial frying oil. The initial silicone level (3.7 ppm) in an edible oil dropped in < 1 day to a steady state of about 1 ppm, because of absorption on the cooked food.

2.3. Solvent Extraction Methods

Solvent extraction techniques have found wide application to foods and beverages. A selective extraction method for food and feeds using methyl isobutylketone (4-methylpentane-2-one, hereafter called MIBK) as suggested by Neal (16) did not report any recovery data. Diethyl ether has been used as an extraction solvent by Nishijima et al. (17) for the analysis of silicone in beer, soy sauce, skim milk powder, cultured milk beverages, ketchup, jam, and sugar. The diethyl ether extract is eluted through activated charcoal to remove contaminants, and the residue after evaporation is dissolved in MIBK and quantified by flame AA. Recoveries of 95–100% are claimed over the range 1–20 ppm.

A low-temperature extraction method (18) using petroleum ether for recovery of silicone from cottonseed oil and cake mix was also recommended for cooking oils and fats (19). In the latter case, the oil was absorbed on cellulose powder to present the maximum surface for extraction, which was carried out at − 80 °C. Recovery, however, is critically dependent on the technical skill of the analyst.

If identification as well as quantification of the silicone is desired, extraction followed by IR analysis is useful. Horner et al. (12) used benzene or chloroform as extractant, which was then evaporated and the residue taken up in carbon disulfide for IR analysis. A microcell can be used if necessary. If interfering materials are extracted also, spectral subtraction can be used to minimize the interference.

An extraction method with an IR finish was reported by Sinclair and Hallam (20) who treated their extract with $0.9N$ NaOH to remove interfering materials. This reagent, however, can attack the siloxane chain and its use may account for the low recoveries reported (50–70%).

Fruit juices such as pineapple juice present special problems, as straight-forward solvent extraction gives poor recoveries. Use of silicones up to 10 ppm is permitted (21), but it is necessary to monitor the level. Kacprzak (22) adsorbed the silicone on Florisil (an activated porous magnesium silicate), dried the adsorbant and extracted it with $CHCl_3$. This extract was then evaporated to dryness, redissolved in MIBK, and the silicone determined using AA spectrometry. A detection limit of 0.2 ppm was claimed.

Gooch (23) has discussed some of the problems involved in juice analysis. First, different matrices respond differently to extraction conditions. It is therefore essential to carry out recovery studies that simulate exactly the conditions encountered with actual samples. Second, problems were noted with apparent lack of stability of standards and synthetic samples. Pineapple juice, to which a known amount of silicone was added, showed an apparent decrease in silicone content on successive days when solvent extracted. Part of the difficulty may have been formation of an emulsion during extraction. He found that acidifying the sample gave cleaner and more complete separations. A 15-mL sample was shaken for 2 min with 15 mL of MIBK and 0.5 mL of 6N HCl. The extract was then analyzed using AA. An inter-laboratory collaborative study has been made on this method (24) and showed an average of 92% recovery on spiked juice samples, with a mean relative standard deviation of 4% in the range 4–30 ppm.

Standards were prepared using silicone-free juice, spiked with sorbitol that had been thoroughly mixed with 0.3% by weight of silicone antifoam compound. Addition of 100 mg of a sorbitol mixture to 15 g of juice gave an antifoam level of 19.2 ppm.

The method, with some modification, is applicable to other juices, but its applicability must be confirmed with appropriate recovery studies. Gooch found, for example, grape juice gave incomplete recovery of silicone when extracted directly, but when diluted 1:1 with water, gave 100 +/− 15% recovery. Lemon juice concentrates required prior neutralization with NaOH solution and controlled reacidification to give complete recovery. Frozen orange juice responded to the same treatment as pineapple juice.

2.4. Migration from Packaging Materials

Contamination of food is a subject of great concern. Regulatory au-thorities from several countries now require measurement of the migration of materials from food contact materials such as packaging papers and plastics into food simulants, before approving the use of these materials for food contact. A method for measuring these migrated materials has been devised by Maturi et al. (25). The treated paper or plastic is held in a demountable frame (Figure 4.2) within a cell, and water, saline solution, or heptane is

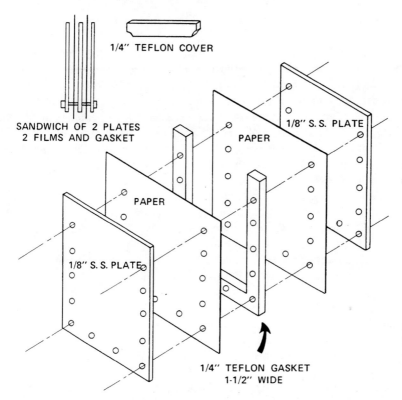

Figure 4.2. Cell for liquid extraction of sheet materials (25). Reprinted with permission from *Journal of Official Agricultural Chemists.*

placed in the cell and held at an elevated temperature for a specified time. The solvent is then analyzed for silicon, and results are reported in terms of milligrams per square centimeter, or weight (mg/cm^2), or weight percent extracted.

Similar tests are specified by European Economic Community (EEC) (26) and German Bundes Gesundheits Amt (BGA) (27) authorities. Here the food simulants used are water, 3% aqueous acetic acid, and ethanol in water. In addition the EEC tests also use fat simulants such as olive oil or HB304 synthetic fat. Legislation of the EEC generally permits migration of < 10 mg of total materials per square decimeter of surface; the German BGA regulations specify that the extracted material must be less than 0.5% of the original plastic or coating. On some occasions the identification or measurement of specific silicone oligomers in the migrated material have been requested. Here techniques such as IR, GC, GC–MS, and HPLC analyses have been applied.

3. SILICONES IN THE ENVIRONMENT

Silicones have found wide industrial application for many years, and ultimately some silicones enter the environment in aqueous effluents, landfills, or as vapors. Because of the nonbiodegradable nature of silicones, concern has been expressed (28) about the possible adverse ecological impact of PDMS. Consequently, studies have been made to evaluate the ecological fate of PDMS (29–32). An overview of these studies has been presented by Frye (32). In brief, although low parts per million concentrations of PDMS have been found in river sediments, the material does not migrate and is not bioconcentrated (33, 34).

Poly(dimethylsiloxane) in contact with certain clay components of common soils has been shown to undergo rearrangement and hydrolysis to form volatile oligomers and water soluble silanols (29). The ultimate hydrolysis product, $Me_2Si(OH)_2$, was shown to undergo demethylation to silicic acid in dilute aqueous solution on exposure to UV radiation in the presence of trace nitrates or nitrites (30, 35). Experiments on volatile PDMS materials in the atmosphere indicated decomposition on exposure to tropospheric UV radiation, with the decomposition proceeding at the same rate as that of n-octane (31). Hence, despite the well-known stability of PDMS in the bulk, the isolated individual molecules do not appear to present any exceptional resistance to environmental chemical degradation processes in the environment (36).

The analysis of these environmental samples can present problems because of high inorganic silicate background from soils, clays, and dust. Extraction procedures are therefore used to separate the organosilicon compounds from other siliceous materials. Solvent extraction with MIBK has proven effective in recovering PDMS from silicate matrices, and the extract solution can be used directly for AA or ICP analysis. If characterization of the organosilicon compound is necessary, it is often possible to evaporate the extraction solvent and apply techniques such as IR, NMR, and GPC to the residue.

3.1. Sampling and Analysis

Sampling is not a trivial or incidental matter. As discussed earlier (Section 1), both loss of silicone and contamination by adventitious sources are matters of real concern. An excellent set of guidelines for working with environmental samples has been developed by the American Chemical Society Committee on Environmental Improvement (2), and should be consulted before environmental studies are undertaken.

Soils should be sampled without drying, if possible, as some dry clays catalyze the decomposition of siloxane polymers (29). If drying is deemed necessary, the samples are placed in suitable containers and freeze-dried, or they can be dried overnight in a desiccator. After the sample is dry, solvent is added, and the container capped and vigorously shaken. The sample is centrifuged to settle the soil particles, and the supernatant liquid sampled directly into the AA or ICP flame.

Water samples are treated similarly; a 1-gal glass jug is a good sample container. At least three 10-mL extractions are made on the sample in its original container. The combined extracts are centrifuged and sampled without transferring. Silicones often are adsorbed on the walls of the container and if, for example, aliquots are taken, recovery will be seriously affected.

3.1.1. Analysis of Soils and Sediments

A survey on sediments from the Potomac River and Delaware Bay has been made by Pellenbarg and co-worker (33, 37), who used diethyl ether as extraction solvent. This solvent was removed by evaporation and the residue dissolved in MIBK for quantification by AA. Concentrations of soluble silicon compounds are reported to be between 0.5 and 3 ppm from the river sediments and between 0.1 and 1.5 ppm from bay sediments. These figures are only approximate, as the phenyl siloxane used to standardize the analysis gives a different response in the AA flame than does PDMS (Chapter 14). Watanabe et al. (38, 39) have made similar surveys on Japanese rivers, analyzing sediments, water, and fish. Petroleum ether was used as an extraction solvent and quantification was achieved by ICP spectrometry on solutions of sample in MIBK. Concentrations of 2–54 ppb in water, 0.3–5.8 ppm in sediments, and 0.4–0.9 ppm in fish are reported. Higher concentrations were found in industrial effluent, domestic waste, and sewage. Unfortunately the fish samples may have been contaminated by improper handling. A better controlled experiment (40) showed no evidence for even the limited bioconcentration inferred from Watanabe's results.

Toluene has been used as extraction solvent on sewage sludge by Tsuchitani et al. (41). The sample is neutralized and then dehydrated by Dean–Stark azeotropic distillation. The toluene extract is cleaned by passage through an activated charcoal column. The solvent is evaporated to dryness and the residue dissolved in MIBK for AA analysis. Nuclear magnetic resonance and GPC are used to identify the type of silicone. Recoveries around 88% are reported.

3.1.2. Analysis of Water

Chlorinated solvents, aromatic solvents, and MIBK are effective extraction solvents for PDMS and other hydrophobic silicones; however, their efficiency is poor for the extraction of hydrophilic siloxanes such as low molecular weight dimethylsiloxanediols from water. Waste water streams from plants that use or process silicones may contain inorganic silicates, water-soluble organosilicon compounds such as silanols, volatile cyclosiloxanes, and hydrophobic polysiloxanes. It is essential for pollution control to monitor and control the amounts of these various entities, but the analytical problem is not straightforward. A total silicon analysis would be misleading; solvent extraction would not include the hydrophilic silicones, and evaporative procedures can give losses of volatile compounds. It has been found that use of a mixed solvent for extraction, 1:1 1-pentanol and MIBK, effectively extracts both hydrophobic and hydrophilic silicones from water (42). The extracted material is analysed by AA (or ICP) spectrometry. Extraction efficiency is 98–100% as long as the pH of the water falls between 5 and 7. The method applies over the range 0.3–30 ppm Si. The standard deviation at 10 ppm is 0.31.

Porapak Q has been used by Cassidy et al. (43) to collect organosilicon compounds from industrial process water. The sample is eluted through a Porapak Q column where the silicone is absorbed. It is then extracted with MIBK for AA quantification. Liquid chromatography with molecular sieve columns and AA detection is used to assist in identification of organosilicon compounds. Recoveries of 89–96% are reported for a number of materials.

Reverse phase liquid chromatography, GC, and GC–MS have been applied by Bruggeman et al. (44) to quantify linear and cyclic PDMS oligomers extracted from fish. Studies from a 6-week feeding test indicated only marginal retention of silicones.

Low molecular weight silanols and siloxanes in water at the low parts per million level have been determined quantitatively by derivatizing them with Me_3SiO groups, and then determining the individual species by GC. Details are given in Chapter 8, Section 3.7.3. Quantification can be achieved on mono-, di-, tri-, and unsubstituted silicon compounds (45).

Organosilicon materials having significant vapor pressure and low water solubility can be collected and quantified using dynamic headspace (purge-and-trap) methods. The Environmental Protection Agency specifies the use of purge-and-trap methods for analysis of over 70 different organic compounds. Concentration factors of 500–1000 over direct injection methods can be achieved. Thus, impurities can be determined in the fractional parts per billion range with good accuracy and precision. In this method, the sample is swept with an inert gas for a few minutes, and the volatiles are collected in an adsorbant trap (typically Tenax®, a 2,6-diphenylene oxide polymer). After collection, the volatiles are desorbed by brief heating of the trap and introduced into a GC, where they are fractionated and may also be examined by MS or IR.

Such a method has been used to determine trace concentrations of octamethylcyclotetrasiloxane (D_4), and other cyclosiloxanes in water (46). Such materials are widely used in consumer products including antiperspirants, hair and skin care formulations, and cosmetics. They are also the principal products of thermal or chemical decomposition of PDMS. The concentrations of these materials in natural waters is thus of environmental interest.

The solubility of $(Me_2SiO)_4$ in water is low ($\sim$ 70 ppb). Henry's law constant for D_4 is calculated as 7 atm·m^3/mol (benzene's is 0.005) so the tendency of D_4 to volatilize from aqueous solutions is extremely high. The difficulty of formulating and maintaining stable solutions for standardization and recovery studies is formidable.

The method utilized standards prepared from D_4 in methanol, which was injected into a known volume of water. With 5-mL samples of water, 1–200 ng of D_4 could be quantified with an RDS of about 5% (MS detection). A flame ionization detector gave less sensitivity but a greater linear range of response (20–2000 ng). The method was thus operable to 200 ppt. An internal standard (deuterated o-xylene) was used. Analysis conditions were

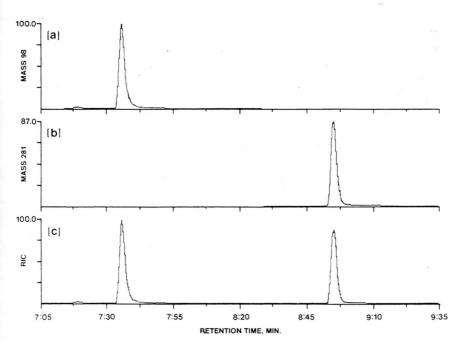

Figure 4.3. Chromatograms (single ion monitoring) for (*a*) perdeuteroxylene, (*b*) octamethyl-cyclotetrasiloxane, and (*c*) typical extract of water sample.

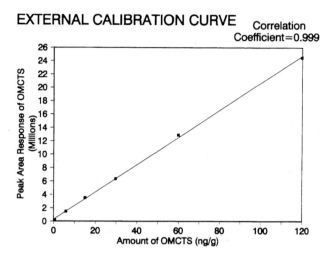

Figure 4.4. Calibration curve for determination of octamethylcyclotetrasiloxane in water.

5-mL sample; 12-min purge plus 4-min after-purge (to reduce the trapped water); desorption 4 min at 180 °C; cryotrapping of sample in a capillary interface to allow plug sample introduction to the GC; 30 m by 0.25 mm GC column with 1.0 μm DB-1 (PDMS) liquid phase; oven program, 2 min at 45 °C, then heat to 220 °C at 15°/min; multiple ion detection by MS. A typical chromatogram is shown in Figure 4.3, and a calibration curve in Figure 4.4.

One of the most severe problems was background contamination. Any solvents or other volatile materials used or stored in the same room inevitably showed in the chromatograms. Glassware had to be rinsed with water and baked in order to eliminate traces of solvent residues. Samples and standards could not be exposed to air without being contaminated.

3.1.3. Analysis of Air

Organosilicon compounds in air may be collected by trapping them as a measured volume of air is drawn through an impinger filled with a suitable liquid. Xylene and MIBK have been used as collection fluids for PDMS and other organosilicon compounds, with analysis by ICP or AA spectrometry. Dust particles must be removed from these solutions prior to analysis. Analyses for specific organosilicon compounds can be made by using special trapping fluids followed by chemical tests: ammonium molybdate solution for $HSiCl_3$ (47); NaOH solution for alkoxysilanes (48); H_2SO_4 solution for silylamines (49) and alkylchlorosilanes (50).

Infrared spectroscopic analysis on chloroform-trapped PDMS has shown acceptable accuracy and of course is specific for the silicone. Standard long path IR gas cells are also applicable, and < 1 ppm of $Me_3SiOSiMe_3$ can be measured in a 10-m cell. Portable IR spectrometers with 10-m sample cells are available for measurement in industrial environments, and wavelengths can be selected to monitor for silicone functionalities such as methoxy. Volatile silicones in air may also be trapped on activated charcoal columns. These can be eluted with carbon disulfide (51) or by thermal desorption for identification and quantification by GC.

A method for determining methyl polysiloxanes associated with airborne particles (aerosols) is described by Weschler (52). He used Teflon® membrane filters with typical collection times of 1 week or more. The filters were subsequently extracted with solvent (Freon 113 or dichloromethane) and the extract placed in a pyrolyzer probe. After evaporation of the solvent at 70 °C for 10 s (two firings), the probe was heated to 980 °C. The volatiles (largely cyclic oligomers) were analyzed using GC–MS. A sensitivity of 0.1 ng was claimed. Not surprisingly, low recoveries were noted for the lower viscosity (more volatile) fluids. The author did not report using any blanks collected at noncontaminated sites, but blanks would seem to be an important part of method validation.

4. SILICONES IN PROCESS MATERIALS

Silicones are frequently added in trace amounts to process materials such as plastics for better mold release, lubricity, or other processing advantage. Sometimes inadvertent contamination by "silicone" (or its omission) is suspected when a manufacturing problem is encountered. Such situations must be approached with awareness of the sampling pitfalls described in Sections 1.2 and 1.3. Because of the varied nature of process materials and problems, no general procedure is applicable, and each analytical problem has to be solved individually. A few examples will illustrate some specific approaches.

Poly (dimethylsiloxane) in plastics additives (53). The additive was first dissolved in *N,N*-dimethylformamide (DMF) and the silicone (350 cS PDMS) extracted using pentane in a continuous extractor. The pentane was evaporated and the residue taken up in carbon disulfide for IR analysis. A sensitivity of 0.2 μg/g of sample was claimed. In cases where interfering materials were simultaneously extracted, their absorptions were eliminated by the use of absorbance subtraction. The method was applied successfully to the determination of siloxanes in several other additives.

Poly(dimethylsiloxane) in polyethylene (54). The plastic was dissolved in hot decalin, and when the solution cooled, it was analysed directly by IR. The SiMe band at 1260 cm^{-1} was used for quantification.

Migration of silicone to the surface of a plastic (54). A silicone polymer that had been mixed into a polyurethane material as a mold release agent was suspected to have migrated to the surface of the plastic after molding. Adhesive strength of a bond to the part was observed to drop significantly with time. A solvent rinse of the surface, followed by concentration of the solvent followed by IR analysis, permitted quantification of the silicone per unit area of surface.

Electronic switch contacts and relay points sometimes develop insulating deposits when silicones are used in their vicinity, especially in enclosures. The phenomenon has been extensively studied (55). Fluid silicones can migrate by creeping (3), but vapors of cyclic and low molecular weight linear oligomers, from silicone rubber, can also cause problems. It is suggested that an acceptable rubber should show less than 0.3% weight loss on heating at 175 °C for 24 h (55). Another test involves Soxhlet extraction of the specimen for 6 h using 1,1,2-trichloro-1,2,2-trifluoroethane (3). The extract is scanned using IR to detect any PDMS. Deposits on contact points have typically been studied using light microscopy, scanning electron microscopy (SEM) and energy dispersive X-ray (EDX) analysis. A more complete discussion of analysis for silicones on surfaces is given in Chapter 5.

5. SILICONES IN MEDICAL AND BIOLOGICAL APPLICATIONS

Whereas organically substituted silicon compounds have never been found in nature, silicon appears to be an essential trace element for the growth and health of most, if not all, plants and animals (56–58). Silicon is well recognized as an essential trace element in the normal metabolism of higher mammals, although its exact mode of action is not well understood. In any case, analytical solutions to problems involving silicones in biological materials must allow for the presence of this endogenous silicon.

5.1. Medical Uses of Silicones

Because PDMS is relatively inert in the human body (59, 60), medical grade elastomers, gels, and fluids are used for a wide variety of medical devices and therapeutic agents. Devices include implants, drug delivery systems, and adjunct devices such as rubber tubing for extracorporeal blood circulation, catheters, and the like. Therapeutic agents include antiflatulent tablets as well as physiologically active materials.

Not all organosilicon materials are as inert as PDMS. Interest in the pharmacological and toxicological aspects of various organosilicon compounds has been strong (61). Such materials are of two categories: organosilicon compounds that are themselves bioactive; and conventional bioactive materials (drugs) that have been modified, either by inclusion of a silicon-bearing moiety, or by direct substitution of Si for C in the molecule. Bioactive and/or toxic silicon compounds include monomers such as $HSi(OMe)_3$ as well as more complex structures such as some of the aryl silatranes (62) and 2,6-*cis*-diphenylhexamethylcyclotetrasiloxane (63). The pharmacology of silanes and siloxanes has been reviewed by Levier et al. (64).

Even PDMS-based materials can evoke a "foreign body" tissue reaction under certain conditions. This reaction seems to be related to particle size and surface texture of the foreign body rather than its chemical constitution.

5.2. Defining the Problem

Analytical problems are likely to be concerned with trace analysis (the analyte is distributed more or less uniformly in the matrix) or micro analysis (the analyte consists of one or a few small particles in a heterogeneous matrix). Unidentified inclusions in tissue are sometimes assumed by pathologists to be "silicone" without any factual evidence. Only when the proper analytical method is used can one be certain of the identity of such contaminants.

It is important to define the nature of the problem before choosing the analytical approach. Is the presence of endogenous Si likely to interfere? Are other potential interferences likely to be present? Is a nonspecific analysis (light microscopy), Si-specific analysis (radiochemical, wet chemical, atomic absorption, gas or liquid chromatography, or XRF) adequate, or should a material-specific analysis (IR, Raman, NMR, or GC–MS) method be employed? What sensitivity is required? Are qualitative, semiquantitative, or quantitative data required? What are the possible sources of contamination? Could the analyte be lost or inadvertently changed during the analysis?

5.3. Analysis of Processing and Synthetic Aids

The use of silicone antifoams at the parts per million level in materials such as fermentation baths is well known, and the necessary precautions for the analysis of the antifoams are discussed in Chapter 2, Section 8.2.1, and Chapter 3, Section 2.4. The $Me_3SiO_{1/2}$ moiety, often used as a blocking reagent in the synthesis of pharmaceuticals (65–67), can be easily characterized and measured by either NMR or IR spectroscopy, where the group shows a strong, distinctive signature.

5.4.   Therapeutic Agents

An example of the skillful use of analytical chemistry is found in a study (68) of the metabolism of 2,6-*cis*-diphenylhexamethylcyclotetrasiloxane (2,6-*cis*), which has been proposed as a treatment for prostate carcinoma (63). As this material is highly lipophilic, the body metabolizes it to a more hydrophilic form in order to excrete it. Such metabolic products were expected to be silanols and other fragments, such as cyclosiloxanes, derived from the original molecule. Special problems were encountered in the analyses. First, some of these smaller molecules were volatile because of their low heat of vaporization. Second, silanols are notoriously unstable, particularly in small molecules. Third, there was little precedent for predicting the distribution of these products in the organism, as very little work had been done on the metabolism of organosilicon compounds.

In addition to 2,6-*cis*, the following materials were used in the study: radiolabeled (^{14}C) 2,6-*cis* to direct the separation of the metabolites; 2,6-*trans* to carry out recovery studies; synthetic reference compounds for identification of metabolites.

Analyses of blood serum, tissues, urine and feces were carried out after solvent extraction. In some cases, up to 10-days extraction (in an ultrasonic bath at room temperature, to minimize losses from volatility) was necessary for complete recovery. Some cross contamination between samples and blanks was noted, because of volatility, and had to be dealt with.

The metabolic products were identified by GC–MS as dimethylsilanediol, methylphenylsilanediol, trimethylphenyldisiloxanediol, and phenol. The silanols were derivatized *in situ* before extraction and analysis to prevent condensation of the silanol groups. Analytical techniques used were GC–MS, mass chromatography, and radio-gas chromatography. Quantitation of the 2,6-*cis* was accomplished using 2,6-*trans* as an internal standard. Sensitivities down to 0.1 μg/g of sample were achieved.

5.5.   Medical Devices

Problems involving silicones in medical devices occasionally arise. Medical grade fluid polymers are sometimes injected in clinical investigations by qualified plastic surgeons to restore facial contours in patients with serious facial disfigurement; however, illicit injections (sometimes using industrial grade fluids) have caused disfigurement and even death among the unfortunate recipients (69). Silicone elastomer implants that serve a dynamic function (such as finger joints) may generate wear particles. There have been reports of small fragments of silicone elastomers from implants migrating into lymph nodes (70). Sometimes silicone from heart-lung machines is transported by the blood into the body (71). Fibrosis and hepatic inflamation has been associated with the presence of particles of silicone in the livers of patients treated by hemodialysis (71). It is important in such cases to identify the source of such contaminants so that the problem can be avoided.

The preferred approach for characterizing foreign materials in tissue or blood is to use a specific method such as IR, Raman, NMR, or mass spectroscopy. Abraham and Etz (72) used a micro Raman technique to identify silicone polymer fragments in a standard tissue section of an enlarged axillary lymph node from a patient that had an implanted silicone finger joint prosthesis. The exciting radiation (the 514.5-nm green line of an Ar–Kr laser) was focused to a small spot, typically 2–20 μm diameter, and the Raman scattered radiation collected and analyzed. Single-crystal sapphire was used as a substrate.

Infrared spectroscopy is also used for microanalysis of specimens as small as 10–20-μm diameter (73). Smahel and Sell (74) used light microscopy to detect and IR to identify small particles in human capsular tissues. Soluble silicones can be detected by the use of standard IR trace analysis methods. Frick and Baudisch detected silicone fluid in tissues from antifoam added to the bubble oxygenator of a heart–lung machine (75). Brain and kidney tissues were extracted with benzene, and subjected to LC. Fractions were then examined by IR, and even small amounts of silicone in organ extracts rich in lipids could be identified.

Spallation and migration of silicone polymer fragments from blood pump tubings in patients on hemodialysis was studied by Leong et al. (71), using light and electron microscopy and AA spectrometry. Liver biopsy samples showed nonstaining, noncrystalline particles, which gave a strong Si response when interrogated with the electron microprobe. Silicon was also found by AA graphite furnace spectroscopy. The silicone was traced to tubing in the roller pump segment of the dialysis machine.

5.6. Summary

It is essential that the analyst be aware of the possibilities for artifacts and misinterpretation of data, as discussed in Sections 1.2 and 1.3. The preferred analytical methods are those that give specific information, as levels of endogenous Si may be variable, and contamination by siliceous materials is an ever-present possibility.

5.7. Analysis of Silicone Materials for Trace Contaminants

1. Catalyst residues in elastomers. Some elastomers are cured using peroxide free radical generators. Benzoyl peroxide and 2,4-dichlorobenzoyl peroxide are typical examples. If the elastomer has not had adequate post-cure, some acid residues (benzoic acid or 2,4-dichlorobenzoic acid) may remain in the material. These acid residues tend to migrate to the surface of the device, where they may appear as a white powdery deposit. These residues also give tissue reaction in implanted devices.

Elastomers can be tested for such acid residues by an extraction–titration procedure. About 1–2 g of the elastomer is sliced into thin sheets using a razor blade, and extracted for 1 h using EtOH in a Sohxlet extractor. The acid is titrated with standardized $0.05N$ KOH in EtOH. If the suspected acid has migrated to the surface, it can be scraped off and characterized by IR, using microsampling methods if necessary.

Other cure systems with organotin salts are also used. Presence of Sn residues or associated products (e.g., 2-ethylhexanoic acid) may be of interest. Extraction by saline solution and cottonseed oil is used to simulate *in vivo* conditions. In one procedure (76), a sample of known surface area (typically $120 \, cm^2$) is extracted for 60 min at 120 °C. The saline solution is then extracted with hexane and analyzed by IR for organic acids. Tin is determined directly on the extraction solvent by XRF. Cyclosiloxanes and low molecular weight linear oligomers may be determined using GC analysis of the hexane and cottonseed oil.

2. Ethylene oxide residues from sterilization in elastomers and mammary gels (77). A solvent extraction–GC method is used to detect ethylene oxide (to 1 ppm), ethylene glycol (to 20 ppm), and 2-chloroethanol (to 10 ppm). About 1.5 g of sample is placed in a $1\frac{1}{4}$-oz vial containing 5 mL of purified acetone and sealed immediately. Extraction is carried out for 3 days on a mechanical shaker. The acetone is then chromatographed (78). Recoveries of 90–110% were demonstrated.

Nanograde acetone is further purified by several passes through a column containing 5 Å molecular sieves with a glass wool plug at the bottom. Standards are prepared fresh daily, as acetone reacts with ethylene glycol if acid is present, and 2-chloroethanol contains traces of HCl. Standard preparation is carried out by partly filling a volumetric flask with acetone, adding ethylene oxide and ethylene glycol to make a 1% solution, filling to the mark, and then adding calculate volumes of this solution to other volumetric flasks partly filled with acetone. Another flask containing 2-chloroethanol is treated similarly, additions to the solutions containing the other analytes is made, and the flasks are filled to the mark. These solutions are sufficiently dilute that they are stable for at least a day.

A nonspecific test for ethylene oxide or catalyst residues involves direct contact cell culture (79). This method is used for quality control of implant devices and materials.

REFERENCES

1. C. L. Frye, "A cautionary note concerning organosilicon analytical artifacts," *Environ. Toxicol. Chem.*, **6**, 329 (1987).

2. American Chemical Society Committee on Environmental Improvement, "Guidelines for data acquisition and data quality evaluation in environmental chemistry," *Anal. Chem.*, **52**, 2242 (1980). "Principles of Environmental Analysis," *Anal. Chem.*, **55**, 2210 (1983).

3. N. M. Kitchen and C. A. Russell, "Silicone oils on electrical contacts—effects, sources, and countermeasures," *IEEE Trans. Parts, Hybrids, Packag.*, **PHP-12**, 24 (1976).

4. S. Felby, "Determination of organosilicon oxide polymers in tissue by atomic absorption spectroscopy using HGA graphite furnace," *Forensic Sci. Int.*, **32**, 61 (1986).

5. V. K. Rowe, H. C. Spencer, and S. L. Bass, "Toxicological studies on certain commercial silicones and hydrolyzable silane intermediates," *J. Ind. Hyg. Toxicol.*, **30**, 332 (1948).

6. R. Dailey, "Methylpolysilicones {in food}," Report FDA/BF-79/29; Order No. PB-289396, 1978; *Chem. Abstr.*, **91**: 37530c (1979).

7. Select Committee of GRAS Substances, "Evaluation of the health aspects of methylpolysilicones as food ingredients," Report, SCOGS-II-14, FDA/BI-81/97; Order No. PB81-229239, 1981; *Chem. Abstr.*, **96**: 50872p (1982).

8. G. W. Monier–Williams, *Trace Elements in Foods*, Wiley, New York, 1949, pp. 396–400.

9. E. J. Underwood, *Trace Elements in Human and Animal Nutrition*, 3rd ed., Academic, New York, 1971.

10. H. J. Horner, private communication, Dow Corning Corp. Midland, MI.

11. E. S. Gladney, P. E. Neifert, and N. W. Bower, "Determination of Silicon in National Institute of Standards and Technology Biological Standard Reference Materials by Instrumental Epithermal Neutron Activation and X-ray Fluorescence Spectrometry," *Anal. Chem.*, **61**, 1834 (1989).

12. H. J. Horner, J. E. Weiler, and N. C. Angelotti, "Visible and infrared spectroscopic determination of trace amounts of silicones in foods and biological materials," *Anal. Chem.*, **32**, 858 (1960).

13. W. G. Doeden, E. M. Kushibab, and A. C. Ingala, "Determination of dimethyl-polysiloxanes in fats and oils," *J. Am. Oil Chem. Soc.*, **57**, 73 (1980).

14. A. Bocca, A. Mazzucotelli, and S. Baragli, "Spectrochemical determination of silicon trace amounts in vegetable oils," *Riv. Ital. Sostanze Grasse*, **61**, 559 (1984); *Chem Abstr.*, **102**: 202684m (1985).

15. D. A. McCamey, D. P. Iannelli, L. J. Bryson, and T. M. Thorpe, "Determination of silicone in fats and oils by electrothermal atomic absorption spectrometry with in-furnace air oxidation," *Anal. Chim. Acta*, **188**, 119 (1986).

16. P. Neal, "Note on the atomic absorption analysis of dimethylpolysiloxanes in the presence of silicates," *J. Assoc. Off. Anal. Chem.*, **52**, 875 (1969).

17. M. Nishijima, M. Kanmuri, S. Takahashi, H. Kamimura, M. Nakazato, and Y. Kimura, "Determination of dimethylpolysiloxane in foods by atomic absorption spectrophotometry," *Shokuhin Eiseigaku Zasshi*, **16**, 110 (1975); through *Chem. Abstr.*, **83**: 145855f (1975).

18. P. Neal, A. D. Campbell, D. Firestone, and M. H. Aldridge, "Low-temperature separation of trace amounts of dimethylpolysiloxanes from food," *J. Am. Oil. Chem. Soc.*, **46**, 561 (1969).

19. J. W. Howard, "Food additives," *J. Ass. Offic. Anal. Chem.*, **55**, 262 (1972).

20. A. Sinclair and T. R. Hallam, "The determination of dimethylpolysiloxane in beer & yeast," *Analyst*, **96**, 149 (1971).

21. Anon., "Canned pineapple and pineapple juice identity standards; listing of dimethylpolysiloxane as optional ingredient," *Fed. Regist.*, **33**(166), 12040 (1968).

22. J. L. Kacprzak, "Atomic absorption spectroscopic determination of dimethylpolysiloxane in juices and beer," *J. Assoc. Off. Anal. Chem.*, **65**, 148 (1982).

23. E. G. Gooch, Private communication, Dow Corning Corp., Midland, MI.

24. R. D. Parker, "Collaborative study on the analysis of polydimethylsiloxane residues in pineapple juice," *J. Assoc. Off. Anal. Chem.*, **73**, 1792 (1990).

25. V. F. Maturi, W. O. Winkler, and W. E. Yates, "Determination of and data on total extractables from paper and film by food-simulating solvents," *J. Assoc. Offic. Agr. Chem.*, **45**, 70 (1962).

26. Anon., "Laying down the basic rules necessary for testing migration of the constituents of plastics materials and articles intended to come into contact with foodstuffs," *Offic. J. Europ. Communities*, No. L297/26, Council Directive of 18 October 1982.

27. B & A Decree Fragenbogen, *Recommendation XV version of the Plastics Commission of the German BGA*, Aug. 1, 1985.

28. P. H. Howard, P. R. Durkin, and A. Hanchett, "Environmental hazard assessment of liquid siloxanes (silicones)," Report No. PB 247778, Natl. Tech. Info. Service, Springfield, VA, 1974.

29. R. R. Buch and D. N. Ingebrigtson, "Rearrangement of poly(dimethylsiloxane) fluids on soil," *Environ. Sci. Technol.*, **13**, 676 (1979).

30. C. Anderson, K. Hochgeschwender, H. Weidemann, and R. Wilmes, "Studies of the oxidative photoinduced degradation of silicones in the aquatic environment," *Chemosphere*, **16**, 2567 (1987).

31. Y. Abe, G. B. Butler, and T. E. Hogen-Esch, "Photolytic oxidative degradation of octamethylcyclotetrasiloxane and related compounds," *J. Macromol. Sci. Chem.*, **A16**, 461 (1981).

32. C. L. Frye, "The environmental fate and ecological impact of organosilicon materials: a review," *Sci. Total Environ.*, **73**, 17 (1988).

33. R. E. Pellenbarg, "Environmental poly(organosiloxanes) (silicones)," *Environ. Sci. Technol.*, **13**, 565 (1979).

34. R. Firmin and A. L. J. Raum, "The environmental impact of organosilicon compounds (O.S.C.) on the aquatic biosphere according to new int'l criteria," *Rev. Int. Oceanogr. Med.*, **85–86**, 46 (1987).

35. R. R. Buch, T. H. Lane, R. B. Annelin, and C. L. Frye, "Photolytic oxidative demethylation of aqueous dimethylsiloxanols," *Environ. Tox. Chem.*, **3**, 215 (1984).

36. C. W. Lentz, "It's safe to use silicone products in the environment," *Ind. Res. Dev.*, **22**(4), 139 (1980).

37. R. E. Pellenbarg and D. E. Tevault, "Evidence for a sedimentary siloxane horizon," *Environ. Sci. Technol.*, **20**, 743 (1986).

38. N. Watanabe, H. Nagase, and Y. Ose, "Distribution of silicones in water, sediment and fish in Japanese rivers," *Sci. Total Environ.*, **73**, 1 (1988).

39. N. Watanabe et al., "Determination of trace amounts of siloxanes in water, sediments and fish tissues by inductively coupled plasma emission spectrometry," *Sci. Total Environ.*, **34**, 169 (1984).

40. R. B. Annelin and C. L. Frye, "The piscine bioconcentration characteristics of cyclic and linear oligomeric permethylsiloxanes," *Sci. Total Environ.*, **83**, 1 (1989).

41. Y. Tsuchitani, K. Harada, K. Saito, N. Muramatsu, and K. Uematsu, "Determination of organosilicones in sewage sludge by atomic absorption spectrometry," *Bunseki Kagaku*, **27**, 343 (1978). Through *Chem. Abstr.*, **89**: 94628y (1978).

42. R. D. Parker, "Determination of organosilicon compounds in water by atomic absorption spectroscopy," *Fresenius' Z. Anal. Chem.*, **292**, 362 (1978).

43. R. M. Cassidy, M. T. Hurteau, J. P. Mislan, and R. W. Ashley, "Preconcentration of organosilicons on porous polymers and separation by molecular sieve and reversed phase chromatography with an atomic absorption detection system," *J. Chromatogr. Sci.*, **14**, 444 (1976).

44. W. A. Bruggeman, D. Weber-Fung, A. Opperhuizen, J. Van der Steen, A. Wijbenga, and O. Hutzinger, "Absorption and retention of polydimethylsiloxanes (silicones) in fish: preliminary experiments," *Toxicol. Environ. Chem.*, **7**, 287 (1984).

45. L. G. Mahone, P. J. Garner, R. R. Buch, T. H. Lane, J. F. Tatera, R. C. Smith, and C. L. Frye, "A method for the qualitative and quantitative characterization of waterborne organosilicon substances," *Environ. Tox. Chem.*, **2**, 307 (1983).

46. J. A. Moore and V. J. Bujanowski, private communication, Dow Corning Corp., Midland, MI.

47. E. Zawadzka, "Determination of trichlorosilane in air," *Chem. Anal. (Warsaw)*, **13**, 1059 (1968), through *Chem. Abstr.*, **70**: 108940r (1969).

48. S. I. Murav'eva, "Determination of organosilicon esters in air," *Nov. Obl. Prom. Sanit. Khim.*, **1969**, 233, through *Chem. Abstr.*, **72**: 58804t (1970).

49. F. D. Krivoruchko, "Colorimetric determination of hexamethylcyclotrisilazane, bis(dimethylhydrazino)dimethylsilane, and bis(methylamino)dimethylsilane in air," *Zavodsk. Lab.*, **29**, 927 (1963), through *Chem. Abstr.*, **59**: 15846h (1963).

50. E. A. Peregud and N. P. Kozlova, "Method of determining alkylchlorosilane

vapors in air," *Zh. Anal. Khim. USSR*, **9**, 47 (1954); through *Chem. Abstr.*, 48: 6917e (1954).

51. J. Diaz-Rueda, H. J. Sloane, and R. J. Obremski, "An infrared solution method for the analysis of trapped atmospheric contaminants desorbed from charcoal tubes," *Appl. Spectrosc.*, **31**, 298 (1977).

52. C. J. Weschler, "Polydimethylsiloxanes associated with indoor and outdoor airborne particles," *Sci. Total Environ.*, **73**, 53 (1988).

53. P. Fux, "Fourier transform infrared spectrometric determination of trace amounts of polydimethylsiloxane in extracts of plastics additives," *Analyst (London)*, **114**, 445 (1989).

54. N. C. Angelotti, private communication, Dow Corning Corp., Midland, MI.

55. A. Eskes and H. A. Groenenduk, "The formation of insulating silicon compounds on switching contacts," *IEEE Trans. Components, Hybrids, Mfg. Technol.*, **11**(1), 78 (1988).

56. T. L. Simpson and B. E. Volcani, Eds., *Silicon and Siliceous Structures in Biological Systems*, Springer-Verlag, New York, 1981.

57. D. Evered and M. O'Connor, Eds., *Silicon Biochemistry* (Ciba Foundation Symposium, Vol. 121), Wiley, Chichester, UK., 1986.

58. E. M. Carlisle, "Silicon as a trace nutrient," *Sci. Total Environ.*, **73**, 95 (1988).

59. P. Vondracek and B. Dolezel, "Biostability of medical elastomers: a review," *Biomaterials*, **5**(4), 209 (1984).

60. M. B. Habal, "The biologic basis for the clinical application of the silicones," *Arch. Surg. (Chicago)*, **119**, 843 (1984). *Chem. Abstr.*, **101**: 157498v (1984).

61. R. Tacke and H. Linoh, "Bioorganosilicon chemistry," in *The Chemistry of Organic Silicon Compounds*, Chapter 18, S. Patai and Z. Rappoport, Eds., Wiley, New York, 1989.

62. M. G. Voronkov, "Biological activity of silatranes," *Top. Curr. Chem.*, **84**, 77 (1979).

63. B. Strindberg, "Biochemical effects of 2,6-*cis*-diphenylhexamethylcyclotetra-siloxane in man," in *Biochemistry of Silicon and Related Problems*, G. Bendz and I. Lindqvist, Eds., Plenum Press, New York, 1978, p. 515.

64. R. R. Levier, M. L. Chandler, and S. R. Wendel, "The pharmacology of silanes and siloxanes," in *Biochemistry of Silicon and Related Problems*, G. Bendz and I. Lindqvist, Eds., Plenum Press, New York, 1978, p. 473.

65. E. Colvin, *Silicon in Organic Synthesis*, Butterworths, Boston, 1981.

66. W. P. Weber, *Silicon Reagents for Organic Synthesis*, Springer-Verlag, New York, 1983.

67. M. Lalonde and T. H. Chan, "Use of organosilicon reagents as protective groups in organic synthesis," *Synthesis*, **1985** (9), 817. *Chem. Abstr.*, **104**: 168509n (1986).

68. J. Vessman, C-G. Hammar, J. Lindeke, S. Stromberg, R. Levier, R. Robinson, D. Spielvogel, and L. F. Hanneman, "Analysis of some organosilicon compounds in biological material," in *Biochemistry of Silicon and Related Problems*, G. Bendz and I. Lindqvist, Eds., Plenum Press, New York, 1978, pp. 535–560.

69. R. Ellenbogen, R. Ellenbogen, and L. Rubin, "Injectable fluid silicone therapy: human morbidity and mortality," *J. Am. Med. Assoc.*, **234**, 308 (1975).

70. A. J. Christie, K. A. Weinberger, and M. Dietrich, "Silicone lymphadenopathy and synovitis," *J. Am. Med. Assoc.*, **237**, 1463 (1977).

71. A. S. Leong, A. P. Disney, and D. W. Gove, "Spallation and migration of silicone from blood-pump tubing in patients on hemodialysis," *N. Engl. J. Med.*, **306**, 135 (1982).

72. J. L. Abraham and E. S. Etz, "Molecular microanalysis of pathological specimens in situ with a laser-Raman microprobe," *Science*, **206**, 716 (1979).

73. M. A. Harthcock, L. A. Lentz, B. L. Davis, and K. Krishnan, "Applications of transmittance and reflectance micro/FT–IR to polymeric materials," *Appl. Spectrosc.*, **40**, 210 (1986).

74. J. Smahel and J. Sell, "Kontamination von menschlichen Geweben mit Silikon," *Mater. Tech.*, **6**(1), 51 (1978). *Chem. Abstr.*, **89**: 53104f (1978).

75. R. Frick and H. Baudisch, "Physico–chemical determination of intravascular silicone in brain and kidney," *Beitr. Pathol.*, **149**, 39 (1973).

76. U. S. Pharmacopeia XXI, "Biological Tests—Plastics," Mack Publishing, Easton, PA, 1985, pp. 1236–1237.

77. P. J. Garner, private communication, Dow Corning Corp., Midland, MI.

78. H. Spitz and J. Weinberger, "Determination of ethylene oxide, ethylene chlorohydrin, and ethylene glycol by gas chromatography," *J. Pharm. Sci.*, **60**, 271 (1971).

79. L. M. Smith, "Direct-contact cell-culture method," *ASTM Spec. Tech. Publ.*, 810, 5 (1983).

CHAPTER

5

ANALYSIS OF SURFACES*

M. J. OWEN

Dow Corning Corporation
Midland, Michigan

1. INTRODUCTION

The solving of analytical problems at surfaces and interfaces typically depends on a combination of techniques to a greater extent than is usual in other areas of analysis. The nature of the problem and the information needed should determine the choice of techniques, although instrument availability often dictates the approach. The surface region in question is similarly defined by such choices. For instance, to the quality control specialist applying protective silicone-alkyd coatings, the surface region is $\sim$ 100-μm coating with the principal concerns being surface smoothness and constant thickness. To the scientist investigating effects of environmental exposure, the surface region of primary interest might be only the top 1–2 μm of that coating and the chemistry occurring therein although, should adhesive failure be involved, attention would focus on the substrate–coating interface as well. This diversity of interests accounts for the wide variety of surface analytical techniques encountered. A "complete" surface analysis not only involves several such techniques but also includes a comprehensive bulk analysis.

---

* Because of the technological importance of semiconductors, silicon has become one of the most studied materials in terms of surface properties and interactions. This field of inorganic silicon surfaces characterization is outside the theme of this chapter but organosilicon materials are sometimes involved in their preparation and study. A comprehensive listing of such articles appears biennially in the surface characterization review of *Analytical Chemistry*. Interested readers are referred to these reviews, for example, G. E. McGuire, "Surface characterization," *Anal. Chem.*, **59**, 294R (1987).

The Analytical Chemistry of Silicones, Edited by A. Lee Smith.
ISBN 0-471-51624-4 © 1991 John Wiley & Sons, Inc.

Many surface analytical techniques are rather specialized and not readily applicable to the various ill-defined, "real" surfaces usually presented to the surface analyst. Selection of the proper technique(s) may be difficult, but is an important part of the problem-solving process. Techniques for surface analysis may be classified as giving information about structural or morphological features; physical properties; or chemical properties (Table 5.1). Of the last, some techniques yield only elemental analysis, whereas others give molecular information. One carefully chosen technique from each of these categories often goes a long way towards solving a given surface analytical problem. For example, the combination of SEM, XPS (also known as ESCA), and contact angle has proved to be very useful in the characterization of silicone polymer surfaces and coatings. Each method has its strong points but the analysis resulting from all three gives a far more complete understanding of the surface than that provided by any one.

Although not mentioned in Table 5.1, chemical methods, such as titration of surface functional groups or the reaction of methylchlorosilanes with hydroxylated surfaces, can provide valuable surface analytical information.

Table 5.1. Techniques for the Analysis of Interfacial Phenomena

Structural Information

Microscopy: optical, electron (SEM, TEM)
Ellipsometry
Low energy electron diffraction (LEED)
Surface profilometry (stylus technique)

Surface Physical Properties

Surface tension
Contact angle
Langmuir trough studies
Heat of adsorption
BET surface area

Chemical Information

XPS, AES, ISS, SIMS, RBS
Electron microprobe analysis (EDS, WDS)
Diffraction: electron, X-ray
Infrared/FT IR: ATR, reflection–absorption, diffuse reflectance, photoacoustic
Inelastic electron tunneling spectroscopy
Solid-state multinuclear NMR

1.1. Defining the Surface

A highly practical definition of what is meant by "surface" is that region of the material sampled by the chosen technique. Thus for XPS, the surface region is 20–50 Å while for attenuated total reflection infrared (ATR–IR) it is several micrometers (μm). One convention regards those techniques that probe 100 Å or less as surface sensitive techniques and treats the rest as relatively thick film analytical methods. When the surface region is different in composition from the bulk, as is often the case, surface sensitivity is less an issue than absolute sensitivity. This is why techniques such as nuclear magnetic resonance (NMR), which are not intrinsically surface sensitive, can be applied effectively to certain surface analytical problems. Sometimes the silicone on a surface is the solution to a problem, as in a release coating to prevent plastics sticking in molds. Sometimes the silicone is the problem, as in the contamination of electrical contacts that causes insulating deposits. Often the presence of silicone is not suspected until the analytical evidence is available. Such surprises derive from the widespread use of silicones as processing aids (antifoams, lubricants, and release agents).

1.2. Sample Handling

Because of the transient nature of many surfaces, the analyst should put a high priority on working with the as-received surface and minimize *in situ* sample preparation and surface modification. Likewise sample selection, handling, and transfer is best done in collaboration with the analyst.

For polymeric solids, surface reorientation and its dependence on different environments introduces further sample timing and storage complications for the analyst. This phenomenon is well recognized but often ignored in polymer surface analysis. Reviews of polymer surface dynamics by Andrade and co-workers (1, 2) show that many polymers, including silicones, exhibit this behavior. Differences in surface polymeric chain orientations are particularly evident at air–water interfaces. A familiar example is provided by silicone elastomers that have been treated with hydrophilic monomers or plasma exposure to give a hydrophilic surface. They maintain their hydrophilic surfaces in contact with water but on exposure to air recover their hydrophobicity. There are a variety of possible mechanisms for this hydrophobic recovery or restructuring of the surface (3), including:

- Reorientation of surface hydrophilic groups away from the surface ("overturn" of polar groups).
- Migration of untreated polymer chains from the bulk to the surface.

- External contamination of the polymer surface.
- Loss of volatile species to the atmosphere.
- Chemical reaction of polar entities (such as surface silanol condensation in the case of plasma-treated silicones).

All these possibilities must be considered by the analyst, but whatever the mechanism of dynamic surface change it will be most obvious in flexible polymers that can reorient easily. Organosilicon polymers, particularly poly-(dimethylsiloxane) (PDMS), are extremely flexible polymers. Poly(dimethyl siloxane) has the lowest glass transition temperature of the common polymers and rotation about the siloxane bond in PDMS elastomers is essentially free, so we should anticipate that this complication will frequently occur with silicone surfaces.

2. INFORMATION PROVIDED BY KEY TECHNIQUES

The diversity of surface analytical techniques is illustrated in Figure 5.1. These sketches are based on those of Lichtman (4), an excellent, older, primary reference to surface analysis. The abbreviations and acronyms used in the figure and in the rest of this chapter are explained in Table 5.2. Figure 5.1 lists most of the important surface analytical techniques although it is not complete and new techniques are constantly evolving. Also there are important techniques whose probes (thermal or high electrical or magnetic field) do not fit this scheme including TDS, IETS, FIM, and FEM.

2.1. Structural Information

Optical microscopy is an almost universal, relatively simple, and inexpensive first step in the characterization of any surface, including organosilicon samples, often indicating which of the more elaborate techniques should next be tried. The application of optical and electron microscopy to the analysis of silicon-containing materials is detailed in Chapter 9 and will not be further discussed here.

Ellipsometry is included in the summary Table 5.1 not only because of its ability to determine thickness and refractive index of coatings but also because of its use in ascertaining the profile of drops spontaneously spreading on solids. The topic of wetting has been rejuvenated in recent years by the theoretical treatment of de Gennes (5). De Gennes and his co-workers have featured carefully purified silicone fluids extensively in experimental tests of these new theories. Such has often been the case in the almost 50 years that

Table 5.2. Abbreviations and Acronyms Used in This Chapter

AEM	Analytical electron microscopy
AES	Auger electron spectroscopy
APS	Appearance potential spectroscopy
ATR	Attenuated total reflectance
BET	Brunauer–Emmett–Teller
EDS	Energy dispersive spectroscopy (electron microprobe)
EIID	Electron induced ion desorption
ELL	Ellipsometry
ESCA	Electron spectroscopy for chemical analysis
ESD	Electron stimulated desorption
ESDIAD	ESD ion angular distribution
FEM	Field emission microscopy
FIM	Field ion microscopy
FT IR	Fourier transform infrared
HPLC	High-performance liquid chromatography
HREELS	High resolution electron energy loss spectroscopy
IETS	Inelastic electron tunneling spectroscopy
ILS	Ionization loss spectroscopy
IMXA	Ion microprobe X-ray analysis
INS	Ion neutralization spectroscopy
IR	Infrared
ISS	Ion scattering spectroscopy
LAMMS	Laser microprobe mass spectrometry
LEED	Low-energy electron diffraction
LRS	Laser Raman spectroscopy
NMR	Nuclear magnetic resonance
PAS	Photoacoustic spectroscopy
PDMS	Poly(dimethylsiloxane)
PIXE	Proton induced X-ray emission analysis
PSD	Photon stimulated desorption
RBS	Rutherford backscattering spectroscopy
RHEED	Reflected high energy electron diffraction
SEM	Scanning electron microscopy
SEMP	Scanning electron microprobe
SERS	Surface enhanced Raman spectroscopy
SEXAFS	Surface extended X-ray absorption fine structure
SIMS	Secondary ion mass spectrometry
SNMS	Sputtered neutral mass spectrometry
TDS	Thermal desorption spectroscopy
TEM	Transmission electron microscopy
TMS	Trimethylsilyl-
UHV	Ultrahigh vacuum
UPS	Ultraviolet photoelectron spectroscopy
WDS	Wavelength dispersive spectroscopy (electron microprobe)
XPS	X-ray photoelectron spectroscopy
XRD	X-ray diffraction
XRF	X-ray fluorescence

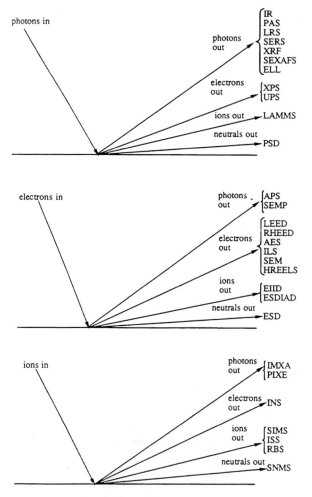

Figure 5.1. Summary of surface analysis techniques [based on a sketch by Lichtman (4)].

silicones have been available as they provide a very convenient liquid polymer. Another example is their use by Israelachvili (6) and others in the surface forces apparatus capable of measuring the forces between surfaces in liquids with a distance resolution of about 1 Å.

2.2. Surface Physical Properties

A variety of silicone properties, closely related to applications of these materials, can be used to rapidly estimate degree of uniformity and extent of surface coverage in instances where the presence of silicone has already been established. Examples of such tests include contact angle of a liquid (usually water), angle of tilt to move a slider, foaming of wash solvent, and response to applied adhesive tape. These tests reflect the familiar silicone properties of hydrophobicity, lubrication, low surface tension, and release behavior that derive from the low intermolecular forces associated with silicones such as PDMS. Other low surface energy materials such as hydrocarbon oils and fluoropolymers give similar responses so such tests cannot be used as the primary indicator of silicone presence. These are the properties that form the second major category of techniques summarized in Table 5.1.

The most useful of these properties is the contact angle and it has been used from the beginning of the silicone industry (7). It probes the outermost molecular surface of a solid and when used expertly is capable of providing considerable information. Neumann and Good (8) provide a full account of contact angle measurement. Careful evaluation of advancing and receding contact angles provides estimates of surface heterogeneity. The use of a series of liquids gives surface activity measurements such as critical surface tension of wetting and solid surface tension. The surface activity of organosilicon polymers has been reviewed (9). The PDMS fluids also provide the analyst with a useful series of contact angle test liquids to characterize low surface energy polymers such as poly(tetrafluoroethylene) (10).

2.3. Chemical Composition

The type of information required and the nature of the sample direct selection of the technique in this category. The minimum need is detection of the major elements present in the surface. Determination of the main functional groups in a surface is also a very important aspect. Rarely is a thin film or surface region characterized to the fullest extent that is possible for a bulk sample. Considerable progress towards this situation has been made in recent years but, for example, information on copolymer constituent sequencing, stereo-regularity, and cross-link distributions is sparse for organosilicon polymer surfaces.

The first set of techniques in this category in Table 5.1 (XPS, AES, SIMS, ISS, and RBS) is what many analysts think of when surface analysis methods are considered. With the exception of RBS they were made possible by the commercial availability of reliable ultrahigh vacuum (UHV) technology in the late 1960's. All of these UHV techniques have been applied to organosilicon materials. The two related electron spectroscopies, XPS and AES, are the most common as befits their status as the most commercially developed and widely applicable of these methods. Basic quantitative elemental composition is virtually guaranteed with these techniques, with oxidation state and some functional group information being a frequent bonus.

Since the early 1970s when the first XPS (11) and AES (12) studies of organosilicon surfaces appeared, hundreds of papers have been published testifying to the now routine use of these methods in surface analysis. The first XPS study featured silane coupling agents on fibers and the first AES study was concerned with silicone contamination, both subjects that continue to be topics of major interest. The XPS and AES studies of organosilicon materials are discussed in Chapter 15 so no further elaboration is included here.

Secondary ion mass spectroscopy is potentially of wide applicability to organosilicon materials as well but it is still developing particularly in polymer surface analysis. It has been used for the past 10 years, usually as a supplementary, more surface sensitive probe along with XPS or AES. Surfaces of interest that have been analyzed by SIMS include silane coupling agent treatments (13, 14) and silicone oils on low density polyethylene (15). Time-of-flight SIMS has also been applied to the analysis of PDMS surfaces (16).

Although ISS is a highly specialized surface analytical technique with a declining use base, it has been used to complement XPS data for various polymer surfaces. The polymers PDMS, poly(phenylmethylsiloxane), and poly(diphenylsiloxane) were analyzed as well as siloxane copolymers and blends (17). Rutherford backscattering spectroscopy has comparatively poor surface sensitivity compared to the other techniques in this section. It has been used to characterize inorganic silicon-containing films, usually in combination with other techniques such as AES, XRD, and SEM, but has not yet, to our knowledge, been applied to the analysis of organosilicon surfaces. A typical application is the analysis of films produced by reactive sputtering in gaseous silane mixtures (18).

The other surface compositional techniques listed in Table 5.1 are all covered in some detail elsewhere in this book so comments here are restricted to those particularly pertinent to surface analysis. The scanning electron microprobe (SEMP), built to high vacuum specifications (10^{-3}–10^{-4} Pa) offers a pre-UHV approach to surface elemental detection. As it depends on X-rays produced in the sample by interaction with the primary electron beam, it is not as surface sensitive as XPS and AES. Often it is information from this thicker region that is required and the electron microprobe techniques have a major advantage over XPS in lateral resolution. The analytical transmission electron microscope (ATEM) can also be operated in the same manner with surface sensitivity limited only by how thin a sample can practically be made. These techniques are described in Chapter 9. A broad comparison of common microanalysis techniques is given in Table 5.3. Each one has its advantages and disadvantages and careful consideration of the problem and sample nature is needed to select the most suitable method.

Table 5.3. Comparison of Microanalysis Techniques

	Element Sensitivity–Resolution			
Technique	Surface Penetration Depth	Lateral Resolution	Elemental Range	Detection Level
SEMP	1 μm	1 μm	B–Hg	0.5%
AEM	1000 Å	500 Å	B–Hg	0.3%
XPS	50 Å	150 μm	All, except H, He	0.2%
SAM	50 Å	500 Å	All, except H, He	0.5%
SIMS	10 Å	500 Å	All	ppm
LAMMS	100 Å	1 μm	All	ppm
ISS	10 Å	10 μm	All, except H	0.5%

Infrared techniques are now applied to organosilicon surface analysis as widely as are XPS and AES. This is no surprise as other methods can provide elemental composition but IR is unsurpassed in the detailed molecular information it can provide. As no vacuum is required, this approach is a rapid, cheap, easier way of obtaining surface chemical information. The specific approach chosen is governed by the nature of the sample. Reflection modes are usually used for surface analysis and customarily probe a depth of several micrometers depending on wavelength and angle of incidence; thus, this technique is significantly less surface sensitive than XPS. Multiple reflections mean lateral resolution is as poor as with XPS but single reflection accessories are now available that can focus on micrometer sized spots of suitable substrates. The advent of Fourier transform infrared (FT IR) with its higher energy throughput and spectral subtraction routines now enable quantities as low as partial monolayers to be detected and identified. Numerous sampling techniques can be used including conventional absorption measurements where a coated powder is pressed into a thin wafer, diffuse reflectance, photoacoustic spectrometry (PAS), reflection = absorption measurements, attenuated total reflectance (ATR), and IR emission spectrometry. The strengths and weaknesses of each sampling technique for FT IR surface analysis have been reviewed by Griffiths and de Haseth (19) and further information on the use of IR techniques with organosilicon materials is given in Chapter 11. Infrared spectroscopy is an excellent primary approach to organosilicon surface analysis. Examples illustrating the diversity of such studies include silane coupling agents on glass and metals (20), silicones on textiles (21) silicone elastomer surfaces (22), silicone contaminants in vacuum

Table 5.4. Comparison of Analytical Techniques for Surface Characterization of Trimethylsilyl-Derivatized Silica

Feature	Technique			
	Gravimetric	NMR	XPS	FT IR = PAS
A. Approximate data acquisition time	Minutes	Several hours	Several hours	Minutes
B. Typical sample requirements	2 g	500 mg	50 mg	25 mg
C. Approximate analysis depths	Bulk	(a)	< 100 Å	< 50 μm
D. Estimated detection limit for TMS:				
$\quad\mu$mol TMS m^{-2}	0.13	0.23	0.45	0.45
$\quad$Monolayers TMS	0.03	0.05	0.10	0.10
$\quad$% silanols reacted	1.4	2.2	4.3	4.3
E. Direct molecular information:				
$\quad$TMS groups	None	^{29}Si (15 ppm)	C 1s (285 eV)	CH$_3$ rock (845 cm^{-1})
$\quad$Silanols	None	^{29}Si (geminal -91 ppm) (single -100 ppm)	SiO$_x$(OH)$_y$ surface stoichiometry (b)	

^{a} Not inherently surface specific, dependent on nature of sample. See original article (26) for discussion.
^{b} Derived from TMS/Si ratios (XPS) versus TMS coverage (gravimetric) and O/Si ratios (XPS). See original article (26) for explanation.

(23), copolymers (24) particularly biomaterials (24), and plasma polymerized films (25).

Inelastic electron tunneling spectroscopy (IETS) is a specialized technique that produces an IR-like spectrum. It is not discussed here (see Chapter 11) but is mentioned because several investigators have applied it to silane coupling agents.

Solid state nuclear magnetic resonance (NMR) using techniques such as cross-polarization and magic angle spinning (MAS) is emerging as a powerful surface analysis tool. The ^{2}H, ^{13}C, and ^{29}Si nuclei can all be conveniently exploited for organosilicon materials. Nuclear magnetic resonance is not an inherently surface specific technique but for suitable samples such as trimethylsilylated silicas, partial monolayers can be studied (26). X-ray photoelectron spectroscopy and FT IR–PAS were also used in this study and Table 5.4 summarizes comparative criteria for these three methods. All gave responses correlating with increasing trimethylchlorosilane coverage but the solid state NMR proved particularly suited to determining reactivity differences between geminal and single silanols. Longer chain bonded silane phases (C_8 and C_{18}) for HPLC columns (27) and other silane coupling agents on silica (28) have also been studied. The ^{31}P isotope has also been exploited for phosphorus-containing silanes (29). The subject of solid state NMR is covered in more detail in Chapter 12.

Numerous other promising probes for surface chemical information have been omitted from this discussion as they have yet to be widely applied to organosilicon surfaces.

2.4. Summary

The key techniques for the surface analysis of organosilicon materials are

Optical–electron microscopy
Contact angle
Infrared (e.g., ATR–FT IR)
XPS–AES
Electron microprobe

Optical microscopy of all samples is recommended followed by SEM or TEM characterization as indicated by the initial optical examination and by the nature of the problem. Contact angle study can be very helpful particularly for flat samples. The contact angle on fibers can also be studied with suitable instruments but powders are difficult. The sample nature similarly influences the choice of surface composition probe. For vacuum sensitive materials, particularly flat reflective films, IR techniques such as reflection–absorption or ATR–IR are recommended. For surface specific analysis of a wide variety of organosilicon surfaces, XPS has become the technique of choice where lateral resolution is not critical. When it is, recourse should be made to electron microprobe or AES depending on the surface sensitivity required and sample type. As stressed earlier, satisfactory solutions to organosilicon surface analysis problems come from appropriate selection of a combination of these and lesser-used, more highly specialized methods by skilled analysts who are well informed about the problem.

3. COMMON SAMPLE TYPES ENCOUNTERED

The form of the sample presented for surface analysis has a major impact on the choice of characterization techniques. For this reason the common types of organosilicon samples encountered are reviewed in this section. The purpose is not to be comprehensive but to select a few examples from each category to illustrate how to tackle such surface analytical problems. These main categories are

- Contaminated surfaces
- Silane coupling agents and primers
- Other applied coatings and thin films
- Polymer surfaces
- Copolymers and blends

3.1. Contaminated Surfaces

Because of their low surface energy, ability to spread over their own adsorbed films, and liquid nature to high molecular weight, PDMS materials can be a perniscious contaminant. Their widespread use as additives results in problems where their presence is not suspected. Problems can arise from a variety of sources including polishes, personal care products, lubricants, stopcock grease, mold release agents, photocopier toner fluid, encapsulants, gaskets, and backstreaming from pump oils. Sophisticated surface analysis is usually not needed to detect significant contamination of this sort. Solvent washing and checking the wash solvent residue by IR for characteristic PDMS bands can be sufficient. For lesser contamination the usual techniques have been used including optical microscopy (in combination with other techniques), SEMP, XPS, AES, ellipsometry, and reflection–absorption IR.

Silicone contamination problems are particularly evident in electrical and adhesive applications. In the electrical area the problem often comes from silicone migrating from encapsulant materials, whereas in adhesion residual silicone release agent often is the culprit. Auger electron spectroscopy is a useful technique for detecting contamination on metal electrical contacts. For example, Haque and Spiegler (30) used AES to examine arcing contacts from circuit packs and Okada and Toda (31) used it to characterize the frictionally polymerized silicones that form insulating layers on the surfaces of metal brush-commutator contacts in small-size dc motors. Other complementary techniques used in silicone contamination detection and identification in the electrical area include SEM (31, 32) SEMP (32), and IR of the solvent extract (32). Recognition of this problem led to the development of nonmigrating silicone encapsulants but difficulties of this sort still arise. Interestingly, the nonsilicone controls of Okada and Toda had low levels of silicone contamination originating from the silicone release liners used with the epoxy adhesive that affixed the rubber pad to the brush to damp out brush chatter.

Numerous workers have examined the effect of silicone contamination on adhesive strength. Aluminum alloys have featured in such studies, as silicone-containing release agents are commonly used for forming parts from such alloys. Mamur et al. (33) used wettability measurements to characterize the silicone-treated alloys in a study with epoxy adhesives. Martin Marietta researchers (34, 35) found that the contamination level responsible for paint adhesive failure was readily quantified by XPS and the type of contaminant identified by reflection = absorption IR. Heptane wiping is an ineffective method for cleaning these contaminated alloy surfaces. Cleaning methods involving abrasion provide a reliable way of removing release agent residue from such surfaces prior to painting.

Silicone contamination is encountered in many unexpected areas. It has even been suggested that some of the evidence implying the existence of polywater may have been biased by silicone stopcock grease contamination (36). A good example of unexpected PDMS surface contamination is given by Schwark and Thomas (37). Their use of XPS to study polystyrene–butadiene AB copolymers showed a 30-Å surface layer of PDMS that was subsequently traced to hydrolysis of dimethyldichlorosilane used to treat the sample casting dishes. Dilks (38) has even shown that in some circumstances silicone release coating contamination on double-sided adhesive tape can migrate through as well as over polymer samples. For this reason, the use of such tape in mounting samples for surface analysis, as is commonly done with techniques such as XPS, should be avoided where low levels of silicone are being studied.

3.2. Silane Coupling Agents and Primers

This is the most prolific category by far. Although silane coupling agents were originally developed to improve glass fiber reinforced plastics, they have been widely applied to other adhesion problems both as integral additives to polymers and as primers at polymer–metal interfaces. Virtually all the techniques discussed in the previous section have been applied to the analysis of silanes, mostly on glass, silica and metal substrates, with an emphasis on modern instrumental techniques, notably XPS–AES, reflection–absorption FT IR, IETS, ^{13}C NMR, and SIMS. A number of reviews of this field have appeared, notably those by Plueddemann (39), Comyn (40), and Boerio and Gosselin (20). Many papers on the analysis of silane surfaces are included in the series of books edited by Leyden on *Chemically Modified Surfaces* (41, 42).

Many silane coupling agent and primer studies are carried out to test the numerous postulated mechanisms of action. Two examples will suffice here, both of which provide good evidence for mechanistic features often hypothesized but rarely demonstrated. Gettings and Kinloch (43) detected $FeSiO^+$ ions by SIMS from steel surfaces treated with γ-aminopropyltriethoxysilane, implying chemical bond formation at the metal oxide–silane primer interface. Chaudhury, Gentle, and Plueddemann (44) used XPS to demonstrate the interdiffusion of an aminofunctional silane into the polymer at a PVC–metal interface. They employed the Solomon et al. (45) technique of depositing a thin layer of metal ($\sim 1500\,\text{Å}$) onto a very thin fluorocarbon release coating ($\sim 100\,\text{Å}$). This is a useful way of using XPS to study buried interfaces without excessive sputtering.

3.3. Other Applied Coatings and Films

This category is an extremely broad one, encompassing systems as different as release coatings, plasma polymerized films, and modified capillary columns. Such varied surface characterization challenges elicit a variety of analytical responses. A few examples are given here to illustrate this diversity and the value of using a combination of techniques.

The commercially important area of silicone-coated papers for organic pressure-sensitive adhesive (PSA) release has been studied by SEMP (46) and contact angle and XPS (47). Differences in surface topography were revealed by SEMP, whereas no differences were apparent by XPS surface analysis. Cross-sectional SEMP showed qualitative differences in silicone soak-in to the paper. X-ray photoelectron spectroscopy was very useful in quantifying the degree of transfer of silicone to the organic PSA.

Wrobel and co-workers (48, 49) have used contact angle to characterize the surface tension of plasma polymerized dimethylsiloxane, dimethylsilazane, and dimethylsilane monomers, and subsequently to monitor their thermal decomposition. Chemical information on these films was obtained by IR–ATR, indicating that silazanes and silanes cross-link more readily than siloxanes under plasma conditions. This is one of the few direct comparisons of the effect on surface behavior of changing the backbone in organosilicon polymers.

The antimicrobial activity of quaternary ammonium compounds has been combined with silane coupling technology to give bound antimicrobial surfaces. Leyden and co-workers (50) have used diffuse reflectance FT IR and ^{13}C and ^{29}Si solid state NMR to probe such surfaces. The former technique was useful in following *in situ* the hydrolysis and condensation on the surface, while the latter technique indicated partial hydrolysis of the methoxy groups and the presence of a quaternary ammonium impurity.

3.4. Polymer Surfaces

Surface analysis of polymers is usually carried out to check the cleanliness of a surface prior to further study or to measure the extent of some type of surface treatment. An example of the former sort, of considerable interest to organosilicon surface analysts, is the XPS polymer surface study of Williams and Davis (51). They investigated the effect of argon ion sputtering on the polymers, using fractured samples of silicone rubber, silicone resin, polyurethane, and vinylidene chloride—vinyl chloride copolymer. Unlike the organic polymers, the silicones showed little change in composition with sputtering depth thus establishing the suitability of this approach for such silicones.

Numerous examples exist of the surface treatment type of study, mostly associated with attempts to make PDMS less hydrophobic. Plasma modification is a common means to this end and, as with other categories of organosilicon surface analysis, a wide range of techniques has been used to study it including contact angle, ATR–FT IR, XPS, and SIMS. A representative study is that of Triolo and Andrade (52) who were concerned with characterizing plasma-modified surfaces of commonly used catheter materials including silicone rubber.

3.5. Copolymers and Blends

This is a very active area. Because they have lower surface energy than most other polymers (with the exception of aliphatic fluoropolymers), silicone-containing polymer hybrids usually have a silicone-rich surface. This need not be the case at other interfaces such as the water–polymer interface. The composition of a copolymer surface formed against a metal can differ from the side cast in air. Polymer surface dynamics must be particularly considered in these cases. Silicones have featured markedly in this type of study because of their widespread applications in polymer surface modification. Some of the more critical of these applications are found in the biomaterial area where the inertness of PDMS is desired.

Block and graft copolymers are of particular importance. Clark et al. (53) pioneered the surface characterization of such materials with their XPS study of AB block copolymers of PDMS and polystyrene. This technique continues to be widely applied. Ratner (54) has reviewed the subject, including his restructuring studies of surface grafted hydrogels on silicone rubber in air and water. Many other silicone-containing biomaterials have been studied. One example is Avcothane, which is a copolymer of polyether, polyurethane, and PDMS that has been characterized by SIMS and ISS as well as XPS (55).

An important nonbiomaterial application is the release coating field. Hsu et al. (56) characterized siloxane-containing methacrylates by contact angle, ATR–IR, and XPS. They provide a cautionary reminder for surface analysts by showing that systems indistinguishable by all three methods gave significant differences in release performance. Inability to differentiate between differently behaving surfaces may mean that a more surface sensitive probe has to be sought, but it may also mean that the solution to the problem lies elsewhere. Lack of correlation of surface analysis with performance is common with silicone applications such as release, adhesion, and lubrication that have a bulk as well as a surface component. Often the value of surface analysis is in redirecting the focus of an investigation.

REFERENCES

1. J. D. Andrade, "Polymer surface and interface dynamics: An introduction," in *Polymer Surface Dynamics*, J. D. Andrade, Ed., Plenum Press, New York, 1988, p. 1.

2. J. D. Andrade, D. E. Gregonis, and L. M. Smith, "Polymer surface dynamics," in *Surface and Interfacial Aspects of Biomedical Polymers*, Vol. 1, J. D. Andrade, Ed., Plenum Press, New York, 1985, p. 15.

3. M. J. Owen, T. M. Gentle, T. Orbeck, and D. E. Williams, "Dynamic wettability of hydrophobic polymers," in *Polymer Surface Dynamics*, J. D. Andrade, Ed., Plenum Press, New York, 1988, p. 101.

4. D. Lichtman, "A comparison of the methods of surface analysis and their applications," in *Methods of Surface Analysis*, (*Methods and Phenomena: Their Applications in Science and Technology*, Vol. 1), A. W. Czanderna, Ed., Elsevier, New York, 1975, p. 42.

5. P. G. de Gennes, "Wetting: statics and dynamics," *Rev. Mod. Phys.*, **57**, 827 (1985).

6. J. N. Israelachvili, "Solvation forces and liquid structure, as probed by direct force measurements," *Acc. Chem. Res.*, **20**, 415 (1987).

7. M. J. Hunter, M. S. Gordon, A. J. Barry, J. F. Hyde, and R. D. Heidenreich, "Properties of polyorganosiloxane surfaces on glass," *Ind. Eng. Chem.*, **39**, 1389 (1947).

8. A. W. Neumann and R. J. Good, "Techniques of measuring contact angles," in *Surface and Colloid Science*, Vol. 11, Experimental Methods, R. J. Good and R. R. Stromberg, Eds., Plenum Press, New York, 1979, p. 31.

9. M. J. Owen, "Siloxane surface activity," in *Silicon-Based Polymer Science*, J. M. Zeigler and F. W. G. Fearon, Eds., Advances in Chemistry Series, Vol. 224, 1990, p. 705.

10. H. W. Fox and W. A. Zisman, "The spreading of liquids on low-energy surfaces. I. Polytetrafluoroethylene," *J. Colloid Sci.*, **5**, 514 (1950).

11. M. Millard and A. Pavlath, "Surface analysis of wool fibers and fiber coatings by x-ray photoelectron spectroscopy," *Text. Res. J.*, **42**, 460 (1972).

12. G. Stupian, "Auger spectroscopy of silicones," *J. Appl. Phys.*, **45**, 5278 (1974).

13. M. R. Ross and J. F. Evans, "Probing the structure of silanized surfaces with secondary ion mass spectrometry and thin film analysis," in *Silylated Surfaces*, Midl. Macromol. Monogr., Vol. 7, D. E. Leyden and W. T. Collins, Eds., Gordon & Breach, New York, 1980, p. 99.

14. F. Garbassi, E. Occhiello, C. Bastioli, and G. Romano, "A quantitative and qualitative assessment of the bonding of 3-methacryloxypropyltrimethoxysilane to filler surfaces using XPS and SSIMS (FABMS) techniques," *J. Colloid Interface Sci.*, **117**, 258 (1987).

15. D. Briggs, "Analysis of polymer surfaces by SIMS. 3. Preliminary results from molecular imaging and microanalysis experiments," *Surf. Interface Anal.*, **5**, 113 (1983).

16. I. V. Bletsos, D. M. Hercules, D. Van Leyen, E. Niehuis, and A. Benninghoven, "TOF–SIMS of polymers in the high mass range," *Springer Proc. Phys.*, **9** (Ion Form. Org. Solids), 74 (1986).

17. T. J. Hook, R. L. Schmitt, J. A. Gardella, Jr., L. Salvati, Jr., and R. L. Chin, "Analysis of polymer surface structure by low-energy ion scattering spectroscopy," *Anal. Chem.*, **58**, 1285 (1986).

18. H. O. Blom, B. Stridh, S. Berg, and J. E. Sundgren, "A comparison of AES and RBS analysis of the composition of reactively sputtered titanium silicide ($TiSi_x$) films," *J. Vac. Sci. Technol.*, A, **1**, 497 (1983).

19. P. R. Griffiths and J. A. de Haseth, *Fourier Transform Infrared Spectrometry*, Chemical Analysis, Vol. 83, Wiley, New York, 1986, p. 536.

20. F. J. Boerio and C. A. Gosselin, "IR spectra of polymers and coupling agents adsorbed onto oxidized aluminum," *Adv. Chem. Ser.*, **203** (Polym. Charact.), 541 (1983).

21. R. S. S. Murthy, D. E. Leyden, and R. P. D'Alonzo, "Determination of poly-dimethylsiloxane on cotton fabrics using Fourier transform attenuated total reflection infrared spectroscopy," *Appl. Spectrosc.*, **39**, 856 (1985).

22. Y. Nakao and H. Yamada, "Enhanced infrared ATR spectra of surface layers using metal films," *Surf. Sci.*, **176**, 578 (1986).

23. J. P. Wightman and R. H. Honeycutt, "Analysis of thin films of DC-704 on aluminum, germanium, and KRS-5 surfaces by infrared reflection spectroscopy," *J. Vac. Sci. Technol.*, **14**, 742 (1977).

24. C. S. Sung, C. B. Hu, E. W. Merrill, and E. W. Salzman, "Surface chemical analysis of Avcothane and Biomer by Fourier transform IR internal reflection spectroscopy," *J. Biomed. Mater. Res.*, **12**, 791 (1978).

25. A. M. Wrobel, "Plasma polymerization of N-silyl-substituted cyclodisilazane," in *Symposium Proceedings—International Symposium on Plasma Chemistry, 7th*, Vol. 4, C. J. Timmermans, ed., Eindhoven, The Netherlands, 1985, p. 1319.

26. R. W. Linton, M. L. Miller, G. E. Maciel, and B. L. Hawkins, "Surface characterization of chemically modified (trimethylsilyl)silicas by ^{29}Si solid state NMR, XPS, and IR photoacoustic spectroscopy," *Surf. Interface Anal.*, **7**, 196 (1985).

27. H. A. Claessens, C. A. Cramers, J. W. DeHaan, F. A. H. Den Otter, L. J. M. Van de Ven, P. J. Andree, G. J. De Jong, N. Lammers, J. Wijma, and J. Zeeman, "Aging processes of alkyl bonded phases in HPLC; a chromatographic and spectroscopic approach," *Chromatographia*, **20**, 582 (1985).

28. A. M. Zaper and J. L. Koenig, "Application of solid state carbon-13 NMR spectroscopy to chemically modified surfaces," *Polym. Compos.*, **6**, 147 (1985).

29. V. D. Alexiev, N. J. Clayden, S. L. Cook, C. M. Dobson, J. Evans, and D. J. Smith, "Solid state ^{31}P NMR spectroscopy of surface-attached triosmium clusters," *J. Chem. Soc., Chem. Commun.*, **12**, 938 (1986).

30. C. A. Haque and A. K. Spiegler, "Investigation of silicone oil contamination on relay contacts using Auger electron spectroscopy," *Appl. Surf. Sci.*, **4**(2), 214 (1980).

31. A. Okada and M. Toda, "Influence of silicone contamination on brush-commutator contacts in small-size DC motors," *IEEE Trans. Components, Hybrids, Manuf. Technol.*, **5**(2), 281 (1982).

32. N. M. Kitchen and C. A. Russell, "Silicone oils on electrical contacts—effects, sources and countermeasures," *Electr. Contacts Proc. Ann. Holm. Semin.*, **21**, 79 (1975).

33. A. Marmur, H. Dodiuk, and D. Pesach, "The effect of contamination on adhesive strength: Wettability characterization by the CSC method," *J. Adhesion*, **24**, 139 (1987).

34. K. A. Barrett, "The effect of release agent transfer on the adhesion of polymers to aluminum," *Natl. SAMPE Tech. Conf.*, **15** (20/20 Vision Mater. 2000), 617 (1983).

35. L. J. Matienzo, T. K. Shah, and J. D. Venables, "Detection and transfer of release agents in bonding processes," *Natl. SAMPE Tech. Conf.*, **15** (20/20 Vision Mater. 2000), 604 (1983).

36. D. Goring, C. Kim, A. Rezanowich, and G. Seibel, "Polywater and silicone grease," *J. Colloid Interface Sci.*, **33**, 486 (1970).

37. D. W. Schwark and E. L. Thomas, "Surface morphology of diblock copolymers," *Polym. Prepr.*, **30**, 377 (1989).

38. A. Dilks, "Polymer surfaces," *Anal. Chem.*, **53**, 802 (1981).

39. E. P. Plueddemann, *Silane Coupling Agents*, Plenum Press, New York, 1982, p. 75.

40. J. Comyn, "Silane coupling agents," in *Structural Adhesives*, A. J. Kinloch, Ed., Elsevier, London, 1986, p. 269.

41. D. E. Leyden, *Silanes, Surfaces and Interfaces*, Gordon & Breach, New York, 1986.

42. D. E. Leyden and W. T. Collins, *Silylated Surfaces*, Midland Macromolecular Monographs, Vol. 7, Gordon & Breach, New York, 1980.

43. M. Gettings and A. J. Kinloch, "Surface analysis of polysiloxane/metal oxide interfaces," *J. Mater. Sci.*, **12**, 2511 (1977).

44. M. K. Chaudhury, T. M. Gentle, and E. P. Plueddemann, "Adhesion mechanism of polyvinyl chloride to silane primed metal surfaces," *J. Adhesion Sci., Tech.*, **1**, 29 (1987).

45. J. S. Solomon, D. Hanlin, and N. T. McDevitt, "Adhesive-adherend bond joint characterization by Auger electron spectroscopy and photoelectron spectroscopy," in *Adhesion and Adsorption of Polymers*, Part A, L. H. Lee, Ed., Plenum Press, New York, 1980, p. 103.

46. J. E. Wilson and H. A. Freeman, "Analysis of silicone-coated papers with the scanning electron microprobe," *Tappi*, **64**, 95 (1981).

47. L. A. Duel and M. J. Owen, "ESCA studies of silicone release coatings," *J. Adhesion*, **16**, 49 (1983).

48. A. M. Wrobel, "Surface free energy of plasma-deposited thin polymer films," in *Physicochemical Aspects of Polymer Surfaces*, Vol. 1, K. L. Mittal, Ed., Plenum Press, New York, 1983, p. 197.

49. A. M. Wrobel, J. Kowalski, J. Grebowicz, and M. Kryszewski, "Thermal decomposition of plasma-polymerized organosilicon thin films," *J. Macromol. Sci. Chem.*, **A17**, 433 (1982).

50. R. S. S. Murthy, G. S. Caravajal, and D. E. Leyden, "DRIFT and NMR spectroscopic studies of an antimicrobial silane," in *Silanes, Surfaces and Interfaces*, D. E. Leyden, Ed., Gordon & Breach, New York, 1986, p. 141.

51. D. E. Williams and L. E. Davis, "Sputter induced compositional changes during ESCA/sputtering of polymers," *ACS Div. Org. Coat. Plast. Chem. Preprints*, **36**, 249 (1976).

52. P. M. Triolo and J. D. Andrade, "Surface modification and evaluation of some commonly used catheter materials. I. Surface properties," *J. Biomed. Mater. Res.*, **17**, 129 (1983).

53. D. T. Clark, J. Peeling, and J. M. O'Malley, "Application of ESCA to polymer

chemistry. VIII. Surface structures of AB block copolymers of polydimethylsiloxane and polystyrene," *J. Polym. Sci. Polym. Chem. Ed.*, **14**, 543 (1976).

54. B. D. Ratner, "Graft copolymer and block copolymer surfaces," in *Surface and Interfacial Aspects of Biomedical Polymers*, Vol. 1, J. D. Andrade, Ed., Plenum Press, New York, 1985, p. 373.

55. S. W. Graham and D. M. Hercules, "Surface spectroscopic studies of Biomer," *J. Biomed. Mater. Res.*, **15**, 465 (1981).

56. T. Hsu, S. S. Kantner, and M. Mazurek, "Surface effects in siloxane-containing methacrylates," *Polym. Mater. Sci. Eng.*, **55**, 562 (1986).

CHAPTER

6

PERSONAL CARE APPLICATIONS

HELEN M. KLIMISCH

Dow Corning Corporation
Midland, Michigan

1. INTRODUCTION

The use of silicones in the personal care market dates back to the early 1950s. The first commercial skin care application was Revlon's "Silicare." This product was marketed as a protective hand lotion based upon the hydrophobic and barrier-forming properties of poly(dimethylsiloxanes). A product identified as a lotion spray, Sudden Date, was marketed by Lanolin Plus for hair care (1). This product was to be sprayed on a tired hairdo that required immediate freshening. The silicone ingredient caused the hair to be shiny and more spirited. These early applications have since expanded to include antiperspirant formulations, shampoos, hair conditioners, sun screen formulations, and cosmetic products such as lipstick, powder, and eye make-up (2–6).

The silicone primarily used has been poly(dimethylsiloxane) (PDMS), whose CTFA (Cosmetics, Toiletries, and Fragrance Association) name is dimethicone. The beneficial properties generally associated with dimethicones are summarized in Table 6.1. More recently, other siloxane polymers and copolymers have been used to a greater degree.

The unique properties of silicones in personal care applications are responsible for the dramatic increase in their use. Although silicones can be difficult to formulate with, because of solubility and compatibility problems, their beneficial properties are manifest at low levels.

The objective of this chapter is to provide guidance for detection of a silicone, identification of the specific siloxane polymer, and, if necessary, quantitative analysis for the silicone in a formulated personal care product or on a surface on which the product has been used. The types of silicone polymers used in personal care applications are discussed. Next, methods

The Analytical Chemistry of Silicones, Edited by A. Lee Smith.
ISBN 0-471-51624-4 © 1991 John Wiley & Sons, Inc.

117

Table 6.1. Why Dimethicones Are Used in Personal Care

Chemical–Physical Property	Product Benefits
Physiological and chemical inertness	Nontoxic, nonirritating and nonsensitizing
Low surface tension	Excellent spreading and film-forming agents
Unique molecular structure	Lubricate without oiliness
	Impart smooth velvety feel to skin
	Water repellency
High refractive index	Imparts gloss and sheen
Incompatibility with aqueous and some organic systems	Detackification of ingredients

useful for detecting silicones, using techniques detailed in other chapters, are highlighted. Very few methods for the analysis of silicones are unique to personal care products and applications.

2. TYPES OF SILICONES USED

The predominant silicone used in personal care applications is dimethicone. Its structure is presented in Figure 6.1 along with the chemical formulas and CTFA designation for other silicones used in skin care products. The dimethyl silicone polymers can exist as linear polymers with n (degree of polymerization or DP) varying from 0 to > 1000, and as cyclic chains or rings (CTFA designation cyclomethicone). The cyclic structures having four or five siloxy units are most frequently used and are referred to as volatile silicones. A variety of other silicone polymers is produced by replacing one of the methyl groups with different organic substituents. Examples are given in Figure 6.1, wherein the R group is phenyl, stearoxy, or polyether. Copolymers, such as dimethicone copolyol, can yield a myriad of compositions as m, n, x, and y are varied. In addition, the polyether chain length can vary, causing further complexity of the copolymer composition.

Aminofunctional polymers are another class of silicones, used primarily in hair care applications. Examples of two aminofunctional polymers with their CTFA designations are illustrated in Figure 6.2. In amodimethicone polymers, end groups are silanol rather than trimethylsiloxy. In addition, internal silanol groups produce potential cross-link sites. Cross-linking gives branch sites within the molecule and ultimately leads to formation of insoluble

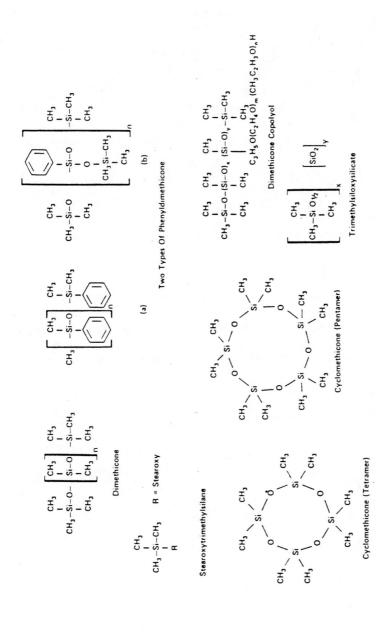

Figure 6.1. Structures and CTFA designations of some silicones used in personal care applications (3). © 1982, *Cosmetic Technology*, Aster Publishing Corp., 859 Willamette St., P.O. Box 10460, Eugene, OR, 97440. All rights reserved.

119

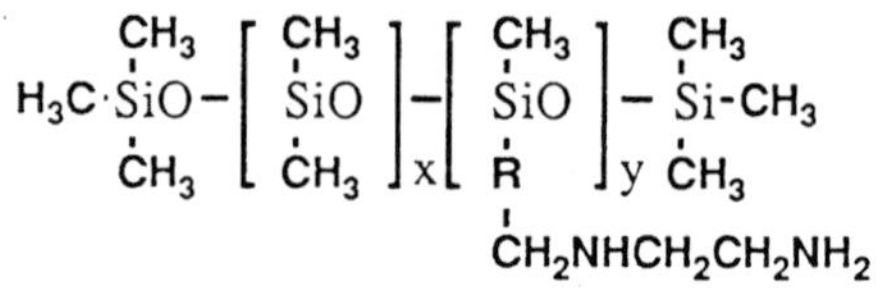

$$\underset{\substack{\displaystyle CH_2NHCH_2CH_2NH_2}}{H_3C \cdot \underset{\overset{\displaystyle |}{CH_3}}{\overset{\displaystyle CH_3}{Si}O} - \left[\underset{\overset{\displaystyle |}{CH_3}}{\overset{\displaystyle CH_3}{Si}O}\right]_x - \left[\underset{\overset{\displaystyle |}{R}}{\overset{\displaystyle CH_3}{Si}O}\right]_y \underset{\overset{\displaystyle |}{CH_3}}{\overset{\displaystyle CH_3}{Si}\text{-}CH_3}}$$

Trimethylsilylamodimethicone

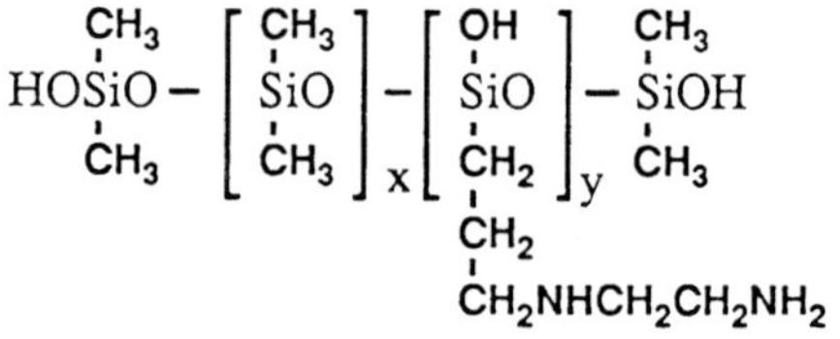

$$HO\underset{\overset{\displaystyle |}{CH_3}}{\overset{\displaystyle CH_3}{Si}O} - \left[\underset{\overset{\displaystyle |}{CH_3}}{\overset{\displaystyle CH_3}{Si}O}\right]_x - \left[\underset{\substack{\displaystyle |\\ CH_2\\ |\\ CH_2\\ |\\ CH_2NHCH_2CH_2NH_2}}{\overset{\displaystyle OH}{Si}O}\right]_y - \underset{\overset{\displaystyle |}{CH_3}}{\overset{\displaystyle CH_3}{Si}OH}$$

Amodimethicone

Figure 6.2. Aminofunctional silicone polymers and their CTFA names.

polymer networks. Network formation causes difficulty in quantitative analysis, especially after deposition of the polymer on a surface.

The following sections contain more detailed descriptions and physical properties of the major silicone types. Additional information can be found in Refs. 2–6. It is important to understand the physical properties of silicones in general and of the specific polymer types, either to identify or quantitatively determine the silicone polymer.

2.1. Dimethicone

As previously stated, the DP, or value of n in Figure 6.1, can vary from 0 to > 1000. Hexamethyldisiloxane, at 0.65 cS ($n = 0$) is completely volatile. The volatility of the polymer decreases as DP increases and approaches a limit around a viscosity of 100 cS. The physical and solubility properties change quite dramatically through this range. The lower viscosity fluids have greater miscibility with mineral oils and solvents such as isopropanol. The fluids with viscosities of 100 cS and above are the most commonly used in personal care applications. These fluids are incompatible with mineral oils and alcohols, but are soluble in aromatic solvents, hexanes, methylene chloride, tetrahydrofuran, and chloroform (cf. Table 3.1).

Dimethicones are noted for their thermal and oxidative stability relative to organic oils. Also unique to dimethicones is their molecular chain flexibility and weak attractive forces. The rotation of the dimethylsiloxy unit around the siloxane bond is nearly free because of the nondirectional character of the SiO bond and the large Si–O–Si bond angle (cf. Chapter 1, Section 5). This flexibility leads to low surface tension and excellent film formation tendencies. The dramatic difference in properties of dimethicones relative to mineral oil is illustrated in Figure 6.3, which illustrates the change in viscosity with molecular weight. The mineral oil viscosity increases rapidly with increasing molecular weight while the dimethicone viscosity increases more slowly. A mineral oil with a viscosity of 20 cS has a molecular weight of about 360 while a corresponding dimethicone has a molecular weight of about 1500. Another way to illustrate the differences between most organic oils and dimethicones is shown in Figure 6.4. The chromatograms in Figure 6.4 are from GPC separations using chloroform as solvent and polystyrene molecular weight standards. The chromatograms compare two mineral oil fluids, Klearol and Kaydol, with petrolatum. The molecular size of the sample increases as the viscosity increases. Also in Figure 6.4, petrolatum is compared to two dimethicone fluids of 20 and 50 cS viscosities. The molecular sizes of the dimethicones are large compared to petrolatum but the viscosities are lower. Another difference between organic oils and dimethicones is the relatively flat viscosity–temperature profile of dimethicones.

Dimethicones have a low surface tension that contributes to film formation properties. The films act as barriers to moisture but allow free passage of gases. The ease of spread and lubricity of the fluids produces the characteristic velvetlike feel of silicones.

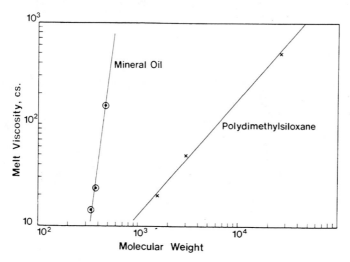

Figure 6.3. Dependence of viscosity on molecular weight for silicone and organic polymers.

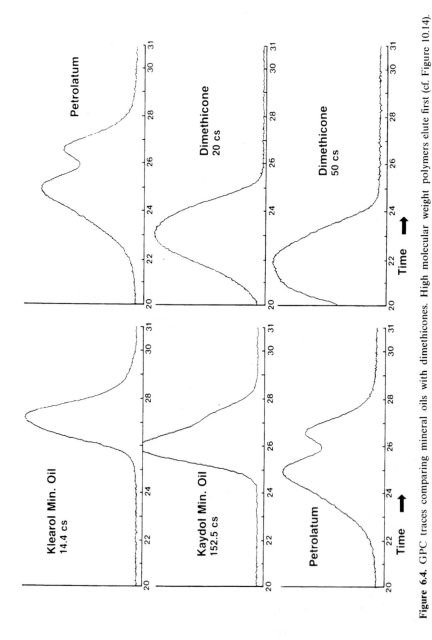

Figure 6.4. GPC traces comparing mineral oils with dimethicones. High molecular weight polymers elute first (cf. Figure 10.14).

2.2.   Cyclomethicones

Two examples of cyclic siloxanes (tetramer and pentamer) are illustrated in Figure 6.1. Cyclomethicones are volatile, low viscosity fluids. They provide the general properties of dimethicones during application such as lubricity, ease of spreading, and detackification but evaporate leaving no residue. Cyclomethicones are compatible with most personal care ingredients including isopropanol and mineral oil. These cyclic fluids have low heats of vaporization, so cooling during evaporation is minimal. This unique property makes the cyclics useful as solvents or additives in antiperspirant products in place of water and alcohol. The heats of vaporization of these materials are presented in Table 6.2. Because of their volatility, cyclomethicones in products may be lost during storage or analysis, and accurate quantitative measurements may be difficult.

Table 6.2. Heats of Vaporization

Material	Heat of Vaporization (cal/g)
Water	539
Ethanol	210
Tetramer component of cyclomethicone	32
Pentamer component of cyclomethicone	32

2.3.   Dimethicone Copolyol

Dimethicone copolyol is the CTFA designation of all silicone glycol (polyether) copolymers, illustrated in Figure 6.1. Replacement of one or more methyl groups along the dimethicone chain with other organic groups produces a change in the physical properties of the silicone. The predominant influence of incorporating polyether chains, or polyoxyalkylated substituents, into the silicone polymer is on its solubility characteristics. Water and polar solvent miscibility increase as the mole% of the polyether moiety increases (value of y in Figure 6.1). Complete water solubility occurs when the ratio of polyether to dimethylsiloxy groups is in the range of 2–4. These copolymers are generally used in predominantly aqueous systems wherein the typical silicone aesthetic properties are required. However, because of their water solubility, these polymers are not as durable or resistant to water and detergent wash-off.

Polyethylene glycol is hygroscopic, and incorporation of this material into the siloxane polymer confers hygroscopicity to the silicone. Most polyethylene glycols are solid above a molecular weight of 600, but such polyether functionality in a siloxane produces a fluid, nonsolid polymer. The flexibility of the siloxane backbone overcomes the polyether character. It is also possible, however, to produce solid, waxy silicone glycol copolymers by proper selection of the amount and chain length of the polyether portion.

The combination of hydrophilic and hydrophobic units in the polymer molecule results in surfactant activity. Aqueous solutions of dimethicone copolyols foam and tend to stabilize foaming products. However, dimethicone copolyols also exhibit inverse temperature solubility. Thus, these polymers act as antifoams above the phase inversion temperature.

2.4. Aminofunctional Polymers

Two types of silicone polymers used predominantly in hair care applications have the CTFA designations amodimethicone and trimethylsilylamodimethicone, illustrated in Figure 6.2. In these materials, the organic group replacement for a methyl group contains an ethylene diamine function. The amine functionality increases the potential for interaction of the silicone polymer with proteinaceous substrates so that enhanced durability of the polymer on hair results.

Both polymer types illustrated in Figure 6.2 are fully protonated in an aqueous media near pH 7. Thus, the molecule possesses multiple positively charged sites for interaction with the negatively charged hair surface. The amodimethicone polymer is delivered only in emulsified form and contains reactive silanol functionality as well. As the emulsion breaks on a surface, the silanol groups react through both chain extension and cross-linking to increase the molecular weight. The resulting deposited film cannot be redissolved or reemulsified. The trimethylsilylamodimethicone polymer contains only amine functionality and no silanol. This polymer is linear and remains linear upon deposition; thus, it is possible to redissolve the polymer. However, the interaction between the amino groups and the hair surface is very strong so that extraction of the polymer is extremely difficult.

3. SILICONES IN FORMULATED PRODUCTS

The analyst is typically asked to determine if a silicone polymer is present in a product, what type of silicone is present, and possibly how much of the silicone is present. It is critical to define the problem before starting any laboratory experiments, and any effort expended at this point is well spent.

Definition of the problem also forces the analyst to think through the reasons for performing the analysis, and possible approaches to the problem. This process is especially important when a sample is submitted for analysis to an analytical service group. The following list of questions is not designed to be all-inclusive but to help direct the thought processes.

1. Is the sample a raw material or formulated product?
2. If a raw material, are recommended methods available from the supplier?
3. If a formulated product, what is the form of the product: aqueous emulsion, solid, dispersion, or solvent based?
4. How much is known about product? What are its ingredients?
5. Why do I need to know if silicone is present?
6. Is it important to identify the type of silicone, if present?
7. Are the other ingredients in the product to be identified?
8. Is the silicone volatile or nonvolatile?
9. Is qualitative information adequate or is quantitative analysis required?

As noted from these questions, it is important to know something about the product itself because the matrix will influence the methods used in the analysis. Chapters 2–4 are recommended reading for additional background information.

3.1.   Separation Techniques

The type or form of the formulated product will strongly influence the separation technique to be used. Personal care products are generally water based or solvent based. Examples of water based products are shampoos, conditioners, sun screen products, and hand and body lotions. Nonwater based products include antiperspirants and deodorants, hair sprays and fixatives, and cosmetic products such as lipstick, eye makeup, and powder. The first step is to answer the easiest question: Is a silicone polymer in the product?

Infrared spectroscopy (Chapter 11) is the best and easiest way to detect silicones. In fact, the IR spectrum can usually be used to identify the type of silicone also. Because water is not a suitable IR solvent, water based products require special sample preparation. The sample can be evaporated to dryness and the nonvolatile portion placed on a salt plate for transmission analysis. Volatile silicones will be lost from the product during the drying step. A second approach is to extract the water based product with a solvent such as

CCl_4 and then analyze the solution. A comparison of the results from the two techniques indicates the presence of volatile components, including silicones, in the original product. Another approach is to use a liquid cell ATR–IR technique (Chapter 11, Section 2). The liquid sample is placed in the cell, the IR spectrum scanned, and the background solvent subtracted to produce a spectrum of the product components. Of course, the minimum detection limit with each technique must be defined if the interest is in trace work. Similar approaches can be used with nonwater based products. Direct sampling of the product on a salt plate or ATR prism permits detection of nonvolatile silicones. For volatile silicones, solvent extraction of the product followed by an IR scan of the solution is preferred. If the amount of silicone present is very small, it may not be detected in the IR scan of the product. In this case, a concentration step is necessary. One recommended procedure (7) for emulsions (or creams) is to shake the emulsion with twice its volume of 1:1 dioxane:toluene, centrifuge, and remove the top layer, which contains the silicone. The solvent is evaporated and the residue examined for silicone using IR.

Chapter 3, Section 2.2.2 contains more detailed separation techniques for a wide variety of mixtures. Additional techniques are presented in Chapter 4.

3.2. Identification

A variety of instrumental and chemical methods can be used to identify the silicone polymer. The chemical methods are discussed in Chapter 8, and will not be expanded upon here. The instrumental techniques are IR, near-IR, Raman, and UV spectroscopy (Chapter 11), NMR spectroscopy (Chapter 12), mass spectrometry (Chapter 13), atomic spectroscopy (Chapter 14), and XRF (Chapter 15). The selection of the method to be used is dependent on the technique available and the information required. The first two categories are applicable to all types and molecular weight silicone polymers and provide specific structural information on siloxane molecules. The methods in the fourth and fifth categories are nonspecific in that they give total Si values, which include silaceous materials as well. Some specificity can be introduced through use of a separation technique in which soluble silicone polymers are physically separated from nonsoluble silicas and silicates. Mass spectrometry is limited by the volatility of the sample and the mass range of the spectrometer but is specific for organosilicon moieties.

The analyst must be aware of the possibilities for contamination of the sample by unexpected sources of silicone. These polymers are found in an ever increasing variety of products, and their use as process aids is also growing. Some potential contamination sources are discussed in Chapter 4, Section 1.2.

3.3   Quantitation

Quantitation requirements are defined by the analysis method selected. Most products will be mixtures of compounds, the complexity of which complicates both the separation and the analysis. Care in the separation technique is the controlling factor in precise and accurate quantitation. One caution concerns the nature of silicones to spread on surfaces. Once present, the silicone is extremely difficult to remove. This behavior can severely hamper quantitation methods, especially where aqueous systems and glass surfaces meet. The use of blanks, synthetic samples, and proper calibration are essential for success.

Silicones in creams and lotions have been determined by shaking the sample with CH_2Cl_2 and aqueous HCl, passing the organic layer through silica gel, and concentrating the effluent for measurement by FT IR. The siloxane absorptions at 1098 and 1015 cm^{-1} were used, after subtracting the spectrum of a blank. Relative standard deviation for samples containing 1% dimethicone was 1.8% (8).

4.   SILICONES ON SURFACES

As with analyzing for silicones in formulated products, definition of the problem is key to the analysis. Is the problem concerned with testing for surface contamination or directed towards proof of deposition?

Testing for contamination of surfaces by silicones can involve simply placing a piece of cellophane tape against the surface. If the tape does not stick, the presence of silicones is suspected. The tape can subsequently be placed on an ATR prism and the IR spectrum scanned. Often some of the silicone will adhere to the tape and a characteristic siloxane IR fingerprint will appear. This method is especially suited for detection of silicones on skin. Washing the surface with a small quantity of solvent, collecting the solvent, and scanning the IR after solvent dry down should show the presence of silicones. Another approach is to check the hydrophilicity of the surface with a water droplet. Silicones produce a hydrophobic surface with minimal water spreadability. This approach is not specific but indicative, and thus further testing is required. The discussion in Chapter 4, Section 1 on contamination, losses, and artifacts in analysis should be consulted.

A number of methods have been used to detect and prove the presence of silicones on surfaces. Chapter 5 on "Analyses of Surfaces" is recommended reading. Most applications concern detection of Si qualitatively rather than quantitatively. The following discussion concerns skin surfaces and fibrous surfaces such as wool and hair. The methods are further classified into total Si versus siloxane specific categories. The total Si methods are nonspecific and thus differentiation between silicates and silicones may be impossible or extremely difficult. For the siloxane specific category, the solubility character- istics of the silicone polymer may aid in the differentiation and identification process.

4.1. Total Si Methods

Auger and ESCA spectroscopy (Chapter 15) are excellent methods for detect- ing the presence of Si in small samples. These are surface techniques and can be used to measure the concentration of Si in the surface volume analyzed, provided the sample dimensions are known. Only *in vitro* analyses of skin samples can be performed in the high vacuum required. Data presented in Ref. 4 illustrate the use of ESCA with treated hair samples. The data presented in Figure 6.5 compare both the deposition level and the durability of two types of silicone polymers. The natural abundance of silicon in hair is very low, thus, the silicon detected by ESCA in this case is assigned to the presence of the silicone polymer. The hair had been treated with 1.75% emulsions of each polymer type. The amodimethicone deposited about twice the level of silicone as the dimethiconol polymer. The treated hair samples were subjected to a series of shampoos with a sodium lauryl sulfate-based

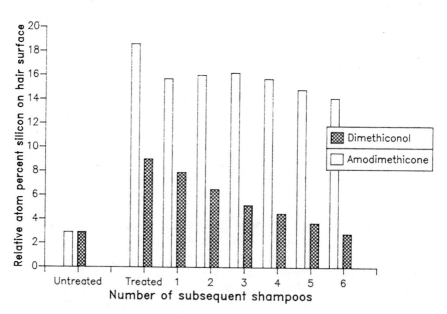

Figure 6.5. Comparative durability of amodimethicone and dimethicone on hair.

formulation. The dimethicone concentration on the hair decreased with the number of shampoos while the amodimethicone level remained relatively constant. These data clearly illustrate the affinity of the amine function for a keratin surface.

One advantage of the ESCA technique is that all silicone on the surface is detected irrespective of molecular weight, cross-linking, and functionality. However, results from systems that contain both silicas and silicones may be difficult to interpret. Success depends on the resolution of the instrument as well as appropriate use of controls and simulated samples.

X-ray fluorescence is a technique that measures all Si present irrespective of form and type. The analysis is nonspecific for silicones but can be quantitative based upon the bulk sample. Fibrous samples must be homogenized, as with a cryogenic grinding apparatus, for placement in the instrument. The background Si concentration must be carefully evaluated. Chapter 15 provides additional information on these techniques.

4.2. Siloxane Specific Methods

Atomic spectroscopy (AA or ICP) is useful for quantitation of total Si, provided the silicone can be gotten off of the surface and into solution. A simple solvent extraction of the substrate may not be sufficient to remove the silicone quantitatively. Sonication improves recovery in solvent extraction over mechanized agitation. An interesting technique has been developed (9) that involves enzymatic breakdown of the hair structure followed by solvent extraction and detection of Si by AA. The enzyme used was papain with sodium sulfite as catalyst. An aqueous solution of the reagents is mixed with cut-up hair samples and heated at 65 °C for 3 days. A liquid–liquid extraction of the mixture is conducted with methyl isobutyl ketone (MIBK) and a small quantity of hydrochloric acid. The MIBK dissolves the silicone polymer and this layer is assayed for Si content. Reagent controls and untreated hair samples must be carried through the procedure along with the treated hair samples to define the background contribution. Success with this method requires that the silicone polymer be soluble in the solvent. Cross-linked polymers are not quantifiable by this technique. It is important that when a specific method is selected, synthetic standards be prepared that simulate real-life samples, to check the analyst and the technique.

Infrared spectroscopy can be used for detection and quantitation of silicones on surfaces. A simple solvent extraction with sonication of the substrate followed by the IR scan may be sufficient but the efficiency of this approach must be evaluated. Dimethicones, cyclomethicones, and dimethicone copolyols may be amenable to this approach, if proper standardization is used. However, aminofunctional and cross-linked polymers are not as

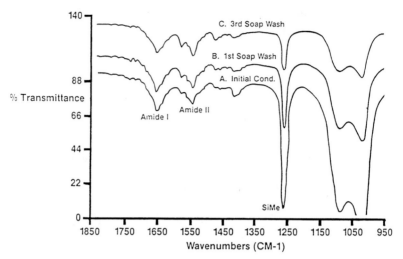

Figure 6.6. IR spectra of treated skin before and after washing (10). Courtesy *J. Soc. Cosmet. Chem.*

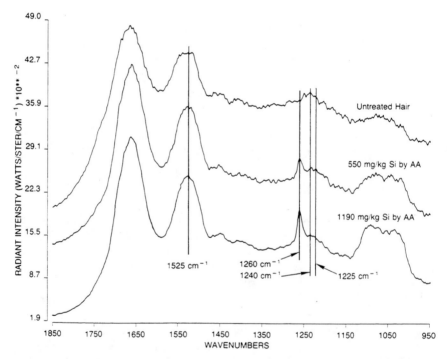

Figure 6.7. IR spectra of silicone treated hair samples, showing use of the band ratio method to quantify treatment levels (11). Courtesy *J. Soc. Cosmet. Chem.*

130

easily analyzed. A quantitative method has been developed for *in vivo* analysis of silicones on skin using ATR with FT IR (10). A literature review on the use of ATR on skin is also given in this paper. Methodology was developed to control skin–prism contact. Quantitation was achieved by using the amide II protein band from skin as an internal standard. The method was used to study the soap wash resistance or substantivity of a variety of dimethicone fluids. The IR spectra in Figure 6.6 illustrate the results obtained with a germanium prism, and the change in the intensities of the SiMe band relative to the amide II band with soap washing. The method was also used to evaluate the influence of the silicone as a substantivity aid for other personal care ingredients. The example given used mink oil as the ingredient. This ATR–FTIR technique is applicable to all nonvolatile polymers regardless of functionality or cross-linking.

Diffuse reflectance IR (DRIFTS) has been used for quantitation of silicones on keratin surfaces (11). A review of the literature pertaining to DRIFTS is also included in this paper. Proper sample preparation is critical for this method and procedures were defined and optimized for both surface and bulk analysis of the fiber. A band ratio technique was used to quantitate

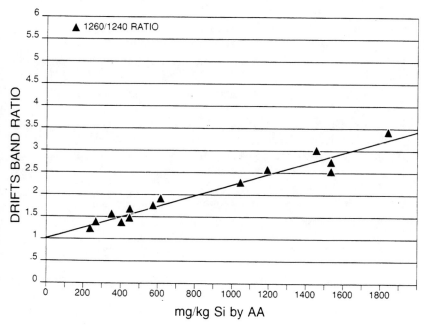

Figure 6.8. Correlation of IR (diffuse reflectance) and AA data for Si (11). Courtesy *J. Soc. Cosmet. Chem.*

the siloxane polymer and is illustrated in Figure 6.7. The band ratios, 1260:1240 and 1260:1225 cm^{-1} were correlated to milligrams per kilogram of Si data generated independently by atomic absorption analysis of hair fibers. The correlation of band ratio with AA data is presented in Figure 6.8. As with ATR, this technique is applicable to all nonvolatile linear or cross-linked polymers.

It is also possible to use GC as a technique for measuring silicones on surfaces. Volatile silicones are immediately quantifiable via GC with head-space analysis. Nonvolatile silicones must be removed from the surface and reduced in molecular weight to increase their volatilities. Examples and further discussion are presented in Chapter 8, Section 3.7.

REFERENCES

1. R. Y. Lochhead, "The history of polymers in hair care," *Cosmetics and Toiletries*, **103**(12), 23 (1988).
2. S. R. Wendel, "Silicones for personal care," *Household and Personal Products Industry*, 53 (Aug. 1982).
3. M. S. Starch and C. R. Krosovic, "Silicones in skin care products," *Cosmetic Technology*, 20 (Nov. 1982).
4. S. R. Wendel and A. J. DiSapio, "Organofunctional silicones for personal care applications," *Cosmetics and Toiletries*, **98**(5), 103 (1983).
5. M. S. Starch, "Silicones in hair care products," *Drug and Cosmetic Industry*, 38 (June 1984).
6. B. Idson, "Polymers in skin cosmetics," *Cosmetics and Toiletries*, **103**(12), 63 (1988).
7. A. J. Senzel, Ed., *Newburger's Manual of Cosmetic Analysis*, 2nd ed., Association of Official Analytical Chemists, Washington, DC, 1977, p. 38.
8. M. Sabo, J. Gross, and I. E. Rosenberg, "Quantitation of dimethicone in lotions using Fourier transform infrared spectral subtraction," *J. Soc. Cosmet. Chem.*, **35**, 273 (1984).
9. E. G. Gooch and G. S. Kohl, "Method to determine silicones on human hair by atomic absorption spectroscopy," *J. Soc. Cosmet. Chem.*, **39**, 383 (1988).
10. H. M. Klimisch and G. Chandra, "Use of Fourier transform infrared spectroscopy with attenuated total reflectance for in vivo quantitation of polydimethylsiloxanes on human skin," *J. Soc. Cosmet Chem.*, **37**, 73 (1986).
11. H. M. Klimisch, G. S. Kohl, and J. M. Sabourin, "A quantitative diffuse reflectance method using Fourier transform infrared spectroscopy for determining siloxane deposition on keratin surfaces," *J. Soc. Cosmet. Chem.*, **38**, 247 (1987).

THE BASIC TECHNIQUES

CHAPTER

7

PHYSICAL PROPERTIES AND POLYMER STRUCTURE

ORA L. FLANINGAM and NEAL R. LANGLEY

Dow Corning Corporation
Midland, Michigan

1. INTRODUCTION

This chapter describes important physical properties of typical organosilicon materials. Section 2 emphasizes measurement techniques for polymers and materials of lower molecular weight. Section 3 addresses typical silicone polymers and their important structural aspects such as molecular weight distribution, branching, cross-linking, and phase composition. Properties and techniques of interest in Section 3 are those that serve to characterize these aspects of polymer structure. The importance of the structure is illustrated by relationships to properties that make silicones especially useful.

2. THE MEASUREMENT OF PHYSICAL PROPERTIES

The physical properties of silicones are useful for their identification, determination of purity, and characterization of structure.

In general the same methods used to measure physical properties of nonsilicone materials can be used for silicones (1–3). We discuss specific techniques only when they differ from established methods, or when the properties of the silicones require variation from conventional practices.

The two major difficulties in measuring physical properties of silicone compounds are first, it is sometimes difficult to obtain a pure sample and second, many of the common organosilicon monomers are reactive and must be handled with care to avoid contamination during routine transfers to the measuring apparatus.

The Analytical Chemistry of Silicones, Edited by A. Lee Smith.
ISBN 0-471-51624-4 © 1991 John Wiley & Sons, Inc.

The chemist used to handling halocarbon compounds or alkoxycarbon compounds may, at first, be unaware that most chlorosilanes and alkoxysilanes are easily hydrolyzed by atmospheric moisture. Thus, any open air transfer of these materials will result in some degree of reaction with production of HCl or an alcohol, and a silanol or siloxane. The silicon–nitrogen bond is similarly subject to hydrolysis with formation of ammonia.

Physical property data on organosilicon compounds can be found in Noll (4). Other sources are Bazant et al. (5–9), Eaborn (10), and Voorvoeve (11). *Lange's Handbook of Chemistry* (12) lists more than 280 organosilicon compounds. The *CRC Handbook of Chemistry and Physics* also lists about 100 organosilicon compounds (13). When complete, the DIPPR Data Compilation (14) will contain evaluated data on about 30 organosilicon compounds.

Most of these sources give only the melting and boiling points, density, and refractive index. The DIPPR data compilation has values for 24 "point properties" and 13 temperature dependent properties. Lange's handbook also gives vapor pressure constants for about 20 organosilicon compounds, and the CRC handbook has about 120 such compounds in its vapor pressure tables.

In many cases, literature values are not very reliable. The reasons for this are inadequate or poorly executed measurement techniques; the difficulty of obtaining pure materials; and the high reactivity of many organosilicon monomers. For this reason, evaluated compilations such as the DIPPR data compilation are valuable resources. Values for physical properties of some silicone monomers and polymers are given in this chapter and in the Appendix. These tables include the most common organosilicon materials, but are not intended to be complete.

2.1. Refractive Index

Refractive index measurements are easily made on organosilicon monomers and polymers using the Abbé refractometer (15) as long as the usual precautions of temperature control and calibration are observed. In addition, if the measurements are to be made on reactive materials, precautions should be taken to ensure that exposure to the atmosphere during loading and measurement does not cause decomposition.

Refractive index measurements of monomers are useful in determining both their structure and their purity. The molar refraction, (M_R), is calculated by summation of the bond (16–20) or group refractions (21–30) in Table 7.1 and comparison with the experimental M_R from Eq. (7.1).

Table 7.1. Bond and Group Refractions for Calculating Molar Refractions[a]

Bond[c]	Refraction (mL/mol)	Bond	Refraction (mL/mol)	Bond	Refraction (mL/mol)
		Bond Refractions[b]			
SiO	1.80				
SiC (al)	2.52	SiF	1.70	SiN	2.16
SiC (ar)	2.93	SiCl	7.11	SiS	6.14
SiSi	5.89	SiBr	10.08	CH	1.674
SiH	3.17	SiI	15.92	CC	1.296

Group	Refraction (mL/mol)	Group	Refraction (mL/mol)	Group	Refraction (mL/mol)
		Group Refractions[d]			
SiMe	7.58	SiO n-Pr	17.65	SiOSi	3.62
SiEt	12.09	SiO i-OPr	17.71	SiCH$_2$Cl	14.21
Si n-Pr	16.82	SiO n-Bu	22.26	SiCH$_2$OH	9.12
Si n-Bu	21.51	SiO i-Bu	22.30	SiOH	3.5
Si i-Bu	21.50	SiO s-Bu	22.23	SiCH$_2$=CH$_2$	11.73
SiPh	27.52	SiO t-Bu	22.29	SiCH$_2$–	5.87
SiOMe	8.28	SiO n-Hex	31.53		
SiOEt	13.00	SiO cyclo-Hex	29.30		

[a] Reference 1 recommends 1.95 for SiF.
[b] See Refs. 10 and 16–20.
[c] al = aliphatic and ar = aromatic.
[d] See Refs. 10 and 21–30.

Agreement of calculated and experimental values constitute partial proof of structure.

$$M_R = (n^2 - 1)MW/(n^2 + 2)d \qquad (7.1)$$

where n = refractive index, MW = molecular weight, and d = density.

The refractive index of a polymer is a function of both the structure and of the molecular weight. The variation of refractive index for poly(dimethylsiloxane) (PDMS) polymer can be seen in Tables 7.2 and 7.3. Hexamethyldisiloxane (MM) is the shortest of the PDMS polymers and has a refractive index of about 1.375. As the molecule is lengthened by insertion of Me$_2$SiO (D) units, the refractive index increases asymptotically to about 1.4035. The

Table 7.2. Physical Properties of Linear Poly(dimethylsiloxanes)[a]

Compound[b]	Refractive Index (n_D^{25})	Density at 25 °C (g/mL)	Viscosity at 25 °C (cS)	Boiling Point (°C)	ΔH_v (kcal/mol)	Melting Point (°C)
MM	1.3748	0.7619	0.65	100.5	7.47	− 68.22
MDM	1.3822	0.8200	1.04	152.5	8.3	− 82
MD$_2$M	1.3872	0.8536	1.53	196.4	9.1	− 68
MD$_3$M	1.3902	0.8755	2.06	229.9	9.8	− 81
MD$_4$M	1.3922	0.8910	2.63	259.7	10.4	− 59
MD$_5$M	1.3940	0.911	3.24	286.8	11.1	− 78
MD$_6$M	1.3952	0.913	3.88	310.4	11.5	− 63
MD$_7$M	1.3959	0.9134	4.46		11.8	< − 75
MD$_8$M	1.3965	0.9191			12.0	− 65
MD$_9$M	1.3971	0.9233	5.94			
MD$_{10}$M[c]	1.3977	0.9245				
MD$_{11}$M[c]	1.3980	0.9305	7.48			
MD$_{12}$M[c]	1.3985	0.9330				
MD$_{13}$M[c]	1.3989	0.9335	9.25			
MD$_{14}$M[c]	1.3991	0.9386				
MD$_{15}$M[c]	1.3995	0.9406	10.60			
MD$_{16}$M[c]	1.3996	0.9242				
MD$_{17}$M[c]	1.3998	0.9441	12.24			
MD$_{18}$M[c]	1.4000	0.9452				
MD$_{19}$M[d]	1.4002	0.9479	13.78			
MD$_{22}$M[d]	1.4039	0.9731				
MD$_{29}$M[d]	1.4035	0.9727				

[a] See Refs. 4, 31–33.
[b] M = $Me_3SiO_{1/2}$; D = Me_2SiO.
[c] These compounds were about 95% pure (see Ref. 33).
[d] These compounds were less than 95% pure (see Ref. 33).

cyclic compounds in Table 7.3 have refractive index maximum of about 1.405 near 12 siloxane units. The phenyl containing polymers have higher refractive indexes and the phenyl content can be correlated with the refractive index. A pure methylphenylsiloxane polymer has a refractive index of about 1.531.

2.2. Density and Specific Gravity

The usual cgs units for density are grams per cubic centimeter (g·cm^{-3}) and the recommended SI units are kilograms per cubic meter (kg·m^{-3}). The specific gravity of a liquid is defined as the ratio of the mass of the material at

Table 7.3. Physical Properties of Cyclic Poly(dimethylsiloxanes)[a]

Compound	Refractive Index (n_D^{25})	Density[b] at 25°C (g/mL)	Viscosity at 25°C (cS)	Boiling Point (°C)	ΔH_v (kcal/mol)	Melting Point (°C)
D_3	(s)[b]	1.12(s)		135.1	8.2	64.5
D_4	1.3940	0.950	2.30	175.4	8.8	17.5
D_5	1.3957	0.9528	3.87	211.0	9.5	− 44
D_6	1.3991	0.9621	6.62	245.0	10.0	− 3
D_7	1.4016	0.9694	9.47	275.6	10.5	− 26
D_8	1.4039	1.177(s)	13.23	303.2	10.8	+ 31.5
D_9	1.4048	0.9756	17.98	325.6	11.3	− 23.5
D_{12}	1.4050	0.9763	24.75	363		
D_{15}	1.4043	0.9737	24.56?	404		
D_{18}	1.4037	0.9736	24.82	∼ 440		
D_{21}	1.4035	0.9724	24.65	∼ 465		
D_{24}	1.4039	0.9731				
D_{27}	1.4036					
D_{30}	1.4035	0.9726				

[a] See Refs. 31, 34, and 35.
[b] (s) = solid.

some stated temperature to the mass of an equal volume of distilled water at a specified reference temperature. Since it is a ratio, specific gravity is dimensionless. A specific gravity of 0.761 25/15 means that the material at 25 °C has 0.761 times the mass of an equal volume of water at 15 °C. Although specific gravity and density are numerically similar, they are not identical in magnitude.

The ASTM compilation (1) lists many methods for making measurements of density and specific gravity (36–39). The two most common methods are by hydrometer and by pycnometer. When hydrometers are used with silicone fluids, the low surface tension of the sample may necessitate an appreciable correction. The specific gravity correction may be as much as ± 0.006 for hydrometers that were calibrated with high surface tension liquids. Measurement of the density of silicone materials using pycnometers are made by the usual methods (40). For materials that are volatile, reactive, or measured over a range of temperatures, it has been found convenient to use a bicapillary pycnometer of the Lipkin type that has been fitted with ground glass joints. The capping of the arms minimizes evaporation and hydrolysis.

Electronic vibrating tube densimeters (ASTM D4052, Mettler/Paar DMA-40 or DMA-45, or Mettler DA-200 or DA-210 or equivalent) have also been found to be very useful. Because the tube can be sealed during the measurement, sample loss or hydrolysis can be avoided. Very reactive materials can be successfully measured. By taking care to avoid bubbles in the sample, one can use the densimeters for fluids having viscosities up to about 15,000 cS. These densimeters can operate over a temperature range for determining temperature coefficients of density or coefficients of thermal expansion.

Densities of monomers and low molecular weight silicone fluids are more informative than they are for higher molecular weight polymers. As can be seen in Tables 7.2 and 7.3, the density of polymers approaches an asymptotic value at relatively low molecular weight.

2.3. Flash Point

Flash point testing is well defined by ASTM methods (41, 42). These methods may be either manual or automated. The automated methods have the advantage that the heating rate may be more closely controlled than for the manual methods. This factor is especially important when performing flash point testing on silicone polymers. Flash testing on polymers actually is done to detect volatile materials. The volatiles require some time to accumulate in the head space above the liquid, and different apparent flash values may be obtained depending on the rate of heating. In some cases, variations of up to 75 °C have been seen. A steady temperature increase up to the flash point will give more reliable values than will rapid heating. An automatic flash point apparatus may not always detect the flash properly. This problem is caused by formation of silica from combustion of silicone materials, which may desensitize the detector. Visual observation of the flash should be the rule.

Aside from the above problems, flash point determinations on silicone materials are similar to those on organic materials.

2.4. Viscosity

The viscosity of organosilicon monomers is usually in the range 0.5–10 cS [viscosity (cS) = viscosity (cP)/density (g/cm^3)]. Values for polymers can range from < 1 cS to more than a million centistokes. Tables 7.2 and 7.3 show the viscosities of some linear and cyclic poly(dimethylsiloxanes). Figure 7.1 shows the relationship of viscosity to molecular weight for the low molecular weight pure PDMS species, and for narrow molecular weight fractions of higher molecular weight polymers (43). The phenylmethylsiloxanes and fluorosilicone polymers have similarly wide viscosity ranges. The polymer viscosities are also affected by molecular weight dispersity and by branching of polymer chains (Section 3.2). Commercial fluids may contain a

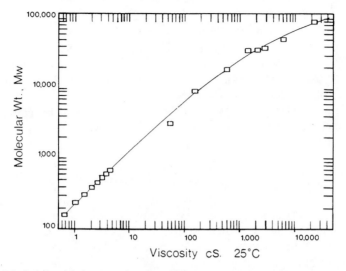

Figure 7.1. Relationship between viscosity of PDMS and weight-average molecular weight.

wide range of molecular weights resulting from the manufacturing process, or from blending of fluids of different viscosities. The relationships between polymer viscosities of blended fluids has been investigated by Kataoka and Ueda (44).

2.4.1. Measurement of Viscosity

Because of the wide range of possible viscosities mentioned above, the choice of which viscometric method to use depends on the material being measured. For materials with viscosities below 100,000 cS, glass capillary viscometers can be used, provided adequate standards are available for calibration. Any type of viscometer can be used, although the Ubbelohde tube, a suspended level viscometer, is preferred.

Recommended procedures for viscosity measurements have been given in detail in the literature and by the ASTM (45, 46). Viscosities above 100,000 cP can be measured by rotational or falling ball viscometers (45). A reverse flow viscometer or a viscometer similar to the Zahn cup can be used to measure viscosity of emulsions or opaque materials.

With glass capillary viscometers, the sample is exposed to air each time it is sucked into the upper bulb for another measurement. With reactive compounds this operation can lead to considerable exposure to moist air and consequent hydrolysis. This problem can be countered by filling the viscometer with dry nitrogen and keeping a nitrogen filled "air bag" connected to the viscometer when such materials are measured. With Ubbelhode visco-meters, the bag is connected to the viscometer by three tubes connected to the same bag through plastic tubing "tees" so that there is no difference in pressure between any of the tubes. Analysis of "before" and "after" samples from tests done over several hours have shown negligible hydrolysis of very reactive chlorosilanes.

Poly(dimethylsiloxane) fluids can be used as standards for calibration of viscometers used with organic fluids, provided the viscometer is well cleaned after calibration.

2.4.2. Cleaning Viscometers

Dimethylsiloxane and phenylmethylsiloxane polymers are soluble in aro-matic and chlorinated hydrocarbon solvents such as toluene, xylene, or chlorothene. Fluorosilicone polymers are more soluble in ketones, so a better solvent for them is acetone. Once the polymer is removed, the cleaning solvent may be removed with acetone and the viscometer can be air dried (see also Chapter 2, Section 8.3).

Glassware in contact with silicone fluid becomes hydrophobic, and in subsequent use with aqueous solutions may display nonwetting tendencies. Viscometers that are so treated give low efflux times and erroneous viscosities when used with aqueous solutions. If less rigorous cleaning methods fail, the glass may be made hydrophilic again by cleaning as suggested in Chapter 2, Section 8.3. The user should be cautioned about the dangers of using corrosive cleaning solutions. They are dangerous to the person, and they act by removing part of the surface of the glass. After any such cleaning recalibration is necessary.

2.4.3. Silicone Fluids as Standards

Silicone fluids are used as secondary standards for calibration of viscometers. The prime prerequisite for such standards is long-term stability. Trimethylsil-oxy end-blocked PDMS fluids with viscosities below about 1000 cS have very good stability. Retesting of silicone standards indicates no drift in viscosity due to oxidation, polymerization, or degradation over a number of years. High viscosity fluids may drift to a higher viscosity if some residual silanol is present, or may drift lower if acidic or basic contaminants are present. If these factors are absent, the fluid should be stable for long periods. Silicone fluids for secondary standards should be carefully chosen and if necessary, sub-jected to heat aging at 200 °C for 16 h to test long-term stability. Little or no viscosity change should be found for a fluid selected as a secondary standard.

Secondary standard fluids should be measured in master viscometers that have been calibrated with oils traceable to the National Institute of Science and Technology.

2.4.4. Flow Behavior of Silicone Fluids

The viscous behavior of poly(dimethylsiloxane) fluids has been investigated in many laboratories (43, 44, 47–52). The rheology of PDMS containing fillers was measured by Currie (49) and Warrick (53). Poly(dimethylsiloxane) exhibits constant viscosity at low shear rates. However, if the shear rate is increased, the viscosity begins to decrease (non-Newtonian behavior). The molecular weight distribution (MWD) of the polymer influences the point at which this change takes place. Poly(dimethylsiloxane) with broad MWD shows deviation from Newtonian behavior at lower shear rates than does narrow MWD material (Figure 7.2). The shear thinning at high shear rates is attributed to polymer disentanglement and alignment in the direction of flow at high flow rates (49). Rheological behavior of filled silicone polymer systems varies with the polymer and filler combination.

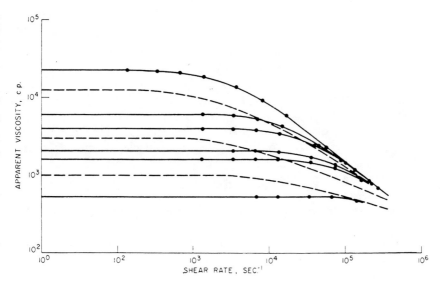

Figure 7.2. Apparent viscosity of PDMS as function of shear rate for (----) commerical grade PDMS and (——) narrow distribution PDMS (43). Courtesy of Journal of Polymer Science.

2.5. Melting Point

The melting point of a pure compound is a well-defined thermodynamic property. Data have been obtained on many organosilicon monomers (see the Appendix). The melting point of organosilicon compounds can be measured by any standard technique. Since, however, many of the melting points are well below room temperature, one of the most convenient methods is that of differential scanning calorimetry (DSC). The ASTM has defined procedures (54) for this measurement. Once a material has been sealed in a DSC pan, it can be easily cooled to low temperatures and the melting point measured. As discussed in Section 3.4, silicone polymers can be supercooled 30–40 °C below their melting point, and if cooled rapidly enough may supercool to a glassy state without crystallizing (55, 56). The melting behavior of silicone polymers is also dependent on their thermal history during the cooling cycle (55, 56). Several "melting points" may be found and their values may vary significantly. True melting points for some PDMS polymers may be found by slow cooling through the − 40 to − 50 °C temperature range, or by holding at these temperatures for several days. Once the polymer has crystallized, it is transferred without warming to a DSC apparatus and scanned in the usual manner.

2.6. Vapor Pressure, Boiling Point, and Heat of Vaporization

2.6.1. Vapor Pressure

Boiling point and vapor pressure of organosilicon compounds may be measured in the standard manner using ebulliometric or isoteniscopic techniques (57–60). The boiling tube of Stull (31, 61) also is useful for vapor pressure measurements on many substances.

Many silicone compounds are very thermally stable and we have measured some vapor pressures above 300 °C. Since temperature measurement is very important, some discussion of it is worth while. If glass thermometers are used without attention to calibration or emergent stem correction, errors of 5–10 °C may be present. Platinum resistance thermometry (31, 61) is the best way to avoid such errors. Of course, calibration of the working thermometers is important in any case. If platinum resistance thermometry is used, the current in the thermometer bulb should be kept as low as possible consonant with the required temperature sensitivity so as to avoid heating the thermometer with the measuring current. About 1 mA has been found sufficient for 25-Ω thermometers. Modern thermometer bridges and null detectors or direct reading resistance thermometers are a great improvement over traditional bridges and galvanometers. Thermocouples have been used, but a single couple is not sensitive enough, and thermopiles involve extra calibration.

Pressure is the other important variable and its measurement should be attended with as much care as is taken with temperature measurement. Use of mercury manometers can give good results if all the usual precautions are taken. In this age of electronics, however, capacitance manometers or similar pressure sensors are much more convenient, but they must also be calibrated and checked. A secondary standard can be used where its boiling temperature is measured simultaneously with that of the unknown and its known vapor pressure–temperature equation is used to calculate the vapor pressure of the unknown (59).

As mentioned earlier, care must be taken with reactive silicon compounds to prevent hydrolysis in sample transfers. Evacuating the apparatus and filling it with dry nitrogen prior to introducing the sample is an excellent way to avoid possible hydrolysis.

In the ebulliometric method, materials with high surface tension and those with polar groups sometimes tend to "bump" when boiled at low pressures and this leads to superheating and erratic temperatures. A continuous read-out of the temperature on a strip chart is invaluable in determining when smooth boiling is occurring. With care, boiling temperatures can routinely be reproduced to within 0.05 °C or better.

The boiling of a substance is the temperature at which the material boils at 1 atm (760 torr or 1.01325×10^5 Pa). If the boiling temperature is known at a pressure near 1 atm, the corrected boiling point may be calculated by standard correction methods (12, 13), or it may be calculated from the Antoine equation (60) or other vapor pressure–temperature equation.

2.6.2. Heat of Vaporization

Heat of vaporization is almost never measured, as it can be calculated from vapor pressure data using the Clapeyron or Clausius–Clapeyron equation (60, 62–64).

2.7. Vapor–Liquid Equilibrium

Vapor–liquid equilibrium (VLE) data are important in the design of distillation columns and other separation processes. Thousands of publications on VLE of various compounds have appeared in the literature and many are of questionable accuracy. As a first step, the compilations of Wichterle et al. (65–68) should be checked to see whether VLE data has ever been determined on a given system. These bibliographies contain many references to VLE of organosilicone materials. If the system has been measured, the original publication is probably listed in these bibliographies. They do not, however, tell whether the data is acceptable, and much of it is not. The VLE data should always be subjected to thermodynamic consistency checks (see below). There are several good compilations of VLE data (69–73), but they contain very little on silicon compounds.

The books of Hala et al. (72, 73) give a good overview of both the theory of VLE and of the methods used for its measurement.

For measurements of VLE of silicone compounds we have had good results with a modified version of the Ellis still (72–74) (Figure 7.3). Our modifications improve handling of the reactive compounds such as the chlorosilanes. Trichlorosilane and silicon tetrachloride cause glass stopcocks in contact with these liquids to "freeze." For this reason all stopcocks were located above the liquid level. Again, because of the sticking problem, all stopcocks were made of Teflon®. The sampling entry above the stopcocks was made to fit Viton® septa. This double seal was found to be effective in preventing loss or reaction of the samples. Sampling is done by long syringe needles. Samples are analyzed by any convenient method, but GC is usually used as the samples can be small. Gas chromatography is also valuable for detecting any decomposition products that may interfere with the VLE measurement or with other analytical techniques.

Because it is difficult to obtain large quantities of high purity samples, the size of the Ellis still was reduced by about $\frac{1}{2}$ so that the amount of liquid required is about 40–60 mL. Pressure and temperature measurements can be made by any suitable method. We have found it convenient to use the same apparatus as is used for vapor pressure measurements.

Besides being reactive, many of the organosilicon compounds in industrial use have similar boiling points and are difficult to separate. This means that more care must be taken to obtain reproducible and thermodynamically consistent results. It is important to subject the data to tests for this consistency (72, 73, 75).

2.8.   Autoignition Temperature

The autoignition temperature methods defined by the ASTM (76) may be used for silicone materials. The internal thermocouple should be placed at the bottom of the flask, since this is where ignition occurs. It has been found that some insulation materials can cause lowering of the autoignition point (77) of silicone polymers and heat transfer fluids. Users of these fluids in high temperature processes should be aware of this problem. Autoignition temperatures (AIT) are included in Table 7.4, which shows the flammability properties of several silicone compounds.

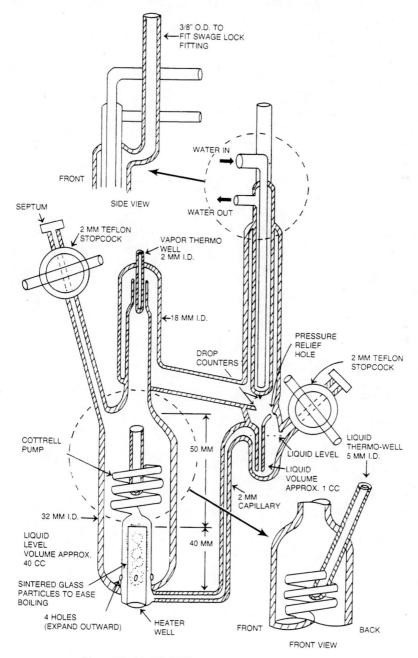

Figure 7.3. Modified Ellis vapor–liquid equilibrium cell.

147

Table 7.4. Flamability Data for Some Silicon Compounds[a]

Compound	AIT[b] (°C)	LFL[b] (v/v%)	UFL[b] (v/v%)	Min. O$_2$ Vol%	Temperature (°C)	Comments
SiH$_4$	100	0.0	100			Stable mixtures of silane in air are not known to exist. Explosions in air can be violent
H$_2$SiCl$_2$	215	4.1	98.8		25	Violent explosions. 78–86 vol% ignites spontaneously
HSiCl$_3$		6.9	>70			
SiCl$_4$						Not flammable
Me$_4$Si	450					
Me$_3$SiH	310					
Me$_2$SiH$_2$	230					Spontaneous explosions may occur
MeSiH$_3$	130					
ViSiH$_3$	90					
Me$_3$SiCl	395	2.0	6.2	10.3	Room temperature	
Me$_2$SiCl$_2$	>400 / 425	>5.5 / 3.4	10.4 / >9.5			
MeSiCl$_3$	395 / 455	7.2 / 7.6	>20 / >70	12.9		
MeHSiCl$_2$	230	4.5 / 3.4 / 5.5	>24 / 17.2	2.9	Room temperature / Room temperature	

	AIT[b]	LFL[b]	UFL[b]		
Me_2HSiCl	263	2.0	83.0		
$ViSiCl_3$	400				
$PhSiCl_3$					
$(Me_2SiO)_4$	400	0.75	7.4	11.8	125
$(Me_3Si)_2O$	340	1.25	18.6	10.0	125
Me_3SiOH	395	1.45	27.5	9.9	125
$(Me_3Si)_2NH$		0.80	16.3	9.7	125
$Me_2Si(OMe)_2$		1.50	14.8	10.4	80
$MeSi(OMe)_3$		1.50	10.4	10.4	100
$(MeHSiO)_4$	239				
$(MeViSiO)_x$	250				
$ViSi(OEt)_3$	268				
$EtSi(OEt)_3$	235				
$MeSi(OEt)_3$	225				
$Bu_2Si(Et)_2$	217				
$PhViSi(OEt)_2$	258				
$ClPrSi(OEt)_3$	230				
$ViSi(OAc)_3$	> 400				

[a] See Refs. 34 and 78–81.
[b] AIT = autoignition temperature.
LFL = lower flammability limit.
UFL = upper flammability limit.

Chloropentamethyldisilane has been reported (82) to give a violent exotherm when distillation under a blanket of N_2 was attempted. Extreme caution is advised during preparation and purification of this and related disilanes.

2.9. Flammability Limits

Although many silicone polymers are not extremely flammable and will self-extinguish because of the silica formed on the burning surface (83), some silicone monomers and oligomers are highly flammable and may pose safety problems if not handled properly. Silanes containing hydrogen bonded to silicon are especially flammable and can be explosive in some concentrations. Some flammability data are given in Table 7.4.

The Bureau of Mines publications (84, 85) of Coward and Zabetakis review flammability characteristics and methods of determining them. Determination of flammability concentration limits is a time-consuming process and can be hazardous if not done properly. The measurements are usually done by introducing the gas or vapor into a tube and mixing it thoroughly with air to obtain a mixture of known composition. Ignition is initiated by a spark at one end of the tube. The process of combustion should be viewed from a safe position, perhaps by a mirror. A test is judged to be "positive" if the flame propagates the length of the vessel. The vessel is usually a glass tube 2–4 in. inside diameter, and $\sim$ 30 in. long.

If the material is not volatile enough to obtain a vapor concentration of about 1% or more by volume, the flammability tube may be placed in an oven in order to reach temperatures that give adequate concentrations for the test.

The lower flammability limit is usually somewhat lower than the stoichiometric composition calculated for complete combustion. There have been a number of papers on estimation of upper and lower flammability limit (86–90). Some of the lower flammability calculation methods appear to be fairly reliable. The upper flammability methods show considerable variability and should be used with caution, if at all.

3. STRUCTURE AND PROPERTIES OF POLYMERS

3.1. Molecular Weight

3.1.1. Significance

Important properties of a polymer usually are dependent on the molecular weights of the polymer molecules. The molecular weights, in turn, depend on the type and extent of the polymerization reaction, and upon subsequent processing such as the removal of components of low molecular weight (MW), or blending. For example, in silicone polymers, low-MW components often are observed. These include cyclic species that exist in equilibrium with the long-chain linear polymer. Average MW values become larger with an increase in the extent of the polymerization reaction, with a decrease in the concentration of chain-terminating species in the monomer composition, or with removal of low-MW cyclic and linear silicone species.

3.1.2. Description

The molecular weight distribution (MWD) of a polymer sample describes the relative numbers of molecules of all MW values. It can be characterized in several ways.

Graphically, a frequency distribution plot gives the number fraction, n, or weight fraction, w, of molecules for all values of MW. For example, Figure 10.14, obtained by GPC, shows relative values of w versus MW for a PDMS sample with the low-MW cyclic species noted above. Different types of distribution plots in the literature arise from the use of number or weight fractions, differential or integral (cumulative) distribution functions, and the use of MW or the degree of polymerization (DP = number of repeat units per molecule) as the independent variable (91). For PDMS, the MW of the $(CH_3)_2SiO$ monomer repeat unit is 74; thus MW = 74 × DP.

Mathematically, empirical equations and their parameters may be found to represent measured MWDs. The equations can be compared with theoretical ones derived for various mechanisms of polymerization. For example, the following reaction shows a hypothetical step-growth polymerization of PDMS from a low-MW linear precursor of DP = x.

$$i\text{H(OSiMe}_2)_x\text{OH} \longrightarrow \text{H(OSiMe}_2)_{ix}\text{OH} + (i-1)\text{H}_2\text{O} \qquad (7.2)$$

It produces the random or most probable" distribution.

$$n_i = (1-p)p^{i-1} \qquad (7.3)$$

where n_i is the fraction of polymer molecules containing i oligomers and p is the fraction of oligomer ends condensed together (91).

Averages of MW, such as the number-average $\bar{M}_n$, the weight-average, $\bar{M}_w$, or the "Z-average" $\bar{M}_z$ are parameters commonly used to describe the general shape of the MWD. They are defined by the first, second, and third moments of MW, respectively,

$$\bar{M}_n = \Sigma \, (MW)n_i / \Sigma n_i \tag{7.4}$$

$$\bar{M}_w = \Sigma \, (MW)^2 n_i / \Sigma \, (MW)n_i \tag{7.5}$$

$$\bar{M}_z = \Sigma \, (MW)^3 n_i / \Sigma \, (MW)^2 n_i \tag{7.6}$$

Higher moments are used to define the averages $\bar{M}_{z+1}$, $\bar{M}_{z+2}$, and so on. Advantages of the use of these averages to describe MWD are that a small number of parameters can be used instead of a complete distribution function or equation, and that the averages can be measured directly by the techniques discussed below. The disadvantage is that a small number of the averages cannot define uniquely the MWD. For the example in Eq. (7.2) of step-growth polymerization (91)

$$\bar{M}_n = M_0 / (1 - p) \tag{7.7}$$

Here $M_0 = 74x$, the increase in MW from each $(OSiMe_2)_x$ growth step. Thus, for 99% condensation of the silanol ends, $p = 0.99$ and $M_n = 7400x$. Conversely, p could be calculated from $\bar{M}_n$ by Eq. (7.7), or from any other MW average by a corresponding equation (91).

3.1.3. Measurement

Methods for measuring MWD and MW averages for silicones are the same as for other polymers, and are summarized in Table 7.5. Special considerations for silicones include the choice of solvent and means of calibrating the methods.

Table 7.5. Methods for Measuring Molecular Weights of Polymers

Methods	Results	MW Range (MW/1000)	Comments
GPC	MWD, MW averages	> 1	Requires calibration
Membrane osmometry	$\bar{M}_n$, χ	15–1500	Slow equilibration
Vapor pressure osmometry	$\bar{M}_n$	< 10	
End-group analysis	$\bar{M}_n$	< 50	Assumes no branching
Light scattering	$\bar{M}_w$, $r_{g,z}$, z	> 10	Requires very clean solutions
Solution viscometry	$\bar{M}_v$, a	> 1	Requires calibration

Gel permeation chromatography (GPC), sometimes called size exclusion chromatography, is probably the most common, convenient and useful method. This technique, described in Chapter 10, gives the MWD and MW averages calculated from the MWD. Gel permeation chromatography is a relative method, usually requiring calibration of the columns with known MW fractions of the same polymer to obtain an absolute MWD. If an additional detector such as a differential viscosity (96) or light scattering (92) detector is used, this calibration is not needed.

Membrane osmometry and *vapor pressure osmometry* are absolute MW methods, giving $\bar{M}_n$ over complementary ranges (91). The concentration dependence from the former method also yields the Flory–Huggins interaction parameter, χ. This parameter expresses the enthalpy of interaction of the polymer and solvent, and increases with the solvating power of the solvent. Although both $\bar{M}_n$ methods are theoretically absolute, results should be checked using known MW standards of some polymer. Methods of measuring $\bar{M}_n$ based on the other colligative properties of boiling point elevation and freezing point depression now are seldom used on synthetic polymers such as silicones.

End-group analysis is another absolute method giving $\bar{M}_n$. It assumes knowledge of the number of end groups per polymer molecule, which is two in the absence of branching. In this case, it follows that the relationships of $\bar{M}_n$ to the end-group concentration E_t in moles of end groups per gram is

$$\bar{M}_n = 2/E_t \qquad (7.8)$$

This equation indicates the high accuracy required for the analysis in order to determine high $\bar{M}_n$ values accurately. If $\bar{M}_n$ is 50,000, the end-group analysis for an unbranched polymer would be 4×10^{-5} and must be determined with the same relative accuracy as that desired for $\bar{M}_n$. Common reactive end groups on silicones include silanol, vinyl, alkoxy, and acetoxy groups, which can be analyzed by chemical or spectroscopic methods. Even the "nonfunctional" Me_3Si- ends can be analyzed by ^{29}Si NMR. The detection limit of the latter is about 0.1 mol% Si (113), corresponding to a degree of polymerization of 2000 and $\bar{M}_n$ of about 150,000 for PDMS. Relative accuracy is about 5% at 1 mol% Si where $\bar{M}_n$ is $15,000 +/-750$.

Light scattering (93) is the most common method of absolute $\bar{M}_w$ measurement. It is based on the theory of scattering by a collection of independent particles, such as the scattering of light from a dilute homopolymer solution. The intensity of the scattered light depends on the scattering angle and polymer concentration. Both of these variables are extrapolated to zero by means of a Zimm plot to calculate $\bar{M}_w$. If a modern laser instrument that measures low-angle scattering is used, the scattering angle extrapolation is not necessary. A measure of the molecular size, $r_{g,z}$, the z-average radius of gyration, also can be obtained from the dependence on scattering angle. Finally, a measure of the shape of the particles can be obtained from the dissymmetry coefficient of scattering, z, at angles 45° and 135°. The accuracy and lower limit of $\bar{M}_w$ are improved by increasing the difference of the refractive indexes of polymer and solvent. For PMDS, therefore, toluene or chloroform are better solvent choices than THF, which has nearly the same refractive index as PDMS (Table 7.6). For poly(methylphenylsiloxane), the order of preference for these solvents is reversed.

Table 7.6. Common Solvents for Siloxane High Polymers

Siloxane Polymer	Solvent	Refractive Index, 25 °C
Dimethyl (PDMS)		1.404
	THF	1.405
	Chloroform	1.446
	Toluene	1.496
Methylphenyl		1.531

Solution viscometry (91) is widely used to determine a "viscosity-average" MW, $\bar{M}_v$, from a measured intrinsic viscosity by the Mark–Houwink–Sakurada equation.

$$[\eta] = K\bar{M}_v^a \tag{7.9}$$

The constants K and a are found from a calibration using the same polymer, solvent, and temperature as for the unknown determination. Values from the literature have been compiled for several types of siloxane polymers (94). For PDMS in the theta solvents bromocyclohexane at 29.0 °C or phenetole at 89.5 °C, $K = 0.074$ mL/g and $a = 0.50$ for $45{,}000 < \text{MW} < 1{,}060{,}000$ (95). The calibration constant K must be corrected if the unknown polymer has a broad MWD (94). The $\bar{M}_v$ thus obtained is related to the MWD by the constant a, and is only slightly less than $\bar{M}_w$ unless the MWD is quite broad (91).

The intrinsic viscosity $[\eta]$ is defined as the limit

$$[\eta] = \lim_{c \to 0} (\eta/\eta_0 - 1)/c \tag{7.10}$$

where η_0 and η are, respectively, the viscosities of the solvent and of a dilute polymer solution in the same solvent at mass concentration c. From the measurements at several values of c (usually < 1 or $2 \, g/100 \, mL$) the extra-polation in Eq. (7.10) to $c = 0$ gives $[\eta]$. The viscometry technique is described above in Section 7.2.4. A differential viscometer (96) also can be used to estimate $[\eta]$ from η/η_0 of a single dilute solution. When used as a GPC detector, the viscometer yields data that permit the Mark–Houwink–Sakurada constants in Eq. (7.9) to be estimated for the polymer, solvent, and temperature of interest. Information on the shape and size of dissolved polymer molecules can be obtained from the constant a. This constant is theoretically related to their shape (coils, spheres, disks, or rods), how the polymer domain sizes scale with MW, and other factors (97).

3.1.4. Effects of MWD on Properties

Performance and processibility of silicones and other polymer products depend on combinations of their basic mechanical, thermal, interfacial, and diffusional properties. These properties, in turn, often depend on the MWD. For example, pumping or extruding of silicone fluids or compounds depends on the silicone viscosity, its shear-thinning behavior, and the viscoelastic storage of energy during extrusion, which may produce "die swell." The strong dependence of the viscosity and shear thinning of undiluted PDMS on $\bar{M}_w$ is discussed in Section 7.2. When $\bar{M}_w$ exceeds about 35,000 for PDMS (its critical or entanglement MW) the viscosity dependence becomes even stronger (98). Here, the viscosity at low shear rate increases in proportion to $\bar{M}_w^{3.4}$ because of the entanglement of polymer coils.

The viscoelastic response also depends on MWD, as well as on the test temperature, time of deformation, and strain level (98). For polysiloxanes, relaxation of viscoelastic deformation is very rapid compared to organic polymers with their carbon chain backbones. Thus, for silicones of lower MW, the deformation time must be short (high dynamic frequency) to capture elastic effects. Lamb and co-workers (99, 100) describe the viscoelastic response at and above audio frequencies for PDMS fluids with viscosities of 100–100,000 cS. Plazek et al. (101) describe viscoelasticity up to audio frequencies for PDMS samples of higher MW. Even for cured elastomers, the MWD before curing can be important. If the polymer ends are functionally active and form the cross-links, then the MWD before curing becomes the distribution of network chain lengths between cross-links, which determines the modulus of rigidity of the elastomer. If the ends are not functionally active, they become "dangling ends," which cannot support deformation stresses, thus lowering the modulus of the elastomer (102). For radiation-cured PDMS, the relation between its equilibrium modulus and the original MWD, cross-link level and scission level has been determined (103).

3.2. Branching

An important structural feature of polymers, in addition to MWD, is the frequency of long-chain branching (LCB). For example, PDMS polymerized from $(Me_2SiO)_x$ cyclics with a few parts per million of $MeSi(OH)O$ impurity may have trifunctional branch points in the polymer at the monomethyl silicon atoms. For a random polymer product of high MW, the concentration of dimethyl (chain propagating) silicon atoms would be orders of magnitude larger than that of the trimethyl (chain terminating) silicons, or of the monomethyl (chain branching) silicons. Thus, the average chain length between branches is many siloxane units long, and each of its branches leads to another branch point or to a chain end. Other LCB structures are the "star" and "comb" shaped polymers (104), which may have many chains that may be quite uniform in length. All of these LCB structures differ from short-chain branched structures. Short branches of only a few chain atoms can arise from intramolecular chain transfer (backbiting) during chain-growth polymerization (105).

3.2.1.1. Significance. Long-chain branching can have dramatic effects on the mechanical properties of polymers. The flow of undiluted polymers depends on intermolecular motion, which may be greatly restricted by branches. The highly interpenetrating, entangled polymer coils may require much longer times to move relative to each other if LCB is present. Elastic energy can be stored for much longer times causing phenomena such as die swell after extrusion or poor flow into molds. For silicones, this effect occurs when reinforcing fillers such as colloidal or fumed silica particles are present. In this case, physical bonding between polymer segments and the particles can create a temporary network of chains joining near-by filler particles. With LCB, the network is established more easily and is more persistent. These phenomena cause processing problems. If the ends of a polymer with LCB are reactive, they may condense to form more branching and, eventually, an elastic network, or they may chemically bond to fillers to form a permanent network. These structural features decrease the stability and shelf life of such elastomer compounds.

If LCB is increased without changing MWD, solution viscosity is decreased because branching makes the polymer coils more compact, lowering their hydrodynamic volumes and radii of gyration. For γ-irradiated PDMS with random branching and an average of one branch per number average molecule, the intrinsic viscosity is equal to that of a linear PDMS of 25% lower $\bar{M}_w$ (106). The viscosity of *undiluted* polymer with LCB also is lower than that of the equivalent linear polymer, unless MW is above a certain value where effects of chain entanglements are large. The viscosity reduction is predicted (98) by viscoelastic theory and observed (107) for star-branched polymers. For the above case of random, radiation-branched PDMS, the bulk viscosity is equal to that of a linear PDMS of 40% lower $\bar{M}_w$ (106). At high MW, entanglements between polymer coils can make the viscosity with LCB *greater* than that of the equivalent linear polymer (107). This effect, however, was not seen in branched PDMS model systems up to $\bar{M}_w = 10^6$ since its viscosity was consistently reduced by LCB (108). Other effects of LCB include broadening of the MWD in random branched polymers, and non-Newtonian viscosity effects at lower shear rates (Figure 7.2).

3.2.1.2.   Measurement. There are two types of methods for characterizing LCB. The first is based on the effect of LCB on molecular size and viscosities, as described above. The second depends on direct analysis of branch sites or or end groups (since branching increases the number of ends per molecule).

Long-chain branching is theoretically related to polymer coil dimensions, such as radius of gyration. This, in turn, is related to *intrinsic viscosity* and *light scattering*, which suggests the use of these techniques to measure branching (105). Intrinsic viscosity often is the more practical method, but its relation to branching is subject to an empirical constant (104), which makes this a relative method. Gel permeation chromatography combined with intrinsic viscosity measurement can determine relative LCB levels (109). When GPC is used with a differential viscometer detector, the apparent relative branching levels can be recorded continuously throughout the MWD (110, 111). Another method is based on the *relationship between dilute-solution viscosity and bulk viscosity* which will, in general, change with branching level, as shown above for PDMS (106). This relationship can be the basis of a relative test for LCB. Each of these relative methods can be calibrated with samples whose LCB levels are determined by their synthesis (104, 112), or by the absolute methods described below.

Absolute methods for branching determination are based on direct analysis of branch sites or end groups. Results reflect both long- and short-chain branches; thus short branches will interfere with LCB measurement. A ^{29}Si *NMR* experiment can distinguish $MeSiO_{3/2}$ branch sites in polysiloxanes (cf. Chapter 12, Section 5.5). Its optimized detection level is as low as 0.01 mol % of Si; thus the detectable branching level is as low as 1 branch per 10,000 siloxane units, or 1 per 740,000 MW for PDMS (113). A *silicon functionality test* (Chapter 8, Section 3.7.3) is capable of even lower detection levels. The $RSiO_{3/2}$ branch units in polysiloxanes (R = alkyl, aryl, or olefinic groups) are derivatized to $RSi(OEt)_3$ by reaction with excess $Si(OEt)_4$ in the presence of KOH or KOEt. The product is quantified by GC (114). The $RSiO_{3/2}$ branch sites can be detected as low as 10–25 ppm by weight (115).

Branching levels also can be deduced by *end-group analysis* since branches add extra ends to polymer molecules. Methods of end-group analysis are discussed above as a means of $\bar{M}_n$ determination for linear (unbranched) polymers. The number of ends per molecule is $(2 + B_3)$, where B_3 is the number of trifunctional branches per molecule. If $\bar{M}_n$ is known independently, then, generalizing Eq. (7.8).

$$B_3 = E_t \bar{M}_n - 2 \tag{7.11}$$

A tetrafunctional branch B_4 can be seen (103) to be equivalent to two B_3 branches, and a B_5 branch (as in a "star" polymer) is equivalent to three B_3 branches, and so on. Thus, B_3 in this equation can be generalized as $(B_3 + 2B_4 + 3B_5 + \cdots)$ where any of these higher functionality branches are present. Large closed loops caused by LCB are ignored here as they should be quite rare.

3.3. Cross-Linking

Cross-links are junctions of polymer strands in a three-dimensional network. They may be viewed as long-chain branches, which are so numerous that a continuous, insoluble network (gel) is formed. A soft gel forms when there is an average of one intermolecular cross-link per two weight-average molecules (one share of a cross-link per weight-average molecule) (116). Increased cross-linking gives an elastomer with progressively increased stiffness until elasticity is lost when there are only a few chain atoms between cross-links. We are concerned both with cross-links formed by chemical bonding, which usually are permanent, and with cross-links formed by physical bonding, including entanglements between polymer chains, which may be temporary. Chemical cross-links may be formed by bonds between established polymer molecules, or by branching during the polymerization process (117). The high-temperature and room-temperature cure (cross-linking) technology of silicone rubber has been reviewed (118).

3.3.1.1. Significance. Cross-linking affects many mechanical properties of polymers. It increases the modulus (stiffness) and decreases the loss tangent (damping), as reported for PDMS cured at several γ-radiation doses (119). As cross-linking increases over the range of practical interest, the ultimate properties of tensile strength and tear strength generally pass through maxima while the strain at failure and degree of solvent swelling both decrease. Other properties such as the dielectric loss factor (120) and, to some extent, the diffusion coefficient (121) of elastomer networks also depend on the extent of cross-linking. Process and environmental conditions that affect the extent of cross-linking clearly affect all of these properties.

3.3.1.2. Measurement. The concentration of cross-links is expressed alternatively in terms of moles per unit volume or mass, number per average molecule or repeat unit of the polymer, or (inversely) as average polymer MW per cross-link, which is related to the average MW of strands between cross-links. The methods below usually are applied to unfilled elastomers; however, results with filled systems may be relevant to the polymer matrix and its bonding to the filler. For the following methods of measurement, the order progresses from those more useful for relative results to those that can give absolute concentrations. The swelling and modulus methods measure the combined concentrations of chemical and effective physical cross-links, while the other methods measure only the concentrations of chemical cross-links.

Solvent swelling is a method based on the penetration and enlargement of cross-linked networks by solvents that would readily dissolve the same polymer if it were uncross-linked. The swelling (relative volume increase) slowly reaches equilibrium at a state of minimum free energy where the negative enthalpy change of further mixing of polymer and solvent is balanced by the negative entropy change of further extending the network strands (122). The Flory–Rehner equation, which results from these considerations, relates the observed volume increase to the cross-link concentration and the Flory–Huggins polymer-solvent interaction parameter, χ. This parameter depends on the polymer, solvent, temperature, and degree of swelling. Application of the Flory–Rehner equation to unfilled PDMS networks (119) gave calculated cross-link concentrations that were in reasonable agreement with those based on equilibrium modulus measurement (below). However, swelling often is used as a relative method since χ values may not be known independently.

The swelling measurement usually is done by immersion in the solvent for several days, briefly draining or blotting excess solvent from the specimen, and weighing the swollen sample in a closed container. The weight gained from swelling and the solvent density give the volume increase for use in the Flory–Rehner equation.

Elastic modulus is related by the theory of rubber elasticity (123, 124) to the concentration of network strands, v, formed by chemical and physical cross-links. Elastic modulus is a stress/strain ratio. It can be measured by uniaxial elongation to give Young's modulus, E, by the relation

$$E = 3 \, (F/A_0)/(\alpha - \alpha^{-2}) \tag{7.12}$$

where the stress, F/A_0, is the applied force per initial cross-section area of sample and the strain, α, is the extension ratio L/L_0 of extended-to-original sample length. Modulus also can be measured by shear to give the shear modulus, G, by the relation

$$G = \sigma/\gamma \qquad (7.13)$$

where σ and γ are the shear stress and strain (98), respectively. A limiting modulus value is measured as strain approaches zero and time approaches infinity. A relative value of v can be estimated from this linear, equilibrium modulus by the simplest form of the theory (125):

$$v = G/RT = E/3RT \qquad (7.14)$$

where R is the gas law constant and T is the absolute temperature. More rigorous equations relate modulus to v, or to the concentrations of chemical cross-links and of chain entanglements that are trapped by chemical cross-links (125). The parameters needed to use these equations have been determined for PDMS networks (103, 126).

Sol fraction, w_s, is the weight fraction of a cross-linked polymer that is soluble in the form of polymer molecules not chemically cross-linked to the continuous gel network. If the network is not highly cross-linked, the network can be swollen sufficiently to extract these soluble molecules and w_s can be measured as the weight fraction dissolved. At degrees of cross-linking beyond the gel point, w_s decreases rapidly from unity to small values as the probability that a molecule is unattached becomes small. Thus w_s is statistically related to the degree of intermolecular chemical cross-linking and to the MWD before cross-linking. For a uniform initial MWD (all molecules of equal size), the number of chemically cross-linked units per number-average molecule, γ_c, is given by the relation (127).

$$\gamma_c = -\ln w_s/(1 - w_s) \qquad (7.15)$$

For a random or "most probable" initial MWD [Eq. (7.3)] (128)

$$\gamma_c = 1/(w_s + w_s^{1/2}) \qquad (7.16)$$

Other, more complex expressions exist for more general initial MWDs and cases where chain scission accompanies cross-linking (102, 129). Expressions also have been derived for networks formed by linking of reactive polymer ends to cross-linker molecules (130) and by branching during the polymerization process (117). Cross-link concentrations of PDMS networks have been characterized using these relationships (103, 119, 130). Results show that sol fraction analysis can be an absolute method of determining the concentration of chemical cross-links.

Nuclear magnetic resonance and the *silicon functionality test,* which were discussed above as techniques for branching analysis, are other absolute methods for chemical cross-link determination. The optimum limits of detection and precision reported above are adequate for measurement of cross-link concentrations of typical commercial silicone elastomers.

3.4. Phase Composition

3.4.1.1. Phases in Polymer Systems. Common silicone elastomers remain flexible and fluids remain pourable at unusually low temperatures compared to other polymers. Their siloxane backbone structure changes conformation easily, giving an unusually low glass transition temperature (about $-125\,°C$ for PDMS) (131). Crystallization can occur below $-40\,°C$ for PDMS (55, 132), depending on rate of cooling. Like other polymers (93), silicone polymers may display a glassy or flexible amorphous phase and a crystalline phase, depending on the temperature and thermal history. A block copolymer may have these phases for each of its constituents. Practical silicone elastomer formulations usually include reinforcing inorganic fillers such as amorphous silica and/or nonreinforcing fillers such as crystalline quartz or $CaCO_3$. Elastomers often contain low molecular weight plasticizers or filler treatments that can constitute a separate phase. The discussion below is confined to glassy and flexible amorphous phases, and to crystalline phases.

3.4.1.2. Methods of Phase Detection. Detecting these phases in silicones usually is done by observing thermal, mechanical or dielectric property changes identified with phase transitions. Property changes and corresponding analytical methods useful for polymer systems in general have been described elsewhere (93). The most common methods for determining the glass transition temperature T_g, the melting temperature T_m, and the crystallization temperature T_c are *thermal methods* such as differential scanning calorimetry (DSC), which determines enthalpy and heat capacity changes during a temperature program. Figure 7.4 shows DSC results for a mixture of high—MW PDMS with cyclics, of MWD described by Figure 10.14.

The sample was cooled rapidly to $-150\,°C$, then heated at $10\,°C/min$. A sudden change of slope near $-120\,°C$ indicates the approximate T_g for this composition. Polymer chains become sufficiently mobile near $-90\,°C$ to align and partially crystallize. This increases the crystalline fraction beyond that which may have formed during the rapid cooling step. The crystallization exotherm continues to about $-55\,°C$ before the endothermic melting

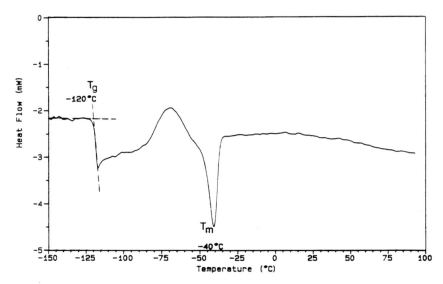

Figure 7.4. Low temperature DSC trace of PDMS gum in He. Heating rate 10 °C/min.

begins and peaks at T_m near -40 °C. The heat of fusion can be estimated from the area of the melting peak compared with that of a material of known heat of fusion. This result can be compared with reported values (55, 132) for complete crystallization (2.72 kcal/mol of Me_2SiO for PDMS) to estimate the degree of crystallization.

The observed T_c and T_g depend on the rate of cooling. Slower cooling gives a higher T_c and lower T_g, indicating kinetic control of these transitions (93, 133). The crystal growth rate at optimum temperatures is relatively rapid for polysiloxanes, compared with polymers having less flexible chains (134). For many polymers, the maximum growth rate occurs (135) at absolute temperature of about 0.80–0.87 T_m, which is consistent with the T_c peak near -70 °C (203 K) and T_m near -40 °C (233 K) for PDMS in Figure 7.4 (also see Section 2.5). Crystallographic data for polysiloxanes have been reported (132). Factors that control T_g of a given polymer such as PDMS include MW, crystallinity, orientation, diluents, and cross-linking (133). The absolute T_g of a polymer usually is about $\frac{2}{3}$ of T_m (133), in reasonable agreement with literature values for PDMS of 146 K for T_g (131) and 233 K for T_m (55, 132). The minimum useful temperature of an elastomer is determined by the "stiffening temperature," which is controlled by T_g, or by T_c if sufficient time is allowed for crystallization. The stiffening temperature of PDMS is lowered to about -110 °C by random copolymerization with approximately 8 mol%

methylphenylsiloxane (92% dimethylsiloxane), which suppresses crystallization without substantially raising T_g (136).

Mechanical and dielectric methods also are useful in detecting phase changes. At T_g the modulus of an amorphous polymer usually changes by several orders of magnitude, and the mechanical and dielectric loss factors typically go through maxima (8, 137). At T_c similar changes of properties are seen, although the modulus change is not as dramatic (138). General instrumental approaches to study these properties and transitions have been reviewed (98, 137). Specific mechanical techniques (139) include dynamic mechanical analysis (140) (DMA), dynamic mechanical thermal analysis (141) (DMTA), thermal mechanical analysis (140) (TMA), and torsional braid analysis (139) (TBA); and methods (142, 143) of "mechanical spectroscopy" and rheology that vary temperature, deformation frequency, amplitude, or time. Dielectric analyzers are available (139–141, 144), which can detect phase changes by scanning temperature, frequency, and time.

3.5. Thermal Stability of the Structure

The above discussion concerns reversible thermal changes in polymer phase structure. Irreversible changes of the MWD, branching or cross-linked structure of silicone polymers inevitably occur at high temperatures because of chain scission or oxidative cross-linking. In the *absence of oxygen*, depolymerization occurs with the loss of volatile products, mostly low-MW cyclic species (145). This depolymerization of silicones is catalyzed by traces of acids, bases, water, hydroxyl groups, and certain organometallic compounds. Depolymerization begins near 400 °C for relatively clean PDMS, but at much lower temperatures for PDMS with residual KOH polymerization catalyst (146). It can be measured by TGA and DSC in inert atmospheres, or by the resulting reduction of the MWD (145) or viscosity. Figure 7.5 includes high-temperature DSC results in helium for the same PDMS gum sample shown in Figure 7.4. An endotherm occurs at 260–350 °C, the same range where TGA, Figure 7.6, shows nearly complete weight loss. The endotherm is attributed to devolatilization of the cyclic products.

In the *presence of oxygen*, thermal depolymerization of silicones is accompanied by oxidative cross-linking. Methyl groups are eliminated and the residual free radicals are believed to cross-link to neighboring molecules (147). A net increase is usually seen in the viscosity of a fluid or in the modulus of a cured elastomer since the effect of cross-linking usually outweighs that of depolymerization. The DSC result in Figure 7.5, as measured in air with the same PDMS polymer, shows a smaller endotherm than for pure depolymerization in helium, which is attributed to exothermic oxidation. Figure 7.6 shows TGA results for this sample in both atmospheres and indicates the

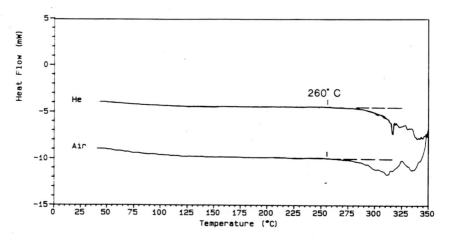

Figure 7.5. DSC trace of a PDMS gum in air and He. Heating rate 10 °C/min.

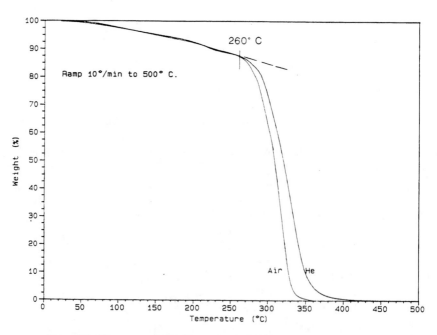

Figure 7.6. TGA trace of PDMS gum in air and He. Heating rate 10 °C/min.

164

relative temperature ranges of thermal degradation. Addition of antioxidants can increase degradation temperatures in oxidative atmospheres (147).

Thus, the overall temperature range of structural and phase stability of silicones can extend from as low as $-110\,^{\circ}C$ (T_g of PDMS copolymerized with about 8 mol% methylphenylsiloxane) to the limit of oxidative stability, about $300\,^{\circ}C$ for most common silicones with heat stability additives. As a result, the physical properties of silicones are quite stable relative to other polymer systems, which undergo structural changes within narrower temperature ranges.

REFERENCES

1. American Society for Testing and Materials, *1989 Annual Book of ASTM Standards*, ASTM, Philadelphia, 1989.

2. A. Weissberger, Ed., *Physical Methods of Organic Chemistry*, 3rd ed., Vol. I, Parts I and II, Interscience, New York, 1959.

3. A. Weissberger and B. W. Rossiter, Eds., *Physical Methods of Chemistry*, Parts I–V, Wiley-Interscience, New York, 1971.

4. W. Noll, *Chemistry and Technology of Silicones*, Academic Press, New York, 1968.

5. V. Bazant, V. Chvalovsky, and J. Rathousky, *Organosilicon Compounds*, Academic, New York, 1965.

6. V. Bazant, J. Hetflejs, V. Chvalovsky, J. Joklik, O. Kruchna, J. Rathousky, and J. Schraml, *Handbook of Organosilicon Compounds. Advances since 1961*, Vol. 1, Dekker, New York, 1976.

7. V. Bazant and J. Hetflejs, *Handbook of Organosilicon Compounds. Advances since 1961*, Vol. 2, Dekker, New York, 1973.

8. V. Bazant and J. Hetflejs, *Handbook of Organosilicon Compounds. Advances since 1961*, Vol. 3, Dekker, New York, 1973.

9. V. Bazant and J. Hetflejs, *Handbook of Organosilicon Compounds. Advances since 1961*, Vol. 4, Dekker, New York, 1973.

10. C. Eaborn, *Organosilicon Compounds*, Butterworths, London, 1960.

11. R. J. H. Voorhoeve, *Organohalosilanes, Precursors to Silicones*, Elsevier, New York, 1967.

12. J. A. Dean, Ed., *Lange's Handbook of Chemistry*, 13th ed. McGraw-Hill, New York, 1985.

13. R. C. Weast, Ed., *CRC Handbook of Chemistry and Physics*, 69th ed., CRC press, Boca Raton, FL, 1989.

14. T. E. Daubert and R. P. Danner, American Institute of Chemical Engineers, *Design Institute for Physical Property Data. Data Compilation*, American Institute of Chemical Engineers, New York, 1989.

15. N. Bauer, K. Fajans, and S. Z. Lewin, "Refractometry," in *Physical Methods of Organic Chemistry*, 3rd ed., A. Weissberger, Ed., Vol. I, Part II, Interscience, New York, 1959, p. 1139.

16. W. T. Cresswell, J. Leicester, and A. I. Vogel, "Bond refractions for compounds of tin, silicon, lead, mercury and germanium," *Chem. Ind. London*, 19, **1953**.

17. A. P. Mills and W. E. Becker, "Silicon–bromine and silicon–carbon(aryl) bond parachors and the silicon–bromine bond refraction," *J. Phys. Chem.*, **60**, 1644 (1956).

18. A. P. Mills and C. A. MacKenzie, "Application of bond parachors to organosilicon chemistry," *J. Am. Chem. Soc.*, **76**, 2672 (1954).

19. A. I. Vogel, W. T. Cresswell, and J. Leicester, "Bond refractions for tin, silicon, lead, germanium and mercury compounds," *J. Phys. Chem.*, **58**, 174 (1954).

20. E. L. Warrick, "The application of bond refractions to organosilicon chemistry," *J. Am. Chem. Soc.*, **68**, 2455 (1946).

21. H. Breederveld and H. L. Waterman, "The calculation of specific refractions of tri-*n* alkylchlorosilanes," *Res. Corresp.*, **6**, 145 (1953).

22. H. Breederveld and H. L. Waterman, "Calculation of specific refractions of tri-*n* alkoxysilanes," *Rec. Trav. Chim.*, **72**, 711 (1953).

23. A. D. Petrov, Yu. P. Egorov, V. F. Mironov, G. I. Nikishin, and A. A. Bugorkova, "Reactivity and molecular optical properties of alkenylsilanes," *Izv. Akad. Nauk SSSR, Ser. Khim.*, **1956**, 50.

24. R. O. Sauer, "Group and bond refractions for organosilicon liquids," *J. Am. Chem. Soc.*, **68**, 954 (1946).

25. B. E. F. Smith, "Bond refractions in organosilicon compounds. I," *Sven. Kem. Tidskr*, **61**, 213 (1949); II, **62**, 146 (1950).

26. H. H. Anderson and A. Hendifar, "*n*-Heptylhalosilanes and *n*-heptylsilane," *J. Am. Chem. Soc.*, **82**, 1027 (1959).

27. B. E. F. Smith, "Identification of liquid organosilicon compounds by refractive constants," *Sven. Kem. Tidskr*, **64**, 330 (1952).

28. B. E. F. Smith, "Calculation of density and refractive index of organosilicon compounds," *Sven. Kem. Tidskr*, **64**, 326 (1952).

29. B. E. F. Smith, "Bond refractions, bond dispersions and ring refractions in cyclopolymethylene silanes," *Acta Chem. Scand.*, **9**, 1286 (1955).

30. B. E. F. Smith, "A preparative, refractometric and spectrophotometric investigation of some phenoxysilanes," *Acta Chem. Scand.*, **9**, 1337 (1955).

31. O. L. Flaningam, "Vapor pressures of poly(dimethylsiloxane) oligomers," *J. Chem. Eng. Data*, **31**, 266 (1986).

32. M. J. Hunter, E. L. Warrick, J. F. Hyde, and C. C. Currie, "Organosilicon polymers, II. Open chain dimethylsiloxanes with trimethylsiloxy end groups," *J. Am. Chem. Soc.*, **68**, 2284 (1946).

33. A. L. Smith, Ed., *Analysis of Silicones*, Wiley, New York, 1974.

34. Dow Corning Corp., unpublished data.

35. M. J. Hunter, J. F. Hyde, E. L. Warrick, and H. J. Fletcher, "Organosilicon polymers. The cyclic dimethyl siloxanes," *J. Am. Chem. Soc.*, **68**, 667 (1946).

36. ASTM Method, D-891, "Test method for specific gravity of liquid industrial chemicals," *1989 Annual Book of ASTM Standards*, Vol. 15.05, ASTM, Philadelphia, 1989.

37. ASTM Methods, D-941, "Test method for density and relative density (specific gravity) of liquids by Lipkin bicapillary pycnometer," D-1298, "Test method for density and relative density (specific gravity) of liquids or API gravity of crude petroleum and liquid petroleum products by hydrometer method," D-1480. "Test method for density and relative density (specific gravity) of viscous materials by Bingham pycnometer," *1989 Annual Book of ASTM Standards*, Vol. 05.01, ASTM, Philadelphia, 1989.

38. ASTM Method, D-3505, "Test method for density and relative density of pure liquid chemicals," *1989 Annual Book of ASTM Standards*, Vol. 06.03, ASTM, Philadelphia, 1989.

39. ASTM Method, D-4052, "Test method for density and relative density of liquids by digital density meter," *1989 Annual Book of ASTM Standards*, Vol. 05.03, ASTM, Philadelphia, 1989.

40. N. Bauer and S. Z. Lewin, "Determination of density," in *Physical Methods of Organic Chemistry*, 3rd ed. A. Weissberger, Ed., Vol. I, Part I, Interscience, New York, 1959, p. 131 ff.

41. ASTM Methods D-56, "Flash point by tag closed tester," D-92, "Flash and fire points by Cleveland open cup," D-1310, "Flash point and fire point of liquids by tag open cup apparatus," *1989 Annual Book of ASTM Standards*, Vol. 05.01, ASTM, Philadelphia, 1989.

42. ASTM Method D-93, "Standard test method for flash point by Pensky-Martin closed tester," *1989 Annual Book of ASTM Standards*, Vol. 05.01, ASTM, Philadelphia, 1989.

43. C. L. Lee, K. E. Polmanteer, and E. G. King, "Flow behavior of narrow-distribution polydimethylsiloxane," *J. Polym. Sci. A2*, **8**, 1909 (1970).

44. T. Kataoka and S. Ueda, "Viscosity of polydimethylsiloxane blends," *J. Polym. Sci. A1*, **5**, 3071 (1967).

45. J. F, Swindels, R. Ullman, and H. Mark, "Determination of viscosity," in *Physical methods of Organic Chemistry*, A. Weissberger, Ed., Vol. I, Part 1, Interscience, New York, 1959, p. 689.

46. ASTM Methods D-445, "Test method for kinematic viscosity of transparent and opaque liquids and calculation of dynamic viscosity," D-446, "Specifications for operating instructions for glass capillary kinematic viscometers," *1989 annual Book of ASTM Standards*, Vol. 05.01, ASTM, Philadelphia, 1989.

47. E. B. Bagley and D. C. West, "Chain entanglement and non-Newtonian flow," *J. Appl. Phys.*, **29**, 1511 (1958).

48. E. B. Bagley, "Power law flow curves of dimethyl siloxane polymers," *J. Appl. Phys.*, **30**, 597 (1959).

49. C. C. Currie and B. F. Smith, "Flow characteristics of organopolysiloxane fluid and greases," *Ind. Eng. Chem.*, **42**, 2457 (1950).

50. G. C. Johnson, "Flow characteristics of linear end-blocked dimethyl polysiloxane fluids," *J. Chem. Eng. Data*, **6**, 275 (1961).

51. R. L. Merker, "Association and entanglement in high polymers. I. Effect on viscometric properties of dimethylpolysiloxanes," *J. Polym. Sci.*, **22**, 353 (1956).

52. R. L. Merker and M. J. Scott, "Viscometric properties of salicyloxymethyldimethyl end-blocked dimethylpolysiloxanes," *J. Polym. Sci.*, **24**, 1 (1957).

53. E. L. Warrick, "Rheology of filled siloxane polymers," *Ind. Eng. Chem.*, **47**, 1816 (1955).

54. ASTM Method E-794, "Melting and crystallization temperatures by thermal analysis," *1989 Annual Book of ASTM Standards*, Vol. 14.02, ASTM, Philadelphia, 1989.

55. C. L. Lee, O. K. Johannson, O. L. Flaningam, and P. Hahn, "Calorimeric studies on the phase transitions of crystalline polysiloxanes. Part I. Polydimethylsiloxane," *Polym. Prepr. Am. Chem. Soc. Div. Polym. Chem.*, **10**(2), 1311 (1969).

56. C. L. Lee, O. K. Johannson, O. L. Flaningam, and P. Hahn, "Calorimetric studies on the phase transitions of crystalline polysiloxanes. Part II. Polydiethylsiloxane and polydipropylsiloxane," *Polym. Prepr. Am. Chem. Soc. Div. Polym. Chem.*, **10**(2), 1319 (1969).

57. W. Swietoslawski, *Ebulliometric Measurements*, Reinhold, New York, 1945.

58. W. Swietoslawski, "Determination of boiling and condensing temperatures", in *Physical Methods of Organic Chemistry*, Vol. I, A. Weissberger, Ed., Interscience, New York, 1945, Chap. II.

59. W. Swietoslawski and J. R. Anderson, "Determination of boiling and condensation temperatures," in *Physical Methods of Organic Chemistry*, Vol. I, Part 1, A. Weissberger, Ed., Interscience, New York, 1959, p. 357.

60. G. W. Thompson, "Determination of vapor pressure," in *Physical Methods of Organic Chemistry*, A. Weissserger, Ed., Vol. I, Part 1, Interscience, New York, 1959, p. 401.

61. D. R. Stull, "Application of platinum resistance thermometry to some industrial physicochemical problems," *Ind. Eng. Chem.*, **18**, 234 (1946).

62. R. C. Reid and T. K. Sherwood, *The Properties of Liquids and Gases*, 2nd ed., McGraw-Hill, New York, 1966.

63. R. C. Reid, J. M. Prausnitz, and T. K. Sherwood, *The Properties of Liquids and Gases*, 3rd ed., McGraw-Hill, New York, 1977.

64. R. C. Reid, J. M. Prausnitz, and B. E. Poling, *The Properties of Liquids and Gases*, 4th ed., McGraw-Hill, New York, 1987.

65. I. Wichterle, J. Linek, and E. Hala, *Vapor–Liquid Equilibrium Data Bibliography*, Elsevier Scientific, New York, 1973.

66. I. Wichterle, J. Linek, and E. Hala, *Vapor–Liquid Equilibrium Data Bibliography—Supplement I*, Elsevier, New York, 1976.

67. I. Wichterle, J. Linek, and E. Hala, *Vapor–Liquid Equilibrium Data Bibliography—Supplement II*, Elsevier, New York, 1979.

68. I. Wichterle, J. Linek, and E. Hala, *Vapor–Liquid Equilibrium Data Bibliography—Supplement III*, Elsevier, New York, 1982.

69. E. Hala, I. Wichterle, J. Polak, and T. Boublik, *Vapor Liquid Equilibrium Data at Normal Pressures*, Pergamon Press, New York, 1968.

70. M. Hirata, S. Ohe, and K. Nagahama, *Computer Aided Data Book of Vapor–Liquid Equilibria*, Elsevier, New York, 1973.

71. Ohe, S., *Vapor–Liquid Equilibrium Data*, Elsevier, New York, 1989.

72. E. Hala, J. Pick, V. Fried, and O. Vilim, *Vapor–Liquid Equilibrium*, Pergamon Press, New York, 1958.

73. E. Hala, J. Pick, V. Fried, and O. Vilim, *Vapor–Liquid Equilibrium*, 2nd ed., Pergamon Press, New York, 1967.

74. S. R. M. Ellis, "A new equilibrium still and binary equilibrium data," *Trans. Inst. Chem. Eng.*, **30**, 58, 1952.

75. D. G. Jordan, Chapter 5, "Vapor–liquid equilibria," in *Chemical Process Development*, Interscience, New York, 1968.

76. ASTM Method E-659, "Autoignition temperature of liquid chemicals," *1989 Annual Book of ASTM Standards*, Vol. 05.03, ASTM, Philadelphia, 1989.

77. R. R. Buch and D. H. Filsinger, "Method for fire hazard assessment of fluid-soaked thermal insulation," *Plant/Operations Prog.*, **4**, 176 (1985).

78. Manufacturing Chemists Association data.

79. R. L. Schalla and G. E. McDonald, "Temperature-composition limits of spontaneous explosion for nine alkylsilanes with air at atmospheric pressure," *Natl. Advisory Comm. Aeronaut. Tech Note No. 3405* (1955). Also R. L. Schalla, G. E. McDonald, and M. Gerstein, "Combustion limits of alkylsilanes," *5th Symposium on Combustion*, Pittsburgh, 1954, p. 705.

80. H. Reuther, "Silicones XVIII. Determination of the ignition temperature of some defined organosilicon compounds," *Chem. Tech. (Berlin)* **5**, 330 (1953).

81. E. W. Balis, H. A. Liebhafsky, and D. H. Getz, "Flammabilities of four chlorosilanes and methyl chlorosilanes," *Ind. Eng. Chem.*, **41**, 1459 (1949).

82. K. A. Horn, "Chemical safety: silane distillation explosion," *Chem. Eng. News*, **68**(23), 2 (1990).

83. J. Lipowitz, "Flammability of poly(dimethyl) siloxanes No. 1, A model for combustion," No. 2, "Flammability and fire hazard studies," *J. Fire Flammability*, **7**, 482, 504 (1976).

84. H. F. Coward and G. W. Jones, "Limits of flammability of gases and vapors," *Bulletin 503*, Bureau of Mines, US Government Printing Office, Washington, DC, 1952.

85. M. G. Zabetakis, "Flammability characteristics of combustible gases and vapors," Bulletin 527, Bureau of Mines, US Government Printing Office, Washington, DC, 1965.

86. L. I. Nuzhda, M. A. Glinkin, E. E. Rafales-Lamarka, and N. F. Tyupalo, "A system method for determining the upper explosive limit," *Sov. Chem. Ind.*, **11**, 230 (1979).

87. S. N. Osipov, "Correlational analysis of the effect of molecular structure of a fuel on the lower limits of explosiveness of vapor–gas–air mixtures," *Sov. Chem. Ind.*, **8**, 20 (1976).

88. Y. N. Shebeko, A. V. Ivanov, and T. MN. Dmitrieva, "Methods of calculation of lower concentration limits of combustion of gases and vapors in air," *Sov. Chem. Ind.*, **15**, 311 (1983).

89. N. Y. Shebeko, A. Y. Korolchenko, A. V. Ivanov, and E. N. Alekhina, "Calculation of flash points and ignition temperatures of organic compounds," *Sov. Chem. Ind.*, **16**, 1371 (1984).

90. N. V. Solov'ev and A. N. Baratov, "Dependence of the lower limit of flammability on molecular structure," *Rus. J. Phys. Chem.*, **35**, 8 (1960).

91. A. Rudin, *The Elements of Polymer Science and Engineering*, Academic, Orlando, FL, 1982.

92. P. J. Wyatt, C. Jackson, and G. K. Wyatt, "Absolute GPC determinations of molecular weight and sizes," *Am. Lab*, 86 (May 1988).

93. H. Elias, *Macromolecules—1, Structure and Properties*, 2nd ed., Plenum Press, New York, 1984.

94. M. Kurata and Y. Tsunashima, "Viscosity-molecular weight relationships," in *Polymer Handbook*, J. Brandrup and E. H. Immergut, Ed., 3rd ed., Wiley, New York, 1989.

95. G. V. Schulz and A. Haug, "Thermodynamics and structure of polydimethylsiloxanes in solutions," *Z. Phys. Chem. (Frankfurt)*, **34**, 328 (1962).

96. M. A. Haney, "A differential viscometer," *Am. Lab*, **17**(3) 41 (March 1985).

97. H. Elias, *Macromolecules—1, Structure and Properties*, 2nd ed., Plenum Press, New York, 1984, p. 360.

98. J. D. Ferry, *Viscoelastic Properties of Polymers*, 3rd ed., Wiley, New York, 1980.

99. A. J. Barlow, G. Harrison, and J. Lamb, "Viscoelastic relaxation of polydimethylsiloxane liquids," *Proc. R. Soc. London*, **A282**, 228 (1964).

100. J. Lamb and P. Lindon, "Audio-frequency measurements of the viscoelastic properties of polydimethylsiloxane liquids," *J. Acoust. Soc. Am.*, **41**, 4 (Part 2) 1032 (1967).

101. D. J. Plazek, W. Dannhauser, and J. D. Ferry, "Viscoelastic dispersion of polydimethylsiloxane in the rubberlike plateau zone," *J. Colloid Sci.*, **16**, 2, 101 (1961).

102. N. R. Langley, "Elastically effective strand density in polymer networks," *Macromolecules*, **1**, 348 (1968).

103. N. R. Langley and K. E. Polmanteer, "Relation of elastic modulus to crosslink and entanglement concentrations in rubber networks," *J. Polym Sci. Polym. Phys. Ed.*, **12**, 1023 (1974).

104. B. J. Bauer and L. J. Fetters, "Synthesis and dilute-solution behavior of model star-branched polymers," *Rubber Chem. Technol.*, **51**, 406 (1978).

105. J. L. Koenig, *Chemical Microstructure of Polymer Chains*, Wiley, New York, 1980.

106. A. Charlesby, "Viscosity measurements in branched silicones," *J. Polymer Sci.*, **17**, 379 (1955).

107. W. W. Graessley, "Effect of long branches on the flow properties of polymers," *Acc. Chem. Res.*, **10**, 332 (1977).

108. E. M. Valles and C. W. Macosko, "Structure and viscosity of poly(dimethyl-siloxanes) with random branches," *Macromolecules*, **12**, 521 (1979).

109. E. E. Drott and R. A. Mendelson, "Determination of polymer branching with gel-permeation chromatography," *J. Polym. Sci. Pt. A-2*, **8**, 1361, 1373 (1970).

110. D. G. Moldovan and S. C. Polemanakos, "A new probe into polymer character-ization using the Viscotek differential viscometer coupled to a 150 C GPC," Waters GPC Symposium, Chicago, IL, 129 (June 1987).

111. F. M. Mirabella, Jr., and L. Wild, "Determination of long-chain branching distributions of polyethylenes by combined viscometry-size exclusion chromato-graphy techniques," *Polym. Mater. Sci. Eng.*, **59**, 7 (1988).

112. L. H. Tung, A. T. Hu, S. V. McKinley, and A. M. Paul, "Preparation of polystyrene with long chain branches via free radical polymerization," *J. Polym. Sci. Polym. Chem. Ed.*, **19**, 2027 (1981).

113. Richard B. Taylor, personal communication, Dow Corning Corp., Midland, MI, 1989.

114. P. J. Garner and R. C. Smith, "Characterization of siloxane polymers by tetraethoxysilane derivatization and subsequent gas chromatography," *Ab-stracts*, 1985 Pittsburgh Conference & Exposition, February 25, 1985, New Orleans, LA, USA, Abstract No. 969 (1985).

115. Paul J. Garner, personal communication, Dow Corning Corp., Midland, MI, 1989.

116. P. J. Flory, *Principles of Polymer Chemistry*, Cornell University Press, Ithaca, NY, 1953, p. 359.

117. P. J. Flory, *Principles of Polymer Chemistry*, Cornell University Press, Ithaca, NY, 1953, Chapter IX.

118. K. E. Polmanteer, "Current perspectives on silicone rubber technology," *Rubber Chem. Technol.*, **54**, 1051 (1981).

119. N. R. Langley and J. D. Ferry, "Dynamic mechanical properties of cross-linked rubbers. VI. Poly(dimethylsiloxane) networks," *Macromolecules*, **1**, 353 (1968).

120. C. P. Wong, "Effect of RTV silicone cure in device packaging," *Polym. Mater. Sci. Eng.*, **55**, 803 (1986).

121. S. P. Chen and J. D. Ferry, "The diffusion of radioactively tagged *n*-hexadecane and *n*-dodecane through rubbery polymers. Effects of temperature, cross-linking, and chemical structure," *Macromolecules*, **1**, 270 (1968).

122. P. J. Flory, *Principles of Polymer Chemistry*, Cornell University Press, Ithaca, NY, 1953, p. 576.

123. P. J. Flory, *Principles of Polymer Chemistry*, Cornell University Press, Ithaca, NY, 1953, Chapter XI.

124. L. M. Dossin and W. W. Graessley, "Rubber elasticity of well-characterized polybutadiene networks," *Macromolecules*, **12**, 123 (1979).

125. J. D. Ferry, *Viscoelastic Properties of Polymers*, 3rd ed., Wiley, New York, 1980, pp. 408–411.

126. K. O. Meyers, M. L. Bye, and E. W. Merrill, "Model silicone elastomers networks of high junction functionality: Synthesis, tensile behavior, swelling behavior, and comparison with molecular theories of rubber elasticity," *Macromolecules*, **13**, 1045 (1980).

127. P. J. Flory, *Principles of Polymer Chemistry*, Cornell University Press, Ithaca, NY, 1953, p. 378.

128. A. Charlesby and S. H. Pinner, "Analysis of the solubility behavior of irradiated polyethylene and other polymers," *Proc. R. Soc. London*, **A249**, 367 (1959).

129. M. Inokuti, "Gel formation in polymers resulting from simultaneous crosslinking and scission," *J. Chem. Phys.*, **38**, 2999 (1963).

130. E. M. Valles and C. W. Macosko, "Properties of networks formed by end linking of poly(dimethylsiloxane)," *Macromolecules*, **12**, 673 (1979).

131. P. Peyser, "Glass transition temperatures of polymers," in *Polymer Handbook*, J. Brandrup and E. H. Immergut, Ed., 3rd ed., Wiley, New York, 1989.

132. R. L. Miller, "Crystallographic data for various polymers," in *Polymer Handbook*, J. Brandrup and E. H. Immergut, Ed., 3rd ed., Wiley, New York, 1989.

133. R. F. Boyer, "Transitions & relaxations in amorphous and semicrystalline organic polymers and copolymers," in *Encyclopedia of Polymer Science and Technology, Supplement Volume 2*, H. F. Mark, Ed., Wiley, New York, 1977, p. 745.

134. K. E. Polmanteer, P. C. Servais, and G. M. Konkle, "Low temperature behavior of silicone and organic rubber," *Ind. Eng. Chem.*, **44**, 1576 (1952)

135. H. Elias, *Macromolecules—1, Structure and Properties*, 2nd ed., Plenum Press, New York, 1984, p. 392.

136. K. E. Polmanteer and M. J. Hunter, "Polymer composition versus low-temperature characteristics of polysiloxane elastomers," *J. Appl. Polym. Sci.*, **1**, 1,3 (1959).

137. J. J. Aklonis and W. J. MacKnight, *Introduction to Polymer Viscoelasticity*, 2nd ed., Wiley, New York, 1983, pp. 207–210.

138. J. D. Ferry, *Viscoelastic Properties of Polymers*, 3rd ed., Wiley, New York, 1980, Chapter 16.

139. R. F. Boyer, "Automated dynamic mechanical testing," in *ACS Advances in Chemistry Series*, No. 203, C. D. Craver, Ed., Vol. 1, 1983.

140. DuPont Instrument Systems, Concord Plaza, Wilmington, DE, 19898.

141. Polymer Laboratories Ltd., The Technology Centre, Epinal Way, Lough-borough, UK, LE11 OQE.

142. Rheometrics, Inc., One Possumtown Road, Piscataway, NJ, 08854

143. Carri-Med Limited, Vincent Lane, Dorking, Surrey, UK, RH43YX.

144. Micromet Instruments, Inc., 21 Erie St., Cambridge, MA, 02139

145. D. W. Kang, G. P. Rajendran, and M. Zeldin, "Kinetics of thermal depolymeriz-ation of trimethylsiloxy end-blocked polydimethylsiloxane and polydimethylsil-oxane-N-phenylsilazane copolymer," *J. Polym. Sci. Part A. Polym. Chem.*, **24**, 1085 (1986).

146. N. Grassie and I. G. Macfarlane, "The thermal degradation of polysiloxanes I. Poly(dimethylsiloxane)," *Eur. Polym. J.*, **14**, 875 (1978).

147. E. M. Acton, K. E. Moran, and R. M. Silverstein, "High-temperature antiox-idants for hydraulic fluids and lubricants. Evaluation and mechanism of pro-tection of a silicone fluid," *J. Chem. Eng. Data*, **6**, 64 (1961).

CHAPTER

8

CHEMICAL ANALYSIS

M. D. GAUL and N. C. ANGELOTTI

Dow Corning Corporation
Midland, Michigan

1. INTRODUCTION

During the early part of this century, instrumental methods were unknown, and all analyses were done by gravimetric or volumetric procedures. Today, such procedures have in large part given way to faster and more definitive instrumental techniques. Even those procedures that rely on chemical reactions or combustion usually utilize automated instrumentation to accomplish the analysis. Nevertheless, it is good to remember that some "wet" chemical procedures are still viable, particularly if a selection of sophisticated instrumentation is not readily available. In some cases, chemical tests give better accuracy, and may be more efficient than an instrumental method. Chemical analyses may also serve as absolute or referee methods where instrument calibration is required. In this chapter we discuss some of the most useful chemical analysis methods for organosilicon compounds. Other more comprehensive discussions are found elsewhere (1, 2).

Many organosilicon compounds require unique reaction parameters or techniques for elemental analysis. Such information is included in the discussions.

2. ELEMENTAL ANALYSIS

2.1. Silicon

As illustrated in Chapter 3, Section 1, analysis for silicon is usually the most important elemental analysis for organosilicon polymers. It is often carried

The Analytical Chemistry of Silicones, Edited by A. Lee Smith.
ISBN 0-471-51624-4 © 1991 John Wiley & Sons, Inc.

175

out by atomic emission or absorption spectroscopy, or X-ray fluorescence analysis, which are usually faster and more convenient than chemical analysis.

2.1.1. Qualitative Tests

Rapid qualitative tests for silicon employ a destructive process to cleave organic substituents and produce a silica residue or soluble silicate. Typical decomposition procedures include Bunsen burner flame, acid digestion, fusion with sodium peroxide, or fusion with sodium carbonate (2). In the latter case, the presence of silicon is indicated by effervescence of the sodium carbonate bead when held over a microburner. With án oxidized sample, further positive proof of silicon may be obtained by volatilization of the residue upon the addition of hydrofluoric acid.

A qualitative test for siloxanes (and silica) has been described (Ref. 3, p. 255). The sample is treated with a hydrofluoric acid–propanol mixture (1:10) and evaporated to dryness in a platinum dish. Siloxanes are converted to fluorosilanes and silica to SiF_4. These products are all readily volatilized.

Very sensitive detection of silica (0.1–1 μg) can be achieved after peroxide or carbonate fusion of silica residues obtained from acid ashing (2). Fusionates are dissolved in water, added to ammonium molybdate on filter paper, and reduced with sodium sulfite, hydroxylamine, Mohr's salt, or hydroquinone to produce a blue color. Phosphorus interferes with the test, but ammonium phosphomolybdate may be destroyed by using oxalic acid before the silicon test.

Reagents

Ammonium molybdate solution. Dissolve 5 g of ammonium molybdate in 100 ml of water; and 35 ml of concentrated HNO_3.

Sodium sulfite solution. Dissolve 15 g of sodium sulfite in 250–300 mL of water.

Apparatus

Platinum loop; Wire gauze and stand.

Procedure

Mix 2 or 3 drops of liquid sample or ~ 0.02 g of solid with a 5:1 Na_2CO_3/Na_2O_2 mixture. Compress a portion of the mixture within a platinum loop and fuse using a microburner. Dissolve the melt in a few milliliters of water and boil for 1 min. Place a drop of the fusionate and a drop of ammonium molybdate on filter paper and warm gently with the filter paper on a wire gauze. Add a drop of Na_2SO_3 solution on the sample spot. A blue color indicates the presence of silicon.

Many silicone products have fillers or pigments that contain silica or silicates. Separation of silicones from these materials for qualitative and quantitative tests may be accomplished using solvent extraction with toluene, acetone, acetone–toluene, or hot piperidine. Organosilicon polymers may be separated from reinforcing silica fillers by an extractive procedure detailed in Chapter 3, Section 6.1.

Thermal depolymerization at 700 to 800 °C of samples under vacuum or nitrogen purge produces volatile silicone products that can be trapped and evaluated for silicon content (Chapter 3, Section 6.2).

2.1.2.   Decomposition Procedures

Many decomposition techniques have been used to oxidize silicones quantitively to silica or soluble silicates. The most popular and generally applicable method, particularly with somewhat volatile samples, is sodium peroxide fusion in nickel Parr-type bombs to give soluble silicates.

2.1.2.1.   Parr Bomb Oxidation (2). Various materials such as sugar, ethylene glycol, sugar and potassium nitrate, sugar and potassium perchlorate, potassium nitrate, and potassium perchlorate have been added with samples to give complete combustion without a carbonaceous residue and to increase the speed of combustion. Samples may be added directly to sodium peroxide, held within gelatin capsules or, if volatile, sealed in silica-free glass tubes. Aqueous samples may be contained in collodion-coated gelatin capsules, although extra caution must be taken if the water content is high. Premature reaction may occur and dangerously high bomb pressures reached. Blanks should be carried through and samples of known Si content run periodically (Chapter 2, Section 8.2).

Reagents

Sodium peroxide, calorimetric grade, Aldrich Chemical Co. Store in a plastic bag in a sealed can.

Apparatus

Blast burner, oxygen-gas, Bethlehem apparatus Company.
Nickel beakers and beaker covers.

Peroxide bomb, 16 or 22 mL, Parr Instrument Company.

Ignition shield, 20 in. length, 6-in. i.d. pipe welded to steel base with $\frac{1}{4}$ in. thick shelf 12 in. from the bottom. The shelf has a hole to allow exposure of the bottom half of the bomb. One side of the pipe is one-quarter cut away to 12 in. from the bottom to position the torch.

Heavy duty tongs.

Size 1 gelatin capsules, Eli Lilly and Co.

Procedure

Weigh between 50–200 mg of sample into a gelatin capsule. Protect the capsule from finger contact with a laboratory tissue. Place the capsule into a bomb crucible. Load the crucible with 10.00 g of sodium peroxide, tapping the crucible on a bench top to settle the peroxide. Wipe any excess sodium peroxide from the crucible lip. Place the crucible lid with lead gasket on and hold in place with iron bolt assembly and securely tighten. Fill the cavity in the assembled unit with ice or water. Lower the bomb assembly into the protective shield from the top with tongs. Adjust the natural gas–oxygen blast lamp flame to just coincide with the bottom of the bomb crucible. Heat the unit until the bottom $\frac{1}{4}$ in. of the crucible is cherry red. Remove the crucible and quench in a metal bucket filled with ice water.

Disassemble the cooled bomb and place the crucible on its side in a 600-mL nickel beaker containing enough water to cover the crucible. Rinse any residue adhering to the cap into the beaker with water from a wash bottle. Cover the beaker with a nickel beaker cover and allow the contents to react to completion. Remove beaker cover and rinse. Remove crucible with a pair of nickel or stainless steel tongs and rinse with water. All rinses are added to the beaker. If the sample is to be analyzed by AA or ICP (Chapter 14), add 37 mL of reagent grade glacial acetic acid to the beaker while stirring. Otherwise, acidify with conc. HCl ($\sim$ 50 mL). Replace beaker cover and allow contents to react. Visually check the beaker contents for unreacted sample. Discard the solution if unreacted sample material is observed. Quantitatively transfer the contents of the beaker to a 500-mL volumetric flask and dilute to volume with distilled water. Degas the solution before analysis.

Nickel crucibles and beakers are used to prevent contamination. Nickel also has a slow erosion rate to molten sodium peroxide, which allows the crucibles to be used a few hundred times before the wall becomes so thin that it poses a hazard. The erosion is monitored by checking the crucible wall thickness with calipers and a micrometer and comparing the value to unused crucibles. Crucibles are discarded when the erosion is 50%.

When a sample fails to completely react in the fusion, it may be necessary to add a fusion accelerator. The sample is weighed without a capsule in a platinum dish and transferred to the bomb crucible. A small amount of sorbitol or mineral oil is then placed directly on the sample and the sodium peroxide added. The procedure is then continued. The added organic matter generates additional heat, which enhances the oxidation of the sample.

A digestion system that uses microwave heating is described in Chapter 14, Section 4.2.1.

2.1.2.2. Acid Decomposition. These methods have an advantage in being somewhat faster and requiring simpler apparatus than peroxide bomb decomposition. Silicon blanks are also generally much lower, and extraneous contamination from the other metals can be minimized. Acid ashing is not generally suitable if volatile silicon-containing compounds are present, however, even when cooling of the sample and crucible are carried out in the initial phase of the decomposition.

Vycor or platinum crucibles and common laboratory reagents are used. Sealed glass containers have been used in the initial oxidation with sulfuric acid at 100–150 °C. Nonvolatile polysiloxanes or resinous samples may be decomposed using fuming sulfuric acid, together with nitric acid when required. (With aromatic and fatty biological materials, a preliminary charring of the sample with sulfuric acid is necessary to avoid formation of explosive organic nitrates.) Nitric acid only, sulfuric acid plus potassium permanganate, sulfuric acid with a drop of mercury, sulfuric acid plus ammonium nitrate, perchloric, sulfuric, nitric, and perchloric, and sulfuric plus perchloric acids have all been used.

An acid ashing procedure for chlorosilanes, resins, polymers, and compounded materials is given below (2).

Reagents

Fuming H_2SO_4 and fuming HNO_3.

Apparatus

Vycor Kjeldahl flasks, 30 mL, neck shortened to about 4 cm. Vycor crucibles, 30 mL, platinum crucibles, 35 mL, muffle furnace, heat lamp. *Note*: All acid decomposition procedures should be carried out in a well-vented fume hood, using good laboratory practices for the handling of concentrated acids.

Procedure

Ignite the Vycor or platinum ware to constant weight at 1000 °C. If
analyzing chlorosilanes, use Vycor and add 0.5 mL of chlorobenzene and
3 mL of NH_4OH to the container. (The NH_4OH neutralizes the chlorosilane, and chlorobenzene reduces frothing.) Add 0.1–0.3 g of sample, using a
weighed capillary dropper for reactive liquid samples, and mix the contents. Weigh the polymer or resin samples directly into the flask or
crucible. Remove water by heating in a vacuum oven at 100 °C with a
slight vacuum or with a heat lamp controlled with a Variac. After cooling,
add 1.5 mL of fuming H_2SO_4 and 0.5 mL of HNO_3 to destroy organic
matter. Heat to SO_3 evolution. Several additions of H_2SO_4–HNO_3 may
be necessary to oxidize all the carbonaceous material as evidenced by a
white silica residue with no black particles. Once the organic material has
been destroyed, heat the flask or crucible with a Meker burner, gently at
first, to remove H_2SO_4 and fully dehydrate the remaining SiO_2. Ignite the
samples in a muffle furnace at 1000 °C to constant weight. Cool and weigh
for silicon determination (see acid dehydration method discussion below).

The use of alkali salts and alkali for sample destruction in quantitative
methods is possible, but loss of a portion of the sample from volatile siloxane
rearrangement products is apt to occur. The main applications of alkali
decompositions have been for solid resinous materials and high molecular
weight polymers. The use of alkali and alcohols together improves recovery
of volatile species, particularly cyclic siloxanes. Fusions of silica with sodium
hydroxide are used to prepare standards or samples for colorimetric analysis.

With some modifications, the Parr calorimetric oxygen bomb may be used
for decompositions. Steel surfaces should be avoided, particularly with
chlorosilanes. Platinum-lined bombs, platinum ignition wire and sample
crucibles are available and should be used. Difficulties may arise in recovering all the silica from the crucible and walls of the bomb. A 5% alkali
solution has been suggested as an acceptor for the oxidized materials.

2.1.3. *Quantitative Silicon Analysis Methods*

Many techniques are given in the literature for determining the silicon
content of organosilicon compounds (4–8). For organosilicon materials that
are soluble in organic solvents, we have found atomic absorption spectrometry to be a rapid and useful technique. The sample is diluted using
4-methyl-2-pentanone, (methyl isobutyl ketone or MIBK) as the solvent of
choice. Other solvents, such as toluene, the chlorinated solvents, or other
ketones may be used, but the MIBK has good solvating properties for a wide

variety of organosilicon materials, along with excellent burning characteristics in the nitrous oxide–acetylene flame. A detailed procedure is given in
Chapter 14, Section 3.1. Gravimetric and colorimetric procedures are most
generally employed for chemical silicon determinations.

2.1.3.1. Gravimetric Methods. The most common gravimetric procedure
involves acid dehydration to SiO_2. In the absence of metals or metalloids, the
acid decomposition procedure described above constitutes a quantitative
method. The silica is dehydrated by repeated ignition with sulfuric or
perchloric acid and ignited to constant weight with a Meker burner or a
muffle furnace at 800–1000 °C.

Percent silicon is calculated using

$$\% \, Si = \frac{\text{residue } SiO_2 \text{ wt (g)}}{\text{sample wt (g)}} \frac{(28.09)}{(60.09)} (100) \tag{8.1}$$

If inorganic materials other than silica are present in the sample, volatilization
of SiF_4 by treatment with hydrofluoric and sulfuric acid may be used after
acid decomposition and dehydration. This procedure distinguishes the silica
present in the residue and thus gives a specific measurement for silicon.

Reagents

Hydrofluoric acid, 45% aqueous, fuming H_2SO_4 and HNO_3.

Apparatus

35-mL Platinum crucibles, heat lamp, and muffle furnace.

Procedure

After fully dehydrating and weighing the SiO_2 produced in the acid ashing
method above, add 1 drop of water, 3 or 4 drops of H_2SO_4, and 7 mL of
HF. Heat gently, using the heat lamp, to SO_3 fumes. Add 1 or 2 drops of
HF, and if a reaction occurs repeat the HF addition and evaporation steps.
When no bubbling is noted, heat with a Meker burner until no SO_3 fumes
evolve. Ignite the crucible to 1000 °C in the muffle furnace to constant
weight or for 1 h. Cool and reweigh the nonsilicious residue in the crucible.
Calculate percent silicon using

$$\% \, Si = \frac{[\text{wt total residue (g)} - \text{wt HF treated residue (g)}]}{\text{sample wt (g)}} \frac{(28.09)}{(60.09)} (100)$$

$$\tag{8.2}$$

2.1.3.2. Colorimetric Silicon Methods. An enormous number of publications has appeared on the determination of silicon using silicomolybdic acid methods. Extensive reviews of this literature have been provided by Shell (9) and by King et al. (10). Both the yellow silicomolybdic acid and blue reduced form have been employed in the analysis of silicone and silicate materials that have been decomposed using peroxide bomb, alkali, or alkali salt fusions. For low levels of silicon ($<5\%$), the reduced method is preferred. Spectrophotometric measurement at 715 nm with sodium sulfite as a reducing agent, and at 815 nm with 1-amino-2-naphthol-4-sulfonic acid, sodium sulfite, or sodium hydrogen sulfite reducing solution have been employed. The latter procedure is roughly five times as sensitive, allowing analysis for 0.005–5% silicon. The effect of fluoride may be masked with boric acid (7). Phosphorus interference may be minimized by the use of oxalic acid, which essentially destroys the phosphomolybdic acid complex before the reduction step. With biological samples, phosphorus and iron interference is critical, and unless special sample treatment is employed, major interference is encountered at low silicon concentrations ($<10\ \mu g/g$).

A typical reduced molybdate colorimetric method (2) is given below.

Reagents

Ammonium molybdate. Dissolve 18.8 g of $(NH_4)_2MoO_4$ in 75 mL distilled water, add 23.0 mL H_2SO_4, cool, and dilute to 250 mL.

1-Amino-2-naphthol-4-sulfonic acid (ANSA).

Reducing solution. Dissolve 2.0 g of anhydrous Na_2SO_3 in 25 mL of water. Add 0.4 g of ANSA and set aside. Dissolve 25 g of $NaHSO_3$ in 200 mL of water. Mix the two solutions and dilute to 250 mL.

Oxalic acid, Eastman. Add 20 g to 80 mL of water.

Poly(dimethylsiloxane), 350 cS.

Apparatus

Spectrophotometer, with good sensitivity at 815 nm. 10- and 50-mm cells, glass or quartz.

Procedure

Decompose samples using the Na_2O_2 bomb procedure described above. Carry a blank throughout. Transfer the fusionates to 500- or 1000-mL volumetric flasks, dilute, adjust to pH of about 4, and dilute to mark. For standards, weigh 5–25 mg of PDMS fluid (37.85% silicon) into gelatin capsules. After Na_2O_2 decomposition, dilute to 1 L. Remove various aliquots to 100-mL flasks to give a concentration range of 6–140 μg silicon/100 mL. Dilute to 80 mL with $(NH_4)_2MoO_4$ solution, mix, and let stand for 30 min. Add 2 mL of oxalic acid solution and shake for 20 s. Add 1 mL of reducing solution, mix, and check the pH. Adjust pH to 1.2–1.7 with acid or base if required. Dilute to volume and allow to stand for 20 min after the addition of the reducing solution. Measure absorbance of the standards and samples at the maximum near 815 nm, using the blank as a reference. Alternatively, water may be used as a reference, and blank values measured and subtracted. Plot the absorbance of the standards versus micrograms of silicon per 100 mL. Read micrograms of silicon for each sample. Calculate percent silicon using

$$\%Si = \frac{(\mu g\ Si)\ (10^{-6})\ (D/A)\ (100)}{\text{sample wt (g)}} \tag{8.3}$$

where D = sample dilution (mL) and A = sample aliquot (mL).

2.1.4.   Unsubstituted Silicon

In many studies it is desirable to determine the silica filler content of compounded products. Silica may be solubilized in the presence of organosiloxane polymers by reaction with saturated sodium butylate (2). Organic substituents bonded to silicon are not cleaved. The gravimetric or colorimetric procedures given above may be used to determine the solubilized silica present.

Reagents

Colorimetric reagents are noted above.
Sodium butylate. Carefully dissolve 6 g of sodium in 100 mL of BuOH. Add small portions to the alcohol contained in a Teflon-lined beaker. Allow to cool, and filter through at 60-mesh stainless steel cone into a polyethylene bottle.

Standards

Silica, dried at 350 °C overnight.
Tetraethylorthosilicate, $Si(OEt)_4$, Eastman.

Apparatus

Teflon® lined, 600-mL, stainless steel beakers.
Nickel reaction vessel (22-mL Parr Na_2O_2 bomb).

60-mesh, stainless steel cone.

Whatman No. 42 filter paper.

Procedure

Add 7–8 mL of saturated NaOBu to the reaction vessel. Weigh a sample (not > 0.5 g) to contain 5–25 mg of silicon. Weigh three standards, containing 5–15 mg of silica or 15–40 mg of $Si(OEt)_4$, with each set of samples. Assemble the reaction vessel, seal, and heat in an oven at 130 °C for 12–16 h. Remove the vessel from the oven and allow it to cool to room temperature. Rinse with water, collecting it in a Teflon® lined beaker. Dilute to 100 mL and place the cup in the beaker. Heat to a gentle boil to dissolve all water-soluble material, cool, and remove the cup, rinsing it inside and out. Collect all washings in the beaker. Add 15 mL of 1:10 HCl to the cup and wait 1 min. Transfer the acid to the beaker and rinse the cup thoroughly. Test the solution with litmus paper and adjust the solution with acid or base so that it is just slightly acid. Filter the solution through Whatman No. 42 paper into a 500- or 100-mL volumetric flask. Rinse the beaker and filter paper with hot water several times. Allow it to cool, dilute the flask to mark, and mix. Transfer an aliquot of the solution to a 100-mL volumetric flask with a pipet for silicon analysis. The 815-nm colorimetric procedure described above may be used at this point. Prepare a calibration curve of micrograms of Si per milliliter versus absorbance from the standards data. Calculate standards concentrations using

$$\mu g \ Si/mL = \frac{(mg \ SiO_2) \ (28.09) \ (10^3 \ \mu g)}{(1000 \ mL) \ (60.09) \ (mg)} \tag{8.4}$$

$$\mu g \ Si/mL = (mg \ SiO_2) \ (0.467) \tag{8.5}$$

or

$$\mu g \ Si/mL = [mg \ Si(OEt)_4] \ (0.135) \tag{8.6}$$

Calculate percent silicon (unsubstituted) as noted in the colorimetric method from the sample concentration–absorbance data.

2.2. Carbon, Hydrogen, and Nitrogen

In the past, oxygen train combustion techniques such as the classical Pregl procedure were used for carbon and hydrogen content (4). A separate determination was made for nitrogen using the Dumas or Kjeldahl procedures (11).

Automated instruments are now available that give analyses for C, H, and N by oxidative degradation of samples with GC separation and measurement of the combustion products. Thus, the carbon dioxide, water vapor, and nitrogen (from the reduction of the nitrogen oxides that had formed), are separated and quantitated.

Typical of the automated C–H–N analyzers are the Carlo Erba Model 1106 and the Perkin Elmer Model 240C elemental analyzers. These instruments are capable of measuring the carbon, hydrogen, and nitrogen content on a single sample, thus eliminating the need for analyzing two separate samples as required for the Pregl and Dumas techniques (12–15).

With the older train-combustion scheme, it was necessary to carry out the combustion of silicone materials at a slow rate to ensure the complete combustion of the sample (4–6, 16, 17). With the Carlo Erba instrument, the combustion of the sample takes place very rapidly. The quartz combustion tube is held in a vertical configuration and is maintained at a temperature of 1030 °C. A constant stream of helium flows through the tube. Samples, held in lightweight tin capsules, are dropped into the tube at preset time intervals. When the samples are introduced, the helium stream is temporarily enriched with pure oxygen. Flash combustion occurs with a rapid rise in temperature in the combustion zone to approximately 1900 °C, primed by the oxidation of the tin container. Quantitative combustion is then achieved by passing the mixture of gases over chromium oxide. The combustion gases are then passed over copper at 650 °C to remove the excess oxygen and to reduce the nitrogen oxides to nitrogen. They then pass through a chromatographic column in which the individual components are separated in the order nitrogen, carbon dioxide, and water. The instrument response is calibrated using standard materials such as benzoic acid, acetanilide, or other primary standards.

This system has proven satisfactory for the determination of carbon and hydrogen in silicone materials, and has also been able to handle such refractory materials as silicon carbide and silicon nitride.

The Perkin-Elmer Model 240C instrument performs similarly, although detection and measurement of the combustion gases is accomplished slightly differently (14).

Many combustion methods have been reported as being capable of analyzing a single sample for several elements, such as carbon, hydrogen, and halogen (18); carbon, hydrogen, halogen, and sulfur (19); carbon, silicon, and aluminum in organoaluminosiloxanes (20) carbon, hydrogen, and fluorine (21); and carbon, hydrogen, and silicon (22). These procedures, however, all suffer from the drawback that analysis conditions are never optimized for a single element. The consequence is that the compromises made for multi-element determinations generally reduce the accuracy and scope of the analysis. Although reasonable results have been reported in the literature for a few simple model compounds, we have not had satisfactory results for the more complex silicone materials.

Nitrogen analysis on organic materials is often done using the Kjeldahl sulfuric acid digestion procedure. This method has found utility in the analysis of organosilicon compounds that contain silazane linkages, or other materials that have nitrogen present in a form easily reducible to ammonia. The nitrogen content of silicon nitrides and some of the aminofunctional silanes used as coupling agents have been analyzed using a modified Kjeldahl procedure with a sealed tube reaction and 80% sulfuric acid. But with polysiloxanes, acids used in the digestion process cause depolymerization and form volatile nitrogen-containing cyclic and linear siloxanes. These oligomers may distill from the digestion flask, giving low recovery for nitrogen.

The micro-Dumas procedure (11) has been found to be satisfactory for nitrogen analysis (except for the analysis of the refractory nitrides). Even samples of glass cloth that have been treated with the aminofunctional coupling agents have been successfully analyzed.

Nitrogen at the low parts per million level is best analyzed using an oxygen tube system. The sample is combusted in a stream of oxygen at a temperature of 800 °C, and the oxidation products are passed over heated platinum gauze maintained at a temperature of 700 °C to oxidize the nitrogen completely to nitrogen dioxide. The nitrogen dioxide is continually absorbed in a solution of sulfanilic acid. After the combustion has been completed, the nitrite, which is coupled to the sulfanilic acid, is diazotized with N-(1-naphthyl)ethylenediamine dihydrochloride to form a red azo dye. The color intensity of the dye is measured at 535 nm, using standards prepared from reagent grade sodium nitrite. This procedure has been used successfully for analysis in the range of 5–1000 ppm.

A qualitative test for nitrogen has been described that uses the oxygen flask decomposition technique with sodium hydroxide as the absorbing solution followed by the nitrite azo dye test procedure (23).

2.3. Oxygen

Oxygen is an element that is difficult to determine in organosilicon materials, and oxygen content is often determined by difference. Combustion techniques cannot be used for obvious reasons. Systems have been described (24) for the determination of oxygen in organometallic compounds that involve pyrolysis of the material in an inert atmosphere at a temperature of 1000 °C. Oxides are converted to carbon monoxide by passing the gas stream over heated carbon, with the carbon monoxide being further oxidized to carbon dioxide by passing it over heated copper oxide. The carbon dioxide is determined gravimetrically by absorption onto ascarite, or titrated iodimetrically.

Neutron activation analysis has been successfully applied to the analysis of some silicone materials. Anders (25) reports that organosilicon compounds may be analyzed by neutron-activation methods to as low as 100 ppm of oxygen with fluorine being the only interferring element.

2.4. Halogens

2.4.1. Qualitative Tests for Halogens

The well-known sodium fusion test (3) and the oxygen flask combustion technique (26–30) have been applied to silicon containing materials for the detection of halides. In the oxygen flask combustion procedure, a dilute solution of sodium hydroxide or sodium carbonate is usually used as absorbing solution. Chloride, bromide, or iodide are detected using either silver nitrate or mercury thiocyanate, while fluoride is most commonly done by the alizarin complexation method.

2.4.2. Quantitative Analysis for Hydrolyzable Halide

Atoms of Cl, Br, or I that are attached directly to a silicon atom react vigorously with water. Such species are considered silicon functional, and their analysis is discussed in Section 3.6.

2.4.3. Total Halide Methods

Halides can often be determined directly by X-ray fluorescence (Chapter 15). Organically bound halide in organosilicon compounds and polymers must be decomposed to form an ionic species before chemical analysis can be undertaken. This decomposition is best accomplished with the Parr peroxide bomb.

Many other decomposition techniques have been described (2), including reaction with sodium biphenyl, the Stepanov reaction of the sample with sodium metal and alcohol, reaction with potassium metal in sealed bombs at high temperatures (850–900 °C), the Carius sealed tube–acid decomposition process, or tube combustion with an oxygen stream and vanadium pentoxide as the catalyst with powdered silver collection of the halide vapors.

Once a suitable decomposition procedure has been established, the preferred finish is the potentiometric titration of the halide with 0.1–0.0025 N silver nitrate solution. The sample is first acidified with nitric acid and the iodate and bromate reduced to the iodide and bromide by the addition of hydrazine sulfate. A potentiometric titration is performed using a silver billet indicating electrode and a silver–silver chloride reference electrode. The internal filling of the electrode (potassium chloride solution) must be replaced with a saturated sodium sulfate solution to ensure correct response of the electrode to the halogen titration. From the potentiometric titration curve, iodide, bromide, and chloride can be distinguished by successive breaks in the titration curve (iodide titrates first, followed by bromide and then chloride).

For fluorine, as found in fluorosilicone polymers, the sample is decomposed in the Parr pressurized oxygen bomb, after which a colorimetric finish can be used. Once the fluoride is in solution, we have found the Technicon AutoAnalyzer system with distillation and SPADNS colorimetric reagent to be a highly reliable method for the analysis. We have also used the fluoride specific ion electrode with potentiometric measurement as the finish, but very careful calibration, pH control and ionic strength adjustment, and masking of interfering cations that precipitate or complex with fluoride have to be employed.

2.5. Sulfur

Sulfur is most commonly determined either gravimetrically as the sulfate, or by titration with barium perchlorate after sample decomposition. The decomposition methods used for sulfur are very similar to those used for halogen analysis. Among these are the Parr oxygen bomb, the oxygen flask technique, Carius sealed tube reaction with concentrated nitric acid, or combustion in tubes with flowing oxygen. Gravimetric procedures are best applied to samples that have a sulfur content of greater than 0.5%. There has also been described in the literature (Ref. 24, p. 55) a procedure for analysis by precipitation with lead in acid solution, redissolving the lead sulfate after filtration in ammonium acetate solution, and titrating the lead with ethylenediaminetetraacetic acid (EDTA).

It is also possible to determine sulfur directly with the use of energy dispersive X-ray fluorescence (Chapter 15, Section 3).

3.   FUNCTIONAL GROUP ANALYSIS

3.1.   Hydroxyl (SiOH)

The analysis for the hydroxyl (OH) group is one of the most important functional group analyses for silicones. Hydroxyl functionality may be present as silanol (SiOH), water (HOH), and possibly as carbinol (COH) or carboxyl (COOH). Amine groups on silicon substituents may also be a source of active hydrogen. The silanol group is particularly important in silicone chemistry because it is through this function that condensation polymerization occurs to produce silicones. It is also an active catalyst for siloxane bond rearrangement. Water is a condensation by-product so the concentration of water in silicones is directly related to the amount of silanol and silane condensation.

The chemical or instrumental analysis for silanol and water in silicones is complicated by several factors. First, the fact that silanol and water generally occur together requires silanol- and water-specific analyses. Unfortunately, there are few if any silanol-specific chemical analyses. Most chemical analyses are nonspecific methods for total active hydrogen; thus silanol and water and other active hydrogens, including carbinols and carboxyl, are determined in total. Second, the reactivity of the silanol group can vary widely depending on the molecular structure or environment. Third, concentrations of interest range from low parts per million to percent levels.

The most common methods for the chemical analysis of silanol and water in silicones are discussed below. Additional information on less common methods is available in the first edition of this book (2). The analysis of functional groups such as carbinol or carboxyl, less commonly found in silicones, can be accomplished with some of the procedures described below. Much more information is available in texts on general organic functional group analysis.

3.1.1.   Total Hydroxyl Methods

3.1.1.1.   Manometric. Manometric methods for active hydrogen involve treating the sample with a reagent that stoichiometrically generates a gas. The quantity of gas is then measured with a manometer. Some reagents used for this type of analysis are discussed below.

A Grignard reagent such as methyl magnesium iodide has often been used (30–32) for the analysis of active hydrogen. As an example, silanol reacts as follows,

$$R_3SiOH + CH_3MgI \longrightarrow R_3SiOMgI + CH_4 \quad (gas) \qquad (8.7)$$

The methane gas can be measured manometrically. Practical use of this procedure for active hydrogen measurement has been limited by a number of problems including methane solubility in solvents, incomplete or nonquantitative reaction, frequent precipitate or gel formation, and complicated apparatus.

Methyllithium has also been used in a manner similar to the Grignard reagent for the determination of silanol groups in silica gels (33). The reaction of the silanol group with the methyllithium reagent also produces methane, which is measured volumetrically.

Lithium aluminum hydride (LiAlH$_4$) has been widely used as a reagent for reaction with active hydrogen (2, 34). This reagent is a powerful reducing agent that reacts with an active hydrogen-containing material as described below:

$$4ROH + LiAlH_4 \longrightarrow 4H_2 + LiAl(OR)_4 \qquad (8.8)$$

The hydrogen gas is measured manometrically with an apparatus such as that shown in Figure 8.1. Lithium aluminum hydride can be used in a variety of solvents including diglyme, THF, ethylether, and toluene. The completeness of reaction and its ease of use makes the lithium aluminum hydride method more generally applicable than the Grignard method. The lithium aluminum hydride method does have some disadvantages, however, most of which result from the powerful reducing ability of this reagent. Volatile silanes can be produced from reductive bond cleavage of siloxanes or

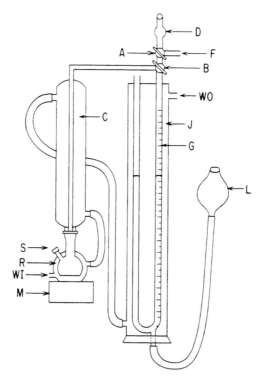

Figure 8.1. LiAlH$_4$ reaction and manometer apparatus. A and B, three-way stopcocks; C, condenser; D, drying tube; F, dry N$_2$; G, gas buret; J, constant-temperature water jacket; L, leveling bulb; M, magnetic stirrer; R, reaction flask; S, rubber septum; WI, water inlet from constant temperature bath; WO, water outlet (2).

functional groups such as R_3SiH, R_3SiCl, or R_3SiOR in the sample. These volatile silanes can alter the gas volume or even result in explosive gas mixtures. A similar explosion hazard exists if various other compounds, such as oxidizing agents, are present in the sample (35). The potential exists for the hydrogen produced to react with unsaturation in the sample. Also, gelation can occur with some samples. For many polymers such as PDMS, however, the method is useful. A typical procedure using lithium aluminum hydride is described below.

Reagents

Diethylene glycol dimethylether (diglyme). Dry over barium oxide.

Lithium aluminum hydride, 95%. Prepare daily 1–3% solution in diglyme. Solutions of lithium aluminum hydride are commercially available.

Xylene. Dry over sodium.

Apparatus

See Figure 8.1.

Procedure

For liquid samples: Add 20 mL of lithium aluminum hydride–diglyme mixture to a dry reaction vessel and purge the apparatus for 5 min with dry nitrogen. Open stopcocks A and B to the atmosphere and adjust manometer to zero with the leveling bulb. Lower the leveling bulb and inject a weighed sample through the septum of the reaction vessel using a syringe. Equalize mercury column heights with the leveling bulb. Determine gas volume 1 min after injection of the sample. Note the atmospheric pressure and calculate percent hydroxyl according to the following equation.

$$\text{net gas volume} = \text{total mL} - \text{sample mL injected} \qquad (8.9)$$

$$\%OH_{total} = \frac{[\text{net gas (mL)}][\text{atm pressure (mm)}](273)(17)(100)}{[\text{sample wt (g)}](760)[\text{ambient temperature (K)}](22{,}400)}$$

$$(8.10)$$

The basic procedure described above will also work with viscous liquid and solid samples. These samples may either be dissolved in dry solvent and injected, or weighed directly into the reaction vessel. Xylene or diglyme is added to dissolve the sample, and the system closed, zeroed, and injected with lithium aluminum hydride reagent. Both of these sample introduction methods require a solvent blank.

3.1.1.2. Condensation Methods. Methods for total hydroxyl that involve silanol condensation take advantage of the fact that most silanols (unless sterically hindered) condense under the proper conditions (36) to produce water.

$$2R_3SiOH \longrightarrow R_3SiOSiR_3 + H_2O \qquad (8.11)$$

The condensation reaction has been used for quantitative hydroxyl determinations when an appropriate catalyst, usually an acid or base, is used. The water produced from condensation is typically removed by azeotropic distillation and measured volumetrically in a Dean–Stark or similar apparatus, or titrimetrically with Karl Fischer reagent.

A catalyst that has been shown effective at promoting complete silanol condensation in a variety of silicone materials consists of a mixture of boron trifluoride, acetic acid, and pyridine (37). This catalyst has been used for rapid and reproducible total hydroxyl determination. Typically, the sample is dissolved in toluene and pyridine and the boron trifluoride–acetic acid catalyst is mixed with the sample solution for 5 min at room temperature. The water produced is then azeotropically distilled to a receiver. This technique can easily be adapted to analyze for free water and silanol separately by distilling the free water from the dissolved sample *before* adding the condensation catalyst. The method is applicable to 0.5–6% levels of hydroxyl but SiH does cause a significant interference (37). The pyridine should be omitted when determining total hydroxyl in monomers.

3.1.1.3. Titration Methods. The most generally applicable titration method for silanol and water uses a strongly basic titrant, lithium aluminum di-*n*-butylamide, to react with active hydrogen compounds as described below (38, 39).

$$LiAl(NBu)_4 + xROH \longrightarrow xBu_2NH + LiOR + Al(OR)_3 \qquad (8.11)$$

The stoichiometry depends on the material being titrated.

ROH	X
Water	2
Silanols	3
2-Naphthol	3
Methanol	3.3

The stoichiometries reported above are empirical and little is known about the actual mechanism of reaction, particularly for the reaction with water. The strongly basic lithium aluminum di-n-butylamide reagent reacts rapidly with virtually all substances with "acidities" equal to or greater than that of the alcohols. The reagent is also air or oxygen sensitive; therefore, reagent storage and buret or titration system design are important considerations. The titration is usually performed in solvents such as THF or a mixture of THF and an amine such as pyridine, and the endpoint is determined colorimetrically with an indicator, 4-phenylazodiphenylamine.

The most important advantages of this total hydroxyl titration are its speed, precision, and broad applicability. The titration can typically be done in 1–2 min and the relative standard deviations for simple and polymeric silanols are 0.5–2% (38). This technique can be applied to a broad range of hydroxyl concentrations from 10 ppm to several percent depending on the reagent normality used. The titration can also be applied to a wide variety of samples including monomeric and polymer silanes and silicone resins. Common functional groups found in silanes and silicones such as silicon hydride (SiH), alkoxy silanes (SiOR), and vinyl groups do not interfere. Finally, liquids, solids, and high viscosity fluids or gums can be analyzed by this method although high viscosity fluids and gums may have to be prediluted in a dry solvent such as xylene.

The lithium aluminum di-n-butylamide titration for total hydroxyl does have disadvantages. The different stoichiometry for water presents some problems although most of the stoichiometry difference is constant and therefore corrections can be made for the amount of water present (38). The acetoxysilane function interferes with this titration and other interferences cause either gel or precipitate formation, or interference with the indicator color change. The procedure (38) for using this titration is described below.

Reagents

THF dried over molecular sieves.

Ethylene glycol dimethylether (Ansul 121, monoglyme) dried over molecular sieves.

Di-n-butylamine.

4-Phenylazodiphenylamine diluted to 0.1% in toluene.

Lithium aluminum hydride, 95%.

2-Naphthol, dried in a desiccator.

Lithium aluminum di-n-butylamide titrant is available commercially from Lancaster Synthesis, England, or made from lithium aluminum hydride and di-n-butyl amine by the following procedure.

Procedure

Place 700 mL of dry ethylene glycol dimethylether in a 1 L, three-neck round-bottom flask fitted with a water condenser and a nitrogen purge of 10 mL/min. Add a few drops of 4-phenylazodiphenylamine indicator and then small portions of the lithium aluminum hydride to the stirred ether to neutralize residual hydroxyl. The indicator should change from yellow to purple when hydroxyl is neutralized. Add an additional 2 g of lithium aluminum hydride and then heat the ether to reflux using a heating mantle and Variac. After establishing a gentle reflux, slowly add 40 mL of di-*n*-butylamine using an addition funnel. Continue reflux for 10 min after the addition of the amine. Allow the solids to settle, then siphon the reagent under nitrogen into the titrant reservoir or appropriate storage bottle. Titrant normality should be approximately $0.05M$ ($0.2N$). The titrant can be diluted to lower concentrations with ethylene glycol dimethyl ether.

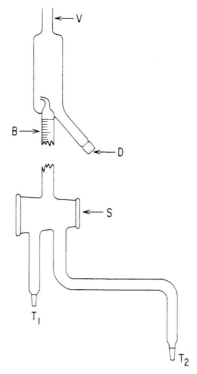

Figure 8.2. Buret of $LiAl(NBu_2)_4$ titration for active hydrogen. B, 2-mL buret, 0.01-mL subdivisions; D, excess reagent drain; S, three-way Teflon stopcock T1, delivery; tip; T2, refill port (T1 and T2 are inner luer type joints, 2-mm i.d.); V, vent to N_2-filled plastic bag (2).

Apparatus

2 or 5 mL auto zeroing buret with inner Luer joints, 2 mm-i.d., on the delivery and refill tips so hypodermic needles can be attached (see Figure 8.2). Hypodermic needles, 1- and 5-in. lengths for insertion into septum top vials.

Vials, 100–125 mL, with Teflon® laminated septum tops.

Procedure

Add 40 mL of dry THF or 1:1 THF–pyridine and 5 drops of the indicator solution to the titration vessel and titrate to a red-purple color persisting for at least 30 s. Standardize the titrant at least four times with primary standard, 2-naphthol. This process serves as a pretreatment for the solvent and indicator. Results from the first one or two titrations should not be included in the average normality calculations. Liquid or solid samples can be titrated using a syringe, or small glass or metal cup, respectively, for sample introduction. Sample weights are obtained by difference. Calculate percent hydroxyl as follows,

$$\%OH = \frac{[EP(mL)](N)(0.017)(100)}{[sample\ wt(g)]} \tag{8.12}$$

where

EP = endpoint volume in mL
N = normality of titrant

This titration procedure has been successfully automated using a Metrohm automatic titrator (40) with a Metrohm dipping probe colorimeter to monitor indicator color change.

The Karl Fischer titration originally developed for water determinations has also been used for total silanol plus water determinations (41, 42). Silanol can react quantitatively with the methanol in the Karl Fischer reagent as follows,

$$R_3SiOH + CH_3OH \longrightarrow R_3SiOCH_3 + H_2O \tag{8.13}$$

$$H_2O + I_2 + SO_2 + CH_3OH \longrightarrow 2HI + CH_3HSO_4 \tag{8.14}$$

Use of this procedure is limited to low molecular weight silanols. Silanol groups of higher molecular weight materials may react incompletely because of stearic hindrance and solubility limitations (2). Also, silanol condensation and alcoholysis of cyclotrisiloxanes are interferences for this method (43).

3.1.2. Silanol Specific Measurements

Only a few silanol-specific chemical methods are available. One method involves the silylation or "capping" of the silanol to form a trialkylsilyl derivative and subsequent analysis of this derivative by GC. This silylation procedure is more commonly used for specific analysis of compounds having a carbinol OH although the procedure does work well for low molecular weight linear and cyclic siloxanes that contain silanol groups. Several silylation reagents are commercially available (44, 45).

Another silanol-specific analysis method involves the direct titration of the more acidic silanols such as triphenyl silanol. Silanols are generally more acidic than carbinols and water with pK_a values in the range of 8–12 (36). Those with pK_a values of 8–10 can be titrated with hydroxide titrants. Good examples are the titration of triphenyl silanol and various hexaorganocyclo-trisiloxanes in pyridine with tetrabutylammonium hydroxide (46, 49).

3.1.3. Water in the Presence of Silanol

Several older analytical methods are used for the determination of water in the presence of silanol, most of which involve a Karl Fischer titration.

One method, described in Section 3.1.1.2, utilizes azeotropic distillation of water with toluene followed by measurement of the water volumetrically or by Karl Fischer titration. The silanol containing sample for this method must be free of condensation catalysts to ensure that the water determined is in fact free water.

Direct Karl Fischer titration of water in the presence of silanol is feasible in spite of the reactivity of silanols toward the Karl Fischer reagents. The water-specific determination uses reagents and titration conditions that minimize both the silanol condensation and silanol–cyclotrisiloxane alcohol-ysis (43, 50, 51). One variation on the standard Karl Fischer titration conditions uses a high molecular weight alcohol such as 2-ethyl-1-hexanol, methyl cellosolve, or dodecanol in place of the methanol as the titration solvent (52–55). Lowering the temperature of the titration slows the inter-fering reactions significantly, and cooling the titration vessel with ice water has allowed water determinations in the presence of short-chain silanols (52). This general methodology has also been extended to the determination of absorbed water on silica surfaces (56) where the titration results compared favorably with the azeotropic distillation and thermogravimetric determina-tion of absorbed water. The advantages of the direct titration of water in the presence of silanol with the modified Karl Fischer reagents include speed, precision, and sensitivity to small amounts of water.

Modified Karl Fischer titration reagents have been developed, which employ substituted alcohols as solvents and amine components other than pyridine (51, 57). These reagents were specifically designed to speed up the reaction with water and slow down or eliminate interfering side reactions, particularly with aldehydes and ketones. These modified Karl Fischer reagents also reduce the interferences caused by silanol and thus allow the determination of water in the presence of silanol (51).

3.2. Silane Hydrogen (SiH)

The hydrogen atom (H) is not normally thought of as a reactive functional group but hydrogen bonded to silicon (SiH) is a special case. The Si–H bond is polarized

$$Si^+H^-$$

depending to some degree on the substituents on the silicon (58). This polarization of the Si–H bond accounts for its reactivity toward nucleophilic or electrophilic attack and also its general reactivity as a reducing agent. The reactivity of the Si–H bond is the basis of several qualitative and quantitative chemical tests for silanes. These chemical tests fall into two categories characterized by [a] the base catalyzed reaction of silanes with alcohols or water and [b] reaction of silanes with easily reduced substances such as Hg(II) or Ag(I) salts. These chemical tests are discussed in detail below.

3.2.1. Hydrogen Evolution Methods

The base catalyzed reaction of SiH with water or alcohols has been known for some time and is described in the equation

$$Si\text{–}H + HOR \xrightarrow{\text{base}} H_2 + ROSi \qquad (8.15)$$

This reaction can be used as a qualitative confirmation of SiH by the observation of evolved gas when the sample is treated with strong base in the presence of alcohol or water. The evolution of hydrogen is rapid, and quantitative procedures can be used to measure the quantities of hydrogen gas produced using an apparatus similar to that shown in Figure 8.1. Several different base catalysts can be used including aqueous or alcoholic NaOH/KOH (59, 60) sodium butoxide (61), or moist piperidine (62).

A method has been developed that uses GC measurement of the hydrogen gas evolved when silanes are treated with alcoholic KOH (30). The reaction of silane with alcoholic KOH is carried out in a sealed septum vial and the gas produced is removed from the vial by syringe and injected into a GC with a thermal conductivity detector. The amount of hydrogen in the sample is calculated from a calibration curve based on reacting a known pure standard silane. This method is very sensitive and capable of measuring parts per billion levels of hydrogen.

Hydrogen gas evolution methods can be generally applied to samples containing silanes as there are few interferences. Silanol and vinyl functionality, and trace amounts of acid or base in the sample do not interfere. Disilane (Si–Si) does release hydrogen in the presence of base and interferes quantitatively with SiH determination (62). Any organic functionality that could release hydrogen (or any gas, if manometric procedures are used) in the presence of strong base also interferes.

A detailed procedure for measurement of silane hydrogen manometrically by reaction with sodium butylate is given below.

Reagents

Butyl alcohol, reagent grade.

Mercury, reagent grade.

Sodium metal, reagent grade. Prepare a 18–20% solution of sodium butylate by dissolving 4.5 g of sodium metal in 100 mL of butyl alcohol. Add small portions at a time.

Xylene, reagent grade

Apparatus

Manometer system as shown in Figure 8.1 or similar apparatus.

Syringes, a variety of sizes of glass hypodermic and plastic disposable syringes.

Procedure

Place about 20 mL of the sodium butylate solution in a dry reaction vessel, attach to the manometer system, and purge the system for 5 min with dry nitrogen. After purging, close the system by placing a rubber septum in the sample entrance tube. Open stopcocks A and B to the atmosphere and adjust the manometer to the zero milliliter mark using the leveling bulb. Close the system to the atmosphere by closing stopcock B and then check the system for leaks by lowering the leveling bulb and observing the mercury level after 1 min. Load a sample into a clean, dry syringe and inject into the reaction vessel recording the sample weight (SW) and sample volume (SV) by difference. Maintain the reaction vessel at 60 °C ± 1 °C with circulating water from a constant temperature bath. Once the evolution of hydrogen stops, equilibrate the reaction vessel temperature to 25 °C ± 1 °C as was done for the manometer. When gas volume has become constant, adjust the leveling bulb to equalize the two mercury columns. Record the gas volume (GV) and the barometric pressure. Then

$$\text{net gas volume, mL} = GV - SV \tag{8.16}$$

$$\text{SiH as } \%H = \frac{(\text{net gas volume})\,(\text{barometric pressure})\,(273)\,(100)}{(SW)\,(760)\,(298)\,(22,400)} \tag{8.17}$$

Solids or viscous liquids can be analyzed by the above method after volumetric dissolution in dry xylene. An alternative procedure for handling solids or viscous liquids is to weigh the sample directly into the reaction vessel, dilute with dry xylene, and then inject the sodium butylate solution. Either of these sample introduction procedures requires that adjustments be made in the gas volume to account for the xylene volume and reaction blank.

Alternatively, an electronic digital manometer (63) can be used in place of hazardous mercury manometer system. This instrument measures the increase in pressure of hydrogen gas liberated in a sealed system by cleavage of the SiH bonds as described above (64). The pressure increase is directly proportional to the volume of silane hydrogen in the sample. Calibration is accomplished by injecting known volumes of air into the reaction vessel with a gas syringe. The meter reading after each injection is plotted versus the volume of air. The calibration curve is checked daily to ensure the method is in control. Calculations for the percent of silane hydrogen are the same as for those using the mercury manometer system.

3.2.2. Mercury(II) Reduction Methods

Several analytical methods for SiH have been developed that depend on the quantitative reduction of mercury(II) salts. In a common example, the reaction of methanolic mercuric acetate with SiH is used

$$\geqslant\text{SiH} + 2\text{Hg(OOCCH}_3)_2 + \text{CH}_3\text{OH} \longrightarrow \geqslant\text{SiOCH}_3 + \text{Hg}_2(\text{OOCCH}_3)_2$$
$$+ 2\text{CH}_3\text{COOH} \tag{8.18}$$

The liberated acid is titrated with alcoholic potassium hydroxide (3, 59). A saturated solution of calcium chloride in methanol is added to react with the excess mercuric acetate (65). Acids, bases, disilanes, and vinyl groups interfere quantitatively with this determination of SiH and must be corrected for if present. A detailed procedure for the mercuric acetate determination of SiH can be found elsewhere (30).

Mercuric chloride reacts with SiH in the presence of water,

$$\geq SiH + 2HgCl_2 + H_2O \longrightarrow \geq SiOH + Hg_2Cl_2 + 2HCl \qquad (8.19)$$

or in a chemically inert solvent

$$\geq SiH + 2HgCl_2 \longrightarrow \geq SiCl + Hg_2Cl_2 + HCl \qquad (8.20)$$

in much the same manner as mercuric acetate.

The hydrochloric acid produced in this reaction can be titrated with alcoholic potassium hydroxide to quantitate the SiH functionality (30). Trace amounts of acids or bases in the product interfere with this determination but the short reaction times minimize interference from vinyl groups. An alternative to the acid determination is measurement of the mercuric chloride formed by filtering and weighing or using a phototurbidimetric procedure (59). The mercurous chloride has also been determined iodometrically by oxidation with iodine (66). Other variations of the mercuric chloride method have been reported that include the amperometric titration of SiH with mercuric chloride (59) and the polarographic determination of excess mercuric ions as iodide complexes (67). The polarographic method is reported to be specific for SiH in the presence of Si–Si bonds and vinyl functionality.

Following is the detailed procedure for the determination of silane hydrogen by reaction with mercuric chloride and titration of liberated acid with alcoholic potassium hydroxide.

Reagents

Calcium chloride, reagent grade. Prepared saturated solution in methanol.
Chloroform, reagent grade.
Mercuric chloride, reagent grade. Prepare a 4% (w/v) solution in 1:1 chloroform–methanol daily.
Methanol, reagent grade.
Phenolphthalein. Prepare a 0.1% solution in ethanol.
Potassium hydroxide, $0.1N$ in methanol.

Procedure

Pipet 20 mL of mercuric chloride solution into 125 mL iodine flasks in duplicate for each sample. Weigh duplicate samples into the flasks recording the weight as SW. Prepare two reagent blanks with each series of samples. Swirl the solutions and allow them to set 5–6 min before adding 15 mL of calcium chloride solution to each flask. Add 15 drops of phenolphthalein indicator solution to each flask and titrate with $0.1N$ alcoholic potassium hydroxide. Record endpoint volume as V_1. Titrate the reagent blanks in the same manner—one at the beginning and one at the end of the sample titrations. Average the blanks, and record as V_2. Acid or base blank (V_3) is determined on the sample in 50 mL of chloroform-methanol, and the result in milliliters is normalized to the sample weight used before being added (basic sample) or subtracted (acidic sample) from V_1. Calculations are shown in Eq. (8.21).

$$\%H = \frac{[(V_1 - V_2) \pm V_3](N_{KOH})\frac{(1.008)}{(2000)}(100)}{\text{sample wt(g)}} \qquad (8.21)$$

where N = normality of basic titrant

3.2.3. Silver(I) Reduction Methods

Silane hydrogen reacts readily to reduce Ag(I) to silver metal. This reaction can be used for detection of SiH by allowing a solution of the sample to react with alcoholic silver nitrate to produce the brown to black suspension of silver metal (68). The intensity of the color of the disperison is indicative of the amount of SiH present.

The reaction of SiH with Ag(I) has also been used for quantitative determinations (69). An excess of silver nitrate is reacted with the sample dissolved in acetone and the excess Ag(I) is titrated with a standard chloride solution. Anything present in the sample other than SiH that reduces or complexes the Ag(I) interferes with the test.

3.2.4. Miscellaneous SiH Reactions

Silane hydrogen is reported to react quantitatively with other oxidizing agents such as bromine (69, 70) and potassium permanganate (71). Bromine reacts with SiH at room temperature and the excess bromine added can be determined iodometrically by adding potassium iodide and titrating the liberated iodine with sodium thiosulfate. The reaction of SiH with potassium permanganate has been followed colorimetrically to measure the decrease in the color intensity of the permanganate solution as it reacts.

A variation on the reaction of SiH with bromine using N-bromosuccinimide has also been reported (72).

3.3.   Disilane (SiSi)

The disilane bond or functional group, Si–Si, similar to SiH in its reactivity toward aqueous or alcoholic base.

$$\text{Si–Si} + 2\text{ROH} \xrightarrow{\text{base}} 2\text{SiOR} + \text{H}_2 \qquad (8.22)$$

This reaction is quantitative for disilanes and the hydrogen produced can be measured manometrically as discussed earlier for SiH. Disilane bonds in organopolysilanes or organopolysiloxanes can be resistant to aqueous or alcoholic alkali if the polymer is not soluble. The solubility and reactivity can be improved by using hot moist piperidine (73) or boiling hexanol–alkali (74). Some hexaorganodisilanes can be resistant to reaction even under these conditions.

The disilane functional group can be determined in the presence of SiH by using the sodium butylate manometric method to measure total Si–Si and Si–H and the mercuric chloride reaction to measure specifically Si–H. The disilane is determined by difference.

3.4.   Vinyl (SiHC=CH$_2$)

Chemical methods for determination of carbon–carbon double bonds in silanes and silicones fall into two groups: addition reactions and cleavage reactions.

3.4.1.   Addition Reactions

Most of the more common quantitative methods for carbon–carbon double bonds involve the use of reagents that add quantitatively to the double bond.

A good example is the determination of total unsaturation using iodine monochloride (75). Excess iodine monochloride in glacial acetic acid is allowed to react with the sample in carbon tetrachloride. The excess iodine monochloride is then determined iodometrically by adding potassium iodide and titrating the liberated iodine with sodium thiosulfate. Silane hydrogen, SiH, is an interference for this test since it reduces the iodine monochloride.

Ozonolysis is another addition reaction used for total unsaturation, which has been applied to silicone polymers (76). The ozonolysis of the vinyl containing silicones dissolved in carbon tetrachloride is carried out with an ozone–oxygen mixture and the change in ozone concentration is measured spectrophotometrically at 254 nm. The data obtained can be represented as a plot of change in absorbance versus time where the calculated "peak" area for a sample corresponds to the reacted ozone. Pure samples of hexamethyldivinylcyclotetrasiloxane and tetramethyldiphenyldivinylcyclotetrasiloxane were used to generate a calibration curve for quantitative analysis. This method has been used to analyze vinyl concentrations reliably down to 0.1 mol%.

The addition of halogens, such as bromine, to carbon–carbon double bonds is well known and has been used for the quantitative determination of vinyl in selected siloxanes (77). Electrogenerated bromine in 90% acetic acid was shown to react quantitatively with some low molecular weight siloxanes depending on structure. The precision was high for this coulometric titration with bromine and analysis time was short.

The reaction mercuric acetate with terminal vinyl groups is an addition reaction that has been known for some time (78) and is widely used (79). Mercuric acetate adds to a terminal vinyl group

$$\equiv\text{Si–CH=CH}_2 + \text{Hg(OAc)}_2 \xrightarrow{\text{MeOH}} \equiv\text{Si–}\underset{\underset{\text{OMe}}{|}}{\text{CH}}\text{–CH}_2\text{–HgOAc} + \text{HOAc} \tag{8.23}$$

to produce one equivalent of acetic acid per equivalent of vinyl. The reaction is usually performed in methanol or a mixture of methanol and chloroform. The acetic acid is titrated with potassium hydroxide after excess mercuric acetate has been reacted with calcium chloride or sodium bromide. The SiH containing materials, acids and bases interfere quantitatively with the test.

3.4.2. Cleavage Reactions

Several degradation reactions have been reported that involve cleavage of a vinyl group bonded directly to silicon to produce ethylene or other easily measured by-products.

A procedure, generally applicable to higher molecular weight silicones, has been reported that involves fusing the silicone with potassium hydroxide followed by GC analysis of the ethylene produced (80). The samples are normally fused in a nickel crucible within a quartz tube purged with nitrogen. The ethylene is collected in a 2–3 L polyethylene bag, butane is added as an internal standard, and the gas is analyzed by GC. This procedure is limited to compounds that boil above 250 °C but is applicable to a concentration range 1–0.001% vinyl.

A standard mixture of potassium hydroxide, dicyclohexyl-1,4,7,10,13,16-hexaoxacyclooctadecane, xylene, and butanol has also been used to cleave vinyl groups bonded to silicon (81). The reaction mixture is heated at 120 °C for 30 min. Ethylene is collected using a glass bulb collection apparatus (82) and analyzed by GC using propyne generated from trimethylsilylpropyne as an internal standard. This procedure is also applicable to a wide concentration range of vinyl, 32–0.004%. The GC analysis is carried out isothermally at 100 °C using a Poropak Q column and a flame ionization detector.

A third example of hydroxide-promoted cleavage of vinyl bonded to silicon has been reported, which measures the evolved ethylene colorimetrically (83). This procedure uses a small reactor with a sodium hydroxide solution to cleave the vinyl groups from silicon. The resulting ethylene is absorbed in a palladium solution. Colorimetric analysis of the solution allows determination of vinyl concentrations from 2.5–0.02%.

Levels of vinyl in the range of 0.1% have been analyzed by bond cleavage reactions employing moist phosphorous pentoxide (82) and 90% sulfuric acid for sample digestion (84). Both methods use GC to analyze for the ethylene from the cleavage reaction.

3.5.   Alkoxy (SiOR)

3.5.1.   Acetylation

A perchloric acid acetylation, originally developed for carbinol analysis (85), has been shown to work well for the determination of SiOR (86–88). The procedure for alkoxysilane analysis involves the addition of the sample to a solution of 1–2M acetic anhydride and 0.06–0.15M perchloric acid in ethylacetate or 1,2-dichloroethane. The reaction mixture is swirled and allowed to stand at room temperature for 10 min or longer. A solution of water, pyridine, and dimethylformamide (DMF) is added to hydrolyze the excess acetic anhydride. The liberated acetic acid is titrated with alcoholic potassium hydroxide. The scheme of reactions suggested for this acetylation procedure (86) is as follows:

$$Ac_2O + H^+ \longrightarrow Ac_2OH^+ \qquad\qquad (8.24)$$

$$Ac_2OH^+ \longrightarrow Ac^+ + AcOH \qquad\qquad (8.25)$$

$$Ac^+ + SiOR \longrightarrow Si\overset{+}{O}R \qquad\qquad (8.26)$$
$$\qquad\qquad\qquad\qquad\quad |$$
$$\qquad\qquad\qquad\qquad\; Ac$$

$$Si\overset{+}{O}R + Ac_2O \longrightarrow SiOAc + AcOR + Ac^+ \qquad (8.27)$$
$$|$$
$$Ac$$

The alkylacetates formed do not hydrolyze under the conditions used so they cause a net consumption of the acetic anhydride. The titration of liberated acid is less than the titration of the reagent blank, and the difference corresponds to the alkoxysilane groups present. The acetoxy silanes do not interfere because they hydrolyze quantitatively to silanol and acetic acid. Water and silanol also do not interfere (with sufficient acetic anhydride) because their product of acetylation either is acetic acid or hydrolyzes to acetic acid. Acids, bases, or alcohols in the sample interfere with the test. This procedure reliably detects alkoxy groups smaller than butoxy at levels of 1% and above.

3.5.2. Zeisel

The Zeisel procedure is an old method originally developed for the analysis of alkoxy groups bonded to carbon in cellulose ethers (89). This method suffers from its length and complexity. In its original form, this method did not work well for alkoxysilanes, but modifications of the original procedure have been developed, which do give satisfactory accuracy (2, 59, 61).

3.5.3. Miscellaneous

A procedure described earlier for vinyl determinations involving alkali cleavage followed by GC analysis has also been shown to work well for the analysis of alcohols generated from alkoxysilanes (81). The alcohols generated from alkoxysilanes during the cleavage reaction are volatilized and collected in a water trap and the water is then analyzed directly by GC using a Porapak Q column. This technique is applicable to a wide concentration range, 1–0.001%. Alkoxy results are generally slightly low but within 90% of theoretical values.

Alkoxy functional group analysis has also been performed with potassium hydroxide in 2-methoxyethanol using sealed ampules held at 80 °C for 1 h (82). The alcohols produced are analyzed by GC. This reaction is applicable to a variety of alkoxysilanes and alkoxy-substituted siloxanes with SiOR concentrations of 90–0.01%.

Finally, a procedure has been reported for the analysis of aryloxysilanes involving bromination of the aromatic ring of the aryloxy group (59). The excess bromine is titrated iodometrically.

3.6. Hydrolyzable Halide (SiX)

The halides chlorine, bromine, and iodine, attached directly to silicon can be hydrolyzed to their corresponding ionic (acidic) form. The hydrolysis can be performed quite easily with aqueous or nonaqueous alkali or even with water or wet organic solvents. Examples of solutions used for hydrolysis include 95% ethanol (61), isopropanol, and mixtures of isopropanol or methanol and toluene (90), acetone (90), and sodium methoxide (91). The halide liberated by hydrolysis may be titrated as an acid, using a base such as alcoholic potassium hydroxide and standard titration procedures. Samples that contain iodine on silicon have to be analyzed by adding the sample to excess alcoholic base and back titrating the acid to prevent the formation of alkyl iodide (92). All the halides except fluoride can also be titrated potentiometrically with silver nitrate with high precision. Silver nitrate titrations can be performed in water or organic solvents such as alcohols, acetone, and mixtures of alcohols and acetone or acetone and water. The organic solvents help solubilize the hydrolyzed silane or sample matrix, sharpen the potentiometric break obtained with the silver nitrate titration, and lower the detection limits, particularly for the titration of chloride (93, 94).

The determination of silicon–bonded halide by silver nitrate titration can be done in the presence of silane hydrogen. Because the silane hydrogen reduces Ag(I) salts, it must first be allowed to react with a base such as sodium methoxide prior to titration with silver nitrate. A detailed titration procedure using sodium methoxide to prevent interference from silane hydrogen is given below.

Reagents

0.5% bromocresol purple or thymol blue in ethanol.

0.1–0.0025N Silver nitrate. The lower concentrations of titrant should use methanol solvent. Standardize the silver nitrate titrant with a standardized HCl solution.

Hydrolyzing solvent: Dissolve 5 g of sodium methoxide in 750 mL of methanol. Dilute the methanol solution with 2700 mL of acetone and add 0.5–1.0 mL of indicator solution.

Apparatus

Automatic titrator.

Silver indicating electrode.

Silver–silver chloride, double junction reference electrode. Inner and outerfill solutions should be 3M potassium chloride and potassium nitrate saturated, 50:50 methanol–water solution, respectively.

Procedure

Weigh accurately 0.5–10 g of sample into a beaker and add 70 mL of the hydrolyzing solvent. Stir for 5 min to neutralize the SiH and then acidify sample to the indicator color change with concentrated nitric acid. Insure the solution is acidic by adding an additional 2–4 drops of concentrated acid. Titrate the solution with standardized silver nitrate titrant.

Calculations

Calculate the percent hydrolyzable chloride using

$$\% \text{ Halide} = \frac{(\text{sample EPV} - \text{blank EPV}) (N) (\text{MEW})}{\text{SW}} \qquad (8.28)$$

EPV = endpoint volume in mL
N = normality of silver nitrate titrant
MEW = milliequivalent weight for the halide of concern
Cl = 0.0355 g/MEW
Br = 0.0799 g/MEW
I = 0.1269 g/MEW
SW = sample weight

With nonaqueous titrations for halide, caution should be used in adding excess concentrated nitric acid to acidify the solution prior to titration. Nitric acid is a strong oxidant and too much of an excess in the presence of organic solvents such as alcohols and ketones can initiate a vigorously exothermic reaction (95). A safer alternative to concentrated nitric acid is glacial acetic acid, which can be added in a large excess without interfering with the potentiometric titration of the halide.

One problem, which can arise when hydrolyzing silicon halides with aqueous or nonaqueous alkali, is the hydrolysis of carbon–halogen bonds if present in a silane or silicone sample. Generally, halide bound to carbon hydrolyzes only under severe conditions of temperature or alkali concentration. There is the potential, however, for some hydrolysis in the presence of sodium methoxide.

The titration of low levels of halide bound to silicon is possible with dilute silver nitrate. Chloride levels as low as 2 ppm can be titrated in some silane or silicon samples but in most situations a more practical detection limit is 10 ppm (96).

Ion chromatography is an alternative analytical technique for determining low levels of ionic or ionizable halide in silicones. Ion chromatography involves the separation of analyte anions or cations on an appropriate ion exchange column, followed by detection of the separated analyte ions with a conductivity detector. Details of equipment and procedures used can be found elsewhere (97–98).

The development and use of ion chromatography for the measurement of low parts per million levels of anions and cations in water was first reported in 1975 (99). Refinement of the technique has continued thru the 1970 and 1980s such that now the detection of parts per billion and subparts per billion levels of cations and anions in water solution has become almost routine. With ion chromatography, procedures have been developed for measuring levels of ionic or ionizable chloride in silicones or silanes down to low parts per billion levels (100).

3.7. Organic Substituents on Silicon

Determination of the organic groups attached to silicon may be required in order to better understand structure–property relationships, for example. Such determinations are usually done using IR or NMR, but if this approach is not practical, chemical reagents can be used to cleave R–Si bonds with the subsequent production of the corresponding hydrocarbon. The identification of the hydrocarbon can then be made by standard techniques, the most important ones being GC, MS, or the combination GC–MS.

3.7.1. Reactions with Acids

Concentrated sulfuric acid cleaves aryl and alkyl groups to form aromatic and aliphatic hydrocarbons. Microreactors have been designed to interface directly with a gas chromatograph and the cleavage carried out using concentrated sulfuric acid coated on a solid support saturated with vanadium pentoxide (101, 102). Cleavage was carried out at a temperature of 175 °C for approximately 15–20 min, after which time the gaseous products were swept into the inlet of the gas chromatograph. The conversion of the R groups was not quantitative; however, it was possible to identify the R groups based on the retention time. Anhydrous HCl has been used to cleave aryl silicon bonds, but the yields are low, ranging from 37–76% recovery of the expected aryl groups. Other acids, such as nitric and phosphoric have been used for the analysis of aryl and vinyl groups, respectively, but again, recovery is low.

3.7.2. Reaction with Bases

Alkali cleavage of the R–Si bond has been more successful and many examples have been reported in the literature. Direct fusion with solid sodium or potassium hydroxide in a variety of crucibles (zirconium, nickel, and gold have been used) yielded the corresponding hydrocarbon. The hydrocarbon was collected by purging the system with nitrogen and sweeping the volatiles into carbon tetrachloride. Analyses were by NMR or GC. We have found it convenient to collect the volatiles in a polyethylene gas bag and sample from it. Methyl, vinyl, silane hydrogen, and phenyl groups have been determined using solid sodium hydroxide as the alkali, with the reaction being in a glass reactor at 200 °C. The outlet of the reactor was fitted with a syringe for collection of the evolved gases. The cleavage products (methane, ethylene, hydrogen, and benzene) were quantitated using GC (103, 104).

Cleavage of phenyl groups to form benzene may be accomplished by heating the sample in a sealed glass ampule for 2 h at 120 °C with 60% aqueous potassium hydroxide in dimethyl sulfoxide. After the reaction is completed, benzene is quantitated using GC with n-propanol as an internal standard. Phenyl contents as low as 0.1% up to as high as 80% are claimed.

3.7.3. Silicon Substituent Analysis

The most accurate process for measuring the amounts of organic substituents in a silicon polymer is to depolymerize the material and analyze for each of the monomers. Cleavage of any RSi bonds must be avoided. Approaches that have been employed for this conversion are discussed below.

3.7.3.1. Alkali Reactions. A scheme for the cleavage of siloxane bonds, using sodium hydroxide, to form sodium salts and liquids that could be separated by solvent fractionation or by distillation has been suggested (Ref. 61, p. 380). Polysiloxanes containing only methyl groups were successfully fractionated by this procedure. It was also possible to separate phenyl containing polymers using this technique. Unfortunately, it is difficult to avoid some C–Si cleavage.

3.7.3.2. Fluorosilane Derivatization. Several procedures have been reported in the literature whereby the siloxane polymers are converted to fluorosilane monomers using either hydrofluoric acid or boron trifluoride as the fluorinating agent (2).

The anhydrous HF procedure is time consuming and hazardous. The reaction is carried out using a specially designed copper reactor, and the products of the cleavage are trapped and then analyzed by analytical fractionation. Boron trifluoride in diethyl ether has also been used for the conversion of organosiloxanes, alkoxysilanes and siloxanes, and silanol functional materials, to the corresponding fluorosilanes. The cleavage reaction occurs fairly readily at room temperature, but the speed of the reaction can be increased by heating the reaction mixture to the boiling point of the etherate reagent (126 °C). This reagent is compatible with glassware, providing a definite advantage over the use of HF. Poly(dimethylsiloxane) has been successfully depolymerized by this reaction to form fluorosilanes. These products were analyzed using GC with flame ionization detection.

An improved version of this reaction was developed by using melting point capillary tubes sealed at one end. A few microliters of the sample are placed in the capillary, cooled, and the boron trifluorideetherate is added. The capillary is flame sealed and allowed to react for $\frac{1}{2}$ at 100 °C. The reaction products are analyzed by GC through the use of a solid sampler injection port, with the glass capillary being crushed in the sample port.

Siloxanes containing silane hydrogen do not maintain functional integrity, since the hydrogen can be partially replaced with fluorine giving the alkyl-fluorosilane instead of the expected alkyl hydrogen fluorosilane.

Trifunctional siloxy structures do not always proceed well with the boron trifluoride system, particularly if they are present at high concentrations. Phenyl groups are generally cleaved to form benzene along with SiF_4. Thus a total phenyl content can be determined, but it is not possible to speciate between monophenyl, diphenyl, triphenyl, and phenylmethyl siloxanes.

3.7.3.3. Alkoxy Derivatization.

In 1959, Voronkov and co-workers (105, 106) described a procedure for the formation of alkoxy derivatives from polysiloxane materials. It consisted of reacting the polysiloxane with tetra-ethoxy silane on a 1:1 molar basis with an alkaline catalyst.

$$2(Me_2SiO)_n + nSi(OEt)_4 \longrightarrow 2nMe_2Si(OEt)_2 + (SiO_2)_n \qquad (8.29)$$

Yields of the dimethyldiethoxysilane were in the range 80–90%, while those of the methyltriethoxysilane and the trimethylethoxy silane varied from 45–80%. Poor recovery restricted this method to semiquantitative analysis.

Two analogous methods have been described for the depolymerization of polysiloxanes. The first utilized the distillation and collection of the ethoxy derivatives in a micro glass apparatus. The second involved reflux of the reaction mixture with subsequent sampling of the products. Analytes in the distillation procedure are limited to methyl ethoxy silanes and excess tetra-ethoxysilane reagent. With the reflux method, high boiling materials such as the phenyl ethoxy silanes and disiloxane and trisiloxane derivatives can be sampled and observed by GC. With such a complex mixture, it is helpful to utilize GC–MS for speciation and quantification of the components. In this manner, corrections can be made for species present as dimers and trimers. Improvements have been made to this procedure, and at present the analysis is performed in the following manner.

Procedure

In a reaction vial (Wheaton Scientific Co., Millville, NJ., Part No. 4502-B17A), weigh 0.2–0.3 g of sample to the nearest 0.1 mg. The internal standard, ~0.1 g, is weighed to the nearest 0.1 mg and chosen from a group of aliphatic hydrocarbons, such that it will elute from the GC in an area that will not interfere with the compounds of interest. We have found dodecane to be a good choice for general use. Add 4.0 ± 0.05 g of tetraethoxysilane, a stirring bar (Teflon® coated), and finally add 0.45 ± 0.05 g of the catalyst. The catalyst is prepared from a 1:1:1 mix of potassium ethoxide, tetraethoxysilane, and ethanol. The vial is stoppered with a 20-mm PTFE faced red butyl rubber stopper (The West Co., Phoenixville, PA, Product No. 1014-6000), and capped with an open top cap (Wheaton Scientific, Part No. 240516). The vial is then placed in a REACTITHERM heating–stirring module (E. Pierce Chemical, Rockford, IL), which has been preheated to 120–125 °C and the stirring started. It is allowed to heat for 1 h, then removed and allowed to cool to room temperature. The vial is opened and gaseous CO_2 is impinged through the mixture to neutralize the alkali. The vial is centrifuged to separate the solids, and an aliquot of the supernate is withdrawn and injected into the GC for analysis. From the chromatogram, the concentration of the various ethoxy silanes that have been formed from the reaction is calculated. Calibration values are obtained by processing well-characterized siloxane polymers or silanes such as are listed in Table 2.3 (Chapter 2).

A large excess of the tetraethoxy silane is used to ensure that the polymer is completely converted to the monomeric form.

Ethoxy silanes are easily analyzed by conventional GC procedures with either thermal conductivity or flame ionization detectors. Often an internal standard is incorporated into the system for the best quantitation. We have successfully analyzed for $Me_3SiO_{1/2}$, Me_2SiO, $MeSiO_{3/2}$, $MeViSiO$, $PhSiO_{3/2}$, $EtMeSiO$, and n-octyl$SiO_{3/2}$ groups.

The equilibrium for the reaction (8.29) is driven far to the right by the large excess of tetraethoxy silane. Yields of monomeric species are greater than 90% and are reproducible. Thus, through the use of response factors derived using well-characterized polysiloxanes, the system is made quantitative. The repeatability of this procedure is typically $\pm 5\%$ relative for major components. The procedure is applied most frequently to macrolevel components and has a detection limit under these conditions of less than 0.1%.

REFERENCES

1. T. R. Crompton, "Analysis of organosilicon compounds," in *The Chemistry of Organic Silicon Compounds*, Chapter 6, S. Patai and Z. Rappoport, Eds., Wiley, New York, 1989.

2. R. C. Smith, N. C. Angelotti, and C. L. Hanson, "Chemical analysis," in *Analysis of Silicones*, A. L. Smith, Ed., Wiley, New York, 1974, pp. 113–156.

3. J. C. B. Smith, in *Identification and Analysis of Plastics*, J. Haslam and H. A. Willis, Eds., Von Nostrand, Princeton, NJ, 1985, pp. 253–263.

4. E. G. Rochow and W. F. Gilliam, "Polymeric methylsilicon oxides," *J. Am. Chem. Soc.*, **63**, 798 (1941).

5. V. A. Klimova, M. O. Korshun, and E. G. Bereznitskaya, "Microelementary analysis of carbon, hydrogen, silicon and halogens," *Doklad. Akad. Nauk SSSR*, **84**, 1175 (1952).

6. V. A. Klimova, M. O. Korshun, and E. G. Bereznitskaya, "Determination of carbon, hydrogen and silicon," *Zh. Analit. Khim. SSSR*, **11**, 223 (1956).

7. H. J. Horner, in *Treatise on Analytical Chemistry*, I. M. Kolthoff, and P. J. Elving, Eds., Part II, Vol. 12, 287, Wiley-Interscience, New York, 1965.

8. J. H. Wetters and R. C. Smith, "Determination of silicon in siloxane polymers and silicon-containing samples employing alkali fusion decomposition methods," *Anal. Chem.*, **41**, 379 (1969).

9. H. R. Shell, in *Treatise on Analytical Chemistry*, I. M. Kolthoff and P. J. Elving, Eds., Part II, Vol. 2, 161, Wiley-Interscience, New York, 1962.

10. E. J. King, B. D. Stacy, P. F. Holt, D. M. Yates, and D. Pickles, "Determination of silicon in the microanalysis of biological material and mineral dust," *Analyst*, **80**, 441 (1955).

11. F. Pregl and H. Roth, *Quantitative Organische Microanalyse*, Springer-Verlag, Vienna, 1958.

12. E. Pella and B. Colombo, "Study of carbon, hydrogen and nitrogen by combustion gas-chromatography," *Microchim. Acta*, **679** (1973).

13. E. Pella and B. Colombo, "Improved instrumental determination of oxygen in organic compounds by pyrolysis-gas chromatography," *Anal. Chem.*, **44**, 1563 (1972).

14. R. Culmo, "Automatic microdetermination of carbon, hydrogen and nitrogen: Improved combustion train and handling techniques," *Microchim. Acta*, **69**, 175 (1969).

15. J. F. Alcino, "The use of platinum in the Perkin-Elmer 240 elemental analyzer, the all platinum ladle," *Microchem. J.*, **18**, 350 (1973).

16. Y. A. Gawargious and A. M. G. MacDonald, "Determination of carbon and hydrogen in organic compounds containing metals and metalloids," *Anal. Chem. Acta*, **27**, 300 (1962).

17. Y. A. Gawargious and A. M. G. MacDonald, "Determination of carbon and hydrogen in silicon containing compounds," *Anal. Chem. Acta*, **27**, 119 (1962).

18. A. Radecki, "Simultaneous determination of carbon and halogens in organoxyhalosilanes by wet combustion," *Chem. Anal. (Warsaw)*, **8**, 607 (1963).

19. V. A. Klimova, T. A. Antipova, and G. K. Mukhina, "Simultaneous microdetermination of carbon, hydrogen, and halogens or sulfur by flash combustion," *Izv. Akad. Nauk USSR Otd. Khim. Nauk*, **1**, 19 (1962).

20. A. P. Terentiev, B. M. Luskina, and S. V. Syavtsillo, "Elementary organic analysis of wet combustion," *J. Anal. Chem. USSR*, **16**, 635 (1961).

21. N. E. Gelman, M. C. Korshun, and K. I. Novozhilova, "Analysis of fluor-organic compounds, microdetermination of fluorine, carbon and hydrogen," *J. Anal. Chem. USSR*, **15**, 342 (1960).

22. E. F. Fedorova, N. A. Nikolaeva, and O. S. Kalita, "Simultaneous determination of carbon, hydrogen and silicon in organosilicon compounds," *Izv. Akad. Nauk SSSR Ser. Khim.*, **1972**, 625.

23. A. D. Campbell and A. M. G. MacDonald, "Oxygen flask in organic analysis—detection of nitrogen," *Anal. Chem. Acta*, **26**, 275 (1962).

24. O. Schwarzkopf and F. Schwarzkopf, in *Characterization of Organometallic Compounds*, M. Tsutsui, Ed., Part I, 35, Interscience, New York, 1969.

25. O. U. Anders, private communication, Dow Chemical Company, Midland, MI.

26. F. W. Cheng, "A rapid method for microdetermination of halogen in organic compounds," *Microchem. J.*, **3**, 537 (1959).

27. R. D. Denny and P. A. Smith, "Comparison of two procedures for the determination of organobromine by the Schoeniger oxygen flask method," *Analyst*, **99**, 1176 (1974).

28. R. A. Lalancette and A. Steyermark, "Collaborative study of the microanalytical determination of bromine and chlorine by oxygen flask combustion," *J. Assoc. Offic. Anal. Chem.*, **57**, 26, 1974.

29. W. Schoeniger, "Eine microanalytishe schnellbestimmung von halogene in organischem substanzen," *Microchem. Acta*, **55**, 123 (1955).

30. R. C. Smith, N. C. Angelotti, and C. L. Hanson, "Chemical analysis," in *Analysis of Silicones*, A. Lee Smith, Ed., Wiley, New York, 1974, pp. 132–156.

31. F. O. Guenther, "Determination of silanol with Grignard reagent," *Anal. Chem.*, **30**, 1118 (1958).

32. J. F. Lees and R. T. Lobeck, "The quantitative determination of active hydrogen by a modification of the Zerewitinoff method," *Analyst*, **88**, 782 (1963).

33. T. Welsch and H. Frank, "A simple method for the quantitative determination of silanol groups," *J. Prakt. Chem.*, **325**, 325 (1983); through *Chem. Abstr.*, **99**, 63632v (1983).

34. G. H. Barnes and N. E. Daughenbaugh, "Diglyme as solvent in the gasometric determination of silanols by the lithium aluminum hydride method," *Anal. Chem.*, **35**, 1308 (1963).

35. L. Bretherick, *Handbook of Reactive Chemical Hazards*, 3rd ed., Butterworths, London, 1985, pp. 41–44.

36. C. Eaborn, *Organosilicon Compounds*, Butterworths, London, 1960, pp. 244–246.

37. R. C. Smith and G. E. Kellum, "Rapid condensation procedure for the determination of hydroxyl in silicon materials," *Anal. Chem.*, **39**, 338 (1967).

38. G. E. Kellum and K. L. Uglum, "Lithium aluminum di-*n*-butyl amide as a direct acid–base titrant for quantitative determination of silanol," *Anal. Chem.*, **39**, 1623 (1967).

39. D. E. Jordan, "Selective hydroxyl group determination by direct titration with lithium aluminum amide," *Anal. Chem. Acta*, **30**, 297 (1964).

40. M. D. Gaul, unpublished procedure, Dow Corning Corporation, Midland, MI.

41. H. Gilman and L. S. Miller, "The determination of silanols with the Karl Fischer reagent," *J. Am. Chem. Soc.*, **73**, 2367 (1951).

42. W. T. Grubb, "A rate study of the silanol condensation reaction at 25° in alcoholic solvents," *J. Am. Chem. Soc.*, **76**, 3408 (1954).

43. R. C. Smith and G. E. Kellum, "Siloxane interference in Karl Fischer reagent titration," *Anal. Chem.*, **38**, 647 (1966).

44. A. E. Pierce, *Silylation of Organic Compounds*, Pierce, Rockford, IL, 1979.

45. *Pierce 1989 Handbook and General Catalogue*, Pierce, Rockford, IL 1989.

46. R. H. Baney and F. S. Atkari, "The potentiometric titration of siloxane structures with tetrabutyl ammonium hydroxide titrant," *J. Organometal. Chem.*, **9**, 183 (1967).

47. R. West and R. H. Baney, "The acid strength of triphenylsilanol," *J. Inorg. Nucl. Chem.*, **7**, 297 (1958).

48. A. P. Kreshkov, V. A. Drozdov, and V. N. Knyazev, "Determination of alkylarylsilanols by potentiometric titration in nonaqueous media," *Plast. Massy* (**1969**); 69; through *Chem. Abstr.*, **70**: 93020q (1969).

49. G. Schott and E. Popowski, "Silanols. VII. Potentiometric titration of substituted triphenylsilanols," *Z. Chem.*, **8**, 113 (1968); through *Chem. Abstr.*, **68**: 104468d (1968).

50. J. Mitchell, Jr., and D. M. Smith, "Aquametry, Part III (the Karl Fischer reagent), A treatise on methods for the determination of water," Wiley, New York, 1980, pp. 658–667.

51. E. Scholz, *Karl Fischer Titration Determination of Water*, Springer-Verlag, Berlin Heidelberg, 1984, p. 92.

52. R. C. Smith and G. E. Kellum, "Determination of water in monomeric and short-chain silanols employing a modified Karl Fischer titration method," *Anal. Chem.*, **39**, 1877 (1967).

53. R. C. Smith and G. E. Kellum, "Modified Karl Fischer titration method for determination of water in the presence of silanol and other interfering materials," *Anal. Chem.*, **38**, 67 (1966).

54. V. Mika and I. Cadersky, "Coulometric determination of microgram amounts of water in silicon compounds," *Fresenius' Z. Anal. Chem.*, **258**, 25 (1972).

55. I. Nordin-Andersson and A. Cedergren, "Coulometric determination of trace water in active carbonyl compounds using modified Karl Fischer reagents," *Anal. Chem.*, **59**, 749 (1987).

56. G. E. Kellum and R. C. Smith, "Determination of water, silanol, and strained siloxane on silica surfaces," *Anal. Chem.*, **39**, 341 (1967).

57. E. Scholz, "Karl Fischer titrations of aldehydes and ketones," *Anal. Chem.*, **57**, 2965 (1985).

58. C. Eaborn, *Organosilicon Compounds*, Butterworths, London, 1960, p. 199.

59. J. Urbanski, W. Czerwinski, K. Janicka, F. Majewska, and H. Zowall; G. Gordon Cameron, transl. Ed., *Handbook of Analysis of Synthetic Polymers and Plastics*, Chapter 20, Ellis Horwood, England, 1979.

60. E. G. Rochow, *An Introduction to the Chemistry of the Silicones*, 2nd ed., Wiley, New York, 1951, pp. 160–167.

61. J. A. McHard, "Silicones," in *Analytical Chemistry of Polymers, Vol. XII, Part I, Chapter XIV*, G. M. Kline, Ed., Interscience, New York, 1959, pp. 361–397.

62. Hans J. S. Winkler, and H. Gilman, "Preparation and reactions of sym-tetraphenyldisilane," *J. Org. Chem.*, **26**, 1265 (1961).

63. Neotronics Corp., P.O. Box 270, Gainesvile, GA 30503.

64. C. J. Dunning, private communication, Dow Corning Corp., Midland, MI.

65. D. N. Ingebrigtson, private communication, Dow Corning Corporation, Midland, MI.

66. G. Fritz, "A method for the quantitative determination of the SiH bond," *Z. Anorg. Allgem. Chem.*, **280**, 134 (1955); *Chem. Abstr.*, **50**, 1522h (1956).

67. J. Cermák and P. Dostál, "Polarographic determination of the Si–H group in the presence of Si–Si in polyorgano-siloxanes", *Collect. Czech. Chem. Commun.*, **28**, 1386 (1963); *Chem. Abstr.*, **59**, 4541f (1963).

68. J. H. Wetters and N. C. Angelotti, unpublished procedure, Dow Corning Corporation, Midland, MI.

69. G. Fritz and H. Burdt, "Determination of the Si–H and Si–phenyl groups," *J. Anorg. Allgem. Chem.*, **317**, 35 (1962); through *Chem. Abstr.*, **57**: 14437g (1962).

70. A. P. Tereut'ev, E. A. Bondareskaya, and T. V. Kirillova, "The determination of small amounts of Si–H groups in organosilicon compounds," *Zavodsk. Lab* **33**, 156 (1967); through *Chem. Abstr.*, **66**: 111323q (1967).

71. V. A. Bork and L. A. Shvyrkova, "Analysis of silicoorganic compounds which contain hydrogen bound to silicon," *Trudy Komiss. Anal. Khim., Akad. Nauk USSR*, **13**, 148 (1963); through *Chem. Abstr.*, **60**, 30a (1964).

72. C. Harzdore, "Titration of silicon hydrogen linkages in polymers," *Fresenius' Z. Anal. Chem.*, **256**, 192 (1971); through *Chem. Abstr.*, **76**, 72817p (1971).

73. C. Eaborn, *Organosilicon Compounds*, Butterworth, London, 1960, p. 354.

74. C. A. Burkhard, "Polydimethylsiloxanes," *J. Am. Chem. Soc.*, **71**, 963 (1949).

75. American Society for Testing and Materials, "Standard test method for iodine value of drying oils and fatty acids," *1989 Annual Book of ASTM Standards, Section 6.03, D 1959–85*, ASTM, Philadelphia, PA, 1989, pp. 277–279.

76. A. A. Lapshova, G. S. Kupreeva, M. P. Strukova, and L. N. Chechetkina, "Determination of vinyl groups in siloxane rubbers by ozonolysis to identify the type of siloxane polymer," *Zh. Anal. Khim.*, **32**, 1816 (1977).

77. M. P. Strukova, G. I. Veslova, and S. I. Kobelevskaya, "Analytical study of siloxanes to determine vinyl groups," *Zh. Anal. Khim.*, **32**, 626 (1977).

78. R. W. Martin, "Rapid estimation of ethylenes," *Anal. Chem.*, **21**, 921 (1949).

79. American Society for Testing and Materials, "Standard Test Method for Vinyl Unsaturation in Organic Compounds with Mercuric Acetate," *1989 Annual Book of ASTM Standards, Section 9.01, E441-74,* ASTM Philadelphia, 1989, pp. 795–797.

80. C. L. Hanson and R. C. Smith, "Determination of alkoxy and vinyl in siloxane materials using alkali fusion reaction and gas chromatography," *Anal. Chem.,* **44**, 1571 (1972).

81. R. D. Parker and W. J. Owen, unpublished procedures, Dow Corning Corporation, Barry, Wales.

82. G. W. Heylmun, R. C. Bujalski, and H. B. Bradley, "Determination of vinyl content of silicone gums," *J. Gas Chromatogr.,* 300 (1964).

83. J. Franc and K. Placek, "Determination of vinyl groups bound to silicon," *Microchimica Acta,* **2**, 31 (1975); *Chem. Abstr.,* **89**, 141522 (1978).

84. E. R. Bissell and D. B. Fields, "Determination of vinyl end groups in siloxane polymers by gas chromatography," *J. Chromatogr. Sci.,* **10**, 164 (1972).

85. J. S. Fritz and G. H. Schenk, "Acid-catalyzed acetylation of organic hydroxyl groups," *Anal. Chem.,* **31**, 1808 (1959).

86. J. A. Magnuson, "Determination of alkoxy groups in alkoxysilanes by acid-catalyzed acetylation," *Anal. Chem.,* **35**, 1487 (1963).

87. J. A. Magnuson and R. J. Cerri, "1,2-dichloroethane as a solvent for perchloric acid catalyzed acetylation," *Anal. Chem.,* **38**, 1088 (1966).

88. P. Dostál, J. Cermák, and B. Novotná, "Determination of silicon bonded alkoxy and aryloxy groups in organosilicon compounds," *Coll. Czech. Chem. Commun.,* **30**, 34 (1965); *Chem. Abstr.,* **62**, 9799a (1965).

89. E. P. Samsel and J. A. McHard, "Determination of alkoxyl groups in cellulose ethers," *Ind. Eng. Chem. Anal. Ed.,* **14**, 750 (1942).

90. J. A. Magnuson, "Nonaqueous titration of trace amounts of HCl and chlorosilanes in siloxane materials," *Chemist-Analyst,* **53**, 15 (1964); *Chem. Abstr.,* **61**, 11315g (1964).

91. Dow Corning Corporation, unpublished results, Midland, MI.

92. H. H. Anderson, D. L. Seaton, and R. P. T. Rudnicki, "Alkyliodosilanes: Ethyl series and monododecyl," *J. Am. Chem. Soc.,* **73**, 2144 (1951).

93. M. D. Gaul, unpublished results, Dow Corning Corporation, Midland, MI.

94. W. Selig, "Lower limits of the potentiometric micro-titration of chloride with silver ions," *Microchemical J.,* **21**, 291 (1976).

95. L. Bretherick, *Handbook of Reactive Chemical Hazards,* 3rd ed., Butterworths, London, 1985, pp. 1100–1129.

96. M. D. Gaul and P. S. Rzyrkowski, unpublished results, Dow Corning Corporation, Midland, MI.

97. J. G. Tarter, Ed., *Ion Chromatography,* Marcel Dekker, New York, 1987.

98. J. Weiss, in *Handbook of Ion Chromatography,* E. L. Johnson, Ed., Dionex Corporation, Sunnyvale, CA, 1986.

99. H. Small, T. S. Stevens, and W. C. Bauman, "Novel ion exchange chromato-graphic method using conductimetric detection," *Anal. Chem.*, **47**, 1801 (1975).

100. M. D. Gaul and G. M. Wyshak, unpublished results, Dow Corning Corporation, Midland, MI.

101. C. A. Brukhard and F. J. Norton, "Analysis of organosilicon compounds and polymers—Determination of hydrocarbon substituents," *Anal. Chem.*, **21**, 304 (1949).

102. J. Franc and J. Dvoracek, "Gas chromatographic structural analysis of organo-silicons," *J. Chromatogr.*, **14**, 340 (1964).

103. J. Frank and K. Placek, "Structural analysis of polyorganosiloxanes by means of gas chromatography – thermal analysis," *J. Chromatogr.*, **48**, 295 (1970).

104. J. Frank and K. Placek, "Determination of methyl groups bound on silicon in organosilicon substances," *J. Chromatogr.*, **67**, 37 (1972).

105. M. G. Voronkov, "Reaction of siloxanes with alkoxysilanes—synthesis of alkoxysilanes and siloxanes," *J. Gen. Chem. USSR*, **20**, 890 (1959).

106. M. G. Voronkov and Z. I. Shabarova, "Cleavage of organosiloxanes by alco-hols," *J. Gen. Chem. USSR*, **29**, 1501 (1959).

CHAPTER

9

MICROSCOPICAL CHARACTERIZATION

H. A. FREEMAN and R. L. DURALL

Dow Corning Corporation
Midland, Michigan

1. INTRODUCTION

The use of microscopical techniques permits the analyst to examine the
surface and interior of a wide range of materials far more extensively than
that possible with the unaided eye. Advances in instrumentation now permit
magnifications up to several million. These techniques, in the hands of
experienced microscopists, provide unparalleled materials characterization
information. Some of the most sophisticated techniques are linked primarily
to research activities, but many have become commonplace in process
studies and quality control efforts, especially those directed toward industries
in which miniaturization is a prime objective. There is an increasing recognition of the importance of surface characteristics, interactions among internal
constituents, composition, and physical properties, as key factors in
improved serviceability of materials.

A competent microscopist recognizes which techniques are most appropriate to use in studying varied materials. No one technique can be universally
applied, for each has its own limitations as well as benefits. Too often,
however, the simple straightforward approach is ignored in favor of one
that has the "glamour" of sophistication and complexity. Recent trends in
instrumentation have engendered technologists who specialize in operating
instruments and interpreting results from the more complex facilities. Increased instrumental sophistication has largely precluded the do-it-yourself
generalist, a trend that jeopardizes the informed application of all microscopical tools. In this chapter, we attempt to provide some guidelines for effective
use of some of these tools as applied to organosilicon materials.

The Analytical Chemistry of Silicones, Edited by A. Lee Smith.
ISBN 0-471-51624-4   © 1991 John Wiley & Sons, Inc.

It is necessary first to define some commonly used terms. The term *morphology* relates to the external form or appearance of material to which such descriptors as size, shape, color, and roughness may be applied. *Microstructure* generally refers to the internal interrelationships that exist between multiphase components. Collectively, these features comprise microscopical characterization.

2. CHOOSING AN APPROPRIATE TECHNIQUE

A moderately well-equipped research laboratory will typically be able to do optical microscopy and scanning electron microscopy for examination of surfaces, microanalysis for elemental composition of microsamples, and transmission electron microscopy for examination and elemental analysis of thin sections. Together with its support and sample preparation equipment and an experienced staff, such a facility is able to provide much of the microscopy required in the modern research environment. Other analysis needs may be filled through the use of surface-specific techniques described in further detail in Chapters 5 and 15. The judicious choice of which technique(s) to use is often critical to the success or failure of a particular investigation.

2.1. Optical Microscopy

Two important concepts to be considered when choosing a microscope are resolution and depth of field. The practical limit of resolution for the optical microscope, based on the wavelength of green light, is approximately 0.5 μm (1, 2). Without the aid of magnification, the average human eye is only able to see particles that are 100 μm in diameter or more. The depth of field for an optical microscope decreases with increasing magnification and NA (numerical aperture) (2). Although the combination of specific optical microscope components may permit attainment of magnifications of up to 2000, the depth of field may be so greatly decreased that another microscopical technique such as TEM or SEM would be more desirable.

The type of illumination used for any given material is first determined by whether the sample is transparent or opaque. Opaque or thick specimens, such as gums and cured elastomers, are generally viewed in reflected light. Transparent or semitransparent materials, including silicone liquids or emulsions, use transmitted light. Depending on the need, specimens may also be examined using polarized light (3).

Sample preparation is relatively simple and numerous techniques are available. As there is little chance for thermal damage or introduction of artifacts common in electron microscopy, samples rarely need special coating techniques. Silicone materials examined with the optical microscope may usually be studied in their original state.

Special objective lenses designed for specific types of information can be chosen. Nomarski DIC (differential interference contrast) objectives are used for obtaining information in reflected and transmitted modes. Phase contrast objectives are used when looking at low contrast silicone emulsions and other fluids in transmitted light (2, 4).

Stereomicroscopes are useful for obtaining three-dimensional images at low magnifications. Their much greater depth of field than the traditional optical microscope renders them extremely useful when delicate sample preparation is needed (1, 2).

The optical microscope is ideally suited to studying features such as the size, shape and distribution of constituents, and the presence and type of contaminants in fluids, emulsions, and gums. Surface and internal features of silicone elastomers, rubbers, resins, coatings, adhesives, and ceramics containing preceramic silicone polymers are commonly examined using an optical microscope.

The limited resolution and restricted depth of field can be disadvantageous to using the optical microscope. These factors may create difficulties when looking at small particles within thick materials such as elastomers without more arduous sample preparation methods. Adequate contrast may also be difficult to achieve with clear silicone materials although contrast may be increased by techniques such as staining or etching, which can enhance subtle differences in texture and constituents.

2.2. Scanning Electron Microscopy

The SEM is a sophisticated, versatile instrument capable of reaching a resolution of greater than two nanometers using a field emission source (Table 9.1). Because of its high resolution and large depth of field, the SEM is used primarily to observe and characterize topographical images of rough surfaces of bulk silicone materials. To create an SEM image, the sample is irradiated with electrons generated by a sharply focused electron beam. The image appears on a cathode ray tube (CRT) as a static spot, or if swept across the sample surface, as a raster pattern (5).

The primary electron beam, accelerated through a 0–30 kV potential, will penetrate a solid and, in the process, lose most of its initial energy by ionizing the atoms within the specimen (5). As the primary beam penetrates a surface, numerous elastic and inelastic events occur producing secondary, backscattered and Auger electrons, and continuum and characteristic X-rays (3, 5, 6). Of these, secondary and backscattered electrons, are of most interest for SEM.

Table 9.1. Summary of Comparative Microscopical Capabilities

	Description	Applications	Suitability	Image Resolution (approx. in nm)	Techniques of Microanalysis
1. Optical microscope	Visible and IR light imaging	Transmitted and reflected light characterization	Comparative low magnification characterization. Wide range of sample types	300	None
2. AEM at 200 keV	Electron beam imaging	Surface, internal morphology and composition combined with microanalysis and phase ID of very limited volumes in suitably thin solid prepared samples	High magnification detail of thin specimens. Control and selection of microanalysis area down to 50 nm in suitable samples		
TEM mode	Scanned electron beam in transmission			0.3	EDS/EELS
STEM mode				1.0	EELS
SEM mode	Scanned reflected			2.0	EDS
3. Dedicated STEM at 100 keV	Scanned electron beam transmitted	Morphology of vacuum stable solid in very thin sections	Similar to AEM	0.3–1.0	EDS/EELS
4. SEM/SEMP at 40 keV	Scanning, usually reflected image	Surface morphology of comparatively large bulk samples and microanalysis of specific constituents	Moderate magnification and controlled microanalysis of bulk specimens	4–6	EDS/WDS
W Source	Conventional heated W electron source		X-rays typically generated from μm-size areas.		
LaB$_6$ Source	High intensity and brightness			3.0	
FEG Source	Cold W source of very high current density			2.0	

222

Elastic scattering creates interactions between the primary beam and the host atom. Elastic scattering events in the sample affect the trajectories of the electron beam without significantly altering its energy. As many as a third of the primary electrons that hit a specimen subsequently escape. The yield of reemerging electrons, called backscattered electrons, increases monotonically with atomic number for elements in the specimen (3, 5).

Inelastic scattering events cause a transfer of energy from the primary beam to the specimen, resulting in the generation of secondary and Auger electrons and characteristic and continuum X-rays (cf. Chapter 15, Section 1). Secondary electrons vary in energy as a result of differences in the sample surface topography as the primary electron beam is swept across a material and interacts with the uppermost atoms of the specimen surface. A relatively high resolution is obtained from secondary electrons because their emission is confined to an interaction volume near the beam impact area (5). As the primary beam travels further into a specimen, it continues to lose more and more energy. The emitted electrons must reach the detector in order to be measured. It is for this reason that secondary electrons are collected from a shallow depth in the sample. Some of the secondary electrons diffuse near the specimen surface, losing their energy by ionizing host atoms. If they reach the surface with residual kinetic energy in excess of the surface barrier energy, they escape the specimen and are collected (5). A detector such as the Everhart–Thornley detector can be used to collect secondary electrons and convert them to a signal that is amplified to produce an image of the specimen on a CRT (5).

One of the greatest strengths of the SEM is its ability to allow examination of samples without time-consuming specimen preparation (3, 5). Silicone materials are usually adhered directly to an aluminum or carbon stub using carbon or silver conductive paint or double stick tape. In cases where small particles are to be examined, the particles are placed in a liquid suspension, dropped onto a stub and allowed to dry. Particles remain adhered to the stub by electrostatic forces. The sample must be grounded to the stub when using these methods. Most silicone materials should be plated with a thin layer of conductive metal such as gold or gold–palladium by either sputter or evaporative techniques (3, 5). Silicones, because they are nonconductive, are especially susceptible to charging and thermal damage effects, which serve to degrade the surface morphology of the sample and reduce the resolution possible in the SEM. Care must always be taken during sample preparation not to introduce surface artifacts that will mislead the interpretation.

2.3. Scanning Electron Microprobe

Choice of the appropriate instrument to use for elemental analysis of microinclusions in a material or characterization of its chemical composition is dictated by the nature of the problem and characteristics of the sample. Because of vacuum constraints, samples for microanalysis in electron beam instruments must generally be solid with little potential for out-gassing. Examination by SEM mode in the analytical electron microscope, AEM, (Section 2.4.4) further requires that samples be less than 3 mm in diameter and 1 mm thick. Much thinner samples, on the order of 100 nm, are required for scanning transmission (STEM) and transmission electron microscope (TEM) modes in this instrument. Samples for the scanning electron micro-probe (SEMP), on the other hand, can be much larger and thicker, depending on the instrument configuration.

Wavelength dispersive spectrometers (WDS) have been in common use on electron microprobe and scanning instruments for several decades. Substantial background literature (5–7) is available describing the theory and application of WDS techniques to materials science and biological specimens.

Organosilicon materials require special care during microanalysis as a consequence of the comparatively high beam currents that are necessary during WDS microanalyses. Evidence of degradation (charring) in the beam renders spectral information suspect. In these situations, the lower current (picoamp range) possible with energy dispersive spectrometer (EDS) modes of analysis may provide more reliable data, though sensitivity may be compromised. Coating organosilicon surfaces with an electrically conductive material such as Al or C is useful in minimizing beam damage and charging. Heavier element coatings such as Au and Pd must be avoided as these elements strongly absorb lighter element X-rays emanating from beneath. Best AEM and SEMP microanalysis results are obtained on stable inorganic surfaces such as those related to some silicone manufacturing processes (Figure 9.1) or materials derived from silicones that have been previously processed at high temperatures (Figure 9.2). Many of these require no prior coating step to provide the necessary stability in the electron beam.

Benefits from use of the SEMP accrue largely from the limited sample preparation effort that is required. Large samples, if solid and vacuum stable, are accommodated in the instrument. Scanning images as well as micro-analysis may be obtained on irregular surfaces to determine morphology and elemental heterogeneity. It must be born in mind, however, that a "surface" in this instrumental context typically includes an area encompassing 2 μm^2 and the classical tear-drop shape 2 μm deep. As such, heterogeneities hidden below the surface may be sampled and contribute to the spectra obtained. This effect becomes more pronounced with higher instrumental accelerating voltages and higher atomic number of those elements contained in the material being examined.

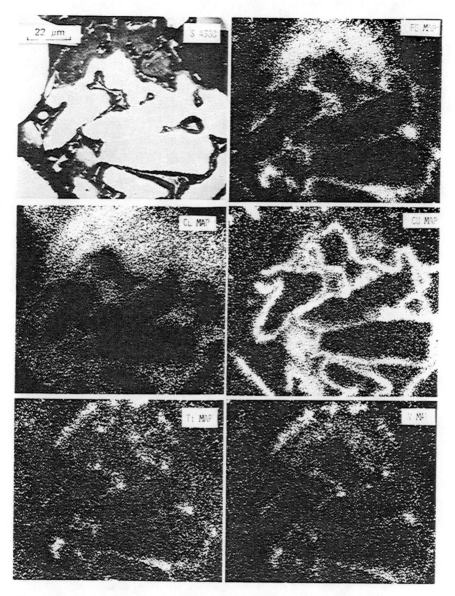

Figure 9.1. SEM micrograph (top left) of polished surface on metallurgical grade silicon metal with corresponding maps of element distributions.

225

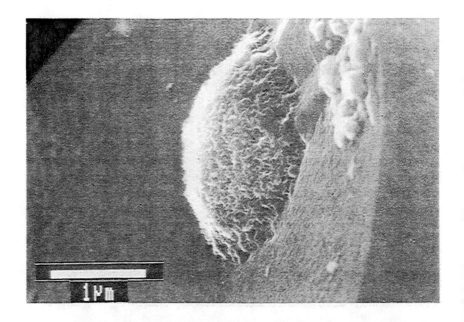

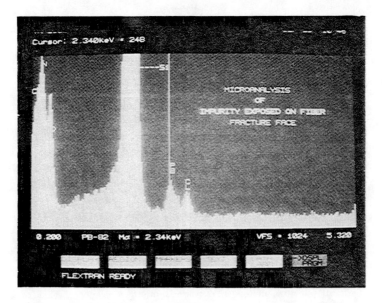

Figure 9.2. Mound-shaped impurity (top) on fracture surface of ceramic fiber derived from a polymeric silicone precursor. AEM microanalysis (lower spectrum) identifies Pb as a prominent contaminant.

Energy dispersive spectrometer detectors have gained favor as attachments on microprobes and conventional scanning electron microscopes. Initially able to detect elements with atomic numbers > 8, newer "light element" detectors now enable microanalysis, additionally, of B, C, N, and O. Innovations in design and spectrometer window compositions have made these advances possible, but have also placed more stringent requirements on vacuum systems, operation, and maintenance of the instruments.

2.4.   Transmission Electron Microscopy

2.4.1.   Usefulness of the Technique

Transmission electron microscopy (TEM) is best applied when meaningful detail at high magnification is required. The technique greatly extends the resolution obtained by microscopical techniques (Table 9.1). Riemer (8) provides a technical discussion of TEM theory and operation; general reviews are also available from other sources (9–11). Section 3 describes selected examples of silicone technology in which the TEM has been profitably used.

2.4.2.   Replication Techniques

Replication techniques were particularly useful for preparing solid specimen surfaces for examination by TEM, before scanning electron microscopes became available. Exact copies of the surface, or replicas, provided a three-dimensional image of the copied surface through application of a heavy element, such at Pt, as a vacuum evaporated thin film on top of a "primary" evaporated carbon film. The heavy element was applied in an evaporator at a comparatively low angle from the horizontal, usually 30°–70°, and provided the impression of surface shadow relief in the TEM.

Preparation of these replicas is exacting (12, 13) and often tedious. Thin plastic films are often applied, followed by evaporated deposition of carbon and heavy element shadow films. Initial plastic films are then dissolved in an appropriate solvent to leave only a heavy element shadowed carbon film that permits transmission of electrons in the microscope. Depending on the replication technique used, replicas often yield excellent surface detail and may be removed intact as extraction replicas (Figure 9.3) for examination of adhering surface components by other means (14). Interpretation of detail, however, from features larger than several micrometers or smaller than 10 nm is often clouded by artifacts generated during the replication process.

Replication methods have been used in a variety of organosilicon surface studies, as in Figure 9.3, but are generally limited to those materials that are

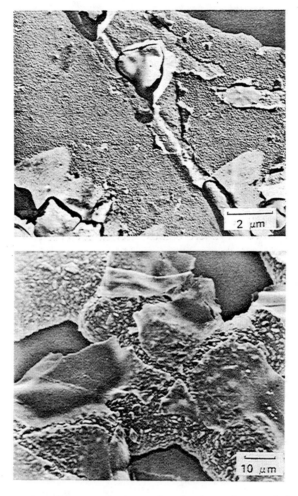

Figure 9.3. Replica from silicone contact lens surface showing surface morphology (top) and by SEM (lower) shown resting on a Cu grid.

unstable in vacuum or are of such size as to preclude examination by other means. The directional shadowing aspect of replicas also enables trigonometric determination of depth and surface film thicknesses, which is often not possible otherwise.

2.4.3. Polymer Characteristics and Artifacts in the TEM

The examination of polymers in the TEM is often difficult. Production of an image requires that different regions in the material produce recognizable variations in image contrast. This contrast differential is produced by scattering of electrons in their passage through the sample prior to their emergence. Dark regions are caused by scattering, whereas light regions are those in which much of the primary beam passes unimpeded. Materials containing heavy elements such as Fe, Cr, and Cu scatter and absorb electrons more efficiently, thereby contributing to a contrasty image. However, since silicone polymers typically contain only atoms of low atomic number, they produce inherently poor contrast, and internal differences in their composition may be poorly discerned.

As in biological materials, polymers may also change quickly in the electron beam. If this change is not recognized, it may lead to erroneous interpretations of the image. The literature discusses radiation damage to inorganic materials (15), even elemental zinc (16), and particularly to organic substances and polymers (4, 17, 18). Although the mechanism by which change occurs is not fully known, degradation of polymers results from scission and breakage of C–C bonds. During irradiation, loss of light elements (19) as well as Na, S, Cl, and K occurs through volatilization from biological specimens (20).

That a beam-sensitive silicone specimen is undergoing change in the electron beam may be recognized in several ways. Dimensional change or shrinking of the specimen upon initial viewing and gradual increase of beam transmission though the thin film (increasing brightness) both indicate a mass loss in the material under examination. Cracks and holes may appear in the film or prepared section. The rate of mass loss generally decreases after short exposure and reaches a steady state. In the meantime, however, some features of interest and their spatial relationships may have been destroyed.

Some carbonaceous materials show a gradual darkening during observation, particularly in the scanning transmission mode (see Section 2.4.4), which presages the buildup of a polymerized carbon deposit. The deposit is often associated with increased fuzziness of an ordinarily sharp, well-focused edge and an apparent growth in the diameter of discrete particles.

The use of TEM to examine silicone polymers may be inadvisable since these materials are often susceptible to the degradation and contamination artifacts described above. If samples are observed to change and contaminate readily in the TEM, different sample preparation techniques may be required. Several approaches may be tried to minimize the deleterious effects of these artifacts. Mass loss and associated instability of highly sensitive specimens may be lessened by using a liquid nitrogen cooled specimen holder that reduces effects of specimen heating (21). Conducting films (e.g., evaporated carbon) applied over the support grid and again on top of the specimen help to "bleed" off the static charge and lessen local beam-induced temperature extremes. A liquid nitrogen "cold finger" has also proven useful in reducing contamination as this device tends to scavenge the volatiles generated from the specimen during examination. Ultimately, however, contamination is best avoided by maintenance of the instrument at high vacuum levels (22). Another technique available on some instruments is that of "low dose" where, by electronic shift of the beam deflection coils, the area of examination is irradiated only during micrograph exposure.

The foregoing discussion should not be construed to indicate that beam-induced changes are all bad. The fact that the electron beam may cause polymerization of some polymers and bases has helped to stabilize thin films and cryotome sections of uncured silicone. Some multicomponent silicone preparations may selectively cross-link in the beam thereby improving the contrast and discrimination of constituent phases.

2.4.4. Electron Diffraction

Techniques of electron diffraction, employed solely in electron beam instruments, have long been a mainstay in defining the nature of crystalline materials (cf. Chapter 15, Section 7) (23, 24). These techniques encompass selected-area diffraction (SAD) in which a beam aperture defines the area from which diffraction information is obtained. Convergent beam electron microdiffraction (CBED) methods produce diagnostic diffraction patterns from far smaller areas than SAD and are often used for the identification of critical microstructural features less than $\sim 0.5\ \mu m$ in diameter. Selected-area diffraction is of value for determining the identity of larger crystals or crystalline regions (24–26).

As polymers are often amorphous or exhibit weak and poorly characterized diffraction patterns, investigations of organic polymers using electron diffraction techniques have been limited to complex orientational and structural determinations (27–29).

Effects of beam damage, especially apparent when using a nonparallel beam of electrons (as in convergent beam and microelectron diffraction), further decrease the likelihood of meaningful crystallographic information being obtained. Even with the use of a parallel beam of electrons, as in selected area diffraction, some crystalline areas of silicone polymers have been observed to become amorphous in a matter of several seconds. Beyond the use of electron diffraction and associated darkfield techniques to identify and locate stable crystalline contaminants in silicone polymers, further advances in this field of characterization will require more reference data and development of less severe instrumental conditions.

Electron diffraction and darkfield techniques are often applied where organosilicon compounds have been pyrolyzed or heated to sufficiently high temperature to develop crystalline assemblages (30). These materials are often less susceptible to detrimental effects of beam heating than their polymeric precursors. Nevertheless, electron beam artifacts of varying severity may be associated with supposedly stable carbonaceous subjects of examination

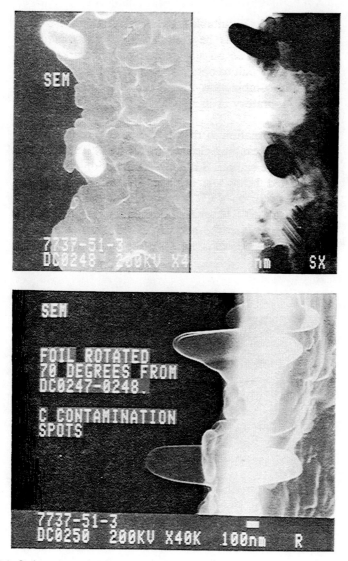

Figure 9.4. Carbon contamination spots shown in different orientations as developed by a focused electron beam.

(Figure 9.4). One such contamination artifact, however, may have useful implications for the determination of thin foil thickness (31).

2.4.5.   Scanning Images in the TEM

Normal TEM methods of examination are significantly broadened by STEM and SEM capabilities. With suitable high intensity or high resolution electron sources, a single sample may be examined in all these modes, providing complementary information on a material or specific area of interest that would otherwise be difficult to obtain. Coupled with energy dispersive X-ray spectrometers, this combination instrument becomes a complete micro-characterization laboratory in its own right, the analytical electron micro-scope (AEM).

Scanning image formation in the SEM mode is accomplished by place-ment of a "reflected" scanning electron image (SEI) detector above and in close proximity to the sample. Scanning transmission electron microscopy images are collected below the sample. The recorded projections, syn-chronized with a raster on a CRT, allow simultaneous viewing of comp-lementary SEM and STEM image information (Figure 9.5). If energy dispersive spectrometers are also part of this AEM installation, selection of areas for X-ray microanalysis is controlled through the SEM and STEM modules.

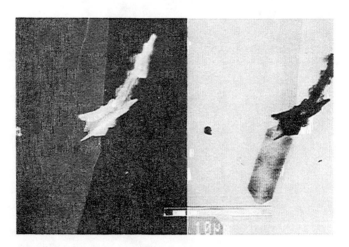

Figure 9.5. SEM (left) and STEM views of particulates resting on transparent film. Comparison shows the benefit of using combined techniques to ensure complete characterization of the preparation.

There is considerable benefit to having this versatility in examination of silicone materials. Unprotected silicone thin films, that is, those without evaporated conductive coatings, may be examined with only minimal beam damage because the SEM/STEM lens configuration significantly decreases the overall incident beam and specimen currents below that characteristic of the TEM mode. This decrease is often sufficient to preserve morphological or microstructural features that are of interest. More concentrated localized doses of damaging current levels may, however, be apparent in positions of stationary spots or raster retrace. Scanning modes also restrict radiation damage to only that area encompassed in the rastered viewing area, as distinguished from the broader irradiation that occurs in the TEM mode.

A prominent attribute of the STEM mode is its ability to provide useful transmitted image information from specimens that are too thick for generation of TEM images. The STEM contrast, moreover, is comparatively enhanced. With the aid of suitable accessories, scanning images may be also combined or juxtaposed with darkfield and backscattered images obtained from the same areas.

Although the STEM configuration associated with the standard TEM instruments has somewhat poorer resolution capabilities than dedicated STEM units with FEG electron sources, the ultrahigh vacuum (10^{-9} torr) required in dedicated STEMs generally rules out the possibility of examining polymers. On the other hand, STEM and SEM units on transmission (AEM) instruments have better resolution (Table 9.1) and meaningful magnification limits than those on scanning electron microprobe (SEMP) or stand-alone SEM instruments. This benefit is the result of high brightness electron sources, improved vacuum, and considerably higher typical accelerating voltages available on the combination AEM instruments.

2.5. Analytical Electron Microscope

Microanalyses of solid organosilicon materials may be performed in the analytical electron microscope (AEM) using thin sections: either thin foils or ultramicrotome preparations. Excellent results are also obtained from manually thinned and polished specimens which, although not thin enough for transmission or STEM examination, can be analyzed with energy dispersive spectrometers in a "reflected" mode in conjunction with scanning imaging. Thin or ultrathin window spectrometers may be used to supplement typical heavy element microanalyses with qualitative microanalyses for C, O, and N. Depending on the AEM configuration, spectrometers positioned to obtain a high take-off angle are preferred. Rotation of the sample holder to accommodate analysis with minimal interference from the holder itself is generally necessary with horizontal detectors. Care should be exercised to ensure that the holders are elementally well characterized, and that spurious X-ray generation from specimen holders and column components is well defined to avoid inadvertent assignment of these extraneous elements to the sample area itself.

Specimens of about 100 nm thickness generally allow direct image generation using TEM and STEM modes. New EDS units installed on AEM systems permit elemental analysis in thin specimen areas as small as 20-nm diameter, subject to instrumental conditions and nature of the sample (32–36).

Limitation of specimen thickness to 50–100 nm improves lateral resolution in light element matrices containing Si during microanalysis, and yet provides tolerable count rates and signal-to-noise (S/N) ratio level in the AEM. Thicker specimens containing light elements and Si may transmit electrons, in the STEM mode, but spatial resolution broadens as the excited X-ray volume becomes greater and encompasses more of the specimen bulk. Absorption, atomic number, and fluorescence effects become more prominent as the excited volume departs (increases) from that of thin film. These effects may be generally ignored in microanalysis of very thin AEM films. Quantitative programs for the AEM, which take these corrections into account, need to be more rigorously applied as the bulk volume situation is approached in thick AEM specimens. Higher instrument accelerating voltages tend to minimize the detrimental effects of electron beam interaction with the excited volume in thin films.

2.6. Image Processing

Modern electron beam instruments have sophisticated interactive equipment and software that control spectrometers as well as acquire STEM and SEM images, X-ray spectra, and generate element distribution maps. Computer programs also process much of the data, allowing selective enhancement of features of interest (37). Energy dispersive X-ray programs are also well developed for the acquisition and correction of data for quantitative microanalysis. Application of these techniques to materials studies are reported frequently in the literature (6).

Image enhancement programs have often been found useful when examining beam-sensitive silicone materials. For example, the judicious accentuation of comparatively low signal levels in elemental distribution maps using computer enhancement techniques tends to avoid beam damage that could otherwise occur at higher current levels during acquisition. As long as acquisition parameters are set appropriately for adequate S/N ratio levels, some computer enhancement of the distribution is effective (Figure 9.6).

Figure 9.6. STEM micrograph (left) of a silicone treated gypsum construction board thin foil. Center map of Si distribution is shown without image enhancement, right map the same, but with moderate computer enhancement of Si distribution in the same area.

3. APPLICATION TO SILICONE MATERIALS

Microscopical examination of silicone materials typically draws upon several different techniques. In order to choose the appropriate technique, the analyst must have good background information and knowledge of what material characteristic is germane to the focus of the investigation.

3.1. Emulsions and Fluid Polymers

These silicones readily permit examination by transmitted optical microscopy. Phase contrast and transmitted Nomarski DIC (Figure 9.7) techniques are often most appropriate to discern "particles" of silicone oil in water or boundaries caused by mixing of other immiscible liquids. Identification of solids and contaminant particles in a liquid medium is often possible using optical microscopy. Filtration or heavy liquid separation techniques are also employed to segregate contaminants for subsequent examination by elemental or X-ray diffraction analyses.

Fine particles $< 1\,\mu m$ in diameter frequently show Brownian motion, though their motion may be slowed by adding a thickener such as glycerine to the supporting medium. Photomicrographs of the rapidly moving particles may be obtained through use of a high intensity illumination source and high speed film. For particles finer than can be adequately resolved by optical techniques, electron microscopy becomes necessary to describe the ultimate particles and the degree of aggregation that they display. Silicone emulsions are often incorporated in a methylcellulose thickener and then dried to preserve the size and shape of emulsion particles for TEM examination (38).

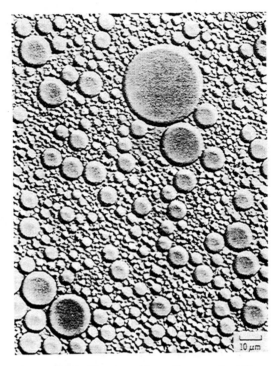

Figure 9.7. Photomicrograph of a silicone emulsion using transmitted Nomarski DIC illumination. Better particle edge definition is achieved with this technique than with phase contrast.

Relatively coarse emulsion particles may deform under pressure of the cover glass during optical microscope examination. The particle size may be preserved by the use of an etched-well glass slide covered with a cover slip (Howard Counting Chamber, Catalog No. 4100-A, from A. H. Thomas Co., Philadelphia, PA). With suitably prepared samples, an estimate of relative particle sizes in a sample may be obtained by visual inspection. More quantitative particle size distribution data is also possible with use of specialized optical measuring equipment interfaced to the optical microscope.

Silicone polymers and copolymers are advantageously examined in either very thin sections or as thin films cast from solvents. Depending on the compositions involved, optical microscopy with bright field and phase contrast or TEM approaches are viable in discerning complex heterogeneity of blended systems (39).

Transmission electron microscope and SEM methods may be employed to characterize and identify extraneous materials and contaminants in silicone

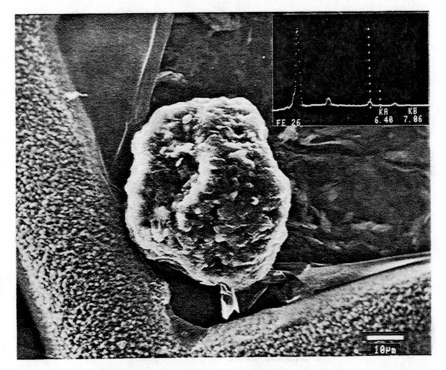

Figure 9.8. Contaminant particle supported in corner of Cu grid by a thin carbon film. Particle was removed from a fluid and analyzed by EDS in the SEMP to reveal (inset) the elements Si, Al, Ca, and Fe as plausible impurities.

fluids, emulsions, and polymers. Some, as in Figure 9.8, may be sufficiently large to physically isolate from their fluid surroundings and analyze by SEMP or AEM to determine their origin. Polymeric contaminants and silicone gels may require IR microscopy (Chapter 11) to assist in this identification.

3.2. Elastomers

3.2.1. Filler Characterization

Many physical and electrical properties of elastomers are determined in part by the particle size, extent of aggregation, and particle shape of filler materials that have been incorporated with silicone polymers during manufacture. These features strongly influence the serviceability of the end product in its intended application. Dry filler powders are typically characterized by optical, SEM, and AEM techniques.

Varying degrees of powder disaggregation may be achieved by dispersing the powder in clean alcohol through use of ultrasonics. Excellent results in the AEM have also been obtained by briefly ultrasonicating a noncatalyzed filled silicone gum in toluene, followed by drying this preparation on a carbon

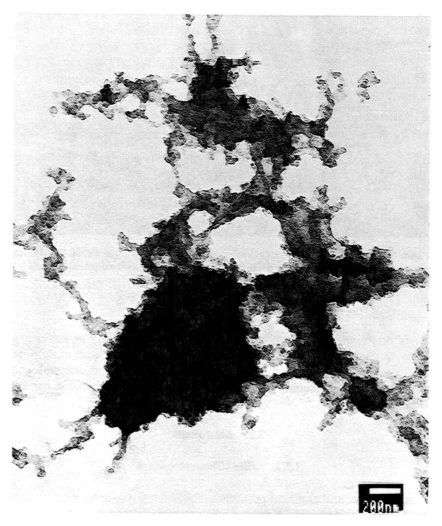

Figure 9.9. Carbon black commercial filler for use in silicone elastomer formulations. TEM micrograph, carbon film support.

coated TEM grid. Variations of these techniques allow visualization of particle affinities in such fillers as carbon black (Figure 9.9), and fumed silica (Figure 9.10). Care must be exercised to scan the dispersed sample thoroughly in order to insure that areas chosen for micrographs are representative of the filler morphology in the specific preparation.

Figure 9.10. Typical fumed silica filler aggregates and chainlike morphology. TEM micrograph, carbon support film.

3.2.2. *Cured Silicone Elastomers*

Once cured, the silicone elastomer and their reinforcing fillers are best studied in the AEM using carefully prepared thin sections (Figure 9.11). Specially

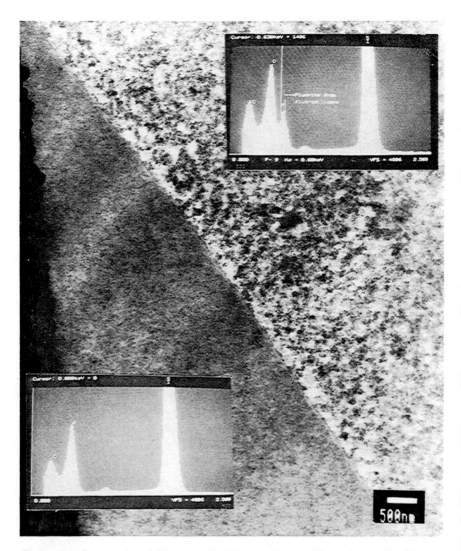

Figure 9.11. Cross-section of silicone prosthetic implant showing different fumed silica distribution in the two tightly adhering portions of the implant. Insets show typical EDS spectra for dimethylsiloxane (lower left) and fluorosilicone polymers (upper right). TEM micrograph, ultracryotome section, ultrathin window EDS detector.

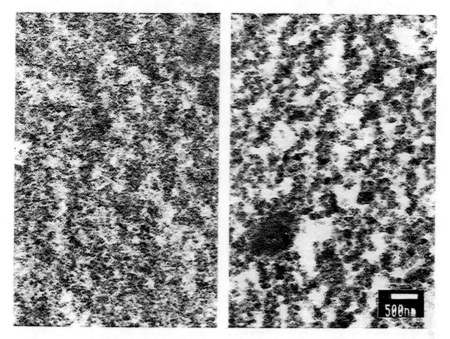

Figure 9.12. Silicone elastomer formulated with fumed silica containing 0% moisture (left) and the same silica containing 20% moisture (right). TEM, ultracryotome section.

equipped ultracryotome instruments (Ivan Sorvall, Inc., Newton, CT; LKB Instruments, Inc., Rockville, MD), are designed to cut sections of about 100 nm thickness while maintaining the sample and diamond knife below the glass transition temperature of the elastomer. Excellent results have been obtained by sectioning silicone elastomers at $-140\,^{\circ}C$ in order to avoid the substantial artifacts that accompany embedment techniques. Meaningful comparisons between elastomer compositions are more readily established (Figure 9.12) if low sectioning temperature and uniform section thicknesses are achieved.

3.3. Resins

Silicone-containing epoxy resins have become extremely useful in the aerospace industry. Adding silicones to resins can enhance desirable features such as fracture toughness, thermal stability, and adhesive strength (39–41). Epoxy resins containing silicones are often examined by SEM for features such as particle distribution, origin and mode of failure and porosity (Figure 9.13). Sample preparation consists of fracturing the material to obtain a fresh

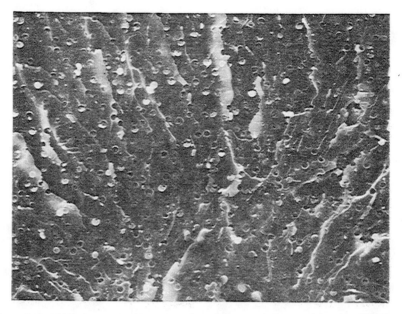

Figure 9.13. SEM micrograph of fracture surface of an epoxy resin containing spherical silicone particles. Fracture section. Bar scale represents 10 μm.

surface and subsequently coating that surface with a thin layer of metal. Image processing and microanalysis techniques are also applied to these materials when the elemental distribution of components in the matrix is required or more detailed analysis concerning the separation of phases is needed. X-ray mapping by both EDS and WDS methods for the distribution of Si provides valuable information regarding the distribution, relative abundance and morphology of the silicone component in the surrounding matrix.

3.4. Special Applications of Microscopy

3.4.1. Coatings

Microscopical characterization is often used to determine the presence and location of silicone coatings on substrates. Optical microscopy and electron beam instruments are commonly employed.

Silicone coatings on paper substrates are well defined by reflected optical Nomarski DIC when coated and uncoated papers that have been prepared in the same manner are compared. The surface appearance of both may be enhanced by evaporated or sputtered coatings of carbon and gold or gold–palladium alloy since the heavy elements, in particular, minimize light diffusion and scattering within the paper interior. Other approaches, such as mapping the paper surface for Si distribution by EDS or WDS (Figure 9.14), also provide evidence of silicone concentrations if Si-containing clays have not also been used in the paper manufacture. Cross-sectional mapping of papers for evidence of relative silicone penetration as a function of paper manufacturing technique has provided valuable information (42).

Many of the same techniques may be employed in demonstrating the microstructure and surface morphology of paints, the location of silicone that imparts water resistance to construction wallboard (Figure 9.6), and continuity of protective coatings on a variety of substrates ranging from automobile windshields to high voltage electrical insulators.

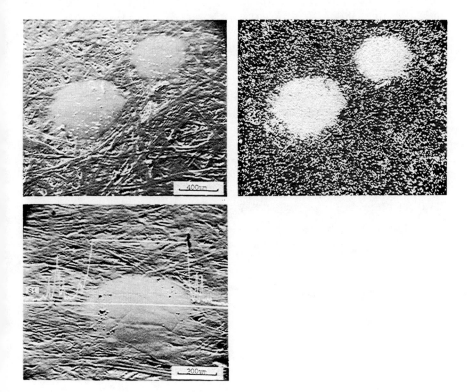

Figure 9.14. Silicone coated paper surface (upper left) showing Si concentrations by WDS mapping (upper right) and Si line scan techniques (lower left).

3.4.2. Electronic Materials

Both SEM and optical microscopy are used extensively by the semiconductor industry for the evaluation and characterization of electronic circuits (43–46). Silicones are used as coatings to provide water repellency, resistance to abrasion and corrosion, and environmental protection (Figure 9.15). Silicone polymer precursors coating the surface of an electronic device help protect it from harsh environmental conditions. Thin silicone films derived from TMS (tetramethylsilane), for example, have been found to be highly resistant to chemical attack (46). Cross sections of the electronic devices are used to establish thickness and planarity of the coating over the rough circuit surface (Figure 9.16). Cross-sectioning techniques for electronic circuits are thoroughly discussed in the literature (43–48). The integrity of coatings is often evaluated by SEM and optical microscopy for such features as surface corrosion and abrasion. Both EDS and WDS techniques are used to gain insight into the effects of heat treatments and corrosion created by environmental testing (48).

3.4.3. Fibers and Composites

Reinforcing whiskers and fibers of inorganic materials made by pyrolysis of polymeric precursors, including organosilicon materials, are being widely used in many commercial applications (49). Because of their high strength,

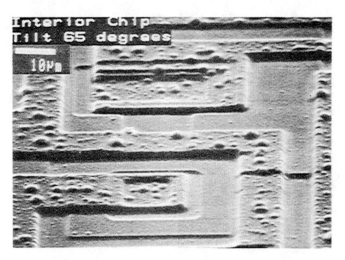

Figure 9.15. SEM micrograph of an electronic device coated with a thin silicone-derived surface film.

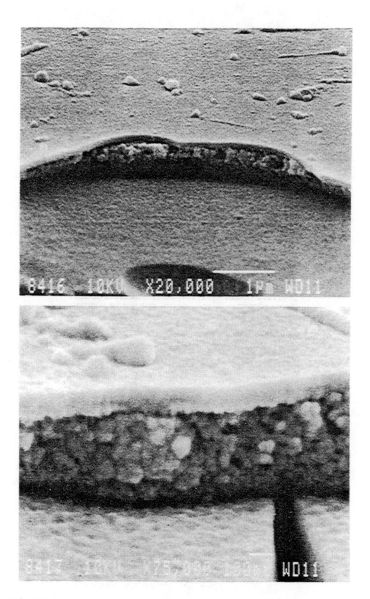

Figure 9.16. SEM micrographs of a silicone protected cross-sectioned electronic device. Each layer shown represents a specific silicone-derived coating.

245

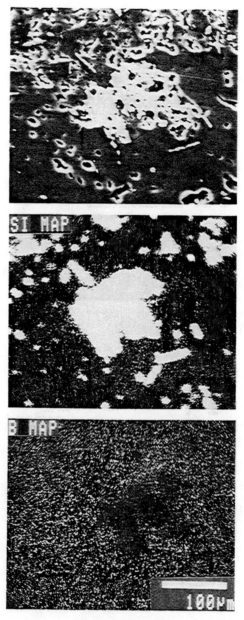

Figure 9.17. Ceramic composite containing SiC grains in a B₄C. WDS X-ray mapping shows lack of B where Si is present. Polished section.

246

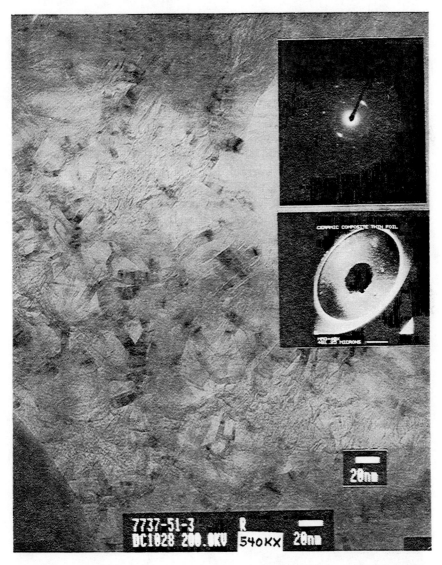

Figure 9.18. Graphite matrix (2H polytype) in ceramic monolith prepared from silicone polymeric precursors. Electron diffraction pattern (top right) and low magnification view of the thin foil sample preparation, perforated at the center.

they impart toughness to composites that are used to fabricate a wide range of materials from golf club shafts to airplane wings. Elimination of defects and blemishes that compromise their strength is central to their use (Figure 9.2).

Composites produced by combining silicone polymers with granular filler materials, and sometimes fired to high temperature, often present an intricate microstructure and fracture surface morphology when viewed by microscopical techniques (50). Polished sections may be prepared for microanalysis in the SEMP and reflected light examination in the optical microscope. Mapping of elements present in the composites by WDS often reveals valuable information regarding interrelationships of the various components (Figure 9.17). Fractured surfaces may be examined by SEM for evidence of preferred fracture patterns as well as internal defects such as porosity and localized grain growth. Sections or foils prepared from these materials by dimpling and ion thinning may be suitable for imaging and microanalysis in the AEM to describe matrix heterogeneity and grain boundary characteristics (Figure 9.18).

WRIST JOINT IMPLANT (LEFT)

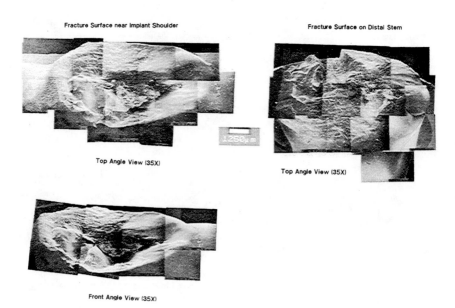

Fracture Surface near Implant Shoulder

Fracture Surface on Distal Stem

Top Angle View (35X)

Top Angle View (35X)

Front Angle View (35X)

Figure 9.19. Mosaic of SEM micrographs showing different views of a wrist joint implant fracture surface.

3.4.4. Biomaterials

Silicone materials are often used in surgical replacement procedures. Optical and electron beam microscopy have been used to characterize surfaces of implants such as fractured wrist joints to determine the mode and origin of failure resulting from a fracture of the prosthetic device (Figure 9.19). Similar techniques have also been employed to evaluate finger joints. One such study involved a patient who was experiencing swelling of the lymph nodes following deterioration of a silicone finger joint. The lymph tissue was studied by SEMP and SEM to determine a possible cause for the swelled tissue. X-ray mapping of the lymph nodes revealed areas of Si concentration within the tissue (Figure 9.20). As the lymph nodes act as receptors for debris in the body, it was determined through additional analyses that the swelling was brought on by granulation of the silicone finger joint and concentration of that material in the lymph nodes.

Microscopy has also been used to study contaminants found on the surface of finger joints. One such example involved the surface of a joint that had been removed from a sterile package and then examined by SEM. The sterile finger joint was then lightly rubbed against the sterile surgical gloves and drapes typically used by doctors. The SEM evaluation of the implant

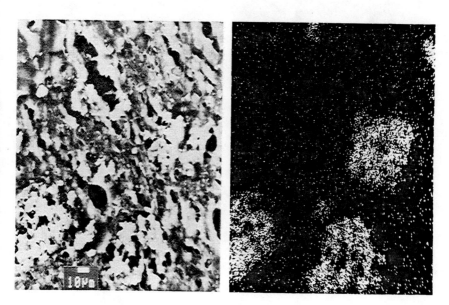

Figure 9.20. Lymph node tissue thin section (left) revealing localized Si concentrations by WDS distribution mapping (right).

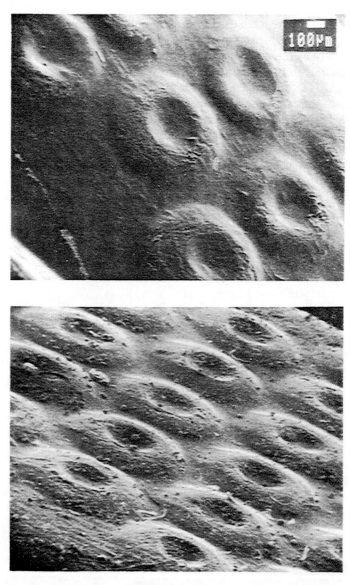

Figure 9.21. SEM micrographs illustrating contamination of silicone finger implant surface. Appearance after removal from sterile packaging (top) and after contact with surgical gloves and drapes.

surface showed that particles from sterile drapes and gloves had transferred to the implant (Figure 9.21).

REFERENCES

1. T. C. Rochow and E. G. Rochow, *An Introduction to Microscopy by Means of Light, Electron, X-rays, or Ultrasound,* Plenum Press, New York, 1978.

2. J. G. Delly, *Photography through the Microscope,* Kodak, Rochester, NY, 1988.

3. W. C. McCrone, L. B. McCrone, and J. G. Delly, *Polarized Light Microscopy,* Ann Arbor Science, Ann Arbor, MI, 1979.

4. L. C. Sawyer and D. T. Grubb, *Polymer Microscopy,* Chapman & Hall, London, 1987, pp. 62–69.

5. J. I. Goldstein, D. E. Newberry, P. Eichlin, D. C. Joy, C. Fiori, and E. Lifshin, *Scanning Electron Microscopy and X-ray Microanalysis,* Plenum Press, New York, 1981.

6. *Proceedings* of the Electron Microscopy Society and Microbeam Analysis Society, published annually by San Fancisco Press, San Francisco.

7. D. E. Newberry, D. C. Joy, P. Echlin, C. E. Fiori, and J. I. Goldstein, *Advanced Scanning Electron Microscopy and X-ray Microanalysis,* Plenum Press, New York, 1986.

8. L. Riemer, *Transmission Electron Microscopy,* Springer-Verlag, Berlin, 1984.

9. B. M. Siegal, *Modern Developments in Electron Microscopy,* Academic, New York, 1964.

10. M. A. Hayort, *Principles and Techniques of Electron Microscopy,* Van Nostrand, New York, 1970.

11. P. Doig, Ed., *Electron Microscopy and Analysis, 1983,* The Institute of Physics, Bristol, 1983.

12. D. H. Kay, *Techniques for Electron Microscopy,* Davis Publishing, Philadelphia, 1965.

13. D. M. Hall, "The preparation of electron microscope replicas from rough porous surfaces," *Br. J. Appl. Phys.,* **8**, 1965, p. 295.

14. H. A. Freeman, *Proceedings, 35th Annual E.M.S.A. Meeting,* G. W. Bailey, Ed., Claitor's Publishing, Baton Rouge, LA, 1977, pp. 174–175.

15. S. S. Breese, Ed., *Electron Microscopy, Fifth International Congress for Electron Microscopy, Vol. 1,* Academic, New York, 1962, pp. F-1–G-14.

16. C. R. Bradley and N. D. Zaluzec, "Observation of transmission electron sputtering at 100 and 120kV," in *Proceedings of the Electron Microscopy Society of America,* G. W. Bailey, Ed., San Francisco Press, San Francisco, 1988, pp. 644–645.

17. M. Isaacson, et al., in *Electron Microscopy and Analysis, 1983,* P. Doig, Ed., The Institute of Physics, Bristol, 1983, pp. 1–30.

18. R. M. Glaeser, "Radiation damage with biological specimens and organic materials," in *Introduction to Analytical Electron Microscopy*, J. J. Hren, J. I. Goldstein, and D. C. Joy, Eds., Plenum Press, New York, 1979, pp. 423–432.

19. G. Hughes, *Radiation Chemistry*, Clarendon Press, Oxford, 1973.

20. T. A. Hall and B. L. Gupta, in *Introduction to Analytical Electron Microscopy*, J. J. Hren, J. I. Goldstein, and D. C. Joy, Eds., Plenum Press, New York, 1979, p. 182.

21. P. Echlin, "Cryomicroscopy and analysis: Procedures, problems, and prospects," in *Proceedings of the Microbeam Analysis Soc.*, P. E. Russell, Ed., San Francisco Press, San Francisco, 1989, pp. 79–84.

22. *Manual on Electron Metallography Techniques*, ASTM Special Technical Publication No. 547, American Society for Testing Materials, Philadelphia, 1973, pp. 41–72.

23. B. E. P. Beeston, "An introduction to electron diffraction," in *Practical Methods in Electron Microscopy*, A. M. Graver, Ed., North-Holland, Amsterdam, 1973, pp. 195–323.

24. P. Hirsh, A. Howie, R. Nicholson, O. Pashley, and M. Whelem, *Electron Microscopy of Thin Crystals*, Kreiser Publications, New York, 1977.

25. D. Williams, *Practical Analytical Electron Microscopy in Materials Science*, Philips Electronic Instruments, Inc., Electron Optics Publishing Group, Mahway, NJ, 1984, pp. 117–146.

26. F. Brissa and R. Marchessaulf, "Fiber diffraction methods," A. D. French and K. H. Gardner, Eds., American Chemical Society Symp. Ser. 141, ACS, Washington, DC, 1980.

27. M. Boer, Ed., "Symposium on electron crystallography of macromolecules," in *Ultramicroscopy*, Vol. 13, Nos. 1–2, 1983.

28. D. R. Petersen, D. R. Carter, and C. L. Lee, "Analysis of X-ray and electron-beam diffraction patterns from poly(dipropylsiloxane)," *J. Macromol. Sci.*, **B3**, 519 (1969).

29. D. Dorset, "Fractionation of polymethylene chains: An electron crystallographic investigation," *Proceedings of the Electron Microscopy Society of America*, San Francisco Press, San Francisco, 1986, pp. 794–795.

30. J. Hyacha, S. Bonnamy, X. Bourrat, A. Duerbergue, Y. Maniette, and A. Oberlin, "Characterization of some pyrolyzed polycarbosilanes by transmission electron microscopy," *J. Mater. Sci. Lett.*, **7**, 885 (1988).

31. D. Williams, *Practical Analytical Electron Microscopy in Materials Science*, Philips Electronic Instruments, Inc., Electron Optics Publishing Group, Mahway, NJ, 1984, p. 77.

32. D. R. Beaman, "Analytical transmission electron microscopy and its application in environmental science," in *Environmental Pollutants*, T. Toriburri, J. Coleman, B. Dahneise, and I. Feldman, Eds., Plenum Press, 1978, pp. 255–294.

33. J. I. Goldstein and N. J. Zaluzec, in *Introduction to Analytical Electron Microscopy*, J. J. Hren, J. I. Goldstein, and D. C. Joy, Eds., Plenum Press, New York, 1979, pp. 55–89.

34. J. I. Goldstein and N. J. Zaluzec, in *Introduction to Analytical Electron Micro-scopy*, J. J. Hren, J. I. Goldstein, and D. C. Joy, Eds., Plenum Press, New York, 1979, pp. 83–167.

35. J. I. Goldstein and N. J. Zaluzec, in *Introduction to Analytical Electron Micro-scopy*, J. J. Hren, J. I. Goldstein, and D. C. Joy, Eds., Plenum Press, New York, 1979, pp. 63–90.

36. J. I. Goldstein and N. J. Zaluzec, in *Introduction to Analytical Electron Micro-scopy*, J. J. Hren, J. I. Goldstein, and D. C. Joy, Eds., Plenum Press, New York, 1979, p. 4.

37. J. I. Goldstein and N. J. Zaluzec, in *Introduction to Analytical Electron Micro-scopy*, J. J. Hren, J. I. Goldstein, and D. C. Joy, Eds., Plenum Press, New York, 1979, pp. 211–217.

38. A. F. Kolb and L. A. Hulce, in *Analysis of Silicones*, A. L. Smith, Ed., New York, pp. 171–181.

39. A. K. St. Clair, T. L. St. Clair, and S. A. Ezzell, "Polyimide adhesives: Modified with ATBN and silicone elastomers," *Polym. Sci. Tech.*, **29**, 467 (1984).

40. A. K. St. Clair and T. L. St. Clair, "Addition polymide adhesives containing aromatic amine-termined butadiene acrylonitrile (ABTN) and silicone elasto-mers," *Int. J. Adhes. Adhes.*, **1**, 249 (1981).

41. E. M. Yorkgitis, N. S. Eiss, Jr., C. Tran, G. L. Wilke, and J. E. McGrath, "Siloxane-modified epoxy resins," *Adv. Polym. Sci.*, **72**, 79 (1985).

42. J. E. Wilson and H. A. Freeman, "Analysis of silicone-coated papers with the scanning electron microprobe," *TAPPI*, **64**(2), 1981, p. 95.

43. R. L. Belcher, G. P. Hart, and W. R. Wade, "Novel sample preparation for microanalysis," Institute of Electrical and Electronics Engineers/Int. Reliability Phys. Sym., 1984, pp. 83–94.

44. R. A. Young and R. V. Kalin, "Scanning electron microscopic techniques for characterization of semiconductor materials," in *Microelectronics Processing: Inorganic Materials Characterization*, L. A. Casper, Ed., ACS Ser. 295, Washing-ton, DC, 1986, pp. 49–74.

45. S. K. Gupta and R. L. Chin, in *Microelectronics Processing: Inorganic Materials Characterization*, L. A. Casper, Ed., ACS Ser. 295, Washington, DC, 1986, pp. 349–365.

46. R. Szeto and D. W. Hess, "Correlation of chemical and electrical properties of plasma-deposited tetramethylsilane films," *J. Appl. Phys.*, **52**, 903, 1981.

47. S. Thomas, "Scanning electron microscope analysis of fracture cross-sections in intergrated circuits process characterization," in *Scanning Electron Microscopy*, O. Johari, Ed., 1983, p. 1585.

48. A. J. Denboer, "From product evaluation to failure analysis low voltage/high brightness SEM plays an important role for integrated circuit engineering corporation," *Norelco Reporter*, **31**, 14, 1984.

49. R. H. Baney, "Designing preceramic polymers," *Chemtech*, **18**, 738, 1988.

50. G. T. Burns and G. Chandra, "Pyrolysis of preceramic polymers in ammonia: Preparation of silicon nitride powders," *J. Am. Ceram. Soc.*, **72**, 333, 1989.

CHAPTER

10

CHROMATOGRAPHIC METHODS

R. D. STEINMEYER

Dow Corning Corporation
Carrollton, Kentucky

and

M. A. BECKER

Dow Corning Corporation
Midland, Michigan

1. INTRODUCTION

This chapter is directed toward the practicing chromatographer who wishes to analyze silicones. Therefore, the basic fundamentals of chromatography are not covered in detail. For those with a limited knowledge of the subject, some preliminary study will be required; numerous basic texts are available. Those with experience in the field, however, should find few surprises, as for the most part, the procedures used for the analysis of silicones are simply extensions of those used for more conventional organic compounds. Nevertheless, some minor but crucial modifications may be necessary to assure success.

Subject matter was selected based upon general applicability to a range of materials as opposed to in-depth reviews of product-specific methods. Most silicone vendors will provide the latter upon request. Shatz et al. (1) have also published a review of 248 references dealing with the gas chromatographic analysis of organosilicon compounds.

Although a brief discussion of supercritical fluid (SFC) and liquid chromatography (LC) is included, the major emphasis of this chapter is on gas chromatography (GC), which currently finds the widest utility for the analysis of silicones. As an indication of the relative importance of the various

The Analytical Chemistry of Silicones, Edited by A. Lee Smith.
ISBN 0-471-51624-4 © 1991 John Wiley & Sons, Inc.

chromatographic techniques, GCs outnumber all other chromatographic instruments by a ratio of approximately 40:1 in the author's sphere of acquaintances.

2. BACKGROUND INFORMATION

2.1. Silanes

Silanes are typically volatile enough to permit analysis by GC. They do, however, present a difficult analytical challenge as, with limited exception, they contain at least one reactive silicon bond (e.g., Si–Cl, Si–N, Si–OR, Si–H). In addition, some silanes also contain reactive organic groups, for example, amine, epoxy, or methacrylate functions.

Chlorosilanes are the basic building blocks of the silicone industry and as such warrant much analytical attention; their purity ultimately influences the properties of finished polysiloxanes. Frequently, quality requirements demand methods capable of measuring impurities in chlorosilanes at parts per million concentration levels.

One's first encounter with chlorosilanes is usually sufficient to instill a respect for their reactivity. The briefest exposure to atmospheric moisture when loading a syringe and injecting a specimen into the chromatograph results in the evolution of hydrogen chloride and the appearance of siloxanes in the chromatogram. Silazanes can react similarly, producing ammonia and siloxanes. Acetoxy silanes and other silyl esters form the corresponding acid and siloxanes. In contrast, Si–OR and Si–H groups are reasonably stable in neutral environments.

Chlorosilanes and silazanes also react readily with silanol on the surface of glass sample containers, glass wool plugs, diatomaceous earth solid supports and other similar surfaces. Chromatographers often use this behavior for deactivating such surfaces since the polar silanol group is converted to a nonpolar siloxane. When it occurs during analysis, however, the reaction is not desirable, as a portion of the sample is consumed in forming these siloxane bonds.

Both water and silanol can cause changes in silane sample composition which, superficially, appear out of proportion to the extent of exposure. This is especially true when the sample is composed of more reactive compounds at low concentrations in a less reactive matrix; the more reactive components may be selectively consumed while little change is noted in the matrix. As examples, the complete loss of low concentrations of the highly reactive $SiCl_4$, $HSiCl_3$, $MeHSiCl_2$, and $MeSiCl_3$ from the less reactive Me_2SiCl_2 and Me_3SiCl have been observed.

Even Teflon®, a material that has found wide use because of its inertness, is not immune from reacting with chlorosilanes. Exchange of chlorine and fluorine can occur, resulting in the formation of fluorosilanes or chlorofluorosilanes. Foltz et al. (2) noted this reaction in a heated mass spectrometer system. The author has also observed the same reaction when chlorosilanes contact such common laboratory materials as fluorocarbon bottle cap liners and ground glass joint sleeves, even at room temperature. The resultant fluorinated silanes elute from the chromatograph much earlier than their chlorinated analogs and may be misidentified as impurities when, in reality, they are artifacts of sample handling.

Hydrogen chloride generated from incidental hydrolysis of chlorosilanes can also effect changes in sample composition, both as a reactant and a catalyst. Examples of reactions that have been observed are shown below.

$$R_1R_2CH{=}CH_2 + HCl \longrightarrow R_1R_2CHCl{-}CH_3 \tag{10.1}$$

$${\geq}SiOSiMe_2Cl + MeSiCl_3 \xrightarrow{HCl} {\geq}SiOSiMeCl_2 + Me_2SiCl_2 \tag{10.2}$$

$$Me_3SiSiMe_3 + HCl \longrightarrow Me_3SiCl + Me_3SiH \tag{10.3}$$

$$\text{metals and/or metal oxides} + HCl \longrightarrow \text{Lewis acids} \tag{10.4}$$

$$R{-}Cl + {\geq}SiH \xrightarrow{\text{Lewis acid}} {\geq}SiCl + R{-}H \tag{10.5}$$

$$2Me_2SiCl_2 \xrightarrow{\text{Lewis acid}} MeSiCl_3 + Me_3SiCl \tag{10.6}$$

$$Me_4Si + HCl \xrightarrow{\text{Lewis acid}} Me_3SiCl + CH_4 \tag{10.7}$$

$$PhMeSiCl_2 + HCl \xrightarrow{\text{Lewis acid}} MeSiCl_3 + \text{benzene} \tag{10.8}$$

$$R{-}Cl + \text{aromatic compound} \xrightarrow{\text{Lewis acid}} R\text{-aromatic} + HCl \tag{10.9}$$

It should be emphasized that the equilibrium of some of the above reactions is far to the left under normal circumstances. Indeed, during macroanalyses, evidence of their occurence may not be observed. As method sensitivity is increased, however, a point will eventually be reached where they can influence the validity of results.

2.2. Polysiloxanes

The molecular weight and boiling points of polysiloxanes increase rapidly within a homologous series because of the mass of each additional siloxane unit. This trend, while advantageous for resolving homologs, can limit the range of an analysis. Only the first few members of a given series may be volatile enough to permit analysis by GC. Supercritical fluid chromatography greatly extends the range of GC and with nearly identical resolution. Gel permeation chromatography usually permits elution of the complete series but, with limited exception, fails to resolve individual members. Figure 10.1 presents a comparison of these three techniques as applied to the separation of a mixture of cyclic and linear dimethyl polysiloxanes.

The siloxane bond itself is quite thermally stable and presents no special problems chromatographically; the most labile portions of the polysiloxane molecule are usually the hydrocarbon groups attached to the silicon atoms. In most instances, then, the precautions required for the analysis of polysiloxanes are those used for hydrocarbons containing the same organic functional groups.

Some polysiloxanes do contain silicon functional groups that require special attention. Most Si–OH groups, for example, condense so readily that they must be derivatized before analysis by GC or SFC. Groups such as Si–OR, Si–H, and Si–Cl require the same precautions as apply to silanes containing these moieties.

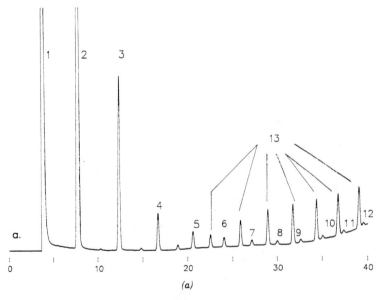

(a)

Figure 10.1. Comparison of (*a*) capillary column GC, (*b*) SFC, and (*c*) GPC chromatograms of cyclic and linear PDMS. Peaks 1–12 represent cyclics D_4 through D_{15}; peaks labeled 13 represent MD_xM oligomers.

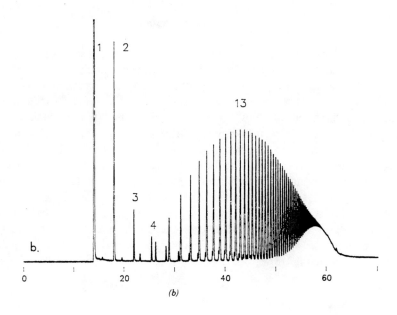

(b)

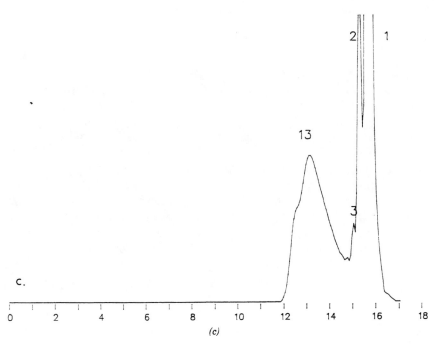

(c)

Figure 10.1 (*Continued*)

259

2.3. Organic Compounds

Conventional organic compounds are used as raw materials, solvents, and processing aids in the silicone industry and, as such, these compounds, their impurities, and their by-products are of interest to the silicone analyst. The direct process reaction of methyl chloride and silicon metal (Section 1.4) generates a large number of by-products including alkanes, olefins, and chlorocarbons. Production of phenyl chlorosilanes, likewise, results in the generation of many aromatic hydrocarbons. In both cases, the boiling points of some of the hydrocarbon by-products are close to those of commercially valuable silane monomers. Separation by fractional distillation, the usual method of purification, is therefore difficult. Not only can these materials interfere with the determination of specific silicones, some may present special concerns of their own. For example, Hanneman has discussed the identification and measurement of incidentally generated polychlorinated biphenyls in silicones (3).

3. SAMPLING

3.1. General Comments

The initial step in the development of any analytical method must be directed toward the sampling procedure. The skilled analyst is fully aware that the data obtained from any analysis can never be better than the samples submitted. Some of the sampling problems encountered with chlorosilanes and other reactive organosilicon materials are discussed in Chapter 2, Section 2.

Time and temperature must be considered enemies. The shorter the interval between sampling and analysis and the lower the temperature the sample encounters, the less likely that detrimental changes will occur. Sampling into dry ice chilled cold traps is sometimes of benefit. Sampling directly into prechilled nonreactive solvent has been successfully used for unstable silanol mixtures (4). This technique takes advantage of both low temperature and dilution. Reactive species can also be converted to less reactive analogs by suitable derivatization procedures during the sampling procedure (5).

If the sampling frequency for in-process material is insufficient, the variability of the material or the process cannot be determined. Application of appropriate statistical sampling plans is therefore highly recommended.

4. SAMPLE PREPARATION

4.1.   General Comments

Most silanes and many siloxanes may be analyzed chromatographically with no special sample preparation required, other than perhaps dilution with a nonreactive solvent to reduce viscosity and aid in syringe injection. Some types of silicones, however, are either so labile or of such high molecular weight that they cannot be chromatographed without some sort of pretreatment. Even in the case of the chromatographically well-behaved silicones, pretreatments are sometimes used in order to increase the sensitivity or to improve the selectivity of an analysis. The commonly used pretreatments can be grouped into three broad catagories: derivatization, decomposition, and extraction. Typical examples are presented in the following sections.

4.2.   Derivatization Schemes

Silanols, almost without exception, are "capped" with a silylating reagent prior to analysis. A procedure that works well with a wide variety of sample types involves dissolving a 1-g specimen in a mixture comprised of 2 mL of diethyl ether and 1 mL of hexamethyldisilizane (HMDZ). Two drops of trifluoroacetic acid are then added to catalyze the reaction. The solution is allowed to stand (gentle heat may be applied) until reaction is complete. Reaction time is initially determined by making periodic analyses to find the point beyond which no further changes occur. N,O-bis(trimethylsilyl)trifluoroacetamide (BSTFA) is sometimes substituted for the HMDZ–ether solution with good result. Reagents containing silyl groups other than trimethyl (e.g., t-butyl dimethyl) may also be used if interferences occur with the trimethylsilyl derivatives. Reagents are even available which contain groups (e.g., chloromethyl dimethyl) that provide specificity with certain detectors (e.g., electron capture). Reviews of silylation techniques by Pierce (6) and Klebe (7) provide other helpful suggestions.

Chlorosilanes may be converted to fluorosilanes in precolumns containing antimony trifluoride. This technique has been used both for increasing the volatility of high boiling silanes and for moving selected silanes away from interfering components (8, 9). Appropriate care should be exercised in using this technique since the reagent and possibly the reaction products are toxic.

Highly reactive amide functional silanes may be converted to their much more stable alkoxy analogs by reaction with alcohols prior to analysis (5).

4.3. Decomposition Schemes

High molecular weight polysiloxane fluids, elastomers, and resins that are
not amenable to direct analysis by GC can be reduced to volatile species by
chemical decomposition. For example, Palamarchuk et al. (10), reported on
the use of NaF to convert multifunctional methyl siloxanes to Me_3SiF,
Me_2SiF_2, and $MeSiF_3$ permitting quantitative determination of moiety
ratios by GC. Others have used BF_3 in diethyl ether for the same purpose
(11, 12). Alkali catalyzed reequilibrations with ethyl orthosilicate quantitatively reduce a wide assortment of silicone polymers to chromatographable
ethoxy silanes (Chapter 8, Section 3.7.3.3). Fluids and resins have also been
thermally decomposed. Capillary column GC–MS allowed the identification
of cyclic oligomers along with small amounts of other compounds formed by
eliminations and/or recombinations of pendant groups (13).

4.4. Extraction Schemes

Free alcohol and Si–OR functionality have been determined in silicone fluids
by extraction with distilled water and aqueous HCl, respectively (14). Agitating a 1-g specimen for a few minutes with 5 mL of distilled water extracted the
alcohol with little effect on the Si–OR. Agitating a separate 1-g specimen
several hours with 5 mL of 5% HCl (aq) converted the Si–OR to the
corresponding alcohol, which partitioned into the acid layer. Analysis of the
aqueous layers allowed determination of the free alcohol and Si–OR content
of the fluid. The sensitivity of the procedure was $\sim 5\,ppm$.

Because of their relatively high permeability, many silicone elastomers are
amenable to solvent extraction. Quantitative determinations of residual
ethylene oxide (used for sterilization) and its reaction products, ethylene
glycol and ethylene chlorohydrin have been made utilizing acetone as an
extraction solvent (Chapter 4, Section 5.7). Low molecular weight dimethylsiloxane oligomers have similarly been extracted with pentane (5).

5. GAS CHROMATOGRAPHY INSTRUMENTATION

5.1. The Chromatograph

Purified silane monomers or siloxane oligomers can usually be analyzed by
use of a simple isothermal GC as they typically span a narrow boiling range.
Even for these materials, however, an instrument with programmed temperature capabilities offers advantages because the column temperature can be
changed with minimal disruption (see Section 5.5). For low boiling silanes,
cryogenic cooling capability is a very useful addition.

A programmed temperature GC is virtually necessary for the analysis of unfractionated silanes and siloxanes because it is not unusual for these materials to span a boiling range of well over 200 °C. Complete analysis of such mixtures by isothermal GC would require combining results from multiple analyses, conducted at a variety of temperatures.

5.2. Materials of Construction

All parts of the apparatus contacting the sample must, of course, be non-adsorptive, nonreactive, and stable at operating temperatures. These requirements eliminate some common materials of construction for use with certain classes of silicones. Copper tubing, for example, has been successfully used for columns for the analysis of chlorosilanes but may react with aminofunctional materials. Aluminum can react with chlorosilanes and chlorocarbons and also has a tendency to oxidize, forming adsorptive surfaces. As mentioned earlier, Teflon® reacts with chlorosilanes. Most other plastics greatly limit operating temperatures.

Stainless steel, glass, and fused silica are the most generally useful materials for sampling and inlet systems, columns, and transfer lines. Prior to use, however, stainless steel should be carefully cleaned by flushing with a succession of solvents (e.g., water, acetone, and chlorothene) and blown dry, preferably with clean nitrogen, to remove oils and other contaminants. The reader is also cautioned that fused silica is sometimes "deactivated" with such materials as Carbowax®, which are incompatible with many reactive silanes.

5.3. Inlet Systems

Beyond the possibility of reacting with certain materials of construction (Section 5.2), silicones place no greater or lesser demands on the chromatograph inlet than most other classes of compounds. All common inlet systems, therefore, find application in the analysis of silicones.

A general review of packed column inlets is provided in most basic texts on GC. Jennings (15) provides an excellent review of capillary column inlets. No further discussion is warranted here.

5.4. The Carrier Gas

Nitrogen, argon, hydrogen, and helium have all been used as carrier gases for the GC analysis of silicones. Helium is the most popular choice because it is compatible with the most commonly used detectors, is readily available in the ultrahigh purity required and is intrinsically safe. Some detectors, the electron capture detector, for example, require the use of other carrier gases.

Even high purity carrier gases contain trace levels of oxygen and moisture, both of which can be detrimental to the chromatographic system and sample. Traps containing oxygen scavengers and efficient drying agents should therefore be incorporated into the supply system and routinely maintained. Electrically heated high performance gas purifiers are highly recommended in addition for those sample types, particularly chlorosilanes, which are highly reactive with moisture (Section 2.1). The use of a single carrier gas offers advantages here since the effluent from one purification system can be manifolded to several instruments.

5.5. System Conditioning

The chromatographic system needs to be conditioned prior to use. Conditioning is accomplished by baking out the column for several hours at elevated temperature (usually near the recommended maximum), with carrier gas flowing but with the detector disconnected. The detector is then connected and the bake-out continued until the detector output stabilizes. This process drives moisture, residual solvents, and other volatiles from the system. It also appears to aid in evenly distributing the column liquid phase (16), resulting in improvement in column efficiency as well as better covering active sites. Repetitive injections of sample are then made under normal operating conditions until consistent results are obtained.

If analyses are not performed for a period of time, the system may need to be reconditioned. Presumably, traces of water, which are still present in the carrier gas, accumulate on the column. Maintaining the column oven at elevated temperatures (typically 150 °C) between analyses is beneficial in minimizing this effect.

Attempts to analyze incompatible samples, for example, aminofunctional silanes and chlorosilanes, on the same instrument can permanently destory system conditioning for both sample types. Cleaning the injection system and cutting a few inches from the head of the column is sometimes effective in removing acitve sites or catalytic deposits that have been introduced. If this fails, both the injection port liner and column probably need to be replaced. A good practice when analyzing potentially reactive materials, therefore, is to dedicate a chromatographic system to each specific sample type for the duration of an investigation.

5.6. Sample Introduction

5.6.1. Syringes

Syringes, without question, provide the most widely used method for introducing silicone specimens into chromatographic instruments. Disposable plastic syringes of $\frac{1}{4}$–5 mL capacity are commonly used to introduce gas samples. To prevent contamination of subsequent samples, the gas syringe is normally discarded after a single use. Glass syringes of 10–25 μL capacity are commonly used for liquid samples. In all analyses, it is good practice to clean these syringes immediately after the specimen is injected to prevent inadvertent contamination of successive specimens. With most silanes, however, immediate and thorough cleaning is an absolute necessity since hydrolysis reactions quickly result in plugged needles and frozen plungers. Syringe cleaning is accomplished by removing the plunger, repeatedly flushing both plunger and syringe bore with solvent (e.g., acetone), and blowing dry with clean nitrogen or air.

Plunger-in-needle type small volume syringes are not recommended for use with highly reactive silanes since immediate and thorough cleaning is difficult. Likewise auto samplers, which rely on syringes for sample introduction, have proven unreliable for such samples as the cleaning procedure is inadequate to prevent eventual freezing of the syringe plunger. Both devices have been used successfully with less reactive samples (e.g., alkoxy silanes and siloxanes).

5.6.2. Sampling Valves

Sample injection valves of appropriate size are used in on-line analyzers. In the laboratory they may also be used in conjunction with sample bombs to prevent contact of the sample with the atmosphere. In the latter case, provisions must be made for cleaning and drying the valve and associated transfer lines to prevent the same problems encountered with syringes. Cleaning is accomplished by solvent flushing followed by nitrogen or air purging.

5.6.3. Headspace and Purge-and-Trap Devices

Headspace and purge-and-trap devices are useful for determining volatile components in less volatile or nonvolatile matrices. Both techniques offer the possibility of extremely low levels of detection as preconcentrated volatiles from a large volume of sample can be injected into the chromatograph. Additionally, they circumvent the introduction of nonvolatile sample components into the inlet and column. Both techniques provide approaches for identifying and quantitating trace levels of residual volatiles, including low molecular weight oligomers, solvents and odorous contaminants, in silicone polymers. Stephan, for example, has used a headspace procedure for quantitating parts per million concentrations of solvents in high molecular weight PDMS (17). Moore and Bujanowski (Chapter 4, Section 3.1.2) have also applied a purge and trap procedure to the determination of parts per billion concentrations of octamethylcyclotetrasiloxane in water. Interestingly, the much higher boiling cyclic siloxane was effectively sparged from the much lower boiling matrix.

5.6.4. Pyrolyzers

Pyrolyzers provide a means for analyzing nonvolatile polymers by thermally reducing them to volatile fragments that can be separated by GC. The author (14), for example, investigated the thermal fragmentation of a variety of siloxane polymers by depositing them on the filament of model engine glow plugs. The glow plugs were then threaded into a specially modified injection port and heated by a current from a dry cell battery. Hanneman has also discussed the pyrolysis of siloxanes (18) as well as preceramic polymers (19). To date, however, pyrolysis techniques have found limited use for the routine analyses of silicones, as they tend to be sample, equipment, and condition dependent. Kleinert and Weschler (20) encountered a typical problem with the technique when using pyrolysis, in combination with GC–MS, to determine trace levels of PDMS in environmental samples. Although their procedure was qualitatively reliable, recovery varied as much as an order of magnitude with changes in the viscosity of the polymer. Chemical decomposition schemes, such as those discussed in Chapter 8, Section 3.7, are usually more capable quantitatively and can be tailored to provide selectivity for specific chemical moieties.

6. GAS CHROMATOGRAPHY COLUMNS

6.1. The Solid Support

The common support materials used to prepare packed columns for the analysis of silicones are diatomites of both the pink and white varieties. The general characteristics and applications of these materials are adequately covered in basic chromatography texts and warrant no further discussion here. The major concern in the use of diatomites for analysis of silicones centers on their potential to adsorb or react with some sample types. Acid washing, to remove possible catalysts, followed by neutralization and silanization to remove surface silanol, are usually necessary pretreatments. Such treated supports are commercially available and require no effort on the analyst's part to prepare.

When properly conditioned, as discussed in Section 5.5, the recommended supports should give few problems. Should problems arise, they usually reveal themselves by the appearance of odd shaped or overly broad "de-composition" peaks or plateaus in the chromatogram. There are occasions, however, when this is not the case. Reaction of a sample component may be so rapid as to produce new peaks of normal appearance. Adsorption may also be so specific that one sample component is removed while the re-mainder of the chromatogram appears normal. Periodic analysis of known standards helps to identify these problems.

6.2. Choosing The Liquid Phase

Identifying a suitable liquid phase for a given analysis can be approached by a process of elimination. Of the several hundred commercially available candidates, many can be excluded from consideration simply by reviewing their chemical and physical properties. Some contain functional groups that react with components of the sample, for example, active hydrogen in the case of reactive silanes. Of those remaining, some have maximum or minimum operating temperatures that are inconsistent with sample requirements. It is best to be conservative; liquid phase loading and detector sensitivity affect maximum operating temperature and, in addition, some vendors are overly optimistic.

After eliminating obviously inappropriate phases, it is advisable to choose several, covering a range of selectivity, for evaluation. Some analysts have attempted to categorize selectivity but none have been totally successful. The system most commonly used, however, is that of Rohrschneider (21), as modified by McReynolds (22). This system is based on the Kovats retention indexes (23) of a set of functionally different test compounds. The differences between the indexes on Squalane, which was chosen as the reference phase, and other phases provide an approximation of phase polarity. Tables of these values, referred to as McReynolds constants, are available in most basic chromatography texts and many vendor catalogs.

McReynolds constants are particularly useful for identifying essentially identical phases produced by different manufacturers, thus avoiding re-dundant evaluation. Unfortunately, they give no direct information on the selectivity of a given phase for any particular separation (excluding, of course, the test compounds). As he peruses these tables, the experienced chromato-grapher, as likely as not, relies on intuition, past experience, and the old adage "like dissolves like" as aids in making his final choice of phases to evaluate. He is also alert to the fact that blends of phases produce intermediate selectivity (24), allowing custom design for specific separations.

If one follows through on the above process in selecting liquid phases for the analysis of silicones, it is almost a foregone conclusion that the phases selected will be silicones. In actual fact, the analyst loses little if any capability if his stock is limited to only three types: a poly(dimethylsiloxane), a fluorosilicone, and a phenyl silicone. These choices are easily justified. Phases less polar than PDMS are less thermally stable. Phases more polar than the fluorosilicones are also probably less thermally stable but more important, they likely contain groups that are reactive with the sample types that benefit from their polarity. Finally, the phenylsilicones provide some selectivity for resolving unsaturated compounds. In our experience, however, the phenylsilicones degrade relatively rapidly when used for chlorosilane analyses. This deterioration presumably is a result of cleavage of phenyl–Si bonds by incidentally generated HCl.

6.3. Designing the Separation

Perhaps the most effective means of discussing separation design is through actual examples. One that is particularly relevant is the development of the packed column that probably finds the greatest use for the analysis of chlorosilanes (14).

The most difficult pairs of chlorosilanes to resolve are compounds of the type $RSiCl_3$ and $RMeSiCl_2$, for example, $MeSiCl_3$ (bp 66 °C) and Me_2SiCl_2 (bp 70.5 °C). The early liquid phases used to effect these separations suffered a number of serious drawbacks: some were too volatile to be truly practical, some were hazardous, and others were, in fact, reactive with chlorosilanes (25–30). Luck intervened in 1961 in the form of a newly commercialized fluorosilicone polymer, Dow Corning® QF-10065 (later called QF-1, now FS-1265).

This fluorosilicone (FS) was more chemically and thermally stable than the previous phases and, importantly, contained the unique electronegative 3,3,3-trifluoropropyl group. It was anticipated that this group would interact more strongly with the $RMeSiCl_2$ molecule than the less polar $RSiCl_3$. A column packed with QF-10065 on Chromosorb-P confirmed this hunch. At column temperatures ranging from 55 to 75 °C, $MeSiCl_3$ and Me_2SiCl_2 were well resolved. At 115 °C, this same column was able to resolve $F_3PrSiCl_3$ from $F_3PrMeSiCl$ and at 190 °C it resolved $PhSiCl_3$ from $PhMeSiCl_2$.

Continuing evaluation of the FS column disclosed that $SiCl_4$ (bp 57.6 °C) and the much lower boiling $MeHSiCl_2$ (bp 41.2 °C) could not be adequately resolved. Previous experience, however, had shown that this separation is easily accomplished using a nonpolar liquid phase, such as PDMS, in a column identical to the FS column. A procedure, independently discovered by Hildebrand and Reilley (31), was then employed. A plot similar to Figure 10.2 was generated by placing retention times obtained on the FS column on the right vertical axis and those obtained on an identical nonpolar column, operated under the same conditions, on the left vertical axis. Connecting

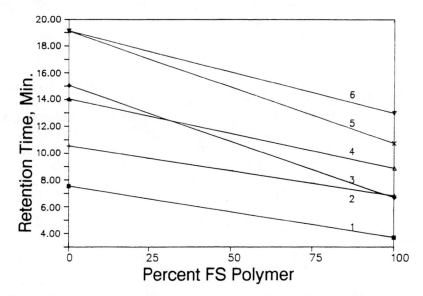

Figure 10.2. Methylchlorosilane retention times as a function of relative column lengths for combination FS/PDMS columns. 1. Me_4Si, 2. $MeHSiCl_2$, 3. $SiCl_4$, 4. Me_3SiCl, 5. $MeSiCl_3$, 6. Me_2SiCl_2.

corresponding points produced a series of lines that allowed the prediction of the relative lengths of the two columns which, when coupled together, would separate the complete sample mixture.

It can be seen in reviewing Figure 10.2 that a ratio of about 70:30 of FS to PDMS appears to be the best compromise for the subject chlorosilanes. It is worth noting that Purnell and Williams (32) have reported that when columns are coupled in this fashion, the apparent length of the upstream column is greater than its true length. This effect, which results from the compressibility of the carrier gas, varies with column head pressure and must be compensated for if the separation is sufficiently critical. Nevertheless, when 70:30 length ratios of FS and PDMS columns are joined, the complete methylchlorosilane mixture is resolved.

Danhaus (33) later used 70:30 blends of FS and PDMS coated supports packed into a single column and obtained essentially identical results to the coupled columns. Codeposits of 70:30 blends of the two liquid phases on the same support (currently the most popular procedure) and experimental copolymers containing a 70:30 ratio of fluorosilicone and dimethylsiloxane units are equally effective.

A plot similar to that described above was also used to design a method for quantitating low levels of $MeSiCl_3$ in purified Me_2SiCl_2. In this case, branched heptane isomers that coelute with MeSiCl$_3$ on FS columns were of concern. Increasing the ratio of PDMS increases the retention times of the heptanes relative to MeSiCl$_3$, as expected, but at the point that they are finally resolved, MeSiCl$_3$ is no longer separated from Me$_2$SiCl$_2$. This problem was solved by coupling the columns by means of a four port column switching valve. An initial separation of the combined MeSiCl$_3$ and heptanes is made from Me$_2$SiCl$_2$ on the FS column. Just prior to the elution of the Me$_2$SiCl$_3$, the valve is switched to direct the effluent from the FS column to the PDMS column. The valve is returned to its original position immediately after the MeSiCl$_3$ enters the PDMS column. The PDMS column easily resolves the MeSiCl$_3$ and heptanes in the absence of the Me$_2$SiCl$_2$. Figure 10.3 presents example chromatograms.

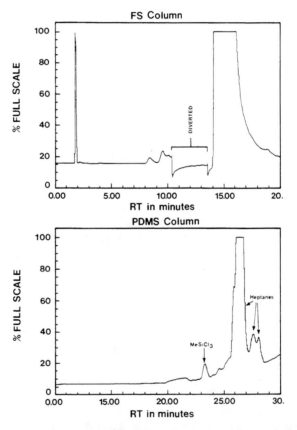

Figure 10.3. Demonstration chromatograms showing column switching technique for separation of MeSiCl$_3$ and heptane isomers.

6.4. Packed Columns for Chlorosilanes

Macroanalysis of the commercially important chlorosilanes may be performed on 3–6 m by $\frac{1}{8}$-in. outside diameter (o.d.) stainless steel columns packed with 20% of the 70:30 FS–PDMS mixed phase on 100:120 mesh, acid washed, dimethyldichlorosilane treated Chromosorb-P or an equivalent support. A thermal conductivity cell is normally used as the detector and the helium carrier gas flow rate is set at approximately 25 mL/min. These columns can be operated over a temperature range from 0 to at least 250 °C.

Complete separation of the methyl chlorosilanes may be obtained on a 6-m column operated isothermally at 60–70 °C. However, a 3.5-m column, programmed at 4 °/min. from 40 to 250 °C, offers adequate separation of the methyl, as well as many higher boiling chlorosilanes. Siloxanes generated during sample handling are also eluted, preventing interference with succeeding analyses. Figure 10.4 presents a typical programmed temperature chromatogram of a mixture of methylchlorosilanes.

The phenylchlorosilanes, through Ph_2SiCl_2, can be separated on a 3.5-m 70:30 column using the temperature program above, but the analysis time may be shortened considerably by increasing the starting temperature to 100 °C. Figure 10.5 presents a chromatogram of a phenylchlorosilane mixture using this latter starting temperature. A slightly higher ratio of FS/PDMS can better resolve $PhSiCl_3$ and $PhMeSiCl_2$.

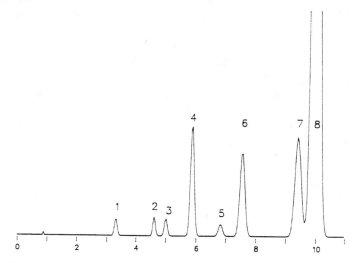

Figure 10.4. Chromatogram of methylchlorosilanes from a 70:30 FS/PDMS packed column.
1. Me_4Si, 2. $HSiCl_3$, 3. Me_2HSiCl, 4. $MeHSiCl_2$, 5. $SiCl_4$, 6. Me_3SiCl, 7. $MeSiCl_3$, 8. Me_2SiCl_2.

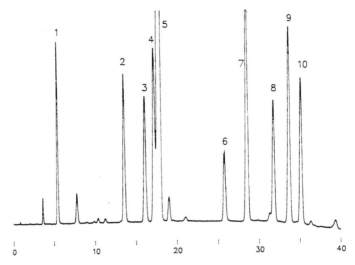

Figure 10.5. Chromatogram of phenylchlorosilanes from a 70:30 FS/PDMS packed column. 1. Chlorobenzene, 2. PhMeHSiCl, 3. PhMe$_2$SiCl, 4. PhSiCl$_3$, 5. PhMeSiCl$_2$, 6. biphenyl, 7. Ph$_2$MeSiH, 8. Ph$_2$HSiCl, 9. Ph$_2$MeSiCl, 10. Ph$_2$SiCl$_2$.

Two additional packed columns complete the battery required for chlorosilanes. Both are identical to the 70:30 columns described above except for the liquid phase: in one case FS is used alone and in the other, PDMS alone. The FS column improves the separation of RSiCl$_3$ and RMeSiCl$_2$ compounds obtained by the 70:30 column and is particularly useful for determining low concentrations of the latter in the former. A column length of 6 m is helpful although shorter columns are adequate in many instances. The PDMS column provides essentially boiling point separation of both chlorosilanes and hydrocarbons, which is required for some specific separations, for example, EtHSiCl$_2$ from Me$_2$SiCl$_2$ and Cl$_2$MeSiSiMe$_2$Cl from Cl$_2$MeSiSiMeCl$_2$. Finally, PDMS and FS columns, when coupled by a switching valve, permit highly specific separations unattainable on any single column. Depending on the particular separation involved, it may be necessary to reverse the order of the columns from that discussed previously.

6.5. Packed Columns for Other Silanes

Peetre (34) discusses the analysis of tetraalkyl and tetraalkoxy silanes on Apiezon and cyanoethylmethylsiloxane liquid phases and correlates their retention behavior with their molecular structure and physical properties. These and many other silanes and silizanes may also be analyzed successfully on 2–6 m by $\frac{1}{8}$-in. o.d. columns packed with 5–10% PDMS phases on 100/120 mesh, acid washed, dimethyldichlorosilane treated Chromosorb-W, or an equivalent support. The advantages of capillary columns, however, coupled with good flame ionization detector response for these compounds, are leading to the rapid elimination of packed columns for such analyses.

6.6. Packed Columns for Siloxanes

Analyses of most volatile siloxanes may be performed on a 2-m by $\frac{1}{8}$-in. o.d. column packed with 10% PDMS fluids or gums on 100/120 mesh, acid washed, dichlorodimethylsilane treated Chromosorb-W or an equivalent support. For very high boiling compounds, the liquid phase loading may be decreased to 1–5% with good effect. Again, capillary columns have virtually replaced packed columns for siloxane analyses.

6.7. Miscellaneous Packed Columns

Three additional packed columns are required if one resorts to the sample pretreatment schemes previously discussed for the analysis of polymers. Hydrogen, for example, is easily resolved from both oxygen and nitrogen by use of a 2–3 m by $\frac{1}{8}$-in. column packed with 80/100 mesh 5-Å molecular sieves. A column temperature of 50–90 °C works well. Because of the low and anomalous thermal conductivity response of hydrogen in helium, argon or nitrogen are used as the carrier gas.

A 2-m by $\frac{1}{8}$-in. column packed with 80/100 mesh Poropak Q, or equivalent, is useful for determining light hydrocarbons generated by alkali cleavage of R–Si. When programmed from 50 to 180 °C at 8 °C/min., this column resolves the important C_1–C_5 saturated and unsaturated hydrocarbons.

A 1–2 m by $\frac{1}{8}$-in. column packed with Poropak T, or equivalent, is useful for determining C_1–C_3 alcohols in aqueous solutions. This column tolerates the 5% aqueous HCl used in the alkoxy extraction procedure better than all others tried to date. A temperature program from 100 to 180 °C at 8 °C/min is suggested. When used in conjunction with a flame ionization detector to minimize response to water, extremely low levels of all three alcohols may be detected (MeOH elutes very near water on this short column).

6.8. Capillary Columns

Capillary columns were used to analyze silicones at least as early as 1960. For example, $MeSiCl_3$ and Me_2SiCl_2 were separated on a stainless steel capillary coated with Apiezon L (14). These compounds eluted in order of molecular weight rather than boiling point. It was not until the commercial introduction of fused silica columns with liquid phases bonded to their walls, however, that capillaries became the columns of choice for many routine silicone analyses.

Fused silica capillaries combine high separation efficiency with the inertness, ruggedness, and flexibility lacking in earlier metal or glass capillaries. Their ruggedness, for example, permits them to be extended directly into the ion source of a mass spectrometer or through a flame ionization detector to the very tip of the flame jet. Such configurations eliminate contact of the sample with any other surface from injection to detection, enhancing the ability to chromatograph labile materials.

The most generally useful fused silica capillary columns for the analysis of lower boiling silanes are 0.25–0.32-mm i.d. by 60 m long with a 1.0–1.5 μm bonded film of PDMS. These columns may be operated from as low as -60 to as high as 300 °C, although extended operation above 250 °C reduces their useful life. Helium carrier gas is used with the column head pressure set to optimize flow rate, usually 15–30 psig, and the splitter flow set to optimize resolution or sensitivity, as required. One major failing of the PDMS capillaries is their inability to separate most of the $RSiCl_3/RMeSiCl_2$ silane pairs, except at low concentrations. Retention indexes on these columns correlate well with boiling points as PDMS is essentially nonpolar. Table 10.1 presents Kovats retention indexes for an assortment of silanes obtained on 60-m PDMS capillary columns, along with their actual and

Table 10.1. Silane Retention Indexes on PDMS Capillary Column

Compound	Retention Index	Estimated bp (°C)	Actual bp (°C)
MeH_2SiCl	406.0	2.0	9.8
Me_3HSi	406.0	2.0	6.7
Me_4Si	476.7	27.9	26.2
Me_2HSiCl	487.0	31.6	35.0
$HSiCl_3$	490.0	32.6	31.9
$MeHSiCl_2$	507.0	38.5	41.2
Me_3SiCl	548.8	52.5	57.6
$ViMe_3Si$	549.3	52.6	54.0
$SiCl_4$	563.0	57.1	57.1
$EtMe_3Si$	574.0	60.6	62.0
Me_2SiCl_2	592.0	66.3	70.1
$MeSiCl_3$	593.6	66.8	66.1
$EtMeHSiCl$	595.0	67.2	69.1
$EtHSiCl_2$	619.0	74.7	75.6

Table 10.1. (*Contd.*)

Compound	Retention Index	Estimated bp (°C)	Actual bp (°C)
Et₂MeHSi	633.0	78.9	74.2
ViMe₂SiCl	642.7	81.8	83.0
AllylMe₃Si	653.0	84.9	86.0
i-PrMe₃Si	655.0	85.5	
n-PrMe₃Si	660.0	87.0	90.0
EtMe₂SiCl	667.2	89.1	90.0
MeViSiCl₂	680.6	93.0	93.8
ViSiCl₃	682.3	93.5	90.7
i-PrHSiCl₂	689.0	95.4	95.0
(ClCH₂)Me₃Si	695.8	97.3	
EtMeSiCl₂	703.7	99.6	101.6
EtSiCl₃	703.7	99.6	98.9
n-PrHSiCl₂	705.0	99.9	
t-BuSiMe₃	706.8	100.4	
i-BuMe₃Si	724.1	105.3	
Me₃SiSiMe₃	730.9	107.2	113.0
AllylMe₂SiCl	736.9	108.8	110.0
i-PrMe₂SiCl	742.0	110.2	
(ClMe)Me₂SiCl	744.6	110.9	
n-PrMe₂SiCl	751.8	112.9	113.0
sec-BuSiMe₃	752.5	113.1	
n-BuMe₃Si	755.6	113.9	
AllylMeSiCl₂	772.5	118.5	119.0
AllylSiCl₃	775.0	119.2	117.5
i-PrSiCl₃	778.0	119.9	120.0
i-PrMeSiCl₂	778.0	119.9	
(ClMe)MeSiCl₂	778.3	120.0	
(ClMe)SiCl₃	784.5	121.7	
t-BuMe₂SiCl	787.0	122.3	124.0
n-PrMeSiCl₂	792.9	123.9	125.0
n-PrSiCl₃	793.8	124.1	124.5
i-BuMe₂SiCl	814.0	129.4	
Et₂SiCl₂	816.0	129.9	130.4
Me₃SiCH₂SiMe₃	817.0	130.2	
Me₃SiSiMe₂Cl	822.0	131.4	129.0
sec-BuMe₂SiCl	840.0	136.0	
(Allyl)₂SiMe₂	847.4	137.9	135.5
n-AmylSiMe₃	850.3	138.6	
n-BuMe₂SiCl	851.9	139.0	
i-BuMeSiCl₂	858.0	140.5	
i-BuSiCl₃	859.0	140.8	142.0

Table 10.1. (*Contd.*)

Compound	Retention Index	Estimated bp (°C)	Actual bp (°C)
$Me_3SiSiMeCl_2$	872.0	144.0	135.0
$(ClCH_2)_2MeSiCl$	877.5	145.4	
$ClMe_2SiSiMe_2Cl$	879.0	145.7	154.0
$PhMeH_2Si$	866.2	142.6	
$n\text{-}BuMeSiCl_2$	891.9	148.9	148.0
$n\text{-}BuSiCl_3$	893.5	149.3	148.0
$Me_3SiCH_2SiMe_2Cl$	910.0	153.3	
$ClMe_2SiSiMeCl_2$	912.0	153.7	
$Cl_2MeSiSiMeCl_2$	926.0	157.1	
$PhMe_2HSi$	932.7	158.7	
$n\text{-}AmylMe_2SiCl$	947.1	162.0	
$n\text{-}HexylSiMe_3$	947.7	162.2	
$Me_3SiCH_2SiMeCl_2$	958.0	164.6	
$AmylMeSiCl_2$	967.8	166.8	
$Me_3SiCH_2SiCl_3$	968.0	166.9	
$AmylMeSiCl_2$	978.5	169.3	
$n\text{-}AmylMeSiCl_2$	990.3	172.0	170.0
$ClMe_2SiCH_2SiMe_2Cl$	993.0	172.6	179.0
$PhSiMe_3$	994.6	172.9	171.0
$n\text{-}AmylSiCl_3$	995.0	173.0	171.0
$PhMeHSiCl$	1014.9	177.5	178.0
$ClMe_2SiCH_2SiMeCl_2$	1029.0	180.6	
$ClMe_2SiCH_2SiCl_3$	1031.0	181.0	
$Cl_3SiCH_2SiCl_3$	1049.0	185.0	185.0
$PhHSiCl_2$	1038.7	182.7	186.0
$Cl_2MeSiCH_2SiMeCl_2$	1054.0	186.1	191.0
$Cl_2MeSiCH_2SiCl_3$	1057.0	186.7	
$ClMe_2SiCH\text{=}CHSiMe_2Cl$	1058.0	186.9	
$PhMe_2SiCl$	1085.5	192.8	195.0
$ClMe_2SiCH_2CH_2SiMe_2Cl$	1086.0	192.9	
$n\text{-}HexylMeSiCl_2$	1091.2	194.0	204.0
$n\text{-}HexylSiCl_3$	1094.1	194.6	192.0
$PhMeSiCl_2$	1127.2	201.6	204.0
$PhSiCl_3$	1130.5	202.2	202.0
$Cl_3SiCH_2CH_2SiCl_3$	1141.0	204.4	202.0
$Cl_2MeSiCH_2CH_2SiMeCl_2$	1161.0	208.5	
Ph_2H_2Si	1328.3	240.6	240.0
Ph_2HSiCl	1506.5	271.7	270.0
$Ph_2MeSiCl$	1661.7	296.6	299.0
Ph_2SiCl_2	1677.5	299.0	304.0

estimated boiling points. The columns were programmed at 2 °C/min from 0 to as high as 300 °C, as appropriate. The reader is cautioned that retention indexes are somewhat temperature dependent and can, therefore, vary as method parameters are changed.

As of this writing, capillary column manufacturers have been unable to produce columns with thick bonded films of fluorosilicone liquid phases. This limitation, coupled with the inability of the bonded fluorosilicone to operate effectively below about 45 °C, results in less than an ideal situation for the analysis of low boiling silanes. On the other hand, the thin film columns that are available permit elution of higher boiling silanes at much lower operating temperatures. Figure 10.6 presents a fluorosilicone capillary column separation of the same phenylchlorosilane mixture anaylzed on the 70:30 packed column (Figure 10.5). This capillary column was 30 m long by 0.25-mm i.d. with a 0.5-μm bonded film of fluorosilicone. The column was programmed at 4 °C/min from 50 to 250 °C. The extended range attainable by this type of analysis when compared to the packed column run is worthy of note.

The most generally useful capillary column for the analysis of higher boiling silanes and siloxanes is 0.25-mm i.d. by 30 m long with a 1-μm film of PDMS bonded to the wall. Typically, the column is temperature programmed at 4 °C/min from 50 to as high as 300 °C. Helium carrier gas is used with the column head pressure set to optimize flow, usually about

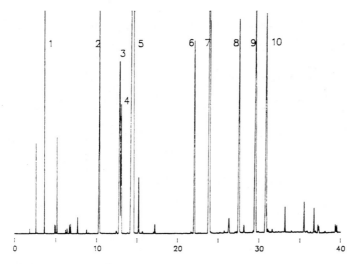

Figure 10.6. Chromatogram of phenylchlorosilanes from an FS capillary column. 1. Chlorobenzene, 2. PhMeHSiCl, 3. PhMe$_2$SiCl, 4. PhSiCl$_3$, 5. PhMeSiCl$_2$, 6. biphenyl, 7. Ph$_2$MeSiH, 8. Ph$_2$HSiCl, 9. Ph$_2$MeSiCl, 10. Ph$_2$SiCl$_2$.

10–15 psig, and splitter flow is adjusted to optimize resolution or sensitivity. Table 10.2 presents Kovats retention indexes for an assortment of siloxanes along with their actual and estimated boiling points. Note the apparent multiple entries for some compounds, indicating the separation of isomeric structures. Again, good correlation of the retention indexes and boiling points is obtained under the conditions used. As mentioned previously, however, these indexes are somewhat temperature dependent and can vary with changes in method parameters.

Table 10.2. Siloxane Retention Indexes on PDMS Capillary Column

Compound	Retention Index	Estimated bp (°C)	Actual bp (°C)
$Me_2HSiOSiMe_2H$	595	67.2	71.5
$(MeHSiO)_3$	675	91.3	93.9
$Me_3SiOSiMe_3$	691	96.0	100.5
$(Me_2SiO)(MeHSiO)_2$	726	105.8	
$(Me_2SiO)_2(MeHSiO)$	778	119.9	
$EtMeHSiOSiEtMeH$	801	126.0	
$Me_3SiOSiCl_3$	805	127.0	
$(Me_2SiO)_3$	824	132.0	135.1
$(MeHSiO)_4$	834	134.5	134.9
$ClMe_2SiOSiMe_2Cl$	836	135.0	138.0
$ViMe_2SiOSiMe_2Vi$	860	141.0	
$ClMe_2SiOSiMeCl_2$	862	141.5	142.0
$(Me_2SiO)(MeHSiO)_3$	881	146.2	69 at 50 mm
$Cl_2MeSiOSiMeCl_2$	882	146.5	144.0
$(Me_2SiO)_2(MeViSiO)$	885	147.2	
$Me_3SiO(Me_2SiO)SiMe_3$	889	148.3	152.5
$(Me_2SiO)_2(MeClSiO)$	895	149.6	
$EtMe_2SiOSiMe_2Et$	898	150.4	
$(Me_2SiO)_2(MeHSiO)_2$	922	156.1	
$(Me_2SiO)_2(MeHSiO)_2$	924	156.6	77.5 at 50 mm
$(Me_2SiO)_2(EtMeSiO)$	924	156.6	
$(Me_2SiO)_2(MeHSiO)_2$	925	156.8	
$(Me_2SiO)_3(MeHSiO)$	962	165.5	165.0
$(MeHSiO)_5$	972	167.8	168.9
$EtMeHSiO(Me_2SiO)SiEtMeH$	999	173.9	
$(Me_2SiO)_4$	1000	174.1	175.4
$(Me_2SiO)(MeHSiO)_4$	1013	177.1	
$ClMe_2SiO(Me_2SiO)SiMe_2Cl$	1036	182.1	184.0
$(Me_2SiO)_2(MeHSiO)_3$	1050	185.2	
$ViMe_2SiO(Me_2SiO)SiMe_2Vi$	1051	185.4	
$(Me_2SiO)_2(MeHSiO)_3$	1053	185.9	

Table 10.2. (*Contd.*)

Compound	Retention Index	Estimated bp (°C)	Actual bp (°C)
$(Me_3SiO)_3SiMe$	1056	186.5	191.0
$(Me_2SiO)_3(EtHSiO)$	1062	187.8	
$(Me_2SiO)_3(MeClSiO)$	1071	189.7	
$(Me_2SiO)_3(MeViSiO)$	1073	190.2	
$Me_3SiO(Me_2SiO)_2SiMe_3$	1074	190.4	194.4
$(Me_2SiO)_3(MeHSiO)_2$	1088	193.4	
$(Me_2SiO)_3(MeHSiO)_2$	1090	193.8	
$(Me_2SiO)_3(EtMeSiO)$	1093	194.4	
$EtMe_2SiO(Me_2SiO)SiMe_2Et$	1094	194.6	
$(MeHSiO)_6$	1108	197.6	93 at 21 mm
$(Me_2SiO)_4(MeHSiO)$	1127	201.5	
$(Me_2SiO)(MeHSiO)_5$	1146	205.4	
$(Me_2SiO)_5$	1161	208.5	211.0
$EtMeHSiO(Me_2SiO)_2SiEtMeH$	1184	213.1	
$(Me_2SiO)_2(MeHSiO)_4$	1186	213.5	
$(Me_3SiO)_4Si$	1194	215.1	222.0
$ClMe_2SiO(Me_2SiO)_2SiMe_2Cl$	1221	220.4	222.0
$(Me_2SiO)_3(MeHSiO)_3$	1223	220.8	
$(Me_2SiO)_4(EtHSiO)$	1227	221.5	
$ViMe_2SiO(Me_2SiO)_2SiMe_2Vi$	1231	222.3	
$(Me_2SiO)_4(MeViSiO)$	1232	222.5	
$(Me_3SiO)_3(Me_2SiO)SiMe$	1233	222.7	
$(MeHSiO)_7$	1236	223.3	
$Me_3SiO(Me_2SiO)_3SiMe_3$	1251	226.2	229.9
$(Me_2SiO)_4(MeSiO_{3/2})_2$	1252	226.4	
$(Me_2SiO)_4(EtMeSiO)$	1255	226.9	
$(Me_2SiO)_4(MeHSiO)_2$	1260	227.9	
$EtMe_2SiO(Me_2SiO)_2SiMe_2Et$	1274	230.5	
$(Me_2SiO)(MeHSiO)_6$	1276	230.9	
$(Me_2SiO)_4(MeSiO_{3/2})_2$	1281	231.9	
$(Me_2SiO)_5(MeHSiO)$	1297	234.9	
$(Me_2SiO)_6$	1339	242.7	245.0
$EtMeHSiO(Me_2SiO)_3SiEtMeH$	1363	246.9	
$ClMe_2SiO(Me_2SiO)_3SiMe_2Cl$	1402	253.9	138 at 20 mm
$(Me_2SiO)_5(EtHSiO)$	1406	254.6	
$ViMe_2SiO(Me_2SiO)_3SiMe_2Vi$	1411	255.4	
$Me_3SiO(Me_2SiO)_4SiMe_3$	1424	257.6	258.6
$(Me_2SiO)_5(MeViSiO)$	1429	258.6	
$(Me_2SiO)_5(EtMeSiO)$	1450	262.2	
$EtMe_2SiO(Me_2SiO)_3SiMe_2Et$	1457	263.4	
$(Me_2SiO)_7$	1518	273.5	275.6

279

Table 10.2. (*Contd.*)

Compound	Retention Index	Estimated bp (°C)	Actual bp (°C)
EtMeHSiO(Me$_2$SiO)$_4$SiEtMeH	1538	276.9	
ClMe$_2$SiO(Me$_2$SiO)$_4$SiMe$_2$Cl	1577	283.2	161 at 20 mm
ViMe$_2$SiO(Me$_2$SiO)$_4$SiMe$_2$Vi	1584	284.4	
Me$_3$SiO(Me$_2$SiO)$_5$SiMe$_3$	1591	285.5	286.8
EtMe$_2$SiO(Me$_2$SiO)$_4$SiMe$_2$Et	1630	291.6	
(Me$_2$SiO)$_8$	1691	301.1	303.2
Me$_3$SiO(Me$_2$SiO)$_6$SiMe$_3$	1755	310.7	310.4
(Me$_2$SiO)$_9$	1855	325.1	325.5
Me$_3$SiO(Me$_2$SiO)$_7$SiMe$_3$	1936	336.3	
(Me$_2$SiO)$_{10}$	2031	349.0	
Me$_3$SiO(Me$_2$SiO)$_8$SiMe$_3$	2100	357.9	

Figure 10.7 presents a portion of a chromatogram of a dimethylcyclosiloxane mixture spiked with approximately 50 ppm each of linears and cocyclics containing Me$_2$SiO, MeHSiO, EtHSiO, MeViSiO, and EtMeSiO moieties. This chromatogram clearly demonstrates the capability of the 30-m PDMS capillary column for detecting detrimental impurities even at these low

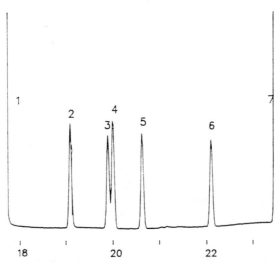

Figure 10.7. Chromatogram showing separation of impurities from dimethylcyclosiloxanes on PDMS capillary column. 1. D$_4$, 2. ⌐(Me$_2$SiO)$_3$(EtHSiO)⌐, 3. ⌐(Me$_2$SiO)$_3$(MeViSiO)⌐, 4. MD$_2$M, 5. ⌐(Me$_2$SiO)$_3$(EtMeSiO)⌐, 6. ⌐(Me$_2$SiO)$_4$(MeHSiO)⌐, 7. D$_5$.

concentration levels. The same column operated under similar conditions has also been used for the analysis of the lower molecular weight methylphenyl-cyclosiloxanes, which, because of the dissimilar R groups, form isomeric structures analogous to the substituted cyclic hydrocarbons. Both cyclic trimer and all four cyclic tetramer isomers can be resolved. The diastereomers of 1,3-dichloro-1,3-dimethyl-1,3-diphenyldisiloxane, formed upon partial hydrolysis of Ph_2SiCl_2, are also resolved.

When new, fused silica columns may even elute low molecular weight silanols as sharp well-defined peaks. This ability can be lost abruptly and without warning, however. The safest procedure for analyzing reactive silanols, therefore, is by use of the derivatization procedure discussed earlier.

7. GAS CHROMATOGRAPHY DETECTORS

7.1. General Comments

The final, but not the least important, element of the chromatograph is the detector, whose purpose is to sense the sample components as they exit the column. Depending on the detector used, quantitative, qualitative, or both types of data may be obtained. Of the dozen or so types available, the thermal conductivity (T/C) and flame ionization (FID) detectors are most widely used for the quantitative analysis of silicones. In addition to these, the mass spectrometer (MS), Fourier transform infrared spectrophotometer (FT IR), and atomic emission spectrophotometers (AES), which utilize microwave induced plasma excitation sources, are able to provide qualitative information as well as functioning as tuneable and highly specific quantitative detectors.

It is generally recognized that, regardless of type, detectors do not respond equally to all compounds. Correction factors must, therefore, be applied in order to obtain truly quantitative results from detector response. These factors are usually determined relative to a reference compound as follows:

because $\quad A_x F_x/A_y F_y = C_x/C_y$ $\qquad\qquad$ (10.10)

then $\qquad\quad F_x = C_x A_y F_y/C_y A_x$ $\qquad\qquad$ (10.11)

where $\qquad x$ = compound of interest
$\qquad\qquad\quad y$ = reference compound
$\qquad\qquad\quad A_x$ and A_y = peak areas of x and y
$\qquad\qquad\quad F_x$ and F_y = response factors for x and y
$\qquad\qquad\quad C_x$ and C_y = concentrations of x and y

Caution must be exercised when determining response factors, for the measured output signal from the detector (i.e., peak area) reflects several possible sources of error. Among these are the following:

1. *Accuracy of the Standard.* The accuracy of the factor can be no better than the purity of the materials used to determine it and the care with which the standard mixtures are prepared. If these materials are reactive, it must be assured that they do not change during preparation, handling, or analysis.

2. *Inertness of the Chromatographic System.* Low or inconsistent response can result from reaction or adsorption of sample components in the chromatograph. Purification of the carrier gas and proper system conditioning, discussed previously, are necessary precautions.

3. *Splitter Linearity.* For those methods using capillary columns, the sample is generally introduced by means of an inlet splitter that may differentiate between high and low boiling sample components. Analysis of a similar boiling range standard mixture with known response factors can be used both to identify this problem and determine correction factors for adjustment. Proper splitter design, injection temperature and split ratio will usually eliminate, or at least minimize, this effect.

4. *Detector Linearity.* All detectors produce response curves that are linear only over a limited range of concentration. The chromatographer must determine, through the use of appropriate standards, the bounds of linear response and design the method to operate within them.

5. *Accuracy of Integration.* Modern electronic integrators are capable of excellent accuracy. If possible, one should design the method to avoid those things that detract from this capability. Of particular concern are noisy and drifting base lines, incompletely resolved peaks, small peaks on the tail of large peaks, low broad peaks late in isothermal runs and inadequate integrator sampling rates for sharp peaks early in the run.

7.2. The Thermal Conductivity Detector

The T/C detector is most commonly used in conjunction with packed columns. Properly designed low volume T/C cells are compatible with larger diameter capillary columns, however, and have found some use in this arena. For example, $SiCl_4$, $HSiCl_3$, and other silicones with few carbon atoms in the molecule provide good response with these T/C detectors while they respond poorly, if at all, with the FID detectors normally used with capillary columns.

Dearlove et al. (35) determined the T/C response of a variety of alkoxy silanes and reported a linear relationship between molar response and

Table 10.3. Relative Molar Response Factor Contributions for Silicon Substituents

Group	RMR Contribution	Group	RMR
Hydrogen	18	Trifluoropropyl	86
Methyl	30	Phenyl	81
Ethyl	47	Chlorine	33
Vinyl	42	Fluorine	23
Propyl	60	Acetoxy	57
Allyl	57	Methoxy	39
n-Butyl	75	Ethoxy	51
n-Pentyl	86	Disilane	0
n-Hexyl	100	Disiloxane	− 12

molecular weight. Rosie and co-workers (36–39) measured the molar response of a large number of organic and inorganic compounds using several T/C detectors and various conditions. Benzene was used as the reference compound and assigned a relative molar response (RMR) of 100. It was concluded that RMR is a physical property of a compound related to the collision diameter of the molecule. Barnes (40) applied this concept to silicones and developed a procedure for estimating their RMR by summing contributions from the groups attached to the silicon atoms. Excellent agreement between estimated and observed response was reported. A portion of this data is presented in Table 10.3.

An example of the use of this data for determining the RMR of Me_2SiCl_2 follows:

$$2 \text{ methyl groups } = 2 \times 30 = \quad 60 \qquad (10.12)$$

$$2 \text{ chlorine atoms } = 2 \times 33 = \underline{\quad 66 \quad}$$

$$RMR = 126 \qquad (10.13)$$

As most analysts report composition as weight rather than mole percent, however, RMR must be converted to weight response factors for practical use. This calculation is accomplished as follows:

$$F_x = (M_x/RMR_x)/(M_b/RMR_b) \qquad (10.14)$$

$$= (M_x \, RMR_b)/(M_b \, RMR_x)$$

$$= (M_x \, 100)/(78.1 \, RMR_x) = 1.28 \, M_x/RMR_x$$

for Me_2SiCl_2 $\quad F_x = 1.28 \, (129/126) = 1.31 \qquad (10.15)$

Table 10.4. Weight Response Factors for Siloxanes

Compound[a]	Factor	Compound[a]	Factor
D_3	1.46	MM	1.17
D_4	1.56	MDM	1.32
D_5	1.65	MD_2M	1.45
D_6	1.73	MD_3M	1.58
D_7	1.82	MD_4M	1.69
D_8	1.89	MD_5M	1.80
D_9	1.97	MD_6M	1.91
D_{10}	2.06	MD_7M	2.01

[a] $M = Me_3SiO_{1/2}$.
$D = Me_2SiO$.

where F_x = weight response factor for compound x
M_x and M_b = molecular weights of x and benzene (b), respectively
RMR_x and RMR_b = relative molar responses for x and benzene (b), respectively ($RMR_b = 100$)

Barne's procedure becomes somewhat more complex in estimating RMR values for polysiloxanes, requiring corrections to be applied depending on the size of the cyclic or linear chain. Weight response factors that have been observed for some cyclic and linear methyl siloxanes, however, are shown in Table 10.4.

These factors appear to fall into a consistent pattern, suggesting the possibility of extrapolating factors for larger oigomers. Kowalski et al. (41), however, reported difficulty when attempting such extrapolations, although the reason was not apparent.

7.3. The Flame Ionization Detector

Much to-do has been made about the behavior of the FID toward silicones. One of their products of combustion is SiO_2, which has a tendency to deposit within the detector. Some designs may be more susceptible to this problem than others. In our experience, however, deposits of SiO_2 cause no apparent detrimental effect so long as they do not bridge between the electrodes or between the electrodes and the instrument chassis. When this happens, wiping or blowing the deposits away (use necessary precautions) will return the detector to normal operation. The frequency of cleaning is directly related to the quantity of Si fed to the flame. Literally hundreds of capillary column analyses, using typical sample sizes and split ratios, have been made between cleanings. With packed column and/or large sample sizes, major peaks, which are beyond the range of quantitation, may be vented away from the detector.

The FID, as other detectors, responds linearly to concentration only over a given range, which may vary with detector design. As concentration is increased, the detector eventually becomes saturated. Many have reported, but have failed to satisfactorily explain, the unusual behavior of silicones at the point of saturation. Peaks become flat topped or M shaped, leading to possible misinterpretation of the chromatogram. This effect is of little practical consequence if the method is designed to operate within the linear range of the detector. In the absence of suitable standards, serial dilution with a noninterfering solvent may be used to define the linear range for specific components in a given sample.

As commonly used, the FID is basically a "carbon counter," responding to the carbon atoms in a molecule. Ongkiehong (42) and Kaiser (43) reported that FID response factors could be closely approximated by dividing the molecular weight of a compound by the weight of carbon it contained.

$$F_x = M_x/(12n_x) \qquad (10.16)$$

where F_x = the weight response factor for compound x
         M_x = the molecular weight of x
         n_x = the number of carbon atoms per molecule of x

Sternberg and co-workers (44) pointed out that not all carbon atoms respond equally. A single halogen atom attached to a carbon was reported to have negligible effect on response. The carbonyl group in aldehydes and ketones and the carboxyl group in acids and esters, however, were generally agreed to give no response. The C–OH group in alcohols and the C–O–C groups in ethers give diminished response. Factors for these compounds reportedly could be estimated as follows:

aldehydes, ketones, acids, and esters $\qquad F_x = \dfrac{M_x}{12(n_x - 1)} \qquad (10.17)$

ethers $\qquad F_x = \dfrac{M_x}{12(n_x - \text{number of ether linkages})} \qquad (10.18)$

primary alcohols $\qquad F_x = \dfrac{M_x}{12(n_x - 0.5)} \qquad (10.19)$

secondary alcohols

$$F_x = \frac{M_x}{12(n_x - 0.75)}$$ (10.20)

tertiary alcohols

$$F_x = \frac{M_x}{12(n_x - 0.25)}$$ (10.21)

If standards are not available, response factors for silicones estimated by the above procedure are sufficiently accurate for many applications. Some examples are presented in Table 10.5.

The estimated factors for the linear siloxanes are amazingly close to the observed factors, whereas the cyclic siloxanes appear to give slightly higher responses (lower factors) than expected. The somewhat larger differences observed for the silanes may be partially from experimental error as these

Table 10.5. FID Response Factors for Silicones

Compound	Estimated Factor	Observed Factor	% Difference
MM	2.25	2.25	0.0
MDM	2.46	2.47	0.4
MD_2M	2.58	2.60	0.8
MD_3M	2.67	2.69	0.8
MD_4M	2.73	2.70	− 1.1
MD_5M	2.77	2.79	0.7
D_3	3.08	2.90	− 5.8
D_4	3.08	2.87	− 6.8
D_5	3.08	2.98	− 3.2
D_6	3.08	3.01	− 2.3
D_7	3.08	3.16	2.6
$(Me_2SiO)_3(MeViSiO)$	2.85	2.83	− 0.7
$(Me_2SiO)_3(EtMeSiO)$	2.87	2.74	− 4.5
$(Me_2SiO)_3(MeHSiO)$	3.36	3.13	− 6.8
$(Me_2SiO)_4(MeHSiO)$	3.30	3.19	− 3.3
Me_4Si	1.83	1.56	− 14.8
Me_2HSiCl	3.94	3.78	− 4.1
$MeHSiCl_2$	9.58	10.13	5.7
Me_3SiCl	3.01	2.53	− 16.0
$EtHSiCl_2$	5.38	6.04	12.3
$EtMe_2SiCl$	2.55	2.28	− 10.6
$ViMe_2SiCl$	2.51	2.40	− 4.4
$ViMeSiCl_2$	3.92	3.51	− 10.5

were determined at concentration levels below 0.1% where weighing and sample handling errors become more significant.

7.4. Hyphenated Methods

Gas chromatography is principally a quantitative technique. In spite of its high resolving power, the only qualitative data that result from an analysis are retention times, hardly sufficient for unequivocable identification. Spectroscopic techniques, on the other hand, yield rich qualitative data sets but provide only composite information on mixtures. Hirschfeld (45) discusses the theoretical advantages of combining GC and spectroscopic techniques into a single, so-called, hyphenated method. The result is synergistic.

7.4.1.   GC–MS

Mass spectra add a second dimension to retention times for purposes of identification. Figure 10.8, for example, presents a portion of the total ion chromatogram of the same cyclosiloxane standard presented in Figure 10.7, run under essentially identical conditions. Figures 10.9 and 10.10 are mass

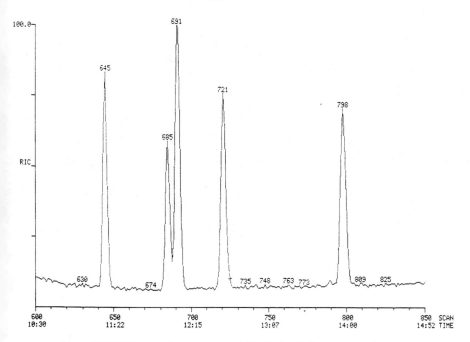

Figure 10.8. Total ion chromatogram of dimethylcyclosiloxane impurities.

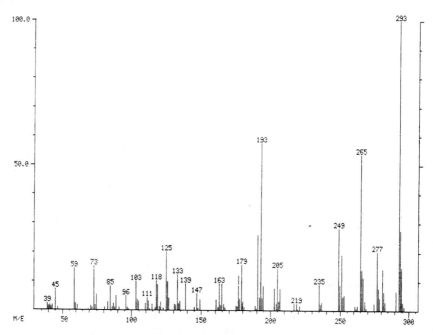

Figure 10.9. Mass spectrum of peak 685 (heptamethylvinylcyclotetrasiloxane).

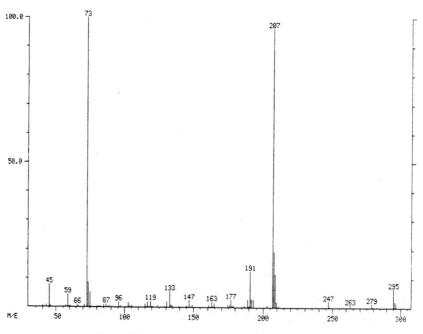

Figure 10.10. Mass spectrum of peak 691 (MD$_2$M).

288

spectra of the peaks that elute at scan numbers 685 and 691. These spectra are characteristic of the cocyclic tetramer, $(Me_2SiO)_3(MeViSiO)$, and the linear tetramer, $Me_3SiO(Me_2SiO)_2SiMe_3$, respectively. The combination of the spectra and the position at which these compounds elute makes these identifications unambiguous.

In addition to its qualitative capability, the mass spectrometer can function as a high sensitivity, selective, and tunable quantitative detector. Single-ions or multiple ions specific to a given compound or group of compounds can be monitored exclusively. Dudding and Sorenson (46), for example, discuss the use of capillary column GC/MS for the determination of trace levels of the potential cross-linker $EtHSiCl_2$ in Me_2SiCl_2. Sensitivity appeared to be similar to that obtained with an FID detector but the linear range of the MS was much more limited.

7.4.2. GC–FT IR

Gas chromatography–IR (Chapter 11, Section 2) is especially useful as a complement to GC–MS as it can often identify specific groups present in a silicone, thus aiding in the interpretation of the mass spectrum. In addition, isomeric silicones generally give identical mass spectra but may well give characteristically different IR spectra. Like GC–MS, GC–IR can also function as a tunable and highly specific quantitative detector, as compounds containing specific functional groups may be monitored to the exclusion of all others.

7.4.3. GC–AES

The final hyphenated technique that has demonstrated utility for the analysis of silicones is GC–AES. This combination, at least theoretically, offers the capability for doing elemental analyses on compounds as they elute from the chromatograph (47). Perhaps a more important use of AES is as a highly selective quantitative detector. For example, by tuning for silicon and chlorine, one can detect chlorosilanes in the presence of interfering hydrocarbons. Dual calibration for both elements in this case aids in assuring that only the anticipated analyte is being monitored.

8. SUPERCRITICAL FLUID CHROMATOGRAPHY

8.1. General Comments

Supercritical fluid chromatography (SFC) is a powerful separation technique. It can accommodate samples that boil higher or are more thermally fragile than can be analyzed by GC, and yet it gives nearly the same resolution and selectivity. Although the application niches for SFC are still being defined, it was shown as early as 1975 that low molecular weight silicone polymers are readily analyzed (48). Chester (49) suggested that "Silicones are almost anomalously well behaved in SFC," and that a properly chosen and purified silicone could be used as a marker in a retention index system for SFC. Interest in the technique was heightened by the adaptation of capillary columns to SFC by Novotny et al. in 1981 (50). A considerable amount of developmental work with both capillary and packed column SFC has led to commercial instrumentation that is suitable for analyzing silicones.

For analysis of PDMS fluids, the resolution and selectivity of capillary SFC is similar to that of capillary GC, except that upper limits for elutable oligomers are greatly extended. For the more volatile components, GC is preferable because of its higher selectivity and efficiency.

Instrumentation for capillary SFC is similar to that for capillary GC, with a few exceptions (51). The injection device is usually a splitter utilizing a fixed volume HPLC loop injector. The columns themselves are usually $50-100\,\mu m$ i.d., and a device must be placed at the end of the column to restrict the flow of supercritical fluid and diminish the pressure drop across the column. The most commonly used restrictor types are integral (52), tapered (53), and porous frit. All of these restrictor types work well for silicones. These restrictors are designed to eliminate spiking in the FID, a phenomenon exhibited by straight-walled capillaries used in early work with capillary SFC.

Both GC and HPLC detectors have been shown to be useful in detecting analytes. The restrictor is placed before the GC detector, allowing the supercritical fluid to be expanded to a gas. Flame ionization detectors and mass spectrometers have proven to be the most useful GC detectors in SFC to date. When HPLC detectors are used, the restrictor is usually placed after the detector, allowing the analyte to be detected in either supercritical or liquid phase, depending on the temperature. The UV–vis and fluorescence detectors have proven to be most useful.

8.2. Applications

For nonpolar silicones, either packed or capillary column SFC is applicable. The inherent inertness of the capillary column lends itself to the analysis of silanol ended polymers. To analyze such polar species on a packed column, one must add mobile phase modifiers such as alcohols or acetonitrile to inhibit interaction with residual silanols on the packing surface.

The physical properties of a supercritical fluid lie between those of a liquid and a gas. The density and solvating power approach that of liquids, but the diffusivity and viscosity approach that of gases. The retention characteristics of an analyte in SFC thus exhibit a dual nature, depending on the temperature and pressure of the mobile phase. An increase in temperature affects retention of the analyte in two opposing ways. The volatility of the analyte is increased, decreasing the capacity factor k', and thereby the retention time. The density, and therefore the solvating power of the supercritical fluid, decreases, causing an increase in the capacity factor. The latter effect can be overcome by increasing the pressure of the supercritical fluid. Figure 10.11 shows the relationship between log k' and reciprocal temperature (54). For thermally stable, volatile analytes such as silicones, it is preferable to stay in the GC-like region by increasing the temperature and the pressure simultaneously (55). Figure 10.12 is a chromatogram of PDMS, obtained using a combined temperature–density program. Excellent separation of oligomers is observed, with oligomers around DP 70 resolved.

All nonionic pendant groups except primary amines are amenable to SFC analysis. However, any pendant group on the silicon that reduces the volatility of the oligomer increases the retention time relative to PDMS. Some aspects of specific pendant groups are

Methyl   Compatible with CO_2, good FID response.
Phenyl   Increases retention dramatically over methyl, excellent FID response.

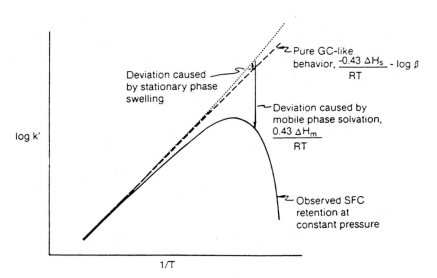

Figure 10.11. Model of observed SFC retention at constant pressure. k' is the chromatographic capacity factor, $(t_r - t_0)/t_0$. [From (54)]. Courtesy *J. High Resolut. Chrom. Chrom. Commun.*

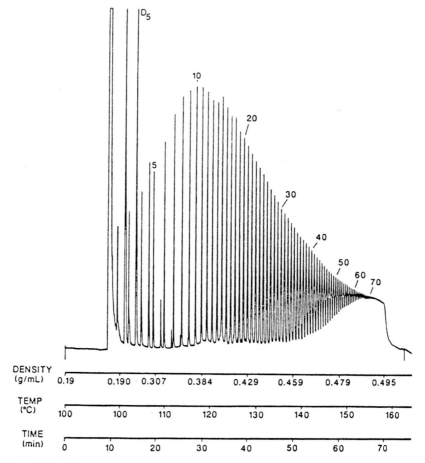

Figure 10.12. SFC chromatogram of PDMS fluid containing a small amount of cyclic siloxanes. [From (55).] Courtesy *J. High Resolut. Chrom. Chrom. Commun.*

H	No FID response, compatible with CO_2.
SiOH	Quite polar, and exhibits some tailing. It is critical when analyzing free silanols to have very dry CO_2, and a well-deactivated capillary column. Silanols can be derivatized with trimethylsilyl groups in order to improve the chromatography.
Trifluoropropyl	Adequate FID response, compatible with CO_2.
Alkoxy	Less polar than SiOH, but somewhat hydrolyzable. CO_2 must be dry.

| Cl | Very hydrolyzable, but compatible with CO_2. No FID response. Dry CO_2 essential. |
| Amine or diamine | Not recommended with CO_2; reacts to form solid material in injector and column. |

Suzuki (56) showed that PDMS with a vinyl in the end group can be analyzed by SFC (Figure 10.13), and that molecular weight distribution (MWD) estimates may be made. This analysis requires that the complete distribution be eluted, and that peak assignments can be made. Poly(dimethylsiloxane) can be eluted at least to 70 monomer units, and peak assignments can be made on the early eluants by the injection of known species. Counting out to the later peaks completes the assignments. The concentration of each oligomer is calculated using area normalization or internal standardization, and the molecular weight of each oligomer can be calculated from its empirical formula. These values are plugged into MWD calculations used in GPC, substituting peak data from the SFC for the slice data from the GPC. This is an absolute method of determination of MWD, as long as there is no sample discrimination from injector or chromatograph or inaccurate peak assignment. Other data available from this analysis include: separate assignments

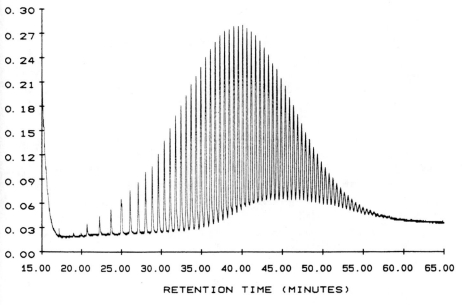

Figure 10.13. Chromatogram of $Me_2ViSiO(SiMe_2O)_xH$ (52). Reproduced from *Polymer*, **30**, 333 (1989) by permission of Butterworth and Co. (Publishers) Ltd. ©.

based on end groups or configuration (such as cyclics and linears) and calculation of end-group concentration. These calculations are easily handled using an electronic spreadsheet.

9. GEL PERMEATION CHROMATOGRAPHY

9.1. General

Gel permeation chromatography (GPC) is commonly used for the analysis of polymers, including silicones. The technique is based on the separation of polymer molecules according to their size in solution, and the system can be calibrated to yield an estimated MWD of a sample. The only basic requirements are that the sample is soluble and stable in a suitable solvent, and that the sample components can be detected by some means in the eluent. Comparison of MWD values or actual distribution curves can provide valuable insight into experimental or process samples. Gel permeation chromatography is invaluable in the optimization of degree of polymerization and in polymer synthesis studies. Performance problems and competitive surveillance often utilize GPC analysis. The theory and practice of GPC are well covered in the monography by Yau et al. (57). Applications to silicone polymer technology are discussed in Chapter 7, Section 3.

9.2. Calibration

Three methods of calibrating the GPC system may be used: narrow MWD standards, broad MWD standards, and use of molecular weight sensitive detectors.

9.2.1. Calibration Using Narrow MWD Standards

The system is calibrated by injecting dilute solutions of narrow dispersity standards of known molecular weight. The retention volume of each standard is plotted against the log MW of the standard. Retention time is often used instead of retention volume when constant flow pumps are used. The resulting data are fitted to an equation for a curve, generally point to point, linear least squares, or cubic.

The narrow standard calibration method is preferred because the entire separation range of the column set is calibrated. Narrow dispersity standards of many different polymer types are commercially available. Narrow molecular weight silicone standards are not, at this writing, commercially available. Standards can be prepared by fractioning a wide dispersity polymer into narrow fractions, and characterizing each fraction by independent methods such as light scattering and osmometry.

Kamide (58) discusses the theoretical and practical aspects of precipitating narrow dispersity fractions of a polymer from solution.

Dodgson et al. (59) used a preparative GPC instrument to fractionate broad dispersity PDMS.

Absolute MWD values are not always necessary, as in a processing study. Relative MWD values can be used to detect changes in the polymer as processing parameters are varied. Calibration in this case is still needed, however, to correct for instrumental instability. This can be done by standardizing with commercially available polymer standards such as narrow molecular weight polystyrene. The MWD values calculated are expressed as relative to polystyrene. Another option is to correct the sample MWD by employing the universal calibration technique. The method requires the measurement of the Mark–Houwink coefficients (57) for both the polystyrene and the sample polymer, using the same GPC solvent and temperature. Once this relationship is established, calibration is done using the polystyrene as secondary standards.

9.2.2. Calibration Using Broad MWD Standards

A standard with a broad MWD that spans the range of interest of molecular weights is characterized by independent means such as light scattering for weight average molecular weights ($\bar{M}_w$), and osmometry for number average molecular weights ($\bar{M}_n$). The standard is then chromatographed and an iterative process used to find a calibration curve that gives the correct $\bar{M}_n$ and $\bar{M}_w$ for the standard. This calibration curve is useful over the same region as covered by the standard, but extrapolation past this region is not recommended. This method allows the GPC to be calibrated for different polymers without the need to obtain narrow MWD fractions. It has been critically reviewed by Kubin (60).

9.2.3. Calibration Using MWD-Sensitive Detectors

Calibration for copolymers and branched polymers presents a special problem. The MWD of the copolymer can be measured accurately only by employing a detector system utilizing both concentration sensitive and molecular weight sensitive detectors such as viscosity and light scattering detectors. Dumelow et al. (61) showed that the block copolymers consisting of polystyrene and poly(dimethylsiloxane) could be analyzed for compositional heterogeneity at each elution volume using LALLS (low angle laser light scattering) detection.

Another method for analyzing copolymers utilizes dual detectors. One detector gives a specific response for one monomer unit, while the other is either specific to the other monomer unit, or gives a nonspecific response. Often one can use a spectroscopic (IR or UV–vis) detector set at two different wavelengths. Comparison of the signals gives an indication of relative distribution of the comonomers in the copolymer. Kohn (62) used this technique to characterize silicone co-polymers by using RI, IR, and UV–vis detectors.

The universal calibration technique, first described by Benoit et al. (63), suggests that a plot of log $[\eta]$ MW versus retention volume is independent of polymer type for a given column set. Thus a detector capable of measuring intrinsic viscosity $[\eta]$ is useful in calibrating the column set. Haney (64) describes the use of a commercially available viscosity detector.

9.3. Calculations

To convert the raw chromatogram to MWD, one must analyze the chromatogram by an appropriate GPC algorithm. This process includes setting a baseline to allow the background to be subtracted and setting summation points that delineate the sample from any extraneous peaks such as solvent impurities or additives. The GPC peak is then divided into time slices, usually correlated to the sampling rate of the A-to-D converter used to collect data from the detector. The retention time of each slice is used to calculate a corresponding molecular weight from the calibration curve generated earlier. The height of the slice (detector response) is then used as a measure of the concentration of the polymer at that molecular weight. The MWD values are calculated according to the following equations:

$$\bar{M}_w = \frac{\sum_{i=1}^{N} (h_i M_i)}{\sum_{i=1}^{N} h_i} \tag{10.22}$$

$$\bar{M}_n = \frac{\sum_{i=1}^{N} h_i}{\sum_{i=1}^{N} (h_i / M_i)} \tag{10.23}$$

Here h_i is height of the ith area slice of molecular weight M_i.

9.4. Application to Silicone Polymers

Figure 10.14 shows the chromatogram of a typical PDMS fluid. The main polymer is highlighted by the vertical lines, and the molecular weight

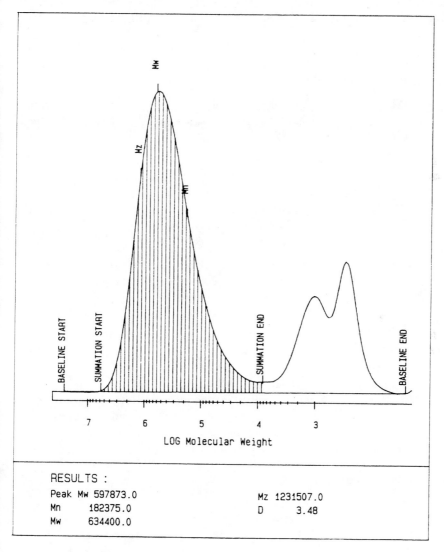

Figure 10.14. GPC tracing of a PDMS fluid (see text).

averages $\bar{M}_n$, $\bar{M}_w$, and $\bar{M}_z$ are labeled. Eluting after the polymer peak are two more peaks from low molecular weight cyclic PDMS. Muller and Opila (65) used GPC as a rapid screening tool for identifying and quantitating low molecular weight components in silicone potting agents. These potting agents are used in the manufacture of electronic components, and low molecular weight components can form insulating deposits on the contacts of nearby relays and switches.

The following general conditions have been found useful.

1. PDMS and Similar Fluids

PDMS of all molecular weights is readily soluble in toluene. The refractometer is very sensitive, allowing dilute samples to be chromatographed.

Column set	Polystyrene–divinylbenzene gel
Solvent	Toluene
Detector	Differential refractometer

Alternative method:

Column set	Same as above
Solvent	Chloroform
Detector	Infrared

This setup allows more flexibility in polymer analysis in that different substituent groups can be selectively detected by using appropriate wavelengths.

2. Fluorosilicones

Column set	Same as above
Solvent	Tetrahydrofuran
Detector	Differential refractometer
Issues	Fluorosilicones are prone to depolymerization in solution, reverting to cyclics. Residual catalyst must be removed, and the sample chromatographed quickly after dilution.

3. Phenyl siloxanes

Column set	Same as above
Solvent	Tetrahydrofuran
Detector	UV 254 nm or differential refractometer

Other detector systems such as fluorescence (66), FT IR (67), mass spectrometry (68), and AES (69), have been used to give direct identification and/or better sensitivity for specific compounds.

9.5. Special Problems and Techniques

Nonsize-exclusion effects can be observed when analyzing polymers containing polar moieties such as amino or carboxylic acid groups. The effect can be reversible adsorption, in which the polymer is retained longer on the column, or irreversible absorption where the polymer does not elute at all. Two ways of alleviating this problem are derivatizing the samples in order to reduce the polarity of the polymer, and changing the chromatographic conditions in order to favor the elution of the sample.

Aminofunctional groups in polymers can be converted to amide groups. An excess of dry acetic anhydride is added to the sample, and allowed to react. The mixture is then diluted and analyzed normally. The amidation of the amine group also permits selective detection of the amide using the IR detector. Relative distribution of the amine group in the polymer is determined by comparing the response of the detector set at an amide-specific wavelength with the response at a siloxane-specific wavelength.

Carboxylic acid functional polymers can be trimethylsilylated using standard derivatizing schemes normally employed in GC (Section 4.2). A useful capping procedure uses hexamethyldisilazane to trimethylsilylate the carboxylic acid groups. Infrared detection specific to the carbonyl group allows the distribution of the acid groups in the polymer fractions to be determined in the same manner as described for the aminofunctional siloxanes.

Changing the chromatographic conditions (rather than the molecules) is somewhat more difficult, because a combination of less active stationary phase and a more polar mobile phase is needed. It is often difficult to find a solvent that dissolves the polymer and also allows detection of the analyte.

REFERENCES

1. V. D. Shatz, R. Y. Sturkovich, and E. Lukevics, "Gas chromatographic analysis of organosilicon compounds," *J. Chromatogr.*, **165**(3), 257 (1979).

2. R. L. Folz, M. B. Neher, and E. R. Hinnekamp, "Applications of mass spectrometry and gas chromatography to the analysis of polymer systems," *Anal. Chem.*, **39**, 1338 (1967).

3. L. F. Hanneman, presented at Capillary Chromatography 1982, Tarrytown, NY, October 1982.

4. L. H. Wood, unpublished work, Dow Corning Corporation, Carrollton, KY.

5. P. J. Garner, unpublished work, Dow Corning Corporation, Midland, MI.

6. A. E. Pierce, *Silylation of Organic Compounds*, Pierce Chemical, Rockford, Il, 1968.

7. J. F. Klebe, "Silylation in organic synthesis," *Adv. Org. Chem.*, **8**, 97 (1972).

8. W. L. Knowlton, unpublished work, Dow Corning Corporation, Midland, MI.

9. C. R. Thrash, D. L. Voisinet, and K. E. Williams, "Analysis of phenylmethyldichlorosilane for phenyltrichlorosilane by reaction chromatography," *J. Gas Chromatogr.*, **3**, 248 (1965).

10. N. A. Palamarchuk, S. V. Syavtsillo, and L. A. Nechaeva, "Analysis of poly-(methylsiloxanes) using a reaction gas chromatograph," *Zh. Anal. Khim.*, **28**(11), 2264 (1973); *Chem. Abstr.*, **80**: 96545g (1973).

11. G. Heylmum and J. Pikula, "Analysis of methylfluorosilanes from methylpolysiloxanes by gas chromatography," *J. Chromatogr.*, **3**, 266 (1965).

12. P. Bailes and D. Rogers, unpublished work, Dow Corning Ltd., Barry, Wales.

13. S. Fujimoto, H. Ohtami, and S. Tsuge, "Characterization of polysiloxanes by high-resolution pyrolysis-gas chromatography-mass spectrometry," *Fresenius' Z. Anal. Chem.*, **331**, 342 (1988); through *Chem. Abstr.*, **109**: 129963 (1988).

14. R. Steinmeyer, unpublished work, Dow Corning Corporation, Carrollton, KY.

15. W. Jennings, *Gas Chromatography with Glass Capillary Columns*, Vol. 2, Academic, New York, 1980.

16. J. Yancey, *Guide to Stationary Phases for Gas Chromatography*, Vol. 12, Analabs Inc., North Haven, CT, 1979.

17. R. W. Stephan, unpublished work, Dow Corning Corporation, Carrollton, KY.

18. L. F. Hanneman, in *Analysis of Silicones*, A. L. Smith, Ed., Wiley, New York, 1974, pp. 239–243.

19. G. Burns, T. Angelotti, L. Hanneman, G. Chandra, and J. Moore, "Alkyl- and arylsilsesquiazanes: Effect of the R group on polymer degradation and ceramic char composition," *J. Mater. Sci.*, **22**, 2609 (1987).

20. J. Kleinert and C. Weschler, "Pyrolysis gas chromatographic mass spectrometric identification of polydimethylsiloxanes," *Anal. Chem.*, **52**, 1245 (1980).

21. L. Rohrschneider, "Method of characterization for gas chromatographic separation of liquids," *J. Chromatogr.*, **22**, 6 (1966).

22. W. O. McReynolds, "Characterization of some liquid phases," *J. Chromatogr. Sci.*, **8**, 685 (1970).

23. E. Kovats, "Gas chromatographic characterization of organic substances in the retention index system," *Advan. Chromatogr.*, **1**, 229 (1965).

24. W. G. Jennings, "Characterization of mixed liquid phases," *Chem. Mikrobiol. Technol. Lebensm.*, **1**, 9 (1971).

25. K. Friedrich, "Partition of methylchlorosilanes by gas–liquid chromatography," *Chem. Ind. (London)*, **1957**, 47.

26. B. Lengyel, G. Garzo, and T. Szekely, "Gas chromatographic analysis of methylchlorosilanes," *Acta Chem. Acad. Sci. Hung.*, **37**, 37 (1963).

27. T. Oiwa, M. Sato, Y. Miyakawa, and I. Miyazaki, "Gas chromatographic determination of methylchlorosilanes," *Nippon Kagaku Zasshi*, **84**, 409 (1963).

28. N. Palamarchuk, S. Syavtsillo, and N. Turkel'taub, "Gas chromatographic determination of impurities in monomeric organosilicon compounds," *Gazov. Khromatogr., Tr. Uses. Konf., Mosc.*, **1962**, 303; through *Chem. Abstr.*, **62**: 7117g (1965).

29. H. Rotzsche, "Silicones. LXXV. Qualitative and quantitative gas chromatographic analysis of the ferrosilicon-methyl chloride reaction products," *Z. Anorg. Allgem., Chem.*, **328**, 79 (1964).

30. I. P. Yudina, L. A. Khokholova, L. P. Sidorova, and A. V. Zimin, "Chromatographic determination of phenyltrichlorosilane impurity in methylphenyldichlorosilane," *Gazov. Khromatogr.*, No. 4, **1966**, 134; *Chem. Abstr.*, **68**: 9149d (1968).

31. G. P. Hildebrand and C. N. Reilley, "Use of combination columns in gas liquid chromatography," *Anal. Chem.*, **36**, 47 (1964).

32. J. H. Purnell and P. S. Williams, "Compressibility effects in the optimization of serially connected gas chromatographic capillary columns by the window diagram technique," *J. High Resolut. Chromatogr. Chromatogr. Commun.*, **6**, 569 (1983).

33. R. Danhaus, unpublished work, Dow Corning Corporation, Midland, MI.

34. I. B. Peetre, Ph.D. Dissertation, Teknisk Analytisk Kemi, Lunds Tekniska Hogskola, 1973.

35. T. J. Dearlove, R. L. Kaas, and R. P. A. Atkins, "Relative molar response factors for the thermal conductometric analysis of silane compounds," *J. Chromatogr. Sci.*, **14**, 448 (1976).

36. D. M. Rosie and R. L. Grob, "Thermal conductivity behavior. Importance in quantitative gas chromatography," *Anal. Chem.*, **29**, 1263 (1957).

37. A. Messner, D. M. Rosie, and P. A. Argabright, "Correlation of thermal conductivity cell response with molecular weight and structure," *Anal. Chem.*, **31**, 320 (1959).

38. E. F. Barry and D. M. Rosie, "Response prediction of the thermal conductivity detector with light carrier gases," *J. Chromatogr.*, **59**, 269 (1971).

39. E. F. Barry, R. S. Fischer, and D. M. Rosie, "Determination and prediction of anomalous response factors for halogenated substances with the thermal conductivity detector," *Anal. Chem.*, **44**, 1559 (1972).

40. G. Barnes, unpublished work, Dow Corning Corporation, Midland, MI.

41. J. Kowalski, M. Schibiorek, and J. Chojnowski, "Correlation of the response factors of thermal-conductivity detector with molecular weight for methylsiloxanes," *J. Chromatogr.*, **130**, 351 (1977).

42. L. Ongkiehong, Ph.D. Dissertation, Technische Hogeschool Eindhoven, Netherlands, 1960.

43. R. Kaiser, *Gas Phase Chromatography*, Vol. 2, Butterworths, London, 1963, p. 99.

44. W. Gallaway, D. Jones, and J. Sternberg, *Gas Chromatography*, Academic, New York, 1962, p. 231.

45. T. Hirschfeld, "The hy-phen-ated methods," *Anal. Chem.*, **52**, 297A (1980).

46. G. F. Dudding and G. T. Sorenson, "Capillary GC/FID and GC/MS determination of ethyldichlorosilane," *J. Chromatogr. Sci.*, **18**, 670 (1980).

47. R. Buffington, *GC–Atomic Emission Using Microwave Plasmas*, Hewlett-Packard Company, Avondale PA, 1988.

48. J. A. Nieman and L. B. Rogers, "Supercritical fluid chromatography applied to the characterization of a siloxane-based GC stationary phase," *Sep. Sci.*, **10**, 517 (1975).

49. T. L. Chester, "Practice and applications of supercritical fluid chromatography in the analysis of industrial samples," *Chromatogr. Sci.*, **45**, 369 (1989).

50. M. Novotny, S. R. Springston, P. A. Peadon, J. C. Fjelsted, and M. L. Lee, "Capillary supercritical fluid chromatography," *Anal. Chem.*, **53**, 407A–414A (1981).

51. J. C. Fjeldsted and M. L. Lee, "Capillary supercritical fluid chromatography," *Anal. Chem.*, **56**, 619A (1984).

52. E. J. Guthrie and H. E. Schwartz, "Integral pressure restrictor for capillary SFC," *J. Chromatogr. Sci.*, **24**, 236 (1986).

53. T. L. Chester, D. P. Innis, and G. D. Owens, "Separation of sucrose polyesters by capillary supercritical fluid chromatography/flame ionization detection with Robot-pulled capillary restrictors," *Anal. Chem.*, **57**, 2243 (1985).

54. T. L. Chester and D. P. Innis, "Retention in capillary supercritical fluid chromatography," *J. High Resolut. Chromatogr. Chromatogr. Commun.*, **8**, 561 (1985).

55. D. W. Later, E. R. Campbell, and B. E. Richter, "Synchronized temperature/density programming in capillary supercritical fluid chromatography," *J. High Resolut. Chromatogr. Chromatogr. Commun.*, **11**, 65 (1988).

56. T. Suzuki, "Preparation of poly(dimethylsiloxane) macromonomers by the 'initiator method': 2. Polymerization mechanism," *Polymer*, **30**, 333 (1989).

57. W. W. Yau, J. J. Kirkland, and D. D. Bly, *Modern Size-Exclusion Liquid Chromatography*, Wiley, New York, 1979.

58. K. Kamide, "Batch Fractionization," in *Fractionation of Synthetic Polymers; Principles and Practices*, L. H. Tung, Ed., Marcell Dekker, New York, 1977.

59. K. Dodgson, D. Sympson, J. A. Semlyen, "Studies of cyclic and linear poly(dimethyl siloxanes). 2. Preparative gel permeation chromatography," *Polymer*, **19**(11), 1285 (1978).

60. M. Kubin, "Calibration of size-exclusion chromatography systems with polydisperse standards," *J. Liq. Chromatogr.*, **7** (*Suppl. 1*), 41 (1984).

61. T. Dumelow, S. R. Holding, L. J. Maisey, and J. V. Dawkins, "Determination of the molecular weight and compositional heterogeneity of block copolymers using combined gel permeation chromatography and low-angle laser light scattering," *Polymer*, **27**, 1170 (1986).

62. E. Kohn, "Size exclusion chromatography (XII). Analysis of silicon-phenyl groups in molecular weight components of polydimethylsiloxanes." Report 1986, MHSMP-86-31; Order No. DE86015001, *Chem. Abstr.*, **106**: 86021 (1987).

63. H. Benoit, Z. Grubisic, P. Rempp, and H. Benoit, "A universal calibration for gel permeation chromatography," *J. Polym. Sci. B.*, **5**(9), 753 (1967).

64. M. A. Haney, "The differential viscometer. II. On-line viscosity detector for size-exclusion chromatography," *J. Appl. Polym. Sci.*, **30**(7), 3037 (1985).

65. A. J. Muller and R. L. Opila, "A new rapid screening method for silicones by size exclusion chromatography," *Electr. Contacts*, 34th, 289 (1988).

66. S. T. Lai, L. Sangermano, and D. C. Locke, "Multimode phenyl-bonded phase liquid chromatography of phenyl-containing room temperature vulcanizeable

silicone raw materials and polystyrene polymers using UV and fluorescence detection," *Separation Sci. Tech.*, **20**, 513 (1985).

67. R. L. White, *Chromatography/Fourier Transform Infrared Spectroscopy and its Applications*, Marcel Dekker, New York, 1990.

68. W. B. Crummett, H. J. Cortes, T. G. Fawcett, G. J. Kallos, S. J. Martin, C. L. Putzig, J. C. Tou, V. T. Turkelson, L. Yurga, and D. Zakett, "Some industrial developments and applications of multidimensional techniques," *Talanta*, **36**, 63 (1989).

69. W. R. Biggs, J. C. Fetzer, and R. J. Brown, "Determination of silicon compounds by gradient liquid chromatographic separation with direct current plasma atomic emission spectrometric detection," *Anal. Chem.*, **59**, 2798 (1987).

CHAPTER

11

INFRARED, RAMAN, NEAR-INFRARED, AND ULTRAVIOLET SPECTROSCOPY

E. D. LIPP and A. LEE SMITH

Analytical Research Department
Dow Corning Corporation
Midland, Michigan

1. INTRODUCTION

Infrared (IR) spectroscopy is a rapid and easily used technique for identification of silicones. Many of the spectral patterns for organic groups attached to silicon are highly specific, and identification of organosilicon materials can be made more readily from an IR spectrum than by any other technique, provided a reference spectrum of that material is available. Even if no exact match to an unknown spectrum can be found, it is usually possible to infer the presence or absence of specific chemical groups and their environment. When combined with NMR and MS data, the IR spectrum provides powerful insights into the identity and structure of unknown materials. A wide selection of sampling techniques provides flexibility in examining solid, liquid, or gaseous samples.

Infrared spectroscopy can also give quantitative information for both molecular species and specific chemical groups. Sensitivity to trace components varies somewhat depending on the absorption intensity of the material sought and on interferences from the matrix. Microgram quantities can usually be detected; special techniques such as GC–matrix isolation can achieve subnanogram sensitivity. Very small molecular effects, such as chain length in high molecular weight polymers, or polymer cross-linking, are usually not amenable to IR analysis.

Many texts covering different aspects of IR spectroscopy are available (1–3). New developments are reported biennially in *Analytical Chemistry* review issues. Because the IR spectra obtained from Fourier transform IR

The Analytical Chemistry of Silicones, Edited by A. Lee Smith.
ISBN 0-471-51624-4 © 1991 John Wiley & Sons, Inc.

(FT IR) spectrometers and dispersive spectrometers are identical (at least in principle), we do not differentiate between them in this chapter except when the difference in capabilities is important.

Raman spectroscopy (4–6) is a useful analytical technique but is found in the laboratory less often than IR because it is somewhat more limited in its applicability. The Raman spectrum is inherently weak and often is overwhelmed by fluorescence, although the use of near-IR laser excitation (7, 8) usually circumvents that problem (but not always—see Ref. 9). The information obtained from the Raman spectrum is often complementary to that from the IR spectrum. Raman has some sampling advantages—water has only a weak spectrum, in contrast to its strong IR absorption—and materials can be sampled through glass or other transparent cells.

Near-IR spectroscopy has developed rapidly as an empirical quantitative technique useful for measuring a variety of properties that may not be related to the spectrum in any obvious way (10, 11). Determination of OH number, total sugar in dry cereals, and iodine value of oils are examples of applications for near-IR spectroscopy. These applications involve statistical programs that compare the spectra of interest with those from a set of known standards, and usually no detailed interpretation of the spectra is attempted. Near-IR spectroscopy can also be used in a conventional fashion where specific group frequencies are correlated with molecular structure, just as in mid-IR spectroscopy.

Ultraviolet (UV) spectroscopy (12) has only limited application to the qualitative analysis of organosilicon compounds because, except for aromatic substituents and polysilanes, most silicones do not absorb UV radiation in the analytically useful region of the spectrum.

1.1. Principles

A molecule can be visualized as a mechanical system consisting of point masses (atoms) connected by springs (chemical bonds). When radiation of the proper frequency impinges on a molecule, a "sympathetic vibration" is set up and the radiation is absorbed. This absorption is displayed as an absorption band when the radiation is dispersed to give a spectrum. Many chemical groups absorb at characteristic positions in the spectrum, so the presence of a band or group of bands at the proper position is good evidence for the presence of a particular chemical group. Such absorptions are known as "group frequencies" and arise because many chemical groups tend to vibrate more or less independently of the total molecule. The number of frequencies depends on the number of masses (atoms) involved, and their exact positions in the spectrum are exquisitely sensitive to small changes in the electronic and geometrical configuration of the molecule.

Whereas such frequencies may vary somewhat in organic compounds, depending on the molecular configuration, many of them tend to be quite constant in organosilicon compounds. This effect results because the silicon atom is large and heavy, and acts as vibrational insulation between the lighter carbon–hydrogen containing portions of the molecule. Sometimes group frequencies vary predictably and thus provide additional clues to the molecular structure. These features make IR the method of choice for the initial screening of unknown compounds, and provide a means for instant recognition of the more common types of organosilicon substituents.

It is important to be aware of the fact that selection rules govern the presence or absence of bands both in IR and Raman spectra (2). Both techniques reflect the vibrational activity of the molecule, but in different ways. In IR, incident radiation that has the same frequency as the vibrational mode of a molecule is absorbed, provided the dipole moment changes during the vibration. The intensity of the absorption depends on the rate of change of the dipole moment with the normal coordinate for the vibration $(d\mu/dQ)^2$, where μ is the dipole moment, and Q is the normal coordinate for the vibration. Thus, symmetrical vibrations such as the Si–Si stretching in $Me_3SiSiMe_3$ do not show any IR absorption.

Raman activity, on the other hand, depends on the change in polarizability of the vibration, $(d\alpha/dQ)^2$, where α is the polarizability tensor. The polarizability may be visualized as a measure of the volume available to the electrons in a molecule. Totally symmetric vibrations, such as the SiSi mode mentioned above, give intense Raman features, whereas intense IR absorptions may be weak or missing in the Raman spectrum.

Comparison of the IR and Raman spectra shown in Figure 11.1–11.5 illustrates the complementary nature of the two techniques. In $Me_3SiOSiMe_3$, for example (Figure 11.1), the SiOSi antisymmetric stretch near $1050\ cm^{-1}$ has a large dipole moment change and a strong IR absorption.

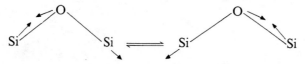

The symmetric stretch, on the other hand, has only a weak dipole moment change and consequently a weak IR absorption. The volume of the electron cloud changes significantly and the vibration shows a strong Raman line.

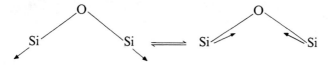

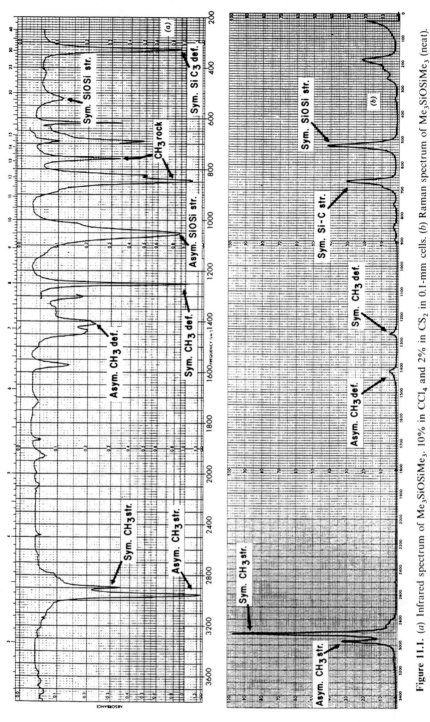

Figure 11.1. (a) Infrared spectrum of $Me_3SiOSiMe_3$, 10% in CCl_4 and 2% in CS_2 in 0.1-mm cells. (b) Raman spectrum of $Me_3SiOSiMe_3$ (neat).

308

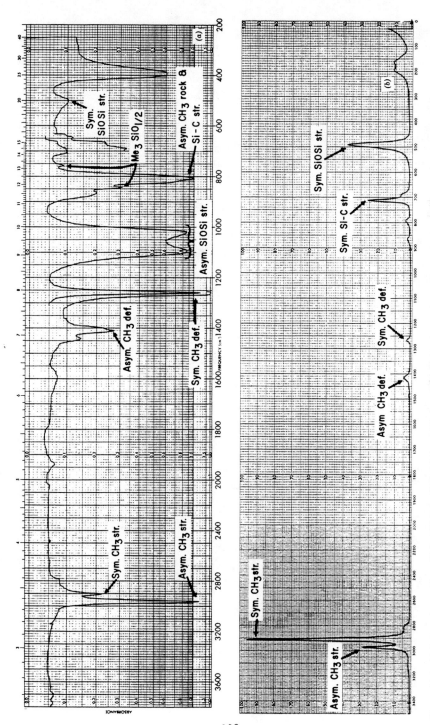

309

Figure 11.2. (a) Infrared spectrum of Me₃SiO(Me₂SiO)ₓSiMe₃, 10% in CCl₄ and 2% in CS₂ in 0.1-mm cells. (b) Raman spectrum of Me₃SiO(Me₂SiO)ₓSiMe₃ (neat).

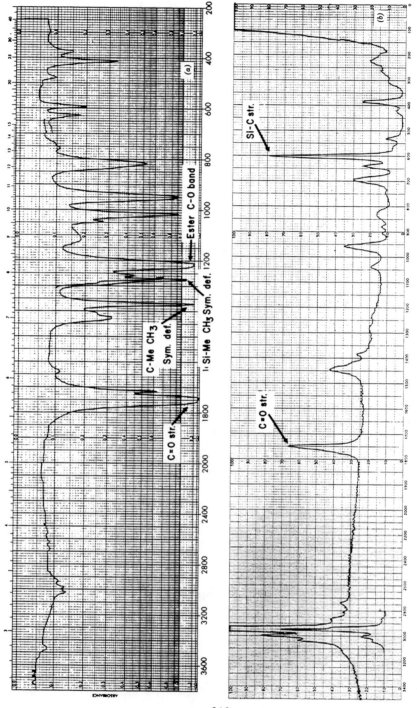

Figure 11.3. (a) Infrared spectrum of MeSi(OAc)$_3$. 5% in CCl$_4$ and 3% in CS$_2$ in 0.1-mm cells. (The band at 1718 cm^{-1} is from HOAc.) (b) Raman spectrum of MeSi(OAc)$_3$ (neat).

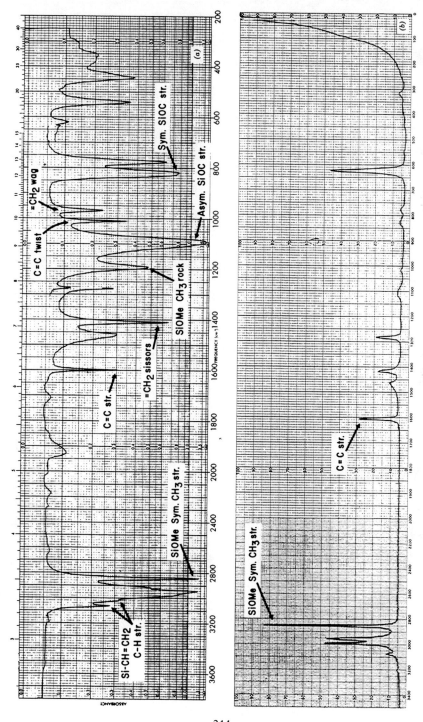

Figure 11.4. (*a*) Infrared spectrum of ViSi(OMe)₃, 10% in CCl₄ and 3% in CS₂ in 0.1-mm cells. (*b*) Raman spectrum of ViSi(OMe)₃ (neat).

311

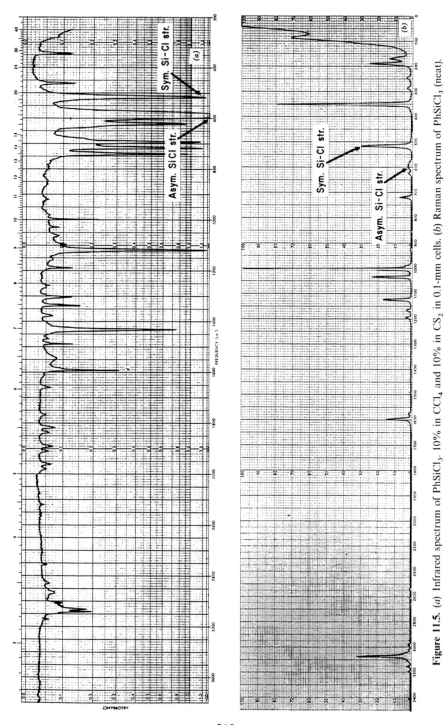

Figure 11.5. (a) Infrared spectrum of PhSiCl₃, 10% in CCl₄ and 10% in CS₂ in 0.1-mm cells. (b) Raman spectrum of PhSiCl₃ (neat).

312

Near-IR absorptions are governed by IR selection rules, but consist almost entirely of vibrational overtones and combinations of hydrogen stretching and deformation modes. The attribution of bands to specific molecular motions may thus be more difficult than in the mid-IR, but often is not necessary. Near-IR excels as an empirical quantitative method when well-characterized materials are available for standards.

Inelastic electron tunneling spectroscopy is a related technique that also gives vibrational and structural information, but it is more difficult experimentally and is used most often as a research tool to address special problems. A typical example is the study of the reaction of silane coupling agents with Al_2O_3 surfaces (13, 14).

Photoelectron spectroscopy (PES), which measures the electronic energy levels, also provides a molecular fingerprint of gaseous atoms and molecules. It has been used to study silanes (15) and can give real-time gas analysis in flow systems. Most studies to date have been structural rather than analytical, however.

With so many potential tools to choose from, selection of the optimum analytical techniques is often difficult. Infrared spectroscopy is particularly powerful, either alone or combined with other techniques, for the solution of problems in identification, molecular structure, or quantitative analysis. The synergism of IR–NMR–MS for solving analytical problems is well known. Both GC–IR and GC–MS are used frequently to identify the components of complex mixtures. Although MS is usually more sensitive to trace components than is IR, it usually cannot distinguish isomers, a task at which IR excels. The "hyphenated techniques" GPC–IR, SFC–IR, and so on, although not fully developed as of this writing, are also useful (16), and some familiarity with their capabilities is essential if one is to choose the best approach to difficult problems.

2. SAMPLING

The basic sampling techniques for IR, near-IR, and Raman spectroscopy of organosilicon materials are generally the same as for organic materials, with only minor modifications needed to accommodate the special characteristics of silicones. Thorough discussions of experimental techniques are given elsewhere (1, 6, 10). Most of the decisions in sampling center on the selection of the proper technique, which depends on the spectral information required and the physical state of the sample.

2.1. Gases

Although only a few organosilicon materials are gases under ambient conditions, many monomers and oligomers are sufficiently volatile to allow spectral measurements. Most IR measurements on gases are performed by transmission, and flexibility in pathlength selection permits analysis over a wide concentration range, including trace measurements using long path lengths (20 m or more), where parts per billion levels can be determined (17) if other atmospheric species do not interfere. Another useful qualitative technique for gases is GC–IR. A plug of air, usually 0.5–3 mL in size, is withdrawn from the sample container and injected directly into the GC. The large volume of air degrades chromatographic resolution, thus making quantitative applications difficult, but the spectra of constituent gases (except monatomic and homonuclear diatomic gases) are obtained within minutes.

2.2. Liquids

Liquids are generally best sampled in transmission, either neat or diluted in the appropriate solvent, because the influence of molecular interactions can be reproducibly controlled and the same molecular environment of the sample can be achieved consistently. This technique produces the high degree of reproducibility necessary for quantitative analysis and for the ability to perform high quality spectral subtractions. Neat measurements employ path lengths anywhere from 2 μm to several centimeters depending on the spectral region of interest and the strength of the absorptions. In the mid-IR region, silicone samples are generally run as 10% (w/v) solutions in CCl_4 for the regions 4000–1350 and 650–200 cm^{-1}, and as 2% solutions in CS_2 for the region 1350–650 cm^{-1} using 0.1-mm path cells. (Vapors of both solvents are toxic; they should be handled only in a fume hood.) This combination of path length and concentration gives appropriate absorbances for the qualitative and quantitative analysis of most silicones. It is noteworthy that the region 1350–650 cm^{-1} is measured as a more dilute solution than is used for most organic materials because of the overall stronger absorbance exhibited by silicones in this region. For samples that have difficulty dissolving in CS_2 or where the use of CS_2 is discouraged, hydrocarbon solvents such as 2,2,4-trimethylpentane or heptane can be substituted. Longer path lengths or more concentrated solutions can be used for minor component or trace analysis in regions devoid of strong fundamental vibrations (e.g., 1800–2600 cm^{-1}). For measurements in the near-IR, where neat samples are generally used with path lengths up to several centimeters, the preferred cell material is fused silica. This material is particularly useful because of its resistance to water and to many corrosive materials. It can be used down to 2300 cm^{-1}.

Raman measurements (4) are generally made on neat samples. Sampling is particularly convenient since only small glass tubes, such as those used in NMR or for melting point determinations, are normally required. The need for solvents in both sample preparation and clean-up is virtually eliminated. Sample size is often a few microliters or less.

Infrared measurements can also be performed on neat liquid samples using attenuated total reflectance (ATR) (18). Common configurations are those where the sampling surface is held horizontal and the liquid poured on top, or of the CIRCLE CELL® variety, which is particularly good for flow-through measurements. This latter cell uses a cylindrical crystal around which the sample flows (19). Attenuated total reflectance provides convenient sampling since no sample preparation is required, and one has some control over the path length through the number of reflections through the ATR crystal. The ATR spectra show smaller intensities at short wavelengths compared to transmission spectra because the depth of penetration of light into the sample is on the order of the wavelength. The spectra are also susceptible to distortions related to polarization and refractive index effects, which vary with the crystal material and angle of incidence of the radiation. An example of these distortions is shown for PDMS in Figure 11.6, where the thick sample spectrum shows intensity changes in the region $1000-1100$ cm^{-1} and a noticeable shift to lower wavenumbers of the 800-cm^{-1} band. Thus, ATR spectra may not match transmission spectra and the identification of subtle spectral details may be more difficult. Measurements on aqueous solutions are feasible with ATR cells, again because very small path lengths are possible.

Some high viscosity polymers do not readily dissolve in solvents or flow into IR cells. Attenuated total reflectance is useful for these high viscosity samples since sample thickness is irrelevant and sample handling is minimized. Devices that control the pressure of the sample against the ATR crystal help to improve the reproducibility of the spectra. For best results, the same region of the ATR crystal should be used, and the sample should cover the whole *width* of the crystal. Poly(dimethylsiloxane) is sometimes difficult to remove completely from KRS-5 crystals.

Mixtures of materials may often be analyzed by any of the techniques for which an IR spectrometer is interfaced to serve as a detector for a chromatographic separation. Of these, GC–IR is by far the most widely used because solvent and background interferences are absent. As an alternative to the light-pipe sample cell, a matrix isolation chamber (20) can be used to trap the eluting species. Spectral measurements in the cryogenic matrix allow a significant sensitivity enhancement through band sharpening and the potential for signal averaging for long periods of time on the frozen samples. Detection limits in the 100-pg range are reported.

Techniques such as LC–IR and GPC–IR are hindered by solvent absorbances that obliterate large portions of the IR spectrum. The SFC–IR technique suffers a similar, though less severe handicap. The most promising

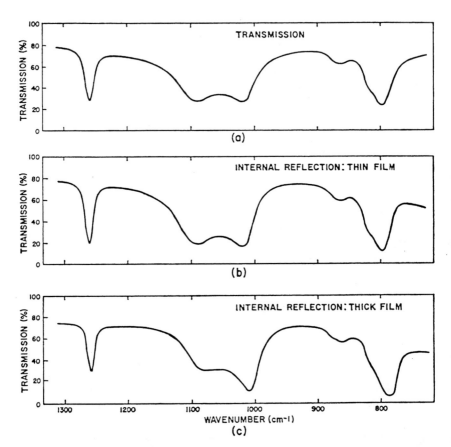

Figure 11.6. Spectra of silicone grease, showing differences between transmission and ATR spectra. Intensity changes at 1000–1100 cm^{-1} and a band shift near 800 cm^{-1} are particularly prominent. From N. J. Harrick, Internal Reflection Spectroscopy, © 1967 John Wiley & Sons.

approaches to these chromatographic–IR techniques involve elimination of the mobile phase. Advances have been made in both LC–IR (21) and SFC–IR (22, 23) measurements. Until the solvent interference problem is overcome, however, these techniques will be restricted to very specific applications.

2.3. Solids

Solids generally present the most difficult case for IR measurements. Consequently, the greatest variety of techniques has evolved for sampling solids in their various forms. Transmission measurements are preferred when information about the bulk of a material, as opposed to surface characteristics, is desired. As with liquids, solution in IR-transmitting solvents is the preferred sampling mode. Other options for transmission measurements include films cast from solvent, thin slices cut from the bulk, or use of a diamond anvil cell to squeeze the specimen down to a useable thickness.

For samples that exist as powders or can be ground to powders, conventional mineral oil or KBr mulls can be used. A transmission measurement is then performed on the suspended particles, although the spectrum is dependent on particle size and shows artifacts if the particle size is not fine enough (1).

An alternative technique for powdery samples is diffuse reflectance (DR) from the powder surface, which can be performed in both the mid- (3, 24, 25) and near-IR (11) regions. Special attachments for holding the sample and collecting the radiation are required; these are generally integrating spheres in the near-IR and off-axis ellipsoidal mirrors in the mid-IR. Samples are normally analyzed neat in the near-IR and either neat or diluted in powdered KBr or KCl in the mid-IR. Compared to KBr pellets and mulls, DR offers the advantages of improved sensitivity, absence of scattering artifacts, and simpler sample preparation. One has to be careful, however, that the absorbance of the bands of interest does not get too large or reflection artifacts (which look like the derivative of the absorbance band) will occur. This problem is generally limited to the mid-IR (near-IR absorbances are too weak for this to occur) and is commonly seen in the extremely intense SiO stretching vibrations near 1100 cm^{-1} in silica and organosilicon materials (Figure 11.7). These artifacts can be eliminated by dilution in an inert matrix. Dilutions of 5% in KBr or KCl are typical. Diffuse reflectance measurements give primarly surface information if grinding or fragmenting the sample is avoided.

Solids that are somewhat pliable (cured elastomers, coated papers, etc.) are effectively analyzed by ATR in the mid-IR. These samples allow good contact with the ATR crystal and clamping devices that produce a consistent pressure help make the spectra more reproducible.

Photoacoustic (PA) spectroscopy (26–28) is less commonly used than DR or ATR, but it offers some definite sampling advantages for samples that are hard or irregularly shaped and cannot be ground. In fact, this acceptance of irregular morphology appears to be the primary advantage of PA detection. Quantitative applications of PA spectroscopy are even more difficult than for ATR or DR, but qualitative applications on otherwise intractable solids are readily handled.

Coatings on flat metallic or highly reflective materials can be analyzed by specular reflection spectroscopy. For coatings in the thickness range of 0.1–10 μm, use of reflectance at near-normal incident angles is a straightforward way of obtaining a spectrum. In this measurement the light passes through the sample twice, and use of a highly reflective substrate gives good

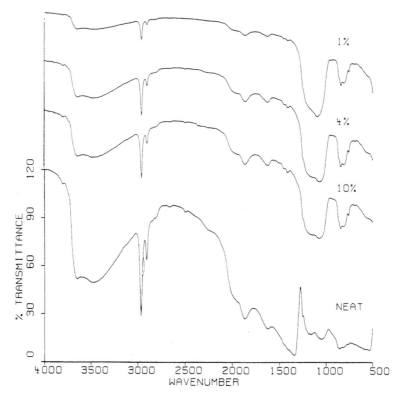

Figure 11.7. Diffuse reflectance IR spectra of treated SiO_2 at different dilutions in powdered KBr. Spectra are displaced for clarity, but the transmittance scale is constant.

energy throughput. Such a measurement contains information from both the bulk and the metal-coating interface.

To concentrate on the interfacial region, one can perform a grazing angle measurement with polarized light. The polarization of the light is set perpendicular, and then parallel, to the plane of incidence. For the former case, no absorption by a dipole at the metal-coating interface can occur. For the latter case, with the polarization parallel to the plane of incidence, absorption can occur. The dipole must also be aligned parallel to the plane of incidence (perpendicular to the surface) for absorption to occur, however. Furthermore, the absorbance will be enhanced relative to a transmission measurement through the material. These absorbance and enhancement effects at the metal surface using polarized light can be explained through imaging of the oscillating dipole by free electrons in the metal. These effects lead to cancellation of absorbance in one case (dipole lying on the surface) and enhanced absorbance in the other (dipole aligned perpendicular to the surface). For molecules away from the surface, there is no perference between the two states of light polarization. Thus, the orientation of molecules at the metal-coating interface can be selectively measured (29).

Raman measurements can also be performed on neat solids. Information about surface rather than bulk properties predominates since most of the light scattering occurs at the surface. Specialized techniques such as surface enhanced Raman spectroscopy (4) are also possible for thin layers adsorbed onto flat metallic surfaces. However, there is much about this technique that is not understood and it is by no means a routine analytical tool.

Microsamples as small as 10 μm in diameter can be analyzed using IR microspectroscopy (17, 30). Inclusions in polymers may be isolated by microtoming, or by sectioning the polymer and removing the inclusion using a needle probe. Special equipment and techniques are required, and FT IR is preferred, because of its energy throughput advantage over dispersive IR.

3.   QUALITATIVE ANALYSIS

3.1.   Group Frequencies

Group frequencies for organic compounds have been studied extensively, and several useful summaries and compilations exist (2, 31, 32). Many chemical groups attached to silicon tend to absorb in narrow spectral regions, so the spectrum can, to a first approximation, be considered to consist of the superposition of absorption peaks from independently vibrating groups, each with its own characteristic spectrum. Thus, from a good knowledge of these group frequencies, the analyst can often make accurate deductions about the molecular characteristics of the sample. In identifying unknowns, of course, the analyst should make full use of sample history as well as any other information available.

Considerable group frequency information is available (33–35) and is summarized in Figure 11.8. Brief discussions of some of the more common group frequencies for silicones are given in this section

One should not infer from the preceding discussion that use of correlation charts and tables is adequate to identify unknown materials from their IR spectra. Attempts to do so invite serious error. Similarly, computer-assisted interpretation schemes (36, 37) may be useful but do not yield unambiguous answers. Identification is achieved only when the unknown spectrum shows a perfect match to an authentic reference spectrum of a well-characterized standard; or, with new materials, when the identification is confirmed by

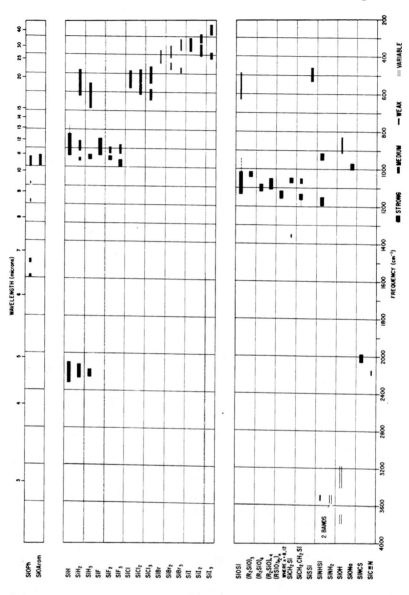

320

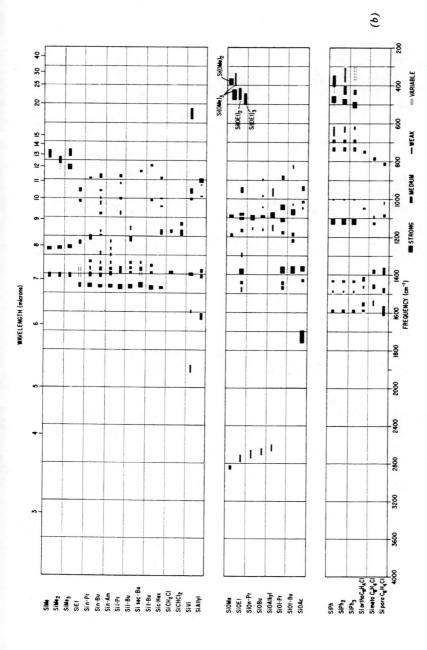

Figure 11.8. Infrared spectra structure correlations for organosilicon compounds (33, 34). Noncharacteristic CH stretching bands are not included.

321

Table 11.1. **Commercially Available Reference Spectra that Include Silicones**[a]

Collection	Silanes	Siloxanes	Range (cm^{-1})	Reference
Atlas of organosilicon compounds	?	?		38
Coblentz society deskbook	29	15	4000–250	39
Coblentz society spectra	163	62	4000–250	40
IR coatings atlas		14	4000–200	41
Atlas of polymers (Hummell)		65	4000–300	42
Aldrich library FT IR spectra	129	11	4600–400	43
Mecke	26			44
Raman–IR atlas (VCH)[b]	6	3	3700–80	45
Sadtler atlas		46	4000–400	46
Sadtler standard IR collection	385	35	varies	46
Sigma library	48	9	4000–400	47

[a] *Note*: in some cases, hydrolyzable materials were apparently exposed to moisture, and hydrolysis products also appear in the spectra. Susceptible materials are chlorosilanes, alkoxysilanes, and silazanes.

[b] Raman spectra also included for the same compounds.

independent techniques such as NMR, MS, and chemical analysis. Thus, in addition to some interpretive skills (which are achieved only by experience), the analyst needs a library of reliable reference spectra and a system for searching them. Some commercial sources for such spectra are listed in Table 11.1. Many search systems are also available.

Small shifts in group frequencies are often predictable, and arise from the inductive effect of neighboring atoms or groups, from mechanical coupling of vibrations, or from steric or geometric effects (48).

3.1.1. The Inductive Effect

An example of the inductive effect is provided by the SiH stretching vibration, nominally 2100 cm^{-1}. The H atom is light and small, and is subject to little or no steric interaction. Very few other bands fall in the SiH region, so mechanical interactions are largely absent. The hybridization of the sp^3 bonds of silicon is, however, influenced by the electron-withdrawing ability of other substituents on the silicon. Slight changes in hybridization are reflected in the SiH bond length and therefore, its vibrational frequency as discussed in Section 3.2.2. The inductive effect of the substituents has an analogous effect on other vibrations as well, as will be discussed later.

3.1.2. Mechanical Effects

When the substituent has about the same mass as the silicon atom, mechanical effects become significant. For example, a single Si–Cl bond absorbs near 480 cm^{-1} (the inductive effect also affects its position). The SiCl$_2$ group shows splitting into symmetric and antisymmetric vibrations that absorb near 470 and 550 cm^{-1}. The SiCl$_3$ group shows corresponding bands near 500 and 600 cm^{-1}. In contrast, SiH$_2$ and SiH$_3$ vibrations interact very little, and if splitting is seen at all, it is usually < 10 cm^{-1}.

3.1.3. Steric Effects

The siloxane band can be used to illustrate steric effects. In the cyclotrisiloxanes, the siloxane framework is constrained, bond angles are well defined, and the SiOSi stretch falls at 1090 cm^{-1}. In the cyclotetrasiloxanes, the structure is less constrained, the Si–O–Si bond angles are larger, and the SiOSi stretch is at 1020 cm^{-1} (cf. Section 3.2.6).

3.2. Interpretation

In interpreting the spectra of organosilicon compounds, the analyst will find it helpful to remember (1) the strong absorptions of the siloxane portion of the molecule can dominate the spectrum even though it may be a minor constituent of the sample; (2) group frequencies are quite reliable, but may be shifted by inductive, mechanical, or steric effects; and (3) much useful information is found in band widths and intensities—features difficult to code in a table or chart.

Representative spectra are shown in this chapter to illustrate the types of absorptions that result from various structures and substituents. We now discuss the group frequencies for a few of the most important substituents commonly found on silicon.

3.2.1. Si–Methyl

Probably the single most characteristic band in the spectra of organosilicon compounds is the symmetrical CH$_3$ deformation (umbrella) mode at 1262 cm^{-1}. This band is intense, sharp, and always falls in the range 1262 ± 5 cm^{-1} (Figures 11.1–11.3). It is accompanied by other absorptions around 800–900 cm^{-1}; SiMe absorbs near 775; SiMe$_2$ near 805; and SiMe$_3$ at 760 and 845 cm^{-1}. The same pattern is followed if one or more of the methyl groups is replaced by an alkyl group; that is, –SiMe$_2$Et gives a similar to that of pattern –SiMe$_3$. The CH stretching absorptions around 2910 and 2970

cm^{-1} are weak in the IR and subject to interference by aliphatic materials; they are not analytically useful. In the Raman spectrum, however, the 2910-cm^{-1} band (symmetric CH_3 stretch) is very intense and polarized. The overtone near 2500 cm^{-1} of the symmetric deformation is also quite distinctive and useful for thick samples.

3.2.2. SiH

Another highly distinctive absorption found in organosilicon compounds is that of the SiH group. It is intense, and falls in a sparsely populated region of the spectrum (2100–2300 cm^{-1}). Its exact position provides a good indication of the inductive effects of the other substituents on the silicon atom (49). Electron-withdrawing groups such as F, O, Cl, and $-CCl_3$ move the absorption to higher frequencies, while electron-releasing groups such as $-SiR_3$ or

Table 11.2. E Values for Substituent Groups

Group	$E(cm^{-1})^b$	Group	$E(cm^{-1})^b$
CH_3COO-	762.1	$CF_3(CH_2)_3-$	708.7
F–	760.8^a	Ph–	708.7
Cl–	752.8	$m\text{-}CH_3C_6H_4-$	708.2
Br–	745.3	$PhCH_2-$	706.5
PhO–	738.6^a	$CH_2{=}CHCH_2-$	706.4
Cl_2HC-	738.3	$p\text{-}CH_3C_6H_4-$	706.3
MeO–	734.4^a	CH_3-	705.9
EtO–	732.0^a	$NC(CH_2)_3-$	703.7
$PhC{\equiv}C-$	731.0	i-Bu–	702.2
i-PrO–	730.2^a	cyclo-C_5H_9-	701.1
ClH_2C-	725.3	$Ph(CH_2)_2-$	700.7
H–	724.8	n-Pr–	700.1
$2,5\text{-}Cl_2C_6H_3-$	722.0	$n\text{-}C_5H_{11}$	700.1
HO–	718.6^a	$NC(CH_2)_4-$	699.7
Me_2N-	715.5^a	$Ph(CH_2)_3-$	699.3
$p\text{-}ClC_6H_4-$	714.0	Et–	699.1
$p\text{-}BrC_6H_4-$	711.9	t-Bu–	696.5
$m\text{-}ClC_6H_4-$	711.3	i-Pr–	694.3
$p\text{-}FC_6H_4-$	711.2	cyclo-$C_6H_{11}-$	691.5
$NC(CH_2)_2-$	710.7	Me_3SiCH_2-	687.6
$CF_3(CH_2)_2-$	709.7	Me_3Si-	684.8
$o\text{-}CH_3C_6H_4-$	709.6		
$CH_2{=}CH-$	709.2		

[a] Multiple groups of the same type not additive.
[b] CCl_4 solution values.

−CHR$_2$ move it to lower frequencies. Its position can be predicted accurately if these substituents are known. Conversely, an accurate measurement of the SiH band position can be used to deduce useful structural information (50).

The calculation is simple; one merely adds the E values given in Table 11.2 for the substituent groups:

$$v = \sum_{i=0}^{3} E_i \qquad (11.1)$$

v is the frequency of the SiH absorption.

Certain substituents such as OMe, OEt, F, and NMe$_2$ that have lone-pair electrons do not follow a simple linear additive law; the relationship here is more complex.

The bending vibrations of the SiH group give useful bands but they are subject to interference from other absorptions. They are summarized in Figure 11.8. The IR spectrum of an SiH-containing polymer is shown in Figure 11.9.

3.2.3. SiOR

Alkoxysilanes are reactive toward water and many other reagents, and are frequently used as intermediates in the preparation of other materials. Their IR absorptions are characteristic and are sometimes predictable. For SiOMe, a sharp intense band at 2840 and a medium intensity absorption at 1190 cm^{-1} are always present. The spectrum of ViSi(OMe)$_3$ shown in Figure 11.4 is typical. The strong 1100 cm^{-1} SiOC stretch is often subject to interference by SiOSi. The SiOEt group has a strong doublet, also partially obscured by SiOS, at 1080 and 1100 cm^{-1} (Figure 11.10). Weaker but more useful bands occur at 960 and 1140 cm^{-1}. Other alkoxy silanes also have distinctive absorptions (Figure 11.8). It is interesting that the exact frequencies of the 960 cm^{-1} band of SiOEt, the 1040 cm^{-1} band of SiOiPr, the 930 cm^{-1} band of SiOPh, the 940 and 1730 cm^{-1} bands of SiOAc can all be correlated with the substituent inductive parameter E given in Table 11.2 (50).

3.2.4. SiOH

When discussing the silanol group, one must be careful to specify the state of the sample, as the spectrum depends strongly on the degree to which the OH group is involved in hydrogen bonding. In the unassociated state (dilute CCl$_4$ solution), the OH stretching vibration of SiOH absorbs at 3695 cm^{-1} (3750 cm^{-1} on silica). When hydrogen bonded to siloxane oxygen atoms, it absorbs near 3580 cm^{-1}. When hydrogen bonded to other silanols, the

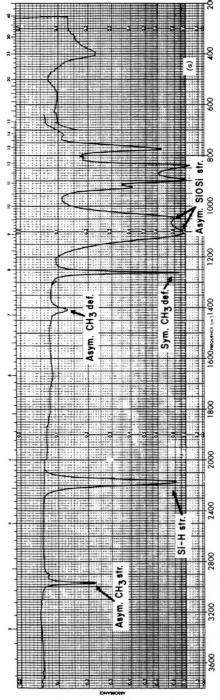

Figure 11.9. Infrared spectrum of $(MeHSiO)_x$. 3% in CCl_4 and 2% in CS_2 in 0.1-mm cells.

326

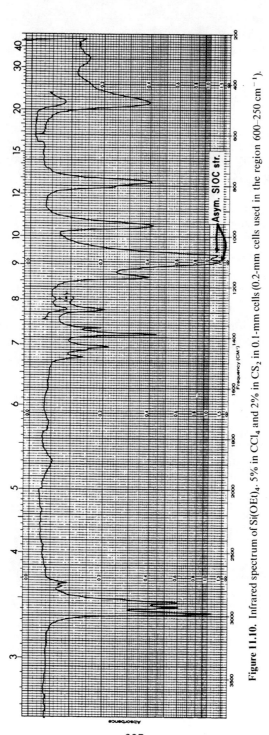

Figure 11.10. Infrared spectrum of Si(OEt)$_4$. 5% in CCl$_4$ and 2% in CS$_2$ in 0.1-mm cells (0.2-mm cells used in the region 600–250 cm^{-1}).

Asym. SIOC str.

327

absorbance is found near 3300 cm^{-1}. These hydrogen-bonded absorbances are very broad and their exact positions are influenced by neighboring functional groups and steric factors. Water also absorbs in the same region (3617 and 3708 cm^{-1} in dilute CCl$_4$ solution) and PDMS itself also shows overtone and combination band activity that overlap the SiOH region (3763, 3700, and 3640 cm^{-1}) (51). Sorting out these absorptions thus becomes an exercise in ingenuity. The SiOH group also shows an absorbance in the region 810–960 cm^{-1}. This band is weaker than the OH stretch and often difficult to identify in siloxanes as it partially overlaps other prominent bands. It is relatively insensitive to hydrogen bonding, however.

A series of organosiloxane silanols has been prepared and studied by IR (52). Infrared and Raman spectra of Me$_2$Si(OH)$_2$ (53), ViSi(OH)$_3$, Ph$_2$Si(OH)$_2$ and PhSi(OH)$_3$ have been reported (54).

3.2.5. Si–Phenyl

Absorptions from this group are very constant, and are similar to those shown by the phenyl group in PhGe, PhSn, and PhPb compounds. Bands characteristic of PhSi are found at 1120 and 1430 cm^{-1} (Figures 11.5 and 11.8).

3.2.6. SiOSi

The siloxane absorption is usually the most intense band in the spectrum of a polysiloxane. Its position, shape, and intensity give us many structural clues. For example, an isolated SiOSi group gives a single strong, rather broad band near 1050 cm^{-1}. The exact position depends on the inductive, steric, and mass effects of the substituents (55). The intensity reflects the large dipole moment change (the bond is partially ionic) and the breadth suggests that the bond angle is somewhat variable; that is, the bonds are less rigidly directional than those in carbon compounds. All these observations are consistent with the concept of partial double bond character and bond shortening described by Pauling (56).

Disubstituted linear siloxane oligomers have one SiOSi absorption for each unit in the chain, but these bands overlap and become less distinct as the chain length increases. When the number of coupled siloxane units becomes >20, the siloxane feature develops the familiar broad doublet pattern (Figure 11.2) with absorption maxima at 1020 and 1090 cm^{-1}. The symmetric stretch at ~500 cm^{-1}, although much weaker, shows a multiplicity that can be observed up to the linear hexamer (34). This band is of medium intensity in the Raman spectrum.

Cyclosiloxanes also show distinctive patterns for the small rings (57), but for rings containing more than about 20 siloxane units, the pattern merges into the same pattern as shown by linear siloxanes.

Monosubstituted siloxanes (silsesquioxanes) can be found in an infinite variety of intriguing structures, including prisms, cubes, larger three-dimensional structures, and ladders (58 and Table 2.1). As might be expected, the smaller, well-defined molecules have correspondingly well-defined IR spectra (59, 60), and show characteristic patterns in the SiOSi region. Silsesquioxanes having poorly defined structure, as might be obtained from incomplete hydrolysis of a trifunctional silane, generally give broad siloxane bands, often with some interferences from residual SiOH groups.

Unsubstituted silicas show considerable variation in both band shape and intensity, depending on the degree of crystallinity and the particle size. "Surface crystallinity" apparently affects the particle size and thus the pore diameter. Surface silanol groups can also vary considerably (absorption near $960 \, cm^{-1}$). Both these factors are important in determining the properties of silicone rubbers in which SiO_2 fillers are used. A study of silicas from various sources led to the conclusion that the sharpness of the $1110\text{-}cm^{-1}$ band was related to the degree of structural order, which is related in turn to bulk density and pore volume (61). Other absorptions in the region $1170–1230 \, cm^{-1}$ of the spectrum of SiO_2 precipitates in Si are attributed to phonon modes (62). Dehydroxylated silica has IR bands at 888 and $907 \, cm^{-1}$, assigned to highly reactive strained surface defects. These defects are probably edge-shared silicate tetrahedra (63). The spectra of silica polymorphs (64) have been discussed. The deconvoluted spectrum of "amorphous" silica is said to cast doubt on the generally accepted random network arrangement (65). A Raman spectral study of gel-derived silicates (66) showed that small siloxane rings are common to high surface–area silicate materials, including leached glasses and Cabosil. Furthermore, these rings form preferentially at the silica surface.

If SiOC or COC groups are present in the sample, the $1100 \, cm^{-1}$ CO stretching absorption will distort the appearance of the SiOSi band. Considerable caution must be used in interpreting the SiOSi region if the presence of alkoxy groups is suspected.

3.3. Ultraviolet

Aromatic substituents and polysilanes comprise the major classes of organosilicon compounds amenable to UV analysis. Other groups showing UV absorption are SiN, just above 200 nm; and SiS, which absorbs at 202 nm in $Me_3SiSSiMe_3$, and 204 and 224 nm in Me_3SiSMe. Other substituents such as

Cl and vinyl, as well as the siloxane bond (67), also show absorption but in the region 160–220 nm, which is not useful for routine UV analysis. The monograph by Ramsey (68) contains a tabulation of data for organosilicon compounds, and a discussion of the theory and practice of UV spectrosopy.

The region has been used extensively to study charge-transfer complexes. Ultraviolet cure of silicone polymers (with a sensitizing agent such as dicumyl peroxide) is common. The effect of UV on rubbers, sealants, and paints is of interest for spacecraft applications (69). Polysilane-based photoresists are of potential interest for lithography (70), but these applications fall outside our primary focus of analytical usage.

4. QUANTITATIVE ANALYSIS

4.1. Principles

Individual components of a mixture may be measured in a traditional manner using IR by measuring an isolated band and comparing its intensity to that of a standard. Sample thickness must be known accurately for both sample and standard. The measurement can be done most conveniently using solvent solutions (CCl_4 or CS_2, Ref. 1). Concentrations are then determined from the Bougher–Beer relationship

$$A = \log 1/T = abc \qquad (11.2)$$

where A is the absorbance of the band, a is a constant characteristic of the band called the absorptivity, b is the path length, and c is the concentration of the analyte. The absorptivity is obtained from a spectrum of pure material (or a mixture containing a known concentration of the analyte). Accuracy of 1% can be had by careful work; 0.1% is possible under special circumstances. If the analyte absorptions are poorly defined, or have interference from other components, more sophisticated multipoint measurements may be used to obtain quantitative data (71).

The PDMS content of lotions and ointments can be determined either directly, or by extraction with CCl_4 or other solvent (Chapter 6). The ATR technique has been used quantitatively to sample thin surface layers of polyurethane for (Me_2SiO) content (72).

Chemical group concentrations can also be determined directly by IR. For example, the amount of $SiMe_3$ in polymers may be calculated from the band at $757\,cm^{-1}$ by comparing it to the same band in $Me_3SiOSiMe_3$. This compound contains 100% $Me_3SiO_{1/2}$, so for a 2% solution in a 0.1-mm cell,

$A = 0.290$ and

$$a = \frac{A}{bc} = \frac{0.290}{\{(2.00)(0.100)\}\,100} = 0.0145 \qquad (11.3)$$

This value for A is used to calculate the concentration of $Me_3SiO_{1/2}$ in the sample shown in Figure 11.2a.

$$c = \frac{A}{ab} = \frac{0.020}{0.0145\{(2.00)(0.100)\}} = 6.9\% \text{ by weight} \qquad (11.4)$$

Groups such as SiH (73) and $SiCH=CH_2$ (74) are easily determined using the same approach. On-line gel permeation chromatography monitoring of SiH and SiOH groups by IR has been reported (75). Near-IR analysis has been used to determine the amount of Me_3Si- in treated silica (76).

Attempts have been made to quantitate the silanol group using IR. In dilute solutions, the SiOH group is largely "free" (not hydrogen bonded), and can be measured in the same manner as any other absorption band (51, 77, 78). Because the solution must be quite dilute, and the SiOH absorption is not very intense, this approach is limtied to samples containing relatively large amounts of silanol. Spectral substraction can be used to reduce the background interference (51).

4.2. Trace Analysis

Quantitative analysis for traces of silicones can be carried out using standard techniques (1, 17). It is necessary to find an absorption of the silicone that has minimum interference from the matrix. Often the 1262-cm^{-1} methyl absorption can be used. In some cases, preconcentration using solvent extraction, gas or liquid chromatography, or supercritical fluid chromatography is helpful. The success of the spectroscopy is critically dependent on attaining the maximum signal-to-noise ratio in the spectrum, for both dispersive and Fourier transform spectrometers. Instrumental parameters are optimized to achieve this objective (1).

4.3. Near-Infrared Spectroscopy

Much progress has been made in automating quantitative analysis by near-IR spectroscopy. It is essential, however, to have a well-characterized set of 40–60 standards covering the extremes of the property or compositions to be determined. The correlation is established by the software once the spectra are measured and the values for the desired property have been input.

Clearly, near-IR spectroscopy is best suited to repetitive analysis involving many similar samples. Applications to quality control and process stream analysis provide many opportunities for exploitation of this technique.

4.4. Ultraviolet

The SiPh group gives two or more moderately intense absorptions in the region 250–270 nm. An analytical method for analysis of Ph_2SiO and $PhSiO_{3/2}$ is based on the difference of their absorption maxima (79). The phenyl content of resins (80) and fluids (81) has also been determined from the UV absorption of that group. Specific detection of PhSi-containing polymers in GPC can be achieved by UV (75).

The UV and ESR spectra of organopolysilanes have been reviewed (82).

5. APPLICATIONS

Applications abound and it would not be prudent to attempt a complete discussion. Rather, in the next few sections, we describe some applications chosen to convey to the reader the wide range of problems in silicone chemistry to which IR spectroscopy has been applied.

5.1. Polymer Analysis

Silicones are often used in copolymer systems to impart thermal and oxidative stability, chemical inertness, easy mold release, resistance to weathering, and chain flexibility. Copolymer systems studied by IR include silicones combined with alkyd (83), urethane (84–86), polyether (85), polyester (87, 88), acrylic (88), and fluorocarbon (89) functional groups. The analysis of such materials generally involves ATR or photoacoustic measurements since the materials are cross-linked solids. The goal of these measurements is to identify spectral changes that may be indicative of changes in the structure resulting from weathering or other environmental stresses. One study involving the weathering of paint (83) by UV light, temperature, and water identified chemical changes from chain scission, generation of COOH, OH, and several carbonyls, oxidation and hydrolysis of binder, and hydroperoxide formation from the IR spectra alone. Infrared spectroscopy has also been useful in the identification of thermal degradation products of some of these copolymer systems (89–91). Studies of strictly silicone based polymers (92, 93) often concentrate on silanol groups, either from the polymer itself or from silica fillers, since many curing reactions take place through hydroxy groups. Attention is usually focused on the presence, changes, or disappearance of silanol moieties, which have significant influence on release properties (93), thermal stability, and inertness of polymers (92). In other cases, spectral changes are used to detect structural differences that relate to elastomer performance (94). There are also numerous studies of the thermal degradation products of silicones as measured by IR spectroscopy (95, 96).

Although IR spectroscopy is not commonly used for end-group analysis of polymers, two applications have been reported where dimethylsiloxanes are endblocked with either SiH (97) or SiMe$_3$ (98) groups and the detection of these species forms the basis for molecular weight (or, equivalently, chain length) determinations. The first case uses the SiH stretching frequency, while the second uses a Fourier deconvolution program to separate dimethyl and trimethyl absorbances in the SiC stretching region. Both methods appear applicable to polymers of molecular weights up to 15,000.

5.2. Silica and Glass

Glass fibers are often treated with silane coupling agents to improve performance in reinforced composite materials, whereas silica is derivatized with various materials for use as a polymer filler or to make chromatographic column supports. Although differing in their applications, silica and glass are similar in the analytical approaches used to characterize them. In both instances it is the surface rather than bulk properties that are of interest. Both have been analyzed by many methods in the past, including transmission IR, but in each case diffuse reflectance appears to be the most effective way of obtaining surface information. Silica and glass both have strong SiO absorbances near 1100 cm^{-1}, which obscure a few hundred wavenumbers of their spectra, but other spectral regions are relatively clean and provide readily accessible information about surface treatments. Silica is especially amenable to diffuse reflectance measurements since it is normally in the form of a fine powder. The identification and effects of many treatments have been studied by IR, including alkoxysilanes (99–101), aminosilanes (102, 103), chlorosilanes (104, 105), epoxides (106), cyclic siloxanes (107), and silazanes (105, 108). Near-IR diffuse reflectance studies of silica (109, 110) have also been helpful in elucidating the roles of silanol and water in surface reactions.

Glass fibers are generally not found in powdery form, and grinding is often inadvisable as this would expose interior regions not affected by surface treatment. Still, diffuse reflectance from the fibers, sometimes with a KBr overlayer, has produced high quality spectra. The surface treatments studied include aminopropyl (111–112), methacrylate (113, 114), vinyl (113), and alkyl (113) functionalities where attachment to the glass occurs through alkoxysilanes.

Information about the course of surface reactions and the effects of environmental factors on the surface treatment have been obtained for both glass and silica. An example of the types of spectra measured in these studies is shown in Figure 11.11a for E-glass (115) before and after treatment with a methacrylate containing coupling agent. Treatment levels are generally quite small so absorptions of the coupling agent are weak compared to the SiO absorptions from the substrate, but there is still sufficient sensitivity to follow changes in the coupling agent during a cure. An expanded view of the carbonyl and olefinic region for this example is shown in Figure 11.11b. The spectral changes observed have been assigned to specific aspects of the cure.

5.3. Biological

Silicones have found application in many areas of biological significance, and IR spectroscopy has been extensively used to analyze silicones in various of these environments. In favorable cases the biological materials can be analyzed directly. Two good examples involve the detection of residual silicone in skin and hair following exposure of these tissues to silicone-containing materials (see Chapter 6, Section 4.2). The resistance of the silicone layer to washing or other exposure was also readily determined in these studies. In many other applications the absorption of biological fluids onto silicone polymers has been determined by ATR measurements. The reaction of the human body to prolonged contact with these materials is the major question of these studies, and the detection of blood components (116–118), proteins (119–121), and lipids (119) is generally the focus of these measurements. A study that demonstrates some of these capabilities concerns the deposition of protein onto polyacrylate contact lenses treated with methoxy silanes (121). It was shown that the deposition could be reduced through replacement of surface silanols with hydrophobic organic groups. Unfortunately, this treatment also decreased wettability and made the lenses less comfortable to wear. The deposition of protein and lipids from milk onto silicone rubber surfaces (122) has been studied in an analogous fashion. The ability of ATR to handle materials with high water content is exploited in some of these studies. Infrared spectroscopy is also used to measure silicone levels in personal care (Chapter 6) or pharmaceutical (Chapter 4) materials, but these measurements are analogous to those for nonbiological applications.

5.4. Other Applications

Infrared spectroscopy has been used to analyze silicone surfaces on many substrates other than those mentioned above. Transmission measurements

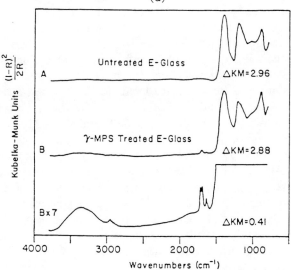

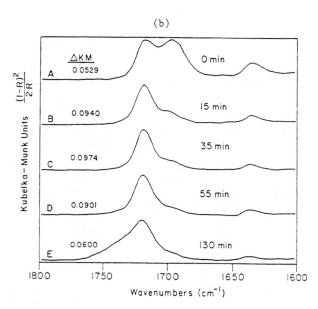

Figure 11.11. (a) Spectra of E-glass before and after treatement with 3-(trimethoxysilyl)-propyl methacrylate. (b) Expanded view of the carbonyl and vinyl region showing spectral changes as a function of cure time at 120 °C. Reprinted by permission from (115). Copyright © 1984 American Chemical Society.

335

through paper have been successful at quantitating silicone coatings applied as release layers on papers of low clay content (123, 124). The paper substrate provides a very small spectral window for transmission measurements, but fortunately it occurs in the area near $800\,\mathrm{cm}^{-1}$ where there is a strong silicone absorbance. Near-IR diffuse reflectance measurements have also been used to quantitate silicone paper coatings (125) and results comparable to XRF measurements (the standard technique) are reported. Sampling by ATR has been used to analyze silicone coatings on more opaque samples, such as clay-filled papers (124) and cotton fabric (126). Measurements on both materials can be performed quantitatively as long as the pressure applied to the sample against the ATR crystal and the surface coverage are reproducible. Also useful for these types of samples is diffuse reflectance with a KBr overlayer (127), although ATR is preferred in most cases. The spectra of silicon-containing materials on metallic surfaces have been measured using ATR (128, 129), external reflectance (130), and reflection–absorption (131, 132) techniques. The latter has been used to give information about the surface reactions and orientation of the absorbed species.

One reaction that has been much studied by IR and other techniques is the hydrolysis and subsequent condensation of alkoxy silanes in aqueous envir-

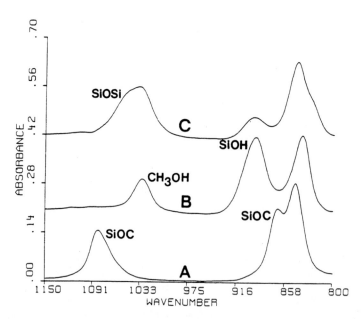

Figure 11.12. Infrared spectra of an 0.025M aqueous antimicrobial methoxysilane in (*a*) unhydrolyzed, (*b*) completely hydrolyzed, and (*c*) partially condensed form. Reprinted by permission from (134). © 1987 Elsevier Science Publishers.

onments and the factors that affect each reaction (133, 134). The kinetics for this reaction of an antimicrobial agent (134) have been delineated in a particularly thorough fashion using an ATR measurement. This reaction was followed in detail through a full range of pH values and the composition of the material at various points was determined from the IR spectra. Figure 11.12 shows spectra of some of the components that could be identified at different points in the reactions. The quality of these spectra is particularly impressive considering that all measurements were made in an aqueous environment.

REFERENCES

1. A. L. Smith, *Applied Infrared Spectroscopy*, Wiley, New York, 1979.
2. N. B. Colthup, L. H. Daly, and S. E. Wiberley, *Introduction to Infrared and Raman Spectroscopy*, 2nd ed., Academic, New York, 1975.
3. P. R. Griffiths and J. A. deHaseth, *Fourier Transform Infrared Spectroscopy*, Wiley, New York, 1986.
4. R. O. Kagel, R. W. Chrisman, and J. A. Roper III, "Raman spectroscopy," in *Treatise on Analytical Chemistry*, 2nd ed., Chapter 10, Part 1, Vol. 8, P. J. Elving, Ed., Wiley, New York, 1986.
5. J. G. Grasselli, M. K. Snavely, and B. J. Bulkin, *Chemical Applications of Raman Spectroscopy*, Wiley, New York, 1981.
6. D. P. Strommen and K. Nakamoto, *Laboratory Raman Spectroscopy*, Wiley, New York, 1984.
7. T. Hirschfeld and D. B. Chase, "FT-Raman spectroscopy: development and justification," *Appl. Spectrosc.*, **40**, 133 (1986).
8. D. B. Chase, "Fourier transform Raman spectroscopy," *Anal. Chem.*, **59**, 881A (1987).
9. F. J. Bergin and H. F. Shurvell, "Applications of Fourier transform Raman spectroscopy in an industrial laboratory," *Appl. Spectrosc.*, **43**, 516 (1989).
10. L. G. Weyer, "Near–infrared spectroscopy of organic substances," *Appl. Spectrosc. Rev.*, **21**, 1 (1985).
11. E. Stark, K. Luchter, and M. Margoshes, "Near–infrared analysis (NIRA): a technology for quantitative and qualitative analysis," *Appl. Spectrosc. Rev.*, **22**, 335 (1986).
12. J. D. Ingle, Jr., and S. R. Crouch, *Spectrochemical Analysis*, Prentice Hall, College Book Division, Old Tappan, NH, 1988.
13. A. F. Diaz, U. Hetzler, and E. Kay, "Inelastic electron tunneling spectroscopy of a chemically modified surface," *J. Am. Chem. Soc.*, **99**, 6780 (1977).
14. J. Comyn, "Inelastic electron tunneling spectroscopy and the adhesive interface," *Adhesion (London)*, **9**, 147 (1984).

15. H. Bock and B. Solouki, "Photoelectron spectra of silicon compounds," in *The Chemistry of Organic Silicon Compounds*, Chapter 9, S. Patai and Z. Rappoport, Eds., Wiley, London, 1989.

16. P. R. Griffiths, S. L. Pentoney, Jr., A. Giorgetti, and K. H. Shafer, "The hyphenation of chromatography and FT–IR spectrometry," *Anal. Chem.*, **58**, 1349A (1986).

17. A. Lee Smith, "Trace analysis by infrared spectroscopy," in *Trace Analysis*, Chapter 3, Gary D. Christian and James B. Callis, Eds., Wiley, New York, 1986.

18. N. J. Harrick, *Internal Reflection Spectroscopy*, Wiley, New York, 1967.

19. Barnes Analytical Division, Spectra-Tech, Inc., 652 Glenbrook Road, Stamford, CT.

20. T. T. Holloway, B. J. Fairless, C. E. Freidline, H. E. Kimball, R. D. Kloepfer, C. J. Wurrey, L. A. Jonooby, and H. G. Palmer, "Performance of a gas chromatographic–matrix isolation-Fourier transform infrared spectrometer," *Appl. Spectrosc.*, **42**, 359 (1988).

21. J. A. DeHaseth, R. M. Robertson, and G. K. Ferguson, "A practical full bandwidth infrared detector for liquid chromatography," *FACSS XVI*, Chicago, Talk No. 182, 1989.

22. S. L. Pentoney, Jr, K. H. Shafer, P. R. Griffiths, and R. Fuoco, "Supercritical fluid chromatography/Fourier transform infrared microspectrometry," *HRC CC, J. High Resolut. Chromatogr. Chromatogr. Commun.*, **9**, 168 (1986).

23. C. Fujimoto, Y. Hirata, and K. Jinno, "Supercritical fluid chromatography-infrared spectroscopy of oligomers: Use of buffer-memory technique," *J. Chromatogr.*, **332**, 47 (1985).

24. M. T. McKenzie and J. L. Koenig, "Further developments in the methodology of surface analysis by FT–IR: quantitative aspects of diffuse reflectance methods," *Appl. Spectrosc.*, **39**, 408 (1985).

25. R. S. Shreedhara Murthy, J. P. Blitz, and D. E. Leyden, "Quantitative variable-temperature diffuse reflectance infrared Fourier transform spectrometric studies of modified silica gel samples," *Anal. Chem.*, **58**, 3167 (1986).

26. D. W. Vidrine, "Photoacoustic Fourier transform infrared spectroscopy of solids and liquids," in *Fourier Transform Infrared Spectroscopy*, Vol. 3, Chapter 4, J. R. Ferraro and L. J. Basile, Eds., Academic, New York, 1982.

27. J. A. Gardella, Jr, G. L. Grobe, III, W. L. Hopson, and E. M. Eyring, "Comparison of attenuated total reflectance and photoacoustic sampling for surface analysis of polymer mixtures by Fourier transform infrared spectroscopy," *Anal. Chem.*, **56**, 1169 (1984).

28. B. S. H. Royce, Y. C. Teng, and J. A. Ors, "Reactions in polymers studied with FTIR–photoacoustic spectroscopy," *Ultrason. Symp. Proc.*, **2**, 784 (1981).

29. W. G. Golden, "Fourier transform infrared reflection–absorption spectroscopy," in *Fourier Transform Infrared Spectroscopy*, Vol. 4, Chapter 8, J. R. Ferraro and L. J. Basile, Eds., Academic, New York, 1985.

30. R. G. Messerschmidt and M. A. Harthcock, Eds., *Infrared Microspectroscopy*, Marcel Dekker, New York, 1988.

31. L. J. Bellamy, *The Infrared Spectra of Complex Molecules*, 3rd ed., Wiley, New York, 1975.

32. L. J. Bellamy, *Advances in Infrared Group Frequencies*, Barnes and Noble, New York, 1968.

33. A. L. Smith, "Infrared spectra–structure correlations for organosilicon compounds," *Spectrochim. Acta*, **16**, 87 (1960).

34. A. L. Smith, "Infrared spectra of organosilicon compounds in the cesium bromide region," *Spectrochim. Acta*, **19**, 849 (1963).

35. A. L. Smith, "Infrared group frequencies for structure determination in organosilicon compounds," in *Chemical, Biological and Industrial Applications of Infrared Spectroscopy*, J. R. Durig, Ed., Wiley, New York, 1985.

36. B. J. Wythoff, C. F. Buck, and S. A. Tomellini, "Descriptive, interactive computer-assisted interpretation of infrared spectra," *Anal. Chim. Acta*, **217**, 203 (1989).

37. S. Moldoveanu and C. A. Rapson, "Spectral interpretation for organic analysis using an expert system," *Anal. Chem.*, **59**, 1207 (1987).

38. V. A. Koptyug, Ed., *Atlas of Spectra of Organic Compounds, No. 33: IR Spectra of Oxygen-Containing Organosilicon Compounds*, Novosibirskii Institut Organicheskoi Khimii, Novosibirsk, USSR, 1986.

39. C. D. Craver, Ed., *The Coblentz Society Desk Book of Infrared Spectra*, 2nd ed., The Coblentz Society, PO Box 9952, Kirkwood, MO, 1982.

40. C. D. Craver, Ed., *The Coblentz Society Collection of Evaluated Infrared Spectra*, The Coblentz Society, PO Box 9952, Kirkwood, MO.

41. Chicago Society for Coatings Technology, *An Infrared Spectroscopy Atlas for the Coatings Industry*, Federation of Societies for Coatings Technology, Philadelphia, 1980.

42. D. O. Hummel, *Atlas of Polymer and Plastics Analysis*, 2nd ed., Verlag, Chemie International, New York, 1978.

43. C. J. Pouchert, *The Aldrich Library of FT–IR Spectra*, ed. I, Aldrich Chemical Co., Milwaukee, WI, 1985.

44. R. Mecke and F. Langenbucher, *Infrared Spectra of Selected Chemical Compounds*, Heyden & Son, Spectrum House, Alderton Crescent, London NW4.

45. B. Schrader, *Raman/IR Atlas of Organic Compounds*, 2nd ed., VCH Publishers, New York, 1989.

46. Sadtler Research Laboratories, *The Infrared Spectra Atlas of Monomers and Polymers*, Sadtler Laboratories, Philadelphia, 1980. *Sadtler Standard Infrared Collection*, Sadtler Laboratories, Philadelphia.

47. R. J. Keller, *The Sigma Library of FT–IR Spectra*, Sigma Chemical Co., St. Louis, MO, 1986.

48. L. J. Bellamy, "The origins of group frequency shifts," in *Spectroscopy*, M. J. Wells, Ed., The Institute of Petroleum, London, 1962, pp. 205–221.

49. A. L. Smith and N. C. Angelotti, "Correlation of the SiH stretching frequency with molecular structure," *Spectrochim. Acta*, **15**, 412 (1959).

50. A. L. Smith, "Are group frequencies obsolete?," *Appl. Spectrosc.*, **41**, 1101 (1987).

51. G. W. Griffith, "Quantitation of silanol in silicones by FTIR spectroscopy," *Ind. Eng. Chem. Prod. Res. Dev.*, **23**, 590 (1984).

52. E. Popowski, N. Holst, and H. Kelling, "Preparation and IR spectroscopic studies of siloxysilanols," *Z. Anorg. Allg. Chem.*, **494**, 166 (1982).

53. T. D. Ho, "Molecular vibrational frequency assignment of dimethylsilanediol," *Appl. Spectrosc.*, **40**, 29 (1986).

54. H. Ishida and J. L. Koenig, "Vibrational assignments of organosilanetriols. I. Vinylsilanetriol and vinylsilanetriol-D_3 in aqueous solutions," *Appl. Spectrosc.*, **32**, 462 (1978); "II. Crystalline phenylsilanetriol and phenyltrisilanetriol-D_3," **32**, 469 (1978).

55. B. I. D'Yachenko, N. K. Vikulova, V. P. Rybalko, V. V. Kireev, S. R. Stepanenko, Y. M. Talanov, A. D. Gut'ko, and E. B. Gubina, "Quantitative determination of the effect of substituent nature on the position of the silicon–oxygen–silicon bond asymmetric valence vibration band in the IR spectra of siloxanes," *Proc. Acad. Sci. USSR*, **271**, 1425 (1983), *Chem. Abstr.*, **100**: 51689x (1984).

56. L. Pauling, *The Nature of the Chemical Bond*, 3rd ed., Cornell University Press, Ithaca, NY, 1960.

57. T. Alvik and J. Dale, "Infrared and Raman spectra of cyclic dimethylsiloxane oligomers," *Acta Chem. Scand.*, **25**, 2142 (1971).

58. J. F. Brown, Jr., L. H. Vogt, Jr., and P. I. Prescott, "Preparation and characterization of the lower equilibrated phenylsilsesquioxanes," *J. Am. Chem. Soc.*, **86**, 1120 (1964).

59. C. L. Frye and W. T. Collins, "The oligomeric silsesquioxanes $(HSiO_{3/2})_n$," *J. Am. Chem. Soc.*, **92**, 5586 (1970).

60. K. A. Mauritz and R. M. Warren, "Microstructural evolution of a silicon oxide phase in a perfluorosulfonic acid ionomer by an in situ sol–gel reaction. I. Infrared spectroscopic studies," *Macromolecules*, **22**, 1730 (1989).

61. J. Gallus-Oleander, B. Franc, and L. Firlus, "Infrared investigations of different kinds of synthetic active silicas," *Spectrosc. Lett.*, **12**, 165 (1979).

62. S. M. Hu, "Infrared absorption spectra of silicon dioxide precipitates of various shapes in silicon: calculated and experimental," *J. Appl. Phys.*, **51**, 5945 (1980).

63. B. C. Bunker, D. M. Haaland, K. J. Ward, T. A. Michalske, J. S. Binkley, C. F. Melius, and C. A. Balfe, "Infrared spectra of edge-shared silicate tetrahedra," *Surf. Sci.*, **210**, 406 (1989).

64. E. Gorlich, K. Blaszczak, and M. Handke, "Infrared spectra of silica polymorphs," *Mineral Pol.*, **141**, 3 (1985), through *Chem. Abstr.*, **105**: 46411z (1986).

65. I. W. Boyd, "Deconvolution of the infrared absorption peak of the vibrational stretching mode of silicon dioxide: evidence for structural order?," *App. Phys. Lett.*, **51**, 418 (1987).

66. D. R. Tallant, B. C. Bunker, C. J. Brinker, and C. A. Balfe, "Raman spectra of rings in silicate minerals," *Mater. Res. Soc. Symp. Proc.*, **73** (Better Ceram. Chem. 2), 261 (1986). *Chem. Abstr.*, **105**: 215776e (1986).

67. E. A. Kirichenko, A. I. Ermakov, N. I. Pimkin, and V. M. Kopylov, "Ultraviolet absorption of siloxane bond in the 186–220 nm region," *Zh. Obshch. Khim.*, **50**, 1576 (1280 CB trans.) (1980).

68. B. G. Ramsey, *Electronic Transitions in Organometalloids*, Academic, New York, 1969.

69. J. C. Guillaumon and J. Guillin, "Paints, potting compounds and silicone varnishes with low outgassing in space environment," *Eur. Space Agency Spec. Publ. ESA SP*, 63 (1979).

70. J. M. Zeigler, L. A. Harrah, and A. W. Johnson, "Self-developing polysilane deep-UV resists—photochemistry, photophysics, and submicron lithography," *Proc. SPIE-Int. Soc. Opt. Eng.*, **539** (Adv. Resist Technol. Processing 2), 166 (1985); *Chem. Abstr.*, **103**: 62418w (1985).

71. G. L. McClure, Ed., *Computerized Quantitative Infrared Analysis*, ASTM Special Technical Publication, American Society for Testing and Materials, Philadelphia, 1987.

72. R. Iwamoto and K. Ohta, "Quantitative surface analysis by Fourier transform attenuated total reflection infrared spectroscopy," *Appl. Spectrosc.*, **38**, 359 (1984).

73. H. Rotzsche, U. Prietz, H. Diedrich, H. Clauss, and H. Hahnewald, "Trace analysis of Si–H and Si–Cl groups in organopolysiloxanes," *Plaste Kautsch.*, **25**, 390 (1978); through *Chem. Abstr.*, **89**: 147244d (1978).

74. K. A. Andrianov, A. Y. Radkina, L. A. Leites, and B. G. Zavin, "Determining the relative content of vinyl groups in various polyvinyl(phenyl)siloxanes by IR spectroscopy," *Vysokomol. Soedin. Ser. (A) (B)*, A17, 682 (1975); through *Chem. Abstr.*, **83**: 28903j (1975).

75. E. Kohn and M. E. Chisum, "Determination of functional groups in molecular components of polydimethylsiloxanes," *ACS Symp. Ser.*, **352**, 169 (1987).

76. S. G. Bush, J. W. Jorgenson, M. L. Miller, and R. W. Linton, "Transmission near-infrared technique for evaluation and relative quantitation of surface groups on silica," *J. Chromatogr.*, **260**, 1 (1983).

77. S. V. Dubiel, G. W. Griffith, C. L. Long, G. K. Baker, and R. E. Smith, "Determination of reactive components in silicone foams," *Anal. Chem.*, **55**, 1533 (1983).

78. V. M. Krasikova, A. L. Klebanskii, L. A. Klimov, Y. A. Yuzhelevskii, V. F. Gridina, L. P. Dorofeenko, B. L. Kaufman, and T. F. Rogozina, "Analysis of the silanol group in organosilicon polymers by infrared spectroscopy," *Zh. Anal. Khim.*, **30**, 1992 (1975); through *Chem. Abstr.*, **84**: 45098p (1976).

79. T. Uriu and T. Hakamada, "Analysis of silicones by ultraviolet absorption spectra," *J. Chem. Soc. Jpn., Ind. Chem. Sect.*, **62**, 1421 (1959); *Chem. Abstr.*, **57**: 9873b (1962).

80. H. Waledziak, "Ultraviolet spectrophotometric determination of the content of the phenyl groups in methyl phenyl siloxane resins," *Chem. Anal. (Warsaw)*, **10**, 579 (1965); through *Chem. Abstr.*, **64**: 9884g (1966).

81. L. A. Efremova and K. K. Popov, "Spectroscopic analysis of chlorosilane mixtures and determination of phenyl groups in methylphenylpolysiloxanes," *Zavodsk. Lab.*, **29**, 708 (1963); through *Chem. Abstr.*, **59**: 8124f (1963).

82. H. Sakurai, "Spectra and some reactions of organopolysilanes," *J. Organomet. Chem.*, **200**, 261 (1980).

83. J. Hodson and J. A. Lander, "The analysis of cured paint media and a study of the weathering of alkyd paints by FTIR/PAS," *Polymer*, **28**, 251 (1987).

84. C. S. Sung and C. B. Hu, "Surface chemical analysis of segmented polyurethanes. Fourier transform IR internal reflection studies," *Adv. Chem. Ser.*, **176**, 69 (1979).

85. F. A. Zhokhova and V. V. Zharkov, "Determination of silicon–hydrogen bond content in foam stabilizers," *Plast. Massy (Plastics)*, **6**, 54 (1983).

86. G. L. Grobe, III, J. A. Gardella, Jr, W. L. Hopson, W. P. McKenna, and E. M. Eyring, "Angular dependent ESCA and infrared studies of segmented polyurethanes," *J. Biomed. Mater. Res.*, **21**, 211 (1987).

87. E. Takeshima, T. Kawano, and H. Mizuki, "Forecast of longevity of paint coated and plastic film laminated steel sheets. Part 3. Analysis of degradation of organic coatings by FT–IR," *Nisshin Seiko Giho*, **47**, 37 (1982).

88. E. W. Kinkelaar, J. T. Rozsa, and L. J. Vavruska, "Cure determination of coil coatings by the use of attenuated total reflectance (ATR) infrared techniques," *J. Coat Technol.*, **46**, 63 (1974).

89. D. J. McEwen, W. R. Lee, and S. J. Swarin, "Combined thermogravimetric and infrared analysis of polymers," *Thermochim. Acta*, **86**, 251 (1985).

90. Z. Il'ina, M. Bryk, V. Kardanov, and O. Kurilenko, "Thermogravimetric and spectroscopic study of epoxide resins and their mixtures with poly-(aluminoethylsiloxane)," *Russ. Book*, **6**, 61 (1974); through *Chem. Abstr.*, **83** 60032k (1974).

91. K. V. Zapunnaya, V. P. Kuznetsova, and A. N. Dobrovol'skaya, "Study of the thermal stability of organosilicon polyurethanes," *Kompoz. Polim. Mater. (Composition Polymeric Materials)*, **7**, 30 (1980).

92. G. Berrod, A. Vidal, E. Papirer, and J. B. Donnet, "Reinforcement of siloxane elastomers by silica. Chemical interactions between an oligomer of poly(dimethylsiloxane) and a fumed silica," *Appl. Polym. Symp.*, **23**, 833 (1981).

93. A. G. Kingsbury, "Poly(dimethylsiloxanes) as release agents," *Pentacol*, **1974**, 52.

94. B. N. Ranganathan, "Fourier-transform infrared spectroscopy evaluation of commercial room-temperature vulcanized silicones," *Anal. Lett.*, **17**, 2221 (1984).

95. A. A. Pashchenko, M. G. Voronkov, and L. A. Shevchenko, "Hydrothermal decomposition of poly(organylsiloxanes)," *Zh. Prikl. Khim. Leningrad (J. Appl. Chem.)*, **49**, 826 (1976).

96. V. S. Papkov, M. N. Il'ina, Y. P. Kvachev, N. N. Makarova, A. A. Zhdanov, G. L. Slonimskii, and K. A. Andrianov, "Features of the thermal degradation of ladder siloxane polymers," *Vysokomol. Soedin. Ser. (A) (B) (Journal of the Theoretical and Experimental Chemistry and Physics of Macromolecular Compounds)*, **A17**, 2050 (1975).

97. P. J. Madec and E. Marechal, "Molecular weight determination of silane-terminated poly(dimethylsiloxane) by infrared spectroscopy," *J. Polym. Sci.*, **18**, 2417 (1980).

98. E. D. Lipp, "Application of Fourier self-deconvolution to the FT–IR spectra of polydimethylsiloxane oligomers for determining chain length," *Appl. Spectrosc.*, **40**, 1009 (1986).

99. J. P. Blitz, R. S. Shreedhara Murthy, and D. E. Leyden, "Studies of silylation of Cab–O–Sil with methoxymethylsilanes by diffuse reflectance FTIR spectroscopy," *J. Colloid Interface Sci.*, **121**, 63 (1988).

100. J. W. DeHaan, H. M. Van den Bogaert, J. J. Ponjee, and L. J. M. Van de Ven, "Characterization of modified silica powders by Fourier transform infrared spectroscopy and cross-polarization magic angle spinning NMR," *J. Colloid Interface Sci.*, **110**, 591 (1986).

101. D. E. Leyden and D. E. Williams, "Spectrochemical characterization of chemically modified surfaces," *Polym. Prepr. Am. Chem. Soc. Div. Polym. Chem.*, **24**, 225 (1983).

102. S. Naviroj, J. L. Koenig, and H. Ishida, "Diffuse reflectance Fourier transform infrared spectroscopic study of chemical bonding and hydrothermal stability of an aminosilane on metal oxide surfaces," *J. Adhesion*, **18**, 93 (1985).

103. S. Kondo, T. Ishikawa, N. Yamagami, K. Yoshioka, and Y. Nakahara, "Surface characterization of silica modified with aminosilane compounds," *Bull. Chem. Soc. Jpn.*, **60**, 95 (1987).

104. S. G. Bush, J. W. Jorgenson, M. L. Miller, and R. W. Linton, "Transmission near-infrared technique for evaluation and relative quantitation of surface groups on silica," *J. Chromatogr.*, **260**, 1 (1983).

105. K. Tsutsumi and H. Takahashi, "Studies of surface modification of solids. IV. Immersional heats and infrared spectra of organosilane-treated silica," *Colloid Polym. Sci.*, **263**, 506 (1985).

106. B. Casal and E. Ruiz-Hitzky, "IR spectroscopic study of the interaction of epoxides with silanol groups on silica surfaces," *An. Quim. Ser. B*, **80**, 315 (1984).

107. R. Lagarde and J. Lahaye, "Surface properties of an aerosil treated with octamethylcyclotetrasiloxane," *Bull. Soc. Chim. Fr.*, **9–10**, 825 (1977).

108. T. Welsch and H. Frank, "A study in high-temperature silylation of silica," *HRC CC, J. High Resolut. Chromatogr. Chromatogr. Commun.*, **8**, 709 (1985).

109. H. Yamauchi and S. Kondo, "The structure of water and methanol adsorbed on silica gel by FT–NIR spectroscopy," *Colloid Polym. Sci.*, **266**, 855 (1988).

110. A. Krysztafkiewicz and B. Rager, "NIR studies of the surface modification in silica fillers," *Colloid Polym. Sci.*, **266**, 485 (1988).

111. M. T. McKenzie, S. R. Culler, and J. L. Koenig, "Applications of diffuse reflectance FT–IR to the characterization of an E-glass fiber/γ-APS coupling agent system," *Appl. Spectrosc.*, **38**, 786 (1984).

112. S. R. Culler, H. Ishida, and J. L. Koenig, "FT–IR characterization of the reaction

at the silane/matrix resin interphase of composite materials," *J. Colloid Interface Sci.*, **109**, 1 (1986).

113. H. Ishida and J. L. Koenig, "A Fourier-transform infrared spectroscopic study of the hydrolytic stability of silane coupling agents of E-glass fibers," *J. Polym. Sci.*, **18**, 1931 (1980).

114. G. Wiedemann, B. Wustmann, H. Maulhardt, and D. Kunath, "FTIR diffuse reflectance spectroscopic analysis of methacrylosilane on glass fiber fabrics," *Acta Polym.*, **35**, 584 (1984).

115. R. T. Graf, J. L. Koenig, and H. Ishida, "Characterization of silane-treated glass fibers by diffuse reflectance Fourier transform spectrometry," *Anal. Chem.*, **56**, 773 (1984).

116. J. H. Kennedy, H. Ishida, L. S. Staikoff, and C. W. Lewis, "Correlation of infrared spectroscopy with platelet morphology in blood compatibility studies of polydimethylsiloxane membranes," *Biomater. Med. Devices Artif. Organs*, **6**, 215 (1978).

117. E. I. Semenenko, A. I. Ivanov, M. A. Markelov, N. B. Dobrova, E. V. Smurova, M. B. Il'ina, and Y. A. Perimov, "Study of changes in polymeric movable elements of artificial heart valves due to the action of blood and its components," *Vysokomol. Soedin. Ser. (A) (B)*, **19**, 1336 (1977).

118. L. M. Seifert and R. T. Greer, "Evaluation of in vivo adsorption of blood elements onto hydrogel-coated silicone rubber by scanning electron microscopy and Fourier transform infrared spectroscopy," *J. Biomed. Mater. Res.*, **19**, 1043 (1985).

119. E. Roggendorf, A. Reklat, and D. Kunath, "Measuring the weakened IR total reflection for testing the biostability of plastics and elastomers," *Plaste Kautsch.*, **23**, 586 (1976).

120. R. Kellner and G. Gidaly, "Fourier-transform infrared versus conventional grating-infrared-attenuated total reflectance spectroscopic investigation of the adsorption of blood proteins on polymer surfaces," *Mikrochim. Acta*, **1**, 119 (1981).

121. X. M. Deng, E. J. Castillo, and J. M. Anderson, "Surface modification of soft contact lenses: silanization, wettability and lysozyme adsorption studies," *Biomaterials*, **7**, 247 (1986).

122. M. Nordman-Montelius and I. von Bockelmann, "Analyses of raw milk deposits on non-heated polymer surfaces," *2nd Int. Conf., Fouling Clean. Food Process.*, **1985**, 276.

123. M. E. Grenoble, "Infrared determination of silicone coating weight," *Course Notes—Solventless Silicone Coat. Short Course*, 28–30 TAPPI: Atlanta, GA, 1979.

124. E. D. Lipp, P. S. Rzyrkowski, and R. F. Geiger, "Comparison of X-ray fluorescence with other techniques for the quantitative analysis of silicone paper coatings," *Tappi J.*, **70**, 95 (1987).

125. C. M. Paralusz, "Near-infrared reflectance analysis for silicone coating weights," *Appl. Spectrosc.*, **43**, 1273 (1989).

126. R. S. Shreedhara Murthy, D. E. Leyden, and R. P. D'Alonzo, "Determination of polydimethylsiloxane on cotton fabrics using Fourier transform attenuated total reflection infrared spectroscopy," *Appl. Spectrosc.*, **39**, 856 (1985).

127. M. T. McKenzie and J. L. Koenig, "Fourier transform diffuse reflectance infrared study of fibers, polymer films and coatings," *Polym. Prepr. (Am. Chem. Soc. Div. Polym. Chem.)*, **25**, 180 (1984).

128. R. W. Phillips, L. U. Tolentino, and S. Feuerstein, "Spacecraft contamination under simulated orbital environment," *J. Spacecr. Rockets*, **14**, 501 (1977).

129. R. H. Honeycutt and J. P. Wightman, "Analysis of thin films of DC-704 on aluminum, germanium, and KRS-5 surfaces by infrared reflection spectroscopy," *J. Vac. Sci. Technol.*, **14**, 742 (1977).

130. F. J. Boerio and C. A. Gosselin, "IR spectra of polymers and coupling agents adsorbed onto oxidized aluminum," *Adv. Chem. Ser.*, **203**, 541 (1983).

131. F. J. Boerio, "Reflection–absorption infrared spectra of gamma-aminopropyl-triethoxysilane adsorbed on bulk iron," in *Applied Polymer Spectroscopy*, E. G. Bramer Jr., Ed., Academic, New York, 1978, p. 171.

132. F. J. Boerio, J. W. Williams, and J. M. Burkstrand, "The structure of films formed by gamma-aminopropyltriethoxysilane adsorbed onto copper," *J. Colloid Interface Sci.*, **91**, 485 (1983).

133. C. C. Lin and J. D. Basil, "Silicon-29 NMR, SEC and FTIR studies of the hydrolysis and condensation of tetraethoxysilane $Si(OC_2H_5)_4$ and hexaethoxy-disiloxane $Si_2O(OC_2H_5)_6$," *Mater. Res. Soc. Symp. Proc.*, **73**, 585 (1986).

134. D. E. Leyden, R. S. Shreedhara Murthy, J. B. Atwater, and J. P. Blitz, "Studies of alkoxysilane hydrolysis and condensation by Fourier-transform infrared spectroscopy with a cylindrical internal-reflection cell," *Anal. Chim. Acta*, **200**, 459 (1987).

CHAPTER

12

NUCLEAR MAGNETIC RESONANCE SPECTROSCOPY

R. B. TAYLOR, B. PARBHOO, and D. M. FILLMORE

Dow Corning Corporation
Midland, Michigan

1. INTRODUCTION

Of all the techniques covered in this edition, the field of nuclear magnetic resonance spectroscopy (NMR) has experienced the most growth and change since the writing of the first edition of this book. Pulsed Fourier transform NMR (FT–NMR) methods and wide-band proton decoupling were known in 1973 but not yet widely available, and were largely limited in application to ^{13}C NMR of low molecular weight or high symmetry organic compounds. Since then, ^{13}C and ^{29}Si NMR of organosilicon compounds have become routine analytical methods in many laboratories. Major technological advances in computers, radio frequency (rf) electronics, and superconducting magnet design have transformed the state of the art in NMR. Dynamic range—the ability to digitize a very small signal in the presence of a very large signal—is no longer limited by the analog to digital converter (ADC); 32-bit word acquisition sizes now permit digitization over a dynamic range of 8 orders of magnitude. Desktop microcomputers have surpassed the power of older, large mainframes to allow fast Fourier transformation of 128k word acquisitions in less than 10 s with 80-bit precision. Powerful pulse programmers have made possible the development of literally hundreds of pulse sequences for specialized one-, two-, and three-dimensional FT–NMR experiments. Applications and breakthroughs in multifilament superconducting materials have resulted in the commercial availability of 600-MHz spectrometers. Taken together, these advancements in instrumentation have lowered the practical concentration limit for ^{13}C and ^{29}Si NMR to approximately 500 M ppm for an overnight experiment. A 200 ppm-wide spectrum

The Analytical Chemistry of Silicones, Edited by A. Lee Smith.
ISBN 0-471-51624-4 © 1991 John Wiley & Sons, Inc.

347

can be acquired with digital resolution exceeding 0.001 ppm. Solid state ^{13}C and ^{29}Si FT–NMR have become practical, and two-dimensional (2-D) multinuclear FT–NMR has become routine.

In this chapter, we cover the application of pulsed ^{1}H, ^{13}C, and ^{29}Si FT–NMR to the analysis of organosilicon compounds as practiced in our laboratories, with pertinent examples from the literature. No attempt has been made to offer a critical literature review of the subject or to reproduce the theoretical details of NMR. Several good reviews and general articles on ^{29}Si NMR may be found in Refs. 1–5. For current awareness of developments in NMR, the reader is referred to the biennial reviews in *Analytical Chemistry* and the *Specialist Periodical Reports* on NMR.

2. PRACTICAL CONSIDERATIONS

Silicon-containing molecules and, in particular, polysiloxanes, are amenable to NMR analytical studies as they are made of atoms having nuclei that resonate in a magnetic field (6). The nuclei ^{1}H, ^{13}C, ^{19}F, and ^{29}Si, have a nuclear spin $I = \frac{1}{2}$ and give sharp resonance signals with frequencies or chemical shifts that are very sensitive to changes in molecular electronic structure (Table 12.1). Other nuclei, such as ^{14}N, ^{17}O, and ^{35}Cl, can also be observed, but the signals are broad because of their nuclear electric quadrupole moments. Chemically similar sites in a molecule cannot be resolved. Therefore, ^{14}N, ^{17}O, and ^{35}Cl are not very informative from the structural point of view. The comparison of the observed ^{1}H, ^{13}C, and ^{29}Si chemical shift range shown in Figure 12.1 illustrates the relative sensitivity of the chemical shift to molecular structure for methylpolysiloxanes and underlines the importance of ^{29}Si NMR (7).

Table 12.1. Properties of ^{1}H, ^{13}C, ^{19}F, and ^{29}Si^a

Isotope	Natural Abundance (%)	Magnetic Moment (μ/μ_n)	Magnetogyric Ratio $\gamma/10^7$ (rad·T^{-1} s^{-1})	NMR Frequency (MHz)	Relative Receptivity
^{1}H	99.985	4.8371	26.7510	100	5.68×10^3
^{13}C	1.108	1.2162	6.7263	25.145	1.000
^{19}F	100	4.5506	25.1665	94.094	4.73×10^3
^{29}Si	4.7	$- 0.9609$	$- 5.3141$	19.867	2.09

^{a} See Ref. 6.

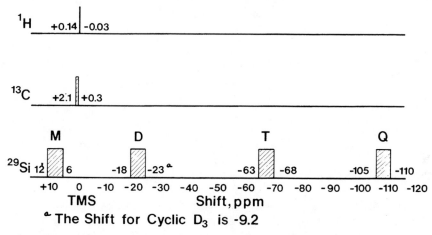

Figure 12.1. Comparison of the chemical shift ranges for methylsiloxanes in ^{1}H, ^{13}C, and ^{29}Si NMR (7).

The origin of the chemical shifts (δ) and of the scalar coupling constants (J) is well understood for protons and is related to molecular electronic structures. The theoretical principles involved are relatively simple as only two electrons can occupy the 1s atomic orbital surrounding the proton nucleus. The chemical shifts of nearly all hydrogen atoms encountered in organic and organosilicon compounds lie in a small range of 0–10 ppm. For ^{13}C, ^{19}F, and ^{29}Si nuclei, more electrons surround the nucleus and fill the higher energy orbitals. The chemical shift then becomes more sensitive to changes in the molecular structure and subtle differences can be observed (2).

Nuclear magnetic resonance experiments performed today are based on a pulsed Fourier transformation (FT–NMR) technique, which allows the observation of a whole spectrum resulting from the application of a pulse of rf energy. The pulse width (PW) is in the order of microseconds and repeated pulses allow the collection of several hundreds of spectra that are coadded to produce a final spectrum with a high signal-to-noise (S/N) ratio. During the rf pulse, the spin system is perturbed from its thermal equilibrium state. After the termination of the pulse the system returns to equilibrium by exponential relaxation processes, which are characterized by spin–lattice and spin–spin relaxation times T_1 and T_2, respectively. During the spin–spin relaxation process, a signal, fluctuating in time and called free induction decay (FID), is detected by the receiver coil of the spectrometer. This FID is a time domain NMR spectrum that after Fourier transformation is converted to a frequency domain NMR spectrum.

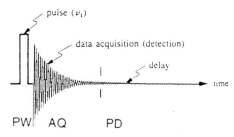

Figure 12.2. The free induction decay following pulse excitation.

The whole sequence of events experienced by the nuclear spin system can be represented by the pulse sequence illustrated in Figure 12.2. In this figure PW represents the pulse width during which the rf field is switched on, AQ is the acquisition time during which the FID data are collected, and PD is the pulse delay that allows the spin system to return to equilibrium before repetition of the entire pulse sequence. These parameters have a direct and profound influence on the spectral characteristics. It is, therefore, important that the chemist understands them.

The events following the excitation pulse are best described by quantum mechanics. However, a vector model with a rotating coordinate system allows basic understanding of the relationship between the sequence parameters and the spectrum information needed by the chemist interested in polysiloxanes.

At equilibrium, the net magnetization M_0 is aligned along the z axis (Figure 12.3). Applying the rf pulse along the x axis tilts M_0 through an angle Θ in the yz plane. It is the projection of the M_0 vector along the y axis that produces the FID. The maximal signal is obtained when the angle Θ is 90°. The system is then allowed to return to equilibrium during the relaxation delay (RD), which comprises the AQ and the pulse delay (PD). The parameters PW and PD can be chosen so as to optimize the S/N ratio of a qualitative spectrum for a given instrumental time. Or by choosing a PW corresponding to 90° and allowing the nuclei of interest to fully relax, one can obtain quantitative spectra. Relating the integrated intensities to the molecular structure proves to be extremely informative for ^{13}C and ^{29}Si NMR, as it is for ^{1}H NMR. Quantitative ^{1}H, ^{19}F, and ^{31}P NMR spectra are easily obtained as the T_1 relaxation times range between milliseconds and seconds. Moreover, the high sensitivity of these nuclei necessitates only a few transients to be acquired and a whole experiment lasts only a few minutes.

In the case of low sensitivity nuclei such as ^{13}C and ^{29}Si, spectra with good S/N ratios are obtained only after accumulation of several hundred transi-

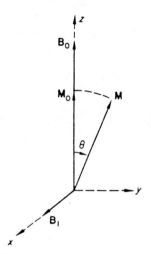

Figure 12.3. Vector model of the nuclear spin system. [From (9)]. Reprinted by permission of John Wiley & Sons, Ltd. © 1980.

ents. The S/N ratio depends on the square root of the number of transients [Eq. (12.1)].

$$S/N \propto (NT)^{1/2} \qquad (12.1)$$

which implies that to improve the sensitivity by a factor of 2, one has to increase the experimental time by a factor of 4. A more practical way of improving sensitivity is to use maximum sample concentration and larger sample tubes for ^{13}C and ^{29}Si NMR. Furthermore, the long spin–lattice relaxation times, T_1, which are on the order of 5–80 s, require relaxation delays of $\geq 5\ T_1$ or 25–400 s, imposing very long experiment times for quantitative spectra. These relaxation times can be greatly reduced by adding a shiftless relaxation reagent to the solution as discussed in Section 5.4.

Another way to reduce the NMR experiment time is to use a higher magnetic field instrument, as the S/N ratio is proportional to the $\frac{3}{2}$ power of the induction field strength B_0.

The theoretical improvement in sensitivity defined as the improvement factor (IF) is reported in Table 12.2 for a constant experimental time as a function of field strength. Alternatively, the time saving realized with higher magnetic fields illustrates the importance of superconducting magnets (Table 12.2).

In ^1H NMR, a single-pulse sequence is used to obtain a spectrum that contains all the information on chemical shifts, coupling constants, and

Table 12.2. Reductions in Experimental Time at Constant S/N Ratios for NMR Experiments Conducted at Magnetic Field Strengths $> 4.7\,T$ (200 MHz)

Field Strength (T)	^{1}H NMR Observation Frequency (MHz)	IF^a	$1/(\text{IF})^2$	Experimental Time (h)
4.7	200	1.00	1.00	24.0
6.3	270	1.57	0.41	9.8
7.1	300	1.84	0.30	7.2
8.5	360	2.42	0.17	4.1
9.4	400	2.83	0.13	3.1
11.8	500	3.95	0.06	1.4

^{a} IF = improvement factor for higher field strengths relative to a 200-MHz spectrometer.

relative intensities of the signals. The same pulse sequence also gives a ^{13}C NMR spectrum. However, the spectrum appears very complex because of ^{13}C–^{1}H spin–spin scalar coupling. This complexity can be eliminated by continuously irradiating the whole proton spectrum while observing the ^{13}C spectrum (Figure 12.4a).

The proton decoupled carbon-13 spectrum, symbolized as ^{13}C{^{1}H}, then shows a series of single lines, each corresponding to a nonequivalent carbon site. With proton decoupling, another advantage emerges: the nuclear Overhauser effect (NOE). The continuous irradiation of the proton nuclei enhances the intensities of ^{13}C signals for those nuclei that are close to hydrogen atoms and decreases the overall accumulation time. Unfortunately, the intensities of the signals are no longer proportional to the number of atoms in the molecule. The NOE can be suppressed, however, by irradiating the proton nuclei only during the acquisition of the ^{13}C FID (Figure 12.4b). One can then obtain ^{13}C{^{1}H} spectra with natural relative intensities. This pulse sequence, used in conjunction with a 90° pulse and with a relaxation delay of $5 \times T_1$ of the longest relaxing nucleus, affords routine quantitative ^{13}C{^{1}H} NMR spectra in a relatively short time. It is noteworthy that the addition of a relaxation reagent also suppresses the NOE.

The structural information deduced from the NMR chemical shift data combined with the integrated intensities of ^{1}H, ^{13}C, and ^{29}Si resonances can provide a detailed description of the molecular structure of organosilicon compounds and particularly of siloxane polymers.

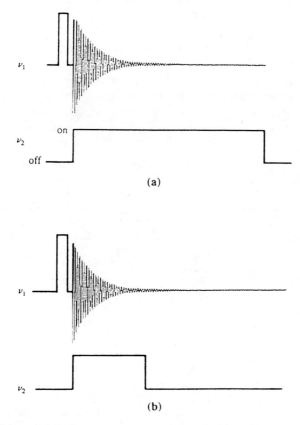

Figure 12.4. Schematic pulse sequences used to observe (*a*) ¹H decoupled HX nuclei with NOE, and (*b*) without NOE. v_1 irradiates X nuclei; and v_2 irradiates the protons.

3. PROTON NUCLEAR MAGNETIC RESONANCE

Many structural groups found as substituents on silicon have distinctive proton resonances, and ¹H NMR is a useful technique for both qualitative and quantitative analysis of organosilicon compounds. When combined with ²⁹Si and ¹³C NMR, ¹H NMR provides valuable synergism that permits solutions to complex microstructural problems.

In this section, we summarize some of the correlations that have been developed for the more common substituent groups on silicon. A more complete discussion has been given elsewhere (10).

3.1. Experimental

Proper solvent selection is important, because chemical shifts vary with the solvent (Table 12.3). Medium induced shifts are important to avoid for most work. Nonpolar solvents are usually safe, except that solvents containing easily polarizable substituents such as bromine may cause large shifts. Aromatic hydrocarbons are especially active. Polar solvents, particularly those that form associative complexes or have large magnetic anisotropy

Table 12.3. Medium-Induced Shifts of Silicon Compounds

Compound	$\Delta\delta$ in Each Solvent[a]					
	C_6H_{12}	$CHCl_3$	C_6H_6	MeOH	Pyridine	Accuracy
$C_6H_{12}{}^{(b)}$	0.019	0.000	− 0.025	0.022	− 0.082	± 0.006
$(Me_3^*Si)_2O^{(c)}$	0.001	0.010	0.052			± 0.016
$Me_3^*SiOSiPh_3{}^{(c)}$	0.021	0.003	0.031			± 0.016
$(Me_2^*PhSi)_2O^{(c)}$	0.017	0.022	0.015			± 0.016
$(Me^*Ph_2Si)_2O^{(c)}$	0.028	0.036	0.040			± 0.016
$Me_3SiCl^{(d)}$	0.05		− 0.24			± 0.02
$MeSiCl_3{}^{(d)}$			− 0.74			± 0.02

Compound	Dilution Shifts[e]	
	$\Delta\delta$ (CH)	$\Delta\delta$ (SiH)
HMe_2SiF	0.00	0.02
HMe_2SiCl	− 0.03	− 0.01
HMe_2SiBr	− 0.09	− 0.04
HMe_2SiI	− 0.09	− 0.08
$(HMe_2Si)_2O$	0.00	0.00
$(HMe_2Si)_2S$	− 0.02	− 0.01
$(HMe_2Si)_2NH$	− 0.01	− 0.02
H_3SiCH_2Cl	0.02	0.10

[a] $\Delta\delta = \delta$ (in above solvent) $- \delta$ (in CCl_4), with 0.5% internal TMS and compound concentration below 0.4 M.

[b] Data from Ref. 11, except $\delta(C_6H_{12}$ in $C_6H_{12})$ is from Ref. 12.

[c] Data from Ref. 13. Error inferred from value cited by author in other similar studies.

[d] Data from Ref. 14, except $\delta(Me_3SiCl$ in $C_6H_{12})$ is from Ref. 15.

[e] $\Delta\delta$ (extrapolated) $= \delta$ (neat) $- \delta$ (100%) C_6H_{12} with C_6H_{12} reference. Accuracy ± 0.02. Data from Ref. 15, except last entry is from Ref. 16.

such as phenyl, vinyl, or carbonyl groups, are likely to cause shifts. Interference of solvent resonances with those of the sample should also of course be avoided.

Hydrolyzable groups such as SiCl, SiOR, SiOAc, or SiN, especially if present in small amounts, react with traces of water in the solvent. Any SiOH groups may condense if acidic or basic impurities are present. Some solvents such as alcohols or dimethyl sulfoxide (DMSO) react with SiCl and other active substituents.

The preferred solvent is $CDCl_3$, which dissolves most organosilicon monomers and polymers (high molecular weight polymers dissolve rather slowly). As purchased, it may contain several hundred parts per million water, so drying with 4-Å molecular sieve spheres is recommended. It should be kept out of the light to prevent photodecomposition. The residual $CHCl_3$ signal is useful as an internal chemical shift reference.

Proper adjustment of instrumental parameters is essential, but is discussed elsewhere (17, 18). Special techniques can be used for trace analysis (19).

Because many organosilicon compounds contain SiMe groups, it may be necessary to record the integration before TMS is added as shift reference.

3.2. Chemical Shifts and Splitting Patterns

Figure 12.5 shows shift ranges for many of the common structural groups found on silicon. Unusual conditions, such as conformational restrictions, ionic substituents, or the use of polar solvents, may cause shifts outside the specified ranges. Many groups also absorb within these ranges when they are attached to elements other than silicon. A summary of the ranges and trends of coupling constants for common substituent groups on silicon is shown in Table 12.4.

The ^{29}Si isotope (natural abundance 4.7%) gives a pair of satellite lines of 2.4% relative intensity symmetrically placed about the main resonance. These resonances are sometimes mistaken for impurities, and may affect integrations for minor constituents if the resonances overlap.

The Si–Me group shows a characteristic sharp single resonance near 0 ppm. It is split only by adjacent hydrogen, that is, Si–HMe. A shift chart for Si–Me resonances is shown in Figure 12.6. Shifts of the SiMe group are additive (10), and can be predicted from a knowledge of the other substituents using the constants listed in Table 12.5.

The Si–alkyl groups usually display the expected shifts and splitting patterns, with the patterns being fairly characteristic because of the shielding effect of silicon. The pattern is, of course, sensitive to the inductive and shielding effects of the other silicon substituents, as illustrated in Figures 12.7 and 12.8.

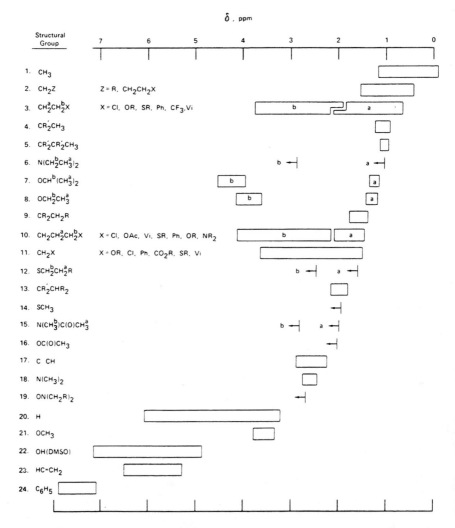

Figure 12.5. Silicon structural group shift correlation chart. Lowercase *a* and *b* denote specific protons. R denotes alkyl, and R' denotes alkyl or H. The SiH and SiOH ranges have more limited applicability. Shift ranges refer to the center of the splitting pattern, except for the SiPh and SiVi groups. The range for these groups includes all members of the splitting pattern at least 5% as strong as the most intense resonance (10).

356

Table 12.4. Ranges and Trends of Coupling Constants[a]

Coupling	Range (Hz)	Comment	References
Range			
$^3J(SiCH_2CH_3)$	7.7–8.0		20
$^3J(H–Si–C–H)$	1.2–4.7	X_2SiHMe; smaller for electronegative X	15, 21
$^3J(H–Si–Si–H)$	2.3–4.0		22
$^4J(H–C–Si–C–H)$	0–0.35	X_2SiMe_2; larger for electronegative X	21
$^1J(^{29}Si–H)$	184–382	Larger for electonegative ligands	15, 21, 23
$^2J(^{29}Si–C–H)$	6.6–9.5	SiMe; Larger for electronegative ligands	15, 21
$^3J(^{29}Si–O–C–H)$	3.5–4.1	$Me_nSi(OMe)_{4-n}$	21
$^1J(^{13}C–H)$	118–126	SiMe	15, 21
$^3J(^{31}P–Si–C–H)$	4.7	Ph_2PSiMe_3	24
$^3J(F–Si–C–H)$	6.6–8.8	Me_nSiF_{4-n}	21
Detailed trends			
$^1J(^{29}Si–H)$	$HSiR_nX_{3-n}$	R = alkyl, H; X = OR, halide	
$n = 0$	298–382		
1	240–289		
2	194–240		
3	184–202		

Coupling constant order: R = H > R = alkyl; X = halide > X = OR: SR similar to OR; NR_2 intermediate between H and OR.

$^2J(^{29}Si–C–H)$	$MeSiR_nX_{3-n}$	R = alkyl, phenyl; X = OR, OAc, Cl, F, H	
$n = 0$	7.9–9.5		
1	7.1–7.8		
2	6.6–7.4		
3	6.6		

Isotope shifts are negligible (< 0.001 ppm) for the 2.5% ^{29}Si satellites.

$^3J(H–Si–C–H)$	Electronegative groups reduce coupling when attached to Si or C; order of reduction is F > OR > Cl > Br > I $\simeq$ SR $\simeq$ NR$_2$ > Me > H

[a] See Ref. 10.

Silicon-functional groups usually show splitting patterns comparable to those of the corresponding organic compounds, except for Si–H and Si–Vi. The Si–H group usually shows a characteristic first-order splitting pattern if coupled with Si–Me protons. It also displays a pair of ^{29}Si satellites located

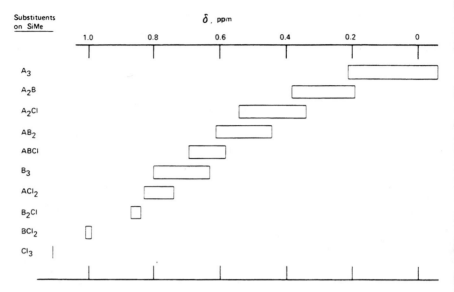

Figure 12.6. SiMe group shift chart. A = $(CH_2)_nH$; $(CH_2)_mX$, $m > 3$, X = any group; NR_2; OR; $OSiMe_3$; OH; H; CH=CH$_2$. B = OC(O)Me; SMe; Ph (10).

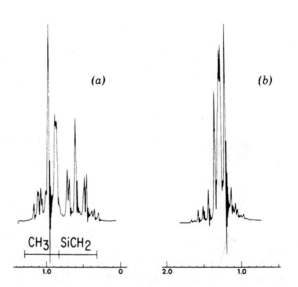

Figure 12.7. (a) Partial spectrum of Et$_4$Si (10). (b) Partial spectrum of EtSiCl$_3$.

358

Table 12.5. Substituent Additivity Constants for SiMe Shifts

Substituent	A_i (Additivity Constant)[a] (ppm)
ONa(DMSO)	-0.20[b]
Cyclopropyl	-0.07[b]
t-Butyl, i-propyl, cyclohexyl	-0.05[b]
OH(DMSO)	-0.02
$(CH_2)_nH$ ($n > 1$)	-0.01[b]
$OSiMe_3$	-0.01
NR_2	0.00
OR	0.02
H	0.05
$CH_2CH_2CF_3$	0.06
Vi	0.07[b]
$ONEt_2$	0.11
ON=CMeEt	0.13[b]
F	0.16
NMeAc	0.22[b]
OAc	0.23
SMe	0.24
Ph	0.24
Cl	0.38
Br	0.52
I	0.78

[a] $\delta = \Sigma_i A_i$, where A_i is the additivity constant for substituent i and the summation is carried over the three substituents. A_i for the methyl group is referenced at zero. All data from Ref. 12.
[b] Additivity constant derived from other additivity constants and measured shifts. All other additivity constants derived from a least-square fit to the above formula in series of the type $L_nMe_{3-n}SiMe$ and $L_nMe_{2-n}XSiMe$.

92–190 Hz from the central resonance (25). The Si–Vi group shows a characteristic pattern except when several electronegative groups are attached to the silicon, in which case it may give only a singlet (26, 27).

The Si–phenyl group shows a complex splitting pattern, which is affected both by the other substituents and by the solvent (Figure 12.9).

Special techniques are often required to observe the Si–OH resonance, which may give a poorly defined band between 4 and 7 ppm, depending on its concentration and the solvent used. An aprotic base solvent such as DMSO (28) may be used to define the resonance, the position of which can then be

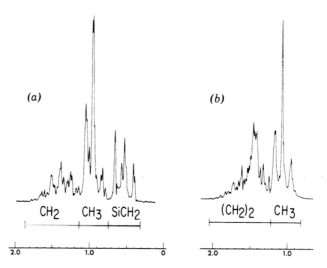

Figure 12.8. (a) Partial spectrum of CH₃*CH₂*CH₂*Si(OMe)₃. (b) Spectrum of *n*-PrSiCl₃ (10).

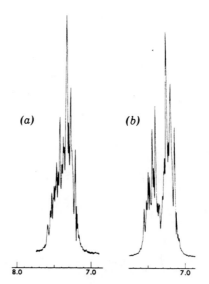

Figure 12.9. (a) Partial spectrum of C₆H₅*SiMe₃ (15% in CCl₄) (10). (b) Partial spectrum of C₆H₅*SiMe₃ (neat) (10).

360

Table 12.6. SiOH–DMSO Additivity Constants[a,b]

Substituent	A_i (Additivity Constant)[a] (ppm)
Cyclohexyl	− 0.11
n-Pr	− 0.08
Et	− 0.06
$Ph(CH_2)_3$	− 0.04
$Ph(CH_2)_2$	+ 0.12
Vi	0.30
Me_3SiO	0.35
H	0.40
Ph	0.63

[a] $\delta = 5.20 + \Sigma_i A_i$, where A_i is the additivity constant for substituent i and the summation is carried over the three substituents. A_i for the methyl group is referenced at zero. Predicted and observed shifts agree to within ± 0.1 ppm. Note that H_2O and DMSO absorb at 3.49 and 2.62, respectively. Data from Refs. 12 and 29.
[b] From Ref. 10.

Table 12.7. ^{19}F NMR Shifts of R_3SiOH–HFA Adducts $Me_nX_{3-n}SiOH$–HFA[a,b]

X	n	Δppm^c
	3	0.13
$Me_3SiO(SiMe_2O)_6$	2	0.41
$Me_3SiO(SiMe_2O)_3$	2	0.39
$Me_3SiO(SiMe_2O)_2$	2	0.36
Me_3SiO	2	0.33
Me_3SiO	1	0.50
Me_3SiO	0	0.76
Ph	2	0.46
Ph	1	0.90
Ph	0	1.5

[a] Samples prepared as $\simeq 2$ wt% sample, 8 wt% HFA, 90 wt% THF, in glass-sealed, medium-walled NMR tubes.
[b] From Ref. 10.
[c] Δ is parts per million downfield from H_2O–HFA adduct (which is ~ 7.3 ppm upfield of free HFA).

predicted with the help of additivity constants (Table 12.6). Hexafluoroacetone forms a 1:1 adduct with a characteristic resonance that can be used for quantitative analysis of Si–OH (10, 29) using ^{19}F NMR. The detection limit is about 0.02 wt% OH. The ^{19}F NMR shifts of various adducts are listed in Table 12.7.

3.3. Applications

The use of ^{1}H NMR for identifying molecular structures is well known and will not be discussed here. Some studies unique to organosilicon materials will be cited to illustrate the breadth of approaches possible.

The reaction kinetics of $(MeO)_4Si$ hydrolysis were studied by Kay and Assink, using high field ^{1}H NMR (30). The product is a sol–gel, $SiO_2 \cdot xH_2O$, but the hydrolysis proceeds stepwise and at different rates for base and acid catalysis. The authors found that condensation is the rate-limiting step in acid catalyzed solution, while hydrolysis is rate limiting in base catalyzed solution. A model was developed that correctly predicts the distribution of product species during the initial hydrolysis process.

The use of ^{1}H NMR is indicated in this hydrolysis study because each intermediate may be observed independently, as contrasted to ^{13}C NMR, where resonances of the intermediate species overlap. Because of its lack of sensitivity in dilute solutions ^{29}Si NMR is of limited use.

Alkyl polysilanes show many interesting properties, such as intense UV absorption in both dilute solution and the solid state. Changes in the conformation of the polymer are thought to be responsible for some of the temperature-dependent changes in the UV spectra, so there has been considerable interest in characterizing the structures of these materials. In one such study, ^{1}H, ^{13}C, and ^{29}Si NMR were used to establish local configurations and confirmed the postulated increase in *trans* conformations along the silicon backbone as the temperature was lowered (31).

A correlation was developed between the ^{1}H NMR spectra and the polarity of Me and PhMe stationary phases used in GC (32). Anisotropic solvent-induced shifts, using benzene as solvent, were used as a quick method to determine molecular weights of Me_3SiO-treated SiO_2 (33).

4. CARBON-13 NUCLEAR MAGNETIC RESONANCE

The ^{13}C nucleus is only 1.1% in natural abundance and its overall sensitivity is 1/5700 that of proton NMR (Table 12.1). However, the availability of broad-band decoupling, the NOE, narrow line widths, and the wide chemical shift range make ^{13}C NMR extremely useful for structural determination.

The purpose of this section is not to cover the basic concepts of ^{13}C NMR, which are explained in several texts (8, 34, 35), but rather to discuss ^{13}C NMR of organosilicon compounds and the effects of silicon on neighboring carbon atoms.

4.1. Chemical Shifts

The range of chemical shifts in ^{13}C NMR is normally 240 ppm for organic compounds. This range is somewhat extended in both directions for organo-silicon compounds (36). Silicon substituents on aliphatic carbon generally result in a shielding effect relative to the analogous organic compound. Deshielding effects are seen, however, for acylsilanes and silaethylenes. Chemical shifts depend on hybridization of the ^{13}C atom and on substituent electronegativity. Upfield shifts may be caused by steric compression and dilution. Hydrogen bonding with polar solvents may cause a downfield shift. The effect of the silicon atom on neighboring carbon atoms is discussed in the following sections.

4.1.1. Si–Methyl

As in ^{1}H NMR, methyl groups on silicon are found upfield in the ^{13}C NMR spectrum and can resonate between $+ 20$ and $- 15$ ppm depending on their environment. Methyl groups on carbon range from 7–32 ppm (35). The Si–Me shift is dependent on the other substituents. The farthest upfield Si–Me chemical shifts ($+ 1$ to $- 16$ ppm) result from the presence of methoxy or ethoxy ligands or other electronegative groups such as chlorine. Chemical shifts around 0 ppm are seen when two OSi ligands are present. When two Me ligands are on silicon, chemical shifts range from 1 to 4 ppm. Downfield chemical shifts are seen with one alkyl and two Cl ligands. Trends in substituent effects are very different for ^{13}C NMR of Si–Me groups as compared to ^{1}H NMR. With the latter, each successive replacement of a methyl group with a chlorine atom causes a downfield shift. However, ^{13}C NMR shifts of Me_4Si, Me_3SiCl, Me_2SiCl_2, and $MeSiCl_3$ are 0, 18, 4, and $- 16$ ppm, respectively (36). This trend is not observed for the analogous carbon compounds, which display consistent downfield shifts with each methyl replacement (36).

4.1.2. Other Substituents

The Si–CH_2 group resonates between 2 and 35 ppm depending on the ligands on the silicon atom and the group next to the carbon atom, referred to as X. When X is a CH_2 group, chemical shifts are normally in the range 3–18 ppm.

However, with three Cl atoms on silicon, the $Si–CH_2$ group resonates at $\sim 24–25$ ppm. When X is a CH group, chemical shifts are commonly 17–25 ppm. The CH_2 group on carbon normally falls in the range 17–52 ppm, so only the upfield range is unique for $Si–CH_2$ (35). Three methoxy and two or three ethoxy ligands give an upfield chemical shift; methyl and OSi ligands fall midrange; and groups such as Cl or X = CH give downfield chemical shifts.

Shifts of organofunctional carbon atoms on silicon follow the same trend as those of $Si–CH_2$ groups in that they depend on the ligand attached to silicon. As one moves farther down the alkyl chain away from the silicon, the chemical shifts of the carbon is less dependent on the Si ligands. At the third or fourth carbon atom any effects of Si are negligible. A vinyl group on silicon resonates at 110–140 ppm for the terminal carbon atom and 130–150 ppm for the internal carbon.

Phenyl groups on silicon have chemical shifts within the same range as they do on carbon. A phenyl group directly bonded to silicon as part of a diphenyl siloxane resonates at 136.3, 134.7, 128.0, and 130.3 ppm starting from the quarternary carbon, as compared with toluene (137.5, 128.9, 128.0, and 125.2) and ethyl benzene (144.3, 127.9, 128.4, and 125.7).

The most common carbon-containing functional groups found on silicon are alkoxy and acetoxy moieties. Methoxy groups on silicon normally appear $\sim 49–50$ ppm. Ethoxy groups resonate at 17–19 and 57–58 ppm. Phenoxy groups on silicon with methyl ligands give carbon resonances at 154.7, 119.6, 129.0, and 121.0 ppm starting from the quaternary carbon. Acylsilanes and silaethylenes experience deshielding compared to their carbon analogs, and have chemical shifts in the range 188–250 ppm for the carbonyl carbon of acylsilanes and 200–215 ppm for the Si=C carbon of silaethylenes (36).

4.2.   Coupling Constants

The normal coupling rules for organics also apply to organosilicon compounds. Additional information on coupling is found in Refs. 8, 33–38. Coupling can occur between ^{13}C and ^{1}H, ^{13}C, ^{29}Si, ^{19}F, or any other nucleus with a net spin. Coupling across one bond is most significant but it can extend over two or three bonds. The magnitude of the coupling constant can be useful in identifying the structural group. Carbon-proton coupling constants can range from 110 to 260 Hz. For a trimethylsilyl group ($Me_3SiO_{1/2}$) the one bond C–H coupling constant is 130 Hz (37).

Poly(dimethylsiloxane) chains have $J_{C–H} = 118$ Hz for dimethyl substituents on silicon (39). Coupling constants for $J_{C–Si}$ have been measured for over 60 silanes and plotted against substituent electronegativity, (38) as shown in

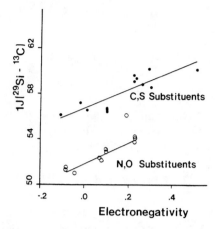

Figure 12.10. Coupling constants for C–Si as a function of the electronegativity of the Si substituents. [From (38)].

Figure 12.10. The plot shows that linear relationships exist when the substituents are divided into two groups; C or S, and N or O. Coupling constants for J_{C-Si} with C or S substituents range from 56 to 60 Hz (38).

4.3. Quantitative Analysis

Quantitative NMR spectra are useful for structure determination and the analysis of mixtures. Typical ^{13}C proton-decoupled NMR spectra are not quantitative because carbon atoms with long relaxation times (T_1 values) may not completely return to thermodynamic equilibrium between pulses. The result is weaker signals for carbon atoms with long T_1 values. Another source of error is the NOE. These topics are discussed in detail in Section 2.

4.4. Applications

4.4.1. Silicone–Glycols

A great deal of information may be obtained from ^{13}C NMR spectra of silicone glycols, such as the type of glycol present, the glycol/siloxane ratio, the glycol block size, the type of capping on the glycol, and whether it is attached to Si through an O or a C link. A typical silicone glycol contains ethylene oxide (EO) groups and propylene oxide (PO) groups as shown in Structure **1**, where m and n may vary from 0–18.

$$Me_3SiO-(Me_2SiO)_x-(MeRSiO)_y-SiMe_3$$

1

$$\text{where } R = -CH_2CH_2CH_2-O-(CH_2CH_2O)_m-(CH_2\underset{\underset{CH_3}{|}}{C}HO)_nOMe$$

The EO and PO units are readily resolved in the ^{13}C NMR spectrum. An EO group resonates at ~ 70 ppm and a PO group at 16.5, 73, and 75 ppm. Figure 12.11 shows a representative ^{13}C NMR spectrum of a silicone glycol.

4.4.2. Hexenyl Hydrolyzate Isomers

Hexenyl hydrolyzates are prepared from the platinum catalyzed addition of an Si–H compound across one double bond of 1,5-hexadiene. However, the platinum also catalyzes rearrangement of the second double bond to form

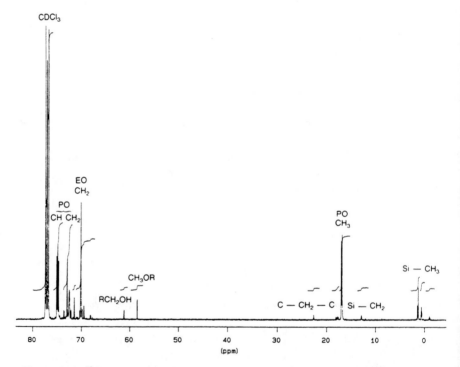

Figure 12.11. ^{13}C spectrum of a silicone–glycol copolymer. The polyol is attached to Si by a $(CH_2)_3$ linkage. MeO– and –OH terminated polyols are present.

several isomers. The ^{13}C NMR can identify the isomers of hexenyl hydrolyzates, Structures 2–4.

$$Si-CH_2-CH_2-CH_2-CH_2-CH=CH_2$$

2

$$Si-CH_2-CH_2-CH_2-CH=CH-CH_3$$

3

$$Si-CH_2-CH_2-CH=CH-CH_2-CH_3$$

4

The terminal species, Structure 2, is the desired product. As shown in Figure 12.12, ^{13}C NMR can be used to measure the level of isomer impurities in hexenyl hydrolyzates.

4.4.3. PC–PDMS Block Copolymers

The ^{13}C NMR approach has been used with ^{29}Si NMR to determine the composition and sequence of polycarbonate (PC)–PDMS block copolymers. Since a two-step reaction sequence is used, it is difficult to predict the average PC block length and the number of isolated bisphenol A (BPA) units. This information is desirable because the physical properties of PC–PDMS block copolymers are linked to the composition. Figure 12.13 shows the ^{13}C NMR spectrum of a BPA–PDMS block copolymer with an average length of five for each block. Assignments were made based on model compounds. The silicone block length can be determined from the ^{13}C spectrum because the Si–methyl carbon atoms at the end of the block resonate apart from internal Si–methyl carbon atoms. Comparing the integrals of the BPA–methyl carbon atoms to the Si–methyl carbon atoms gives the mole ratio of BPA polycarbonate to polysiloxane. The polycarbonate block length may be calculated from the quaternary BPA carbon atoms as shown in Figure 12.14. The NMR analysis provides the only direct analysis of these polymer systems and is the first to allow measurement on isolated BPA units in the polymer (40).

4.4.4. Alkyl-Modified Soluble Silica

The ^{13}C NMR approach is useful for studying alkyl-modified silicas. These materials are the most common stationary phases for LC, and information on solute interaction with bonded alkyl groups is greatly desired (41). In one example, silica powder was modified by adding ^{13}C enriched alkylchlorosilanes. As expected, the bound silanes had broader signals than the solution

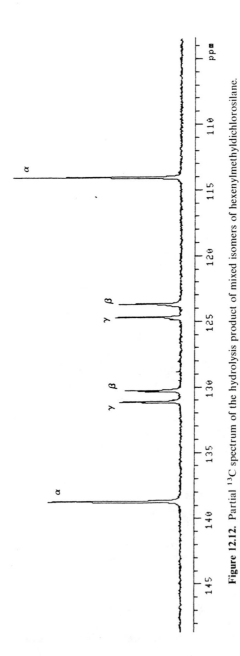

Figure 12.12. Partial ^{13}C spectrum of the hydrolysis product of mixed isomers of hexenylmethyldichlorosilane.

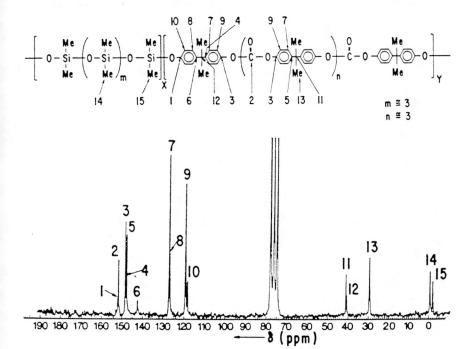

Figure 12.13. ^{13}C NMR spectrum of a BPA–PDMS block copolymer with average block length of five for each block, in which both internal and terminal block units may be seen (40). Reprinted with permission from *Macromolecules*, **10**, 782. © 1977, American Chemical Society.

INTERNAL

TERMINAL

ISOLATED

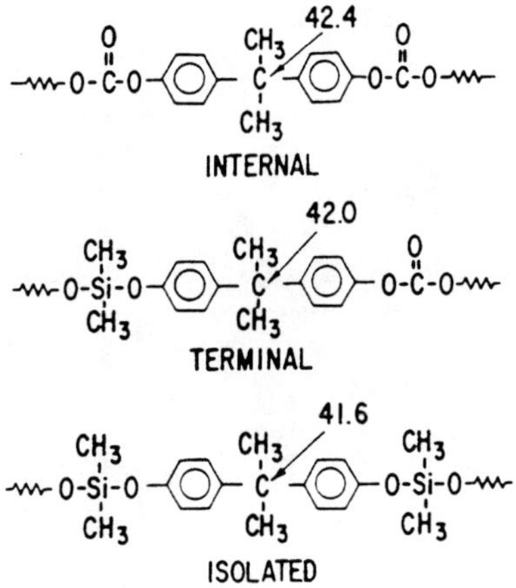

Figure 12.14. Units found in a PBA–PDMS copolymer (39). Average block length is five monomer units. Reprinted with permission from *Macromolecules*, **14**, 1016 © 1981, American Chemical Society.

369

Table 12.8a. Surface Data for Labeled Bound Silanes[a]

Alkylchlorosilane	Chemical Shift[b] (ppm)	Linewidth[c] (Hz)	% Bound Carbon
$-\text{O}-\underset{\underset{\text{CH}_3}{\|}}{\overset{\overset{-\text{CH}_3}{\|}}{\text{Si}}}-(\text{CH}_2)_9{}^*\text{CH}_3$	15.0	100	7.9
$-\text{O}-\underset{\underset{\text{O}}{\|}}{\overset{\overset{\text{O}}{\|}}{\text{Si}}}-(\text{CH}_2)_9{}^*\text{CH}_3$	14.2	160	10.4
$-\text{O}-\underset{\underset{\text{O}}{\|}}{\overset{\overset{\text{O}}{\|}}{\text{Si}}}(\text{CH}_2)_3{}^*\text{CH}_2(\text{CH}_2)_4\text{CH}_3$	33.0	512	8.8
$-\text{O}-\underset{\underset{\text{O}}{\|}}{\overset{\overset{\text{O}}{\|}}{\text{Si}}}{}^*\text{CH}_2(\text{CH}_2)_7\text{CH}_3$	27.8	720	8.0

[a] Obtained in $CDCl_3$ referenced to tetramethylsilane.
[b] Measured at half-height.
[c] Measured at half-height.

Table 12.8b. Solution Data for Labeled Silanes[a]

Alkylchlorosilane	Chemical shift[b] (ppm)	% Enrichment
$Cl(CH_3)_2Si(CH_2)_9{}^*CH_3$	14.1	25
$Cl_3Si(CH_2)_9{}^*CH_3$	14.2	25
$Cl_3Si(CH_2)_3{}^*CH_2(CH_2)_5CH_3$	29.1	15
$Cl_3Si{}^*CH_2(CH_2)_7CH_3$	24.4	50

[a] From Ref. 41. Courtesy Academic Press.
[b] Obtained in $CDCl_3$ referenced to tetramethylsilane.

silanes due to reduced motional freedom and surface inhomogeneities (41). Also, monofunctional bonded silanes had a narrower resonance than trifunctional bonded silanes. Table 12.8 lists chemical shifts and line widths of alkylchlorosilane treated soluble silica in solution. The line width increases for carbon atoms close to the silica particle, because of increasing restrictions on their motion. Solid state ^{13}C NMR of alkyl-modified silicas is discussed in Section 7.3.1.

5. SILICON-29 NUCLEAR MAGNETIC RESONANCE

5.1. General

The first reports of ^{29}Si NMR data in 1956 and 1962 by Lauterbur and co-workers (42, 43) described the basic trends in ^{29}Si chemical shifts and the experimental difficulties involved in the acquisition of ^{29}Si NMR data. The field was quiet until 1968 when Hunter and Reeves (21) prompted new interest with information on both ^{29}Si chemical shifts and relaxation times. With the recognition of the analytical potential and the commercial availability of multinuclear and FT spectrometers in the early 1970s, applications of ^{29}Si NMR grew rapidly.

Today ^{29}Si NMR is routinely utilized in many laboratories as the method of choice for structure determination of organosilicon compounds. Literally thousands of papers have been published reporting use of ^{29}Si NMR for structural and dynamic investigations. In this section we give an overview of the state of ^{29}Si NMR as of 1990 with practical information useful for acquisition and interpretation of ^{29}Si NMR data.

5.2. Chemical Shifts

The sensitivity of ^{29}Si nuclei to their electronic, atomic, and molecular environments gives a chemical shift dispersion range for the analysis of organosilicon compounds that is unsurpassed. The chemical shifts observed for methylpolysiloxanes vary over a range of approximately 120 ppm (Figure 12.1). Increasing the number of oxygen atoms on Si results in greater shielding of the Si nucleus and upfield shifts are observed. Alkyl groups deshield the Si nucleus resulting in downfield shifts. Silicon compounds of the type $(CH_3)_{4-n}SiX_n$, where X is an electronegative group and n is varied from 0 to 4, exhibit a well-known "sagging" pattern (1) (Figure 12.15). This behavior has been discussed in terms of $(p-d)\pi$ bonding (1–4) but a fully satisfactory theory of ^{29}Si shielding has not yet been developed. Liepins et al.

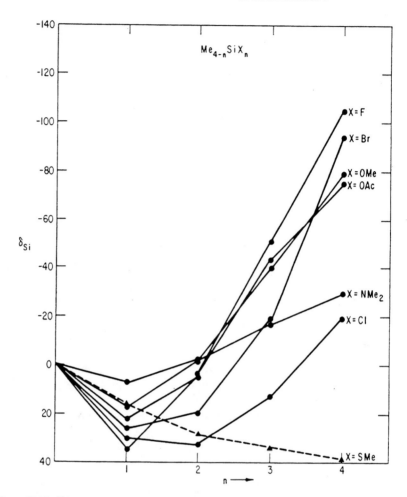

Figure 12.15. ^{29}Si chemical shifts in ppm relative to Me$_4$Si for the methylsilanes Me$_{4-n}$SiX$_n$. Reprinted by permission from Ref. 1. © 1979, Academic Press.

(44) measured ^{17}O, ^{29}Si, and ^{13}C NMR shifts for more than 100 alkoxysilanes and observed that ^{29}Si shifts were more sensitive to inductive than steric effects for all compounds studied except for Si(OR)$_4$. Linear relationships were found between the chemical shifts of ^{17}O, ^{29}Si, and ^{13}C. Unequal transmission of electronic effects along the Si–O bond in different directions were observed and discussed in terms of $(p–d)\pi$ bonding. Radeglia (45) described a simple method to estimate the magnetic screening constant of

tetrahedrally coordinated Si with 4 different ligands that he applied to the calculation of screening constants for 43 substituted silanes. The ^{29}Si NMR shifts calculated from these screening constants agreed semiquantitatively with the experimentally observed chemical shifts. Bellama et al. (46) examined 13 pentacoordinate silatranes, $RSi(OCH_2CH_2)_3N$, and developed an empirical equation to relate their chemical shifts to the corresponding triethoxy silanes. Petrakova et al. (47) reported ^{29}Si and ^{13}C chemical shifts for a series of 30 pertrimethylsilylated oligosaccharides containing the xylopyranosyl unit. They demonstrated the usefulness of ^{29}Si NMR for the analysis of oligosaccharides but were unsuccessful in attempts to establish empirical assignment rules useful for compounds with unknown structures. No relationship between chemical shift and structure was found. This result was attributed to the narrow chemical shift range observed in the oligosaccharides.

Useful ^{29}Si NMR chemical shift correlation charts are presented in Figures 12.16 and 12.17. These charts are compiled from reports in the open literature and supplemented by internally generated data. Other useful compilations may be found in Refs. 1–3, 5, 49; all chemical shifts are given relative to tetramethylsilane (TMS), which is the accepted NMR chemical shift reference standard for ^{29}Si NMR. Examination of these charts reveal the power of ^{29}Si NMR to distinguish between substituents. Substitution of a methyl group in polysiloxanes is denoted by following the M, D, or T notation with the substituent as a superscript, that is, $D^{Vi} = O_{2/2}SiMeVi$, $M^{Ph,H} = O_{1/2}SiMePhH$, and $T^{OH} = O_{3/2}SiOH$. The polysiloxane building units portion of Figure 12.16 shows that replacement of a methyl group by a single phenyl group results in an upfield shift of more than 6, 9, and 9 ppm for M, D, and T siloxanes, respectively. Replacement by a vinyl group gives even larger shifts. Polysiloxanes are sensitive to substitution on neighboring silicons as seen by comparing $M-D^H$ to $M-D$. These differences are useful for determining sequencing of copolymers and will be discussed later in this chapter. The sensitivity of ^{29}Si shifts to bond angles is shown by the cyclic siloxane species D_3, D_4, D_5, and D_6, which show that as ring strain increases the ^{29}Si nucleus is less shielded, resulting in chemical shifts to higher frequency. The selected functional silane compound section of Figure 12.17 shows the sensitivity of the ^{29}Si nucleus to the alkoxy group in a series of Me_3SiOR alkoxysilanes where R includes Ac, Me, Et, and H. This series covers a range of approximately 18 ppm and each silane is resolved from the others. It is worth noting here that while resolution is maintained, the order of appearance of the alkoxy substituents changes from Ac, Me, Et, H to H, Ac, Me, Et on going from Me_3SiOR to $MeSi(OR)_3$ and $Si(OR)_4$. Sequential hydrolysis of $Si(OEt)_4$ results in sequential downfield shifts, which are reduced in magnitude for each consecutive ethoxy group hydrolyzed.

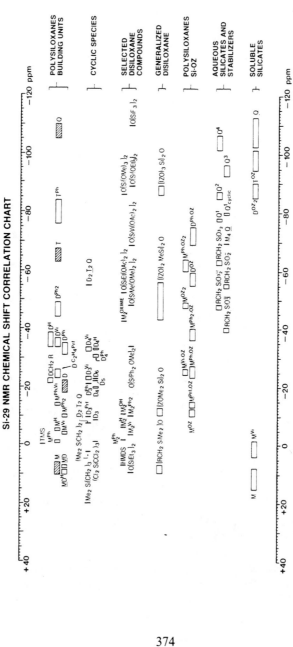

Figure 12.16. ^{29}Si NMR chemical shift correlation chart for selected organosilicon compounds (1, 2, 5, 48). R = H, alkyl, or functional group. Z = H, alkyl, or functional group attached to oxygen. Prf = 1,1,1-trifluoropropyl. M,D,T,Q defined in Figure 1.1, Chapter I.

374

Si-29 NMR CHEMICAL SHIFT CORRELATION CHART

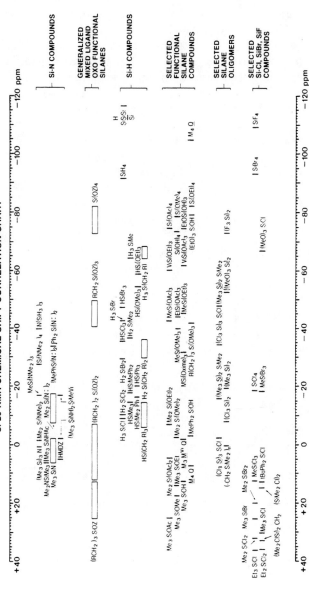

Figure 12.17. ^{29}Si NMR chemical shift correlation chart for functional silane and siloxane compounds (48). R = H, alkyl, or functional group. Z = H, alkyl, or functional group attached to oxygen. Prf = 1,1,1-trifluoropropyl. M,D,T,Q defined in Figure 1.1, Chapter 1.

375

5.3. Coupling Constants

Scalar coupling values can be useful in structure determination and stereo-chemical assignments in ^{29}Si NMR. Reports are available reviewing experimentally determined and calculated ^{29}Si coupling constants (1–3, 5 and references cited therein). Work is continuing on the theory of coupling for silicon, and no attempt to cover the literature in this area is made here. A few trends are worth noting, however. In most cases, the dominant factor in determining the size of the coupling constant is the electronegativity of the substituents. Thus, the $^1J_{Si-H}$ increases from 202.5 to 381.7 in F_nSiH_{4-n} as n increases from 0 to 3 (1). One-bond silicon carbon coupling constants have been reported to be dependent on the s-character of the Si–C bond (3). Two-bond silicon hydrogen coupling constants are dependent on the hybridization of the silicon and the intervening carbon (1). Typical coupling constant ranges and experimentally determined signs are presented in Table 12.9.

5.4. Quantitative Analysis

The quantitative analysis by NMR spectroscopy of ^{29}Si nuclei has requirements similar to those of ^{13}C nuclei and enjoys the same advantages. The major differences are the negative magnetogyric ratio resulting in a negative NOE (4) and long relaxation times of the ^{29}Si nuclei [generally > 30 s (1–3)]. The negative NOE can result in reduced, nulled, or negative signals from heteronuclear proton decoupling experiments that are important for

Table 12.9. Signs and Typical Ranges of Silicon Coupling Constants

Coupling	Sign[a]	Range (Hz)
$^1J_{Si-C}$	−	37–113[b]
$^1J_{Si-Si}$		52–186[b]
$^1J_{Si-H}$	−	147–420[b]
$^1J_{Si-F}$	+	110–488[b]
$^1J_{Si-P}$		7–50[b]
$^2J_{SiCH}$	+	3–12[c]
$^3J_{SiCCH}$	−	1–8[c]
$^4J_{SiCCCH}$		1.0–1.5[c]

[a] See Ref. (1).
[b] See Ref. (50).
[c] See Ref. (2).

^{29}Si NMR. Since two and three bond Si–H coupling constants are appreciable, it is not unusual for ^{29}Si nuclei to be coupled to six or more protons. In a fully coupled spectrum, this leads to complex splitting patterns and concomitantly reduced S/N ratios because the signal is divided among so many peaks. The requirement to wait $5 \times T_1$ for quantitative integrals results in prohibitively long experiment times.

At first glance, this problem may appear to be greater for ^{29}Si than for ^{13}C NMR but the solutions are the same. Addition of small amounts of a paramagnetic relaxation agent (0.01–0.03 M) has been reported to reduce the ^{29}Si relaxation times to a few seconds and effectively reduce or remove any NOE (1–3). This result follows from introduction of an electron-nuclear relaxation mechanism that is much more efficient than the spin–rotation and dipole–dipole interactions normally dominant for ^{29}Si nuclei in organosilicon compounds. Chromium(III) acetylacetonate [Cr(acac)$_3$] is the most commonly used relaxation reagent. It is very soluble in common organic solvents and does not cause paramagnetic chemical shift effects at concentrations up to 0.05 M. The T_1 values for selected compounds are given in Table 12.10 at different Cr(acac)$_3$ concentrations. In our hands, Cr(acac)$_3$ has been observed to slightly broaden ^{29}Si resonances at the higher concentrations, which may be significant for studies such as polymer sequencing where highest resolution is required. A good compromise suitable for general quantitative acquisition is 0.02 M Cr(acac)$_3$ combined with a 45° pulse width, inverse gated decoupling and a relaxation delay of 8 s.

Two other situations are commonly encountered that require different solutions. First, some solvents and some compounds are not compatible with Cr(acac)$_3$ because of limited solubility or chemical reaction. Second, glass NMR tubes, inserts, and dewars give an SiO$_2$ background that extends from approximately -80 to -130 ppm and interferes with signals that resonate in that region (Figure 12.18). Several alternatives have been discussed as solutions to these problems. Levy and Cargioli (3) suggest saturating the solution with dissolved oxygen, which is paramagnetic, but point out that pressures of approximately 10 atm are necessary to reduce T_1 values to a few seconds. Gated decoupling by itself is useful to avoid the negative NOE effects but long acquisition times are required. Harris et al. (2) comment on replacement of glass tubes with tubes made from Teflon®, Kel-F®, or FEP and the use of difference spectroscopy by subtracting the FID of a blank run taken under the same conditions as the sample. Schraml et al. and (49) Dereppe and Parbhoo (51) have investigated the use of the polarization transfer techniques INEPT and DEPT for quantitation of trimethylsilylated compounds. These techniques depend on transfer of magnetization from protons to the Si nucleus. They observed sensitivity enhancement factors ranging from 6.6 to 9.2, which result in direct time-saving factors of 40–80.

Table 12.10. Effect of $Cr(acac)_3$ Concentration on ^{29}Si T_1 Values of 30% Solutions of Selected Organosilicon Compounds in $CDCl_3$[a]

	T_1(s)							
$[Cr(acac)_3]$	TMS	$M\underline{D}^RM$	$MD^R\underline{M}$	$1000\,cS\ MD_nM$	$(PhMeSiOH)_2O$	$MD^H_{60}M$	$M^{Ph_2}\underline{D}^{Ph}M^{Ph_2}$	$\underline{M}^{Ph_2}D^{Ph}M^{Ph_2}$
$0.01M$	5.5	4.5	4.2	5.3	2.1	4.8	4.5	4.3
0.02	3.0	2.4	2.4	2.8	1.2	2.7	2.3	2.2
0.03	2.2	1.6	1.6	2.0	0.8	2.4	1.7	1.7

[a] $R = (CH_2)_3\,NH(CH_2)_2\,NH_2$. Estimated relative error $\pm$ 5%.

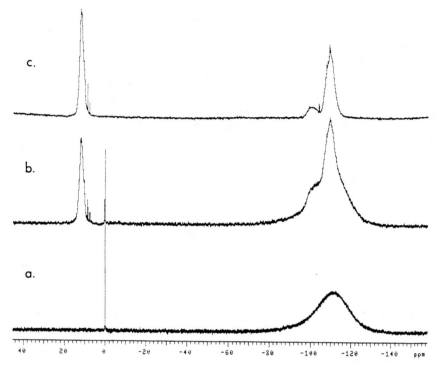

Figure 12.18. ^{29}Si NMR spectra showing the consequence of the SiO$_2$ background signal from glass inserts or sample tubes: (a) spectrum from a CDCl$_3$ sample in a 16 mm glass NMR tube; (b) tube with a 30% solution of an MTQ resin in CDCl$_3$; (c) b in a Teflon® NMR tube and glass-free probe.

Relative error for compounds of unknown structures was 20% but could be improved to 5% by calibration with compounds of known structure. Since these techniques are dependent on ^{1}H relaxation times, which are generally much shorter then ^{29}Si relaxation times, this approach offers another potential savings in experimental time based on a faster repetition rate. It also points out that the compound must have appreciable silicon–proton coupling for polarization transfer to take place. Therefore, the SiO$_2$ background problem is also avoided by polarization transfer methods.

In our laboratory, we use Si-free NMR tubes constructed as shown in Figure 12.19 from standard commercial 15-mm Teflon® NMR tube inserts. These are fitted snugly inside 16-mm o.d. glass tubing by wrapping with Teflon® tape. The assembly is placed in a standard 16-mm spinner. These Si-free tubes give poor resolution ($\sim$ 1 Hz) by comparison with high resolution glass NMR tubes, but this limitation is rarely a significant factor. The

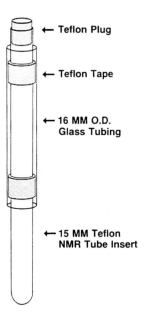

Figure 12.19. A silicon-free NMR tube constructed from a Wilmad 15-mm Teflon® NMR tube liner.

majority of compounds of interest in this region give broad lines anyway, because of the multiplicity of environments present in silicates and resins. These tubes, when used in custom made probes in which glass inserts have been replaced with Kevlar or alumina and the glass dewars removed, allow overnight acquisition of ^{29}Si NMR spectra with no observable background.

Laude and Wilkins (52) have developed a recycled flow method for quantitation of samples with slowly relaxing nuclei and compared it to the addition of relaxation reagents. This method utilizes a flow cell and a premagnetization region such that the excited molecules leave the observed region of the cell after acquisition and are replaced by magnetized but fully relaxed nuclei between pulses. The requirement for repetition delay is no longer based on five times the longest T_1 of the sample but rather on the residence time of the sample in the cell. This approach has been applied (52) to six linear siloxane polymers for molecular weight determination by end-group analysis. Two water soluble samples were also analyzed to demonstrate general applicability to cases where Cr(acac)$_3$ is not an option because of insolubility. They found that recycled flow offered a factor of 3–5 in time savings over Cr(acac)$_3$ relaxed samples, had improved spectral quality, and was fully applicable to aqueous systems. The clear drawback of this method is the requirement for additional nonstandard hardware that is not currently available commercially.

5.5. Structural Applications

The ability of ^{29}Si NMR to distinguish between unsubstituted M, D, T, and Q siloxane units; between substituted $M^{X,Y,Z}$, $D^{X,Y}$, or T^X units within their respective class of M, D, or T siloxy units; and finally between cyclic and linear chains; establishes NMR spectrometry as a powerful tool for structure determination of siloxane polymers and copolymers (1, 2, 4, 53–56).

Polysiloxanes discussed here are classified into four types: linear homopolymers involving M and D units; branched or cross-linked homopolymers involving M, T, and Q units; linear siloxane copolymers; and finally, linear organosiloxane copolymers.

5.5.1. Polymers

Linear homopolymers may be described by the general formula M^X–$D_n^{X,Y}$–M^X, where X and Y are any organic, organofunctional, or inorganic substituents. These polymers are the easiest to observe and to analyze. The chemical shifts of the end-groups M^X and of the chain groups $D^{X,Y}$ are known for the most common substituents, and the deduction of the polymer structure is straightforward.

The ^{29}Si NMR method is unique in that the end groups and the chain units are simultaneously observed. Therefore, the polymer is completely characterized by average parameters deduced from the spectrum. These are the mole percent of a given functionality, the average degree of polymerization, and the absolute number average molecular weight. These parameters are generally independent of impurities, solvents, and additives that may be present in the solution.

Dimethylsilanol terminated PDMS (a) and trimethylsilyl terminated poly-(methyl-1,1,1-trifluoropropylsiloxane) (b) illustrate this class of polymers (Figure 12.20).

The main problem in recording the spectrum of this type of homopolymer comes from the degree of polymerization. For relatively short chains, ~ 50 units, observation of the M and D units does not present any difficulty and the spectrum can be obtained in 15 min. Moreover, a distinct signal arises for the D unit next to the M unit and these M and M–D– signals are of equal intensity, providing two structural probes for the quantitative determination of the terminating units. But as the chain length increases, the ratio of D to the M and M–D– end groups increases and it becomes more and more

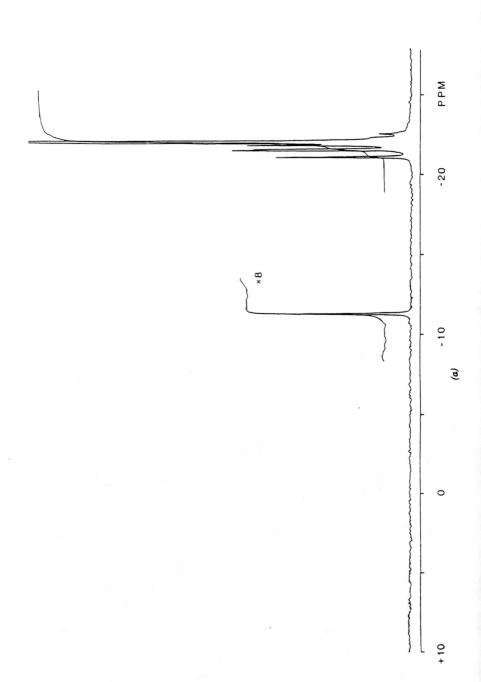

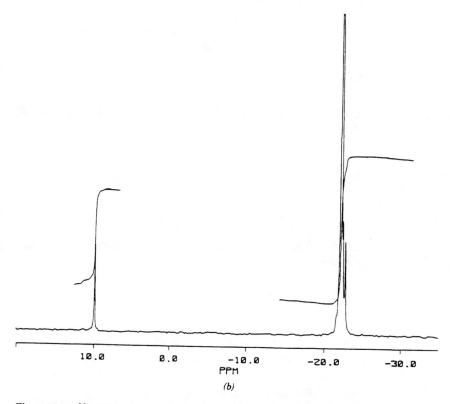

Figure 12.20. ^{29}Si NMR spectra of (*a*) dimethylsilanol terminated PDMS and of (*b*) trimethyl-silyl terminated poly(methyl-1,1,1-trifluoropropylsiloxane) homopolymers.

difficult to detect the end-group resonances. Consequently, the error associated in the integration of the small M signals relative to the large D signal increases significantly and affects the accuracy of the calculated degree of polymerization and of the absolute number average molecular weight.

The average degree of polymerization (DP), the average chain length (CL), and the related number average molecular weight ($\bar{M}_n$) are calculated from the following equations (1, 53, 56):

$$DP = 2I^D/I^M \tag{12.2}$$

$$CL = 2(I^M + I^D)/I^M \tag{12.3}$$

$$\bar{M}_n = 2U^M + (CL)(U^D) \tag{12.4}$$

where I is the integrated intensity of the appropriate unit D or M and U is the unit molecular weight of the siloxy unit D or M.

The same parameters can be calculated for linear copolymers in a similar fashion:

$$DP = 2(I^D + I^{D'})/I^M \tag{12.5}$$

$$CL = 2(I^M + I^D + I^{D'})/I^M \tag{12.6}$$

$$\bar{M}_n = 2U^M + (CL)[(X \cdot U)^D + (X \cdot U)^{D'}] \tag{12.7}$$

where $(X \cdot U)$ is the mole fraction X times the unit molecular weight U of the corresponding D or D' unit. The mole fraction is defined as

$$X^D = I^D/(I^D + I^{D'}) \tag{12.8}$$

Finally, the mole percentage of D or D^X units in cyclics can be determined from the same NMR spectrum by

$$\%D_{cyclics} = 100I_{cyclics}/\sum_i I_i \tag{12.9}$$

where $I_{cyclics}$ is the integrated intensity of D signals in D_n cyclics and $\sum I_i$ is the summation over all observed siloxy units. The extension to cocyclics is straightforward.

Siloxane compounds made from the combination of M and T, D and T, and M and Q siloxy units have been studied and shown to be complex mixtures of highly branched or cross-linked materials (1, 2, 53, 57). The low molecular weight fractions have been extracted and analyzed by NMR and their molecular structures determined. Figure 12.21 collects some of these characteristic compounds made from D and T units. The sensitivity of the

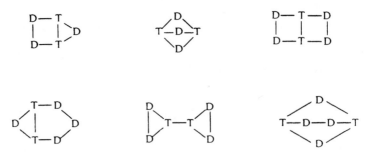

Figure 12.21. Structures found in a DT resin (64).

^{29}Si chemical shift even allows the identification of isomeric structures in DT resins. Similarly, complex structures composing MQ resins have been resolved and are shown in Table 12.11. The degree of trimethysilylation of these resins can be directly determined from the integral ratio of M to Q units.

Table 12.11. ^{29}Si Chemical Shifts^a for Some Silicic Acid Trimethylsilyl Esters^b

	Compounds	M^c	Q^c
M_4Q	M_4Q	7.6 (8.6)	$-105.1 (-104.2)$
M_6Q_2	M_3Q-QM_3	8.0 (9.3)	$-107.5 (-106.5)$
M_8Q_3	$M_3Q-M_2Q-QM_3$	8.0 (8.99) (M_3Q) 8.3 (9.31) (M_2Q)	$-107.7 (-106.7)$ (QM_3) $-110.1 (-109.1)$ (QM_2)
$M_{10}Q_4$	$M_3Q-M_2Q-M_2Q-QM_3$	7.6 (M_3Q) 7.7 (M_2Q)	-108.6 (QM_3) -110.04 (QM_2)
M_6Q_3	$\begin{array}{c} QM_2 \\ M_2Q-QM_2 \end{array}$ (triangular)	6.4	-100.7
M_8Q_4	$\begin{array}{c} M_2Q-QM_2 \\ M_2Q-QM_2 \end{array}$ (ring)	9.2 (10.2)	$-108.8 (-107.8)$
$M_{10}Q_7$	$\begin{array}{c} M \\ Q \\ M_2Q \quad QM_2 \\ MQ-QM \\ MQ-QM_2 \end{array}$ (cage)	9.4 (M_2Q) 10.0 (MQ)	-108.8 (QM_2) -109.8 (QM)
M_8Q_8	$\begin{array}{c} MQ-QM \\ MQ-QM \\ MQ-QM \\ M-M \end{array}$ (cube)	11.7	-109.3
$M_{10}Q_{10}$	$\begin{array}{c} MQ-QM \\ MQ \ MQ-QM \ QM \\ M \\ MQ \quad Q \quad QM \\ Q \\ M \end{array}$ (prismatic cage)	12.4	-110.2

^{a} In ppm relative to Me$_4$Si.
^{b} From Ref. 53. Courtesy Academic Press.
^{c} Values in parentheses are from Ref. 57.

5.5.2. Copolymers

The characterization of copolymer macrostructure involves the determination of the structure of the chain units and the terminating groups, the average degree of polymerization, the molecular weight distribution, and related average molecular weights. To complete this picture, the sequencing of the comonomer units along the polymeric chain or, more precisely, the distribution of sequences of comonomer units, has to be described. A detailed and comprehensive review of the statistical models used for polysiloxane copolymers can be found in Ref. 56.

Structures of pure block copolymers of the type $M-D_m-D'_n-M$ and the block lengths m and n can in principle be directly determined from the ^{29}Si NMR spectrum. One can differentiate between the end groups M–D–D–D– and M–D'–D'–D'–, and also determine the relative numbers of D and D' units. The linkages between the D and D' blocks can also be directly observed as, again, the units next to the mixed linkage will exhibit different chemical shifts:

$$-D \ -D \ -D \ -D \ -D' \ -D' \ -D' \ -D'-$$
$$\gamma \quad \beta \quad \alpha \quad \alpha \quad \beta \quad \gamma$$

The assignment of these units in their position follows the pentad or heptad sequences depending on the strength of the magnetic field. The major limitation of the NMR method applied to this problem comes again from the degree of polymerization. Only relatively short chains, ~ 200 units maximum of pure siloxane block copolymers, can be analyzed.

Similarly, triblock copolymers of the type

$$M-\ D_m-D'_n-D_m-M \qquad \text{or} \qquad M-\ D'_n-D_m-D'_n-M$$

can be determined by careful analysis of the signals in the M, D, and D' regions.

The basic problem for the siloxane block copolymers does not come from the analysis but rather from the synthetic method. The silicon atom has the propensity to undergo rearrangement or transfer reactions, so partial randomization usually occurs. The NMR technique can assess the extent of these side reactions.

5.5.3. Siloxane Random Copolymers

Most siloxane copolymers are synthesized via an equilibration route and ordering of the siloxy units within the linear and cyclic chains is dependent on synthetic conditions (1).

5.5.3.1. **Linear Copolymers.** The ^{29}Si NMR spectra of linear copolymers show two main groups of signals corresponding to D and D' units in the difunctional siloxy unit region, represented by A and B, respectively (56, 58). Each of these patterns is a triplet, as a given unit A or B is influenced in three different ways by its first A and B neighbors (Figure 12.22). These triplets correspond to three triads. With higher field strength, the influence of the second unit can be detected and each of the signals constituting the triplet is further split into three. A symmetric pattern of 9 signals corresponding to 10 pentad sequences is obtained. This sequencing of comonomer units is exactly the same within the chains and at the end of the polymer chains. The terminating units, M, probe the last sequences of the chains at pseudotriad and pseudopentad levels (Figures 12.22 and 12.23) (54).

When the distribution of the D and D' units changes from random to blocky microstructures or when their molar ratios are no longer equal, the relative intensities in the NMR patterns deviate from the symmetric case (Figure 12.24) (59). Furthermore, a splitting of signals leading to a quartet can arise from nonadditivity of the chemical shift resulting from different arrangements of the neighboring units in the pentad sequences.

The assignment of each of the NMR signals to specific n-ads is best performed by analyzing model compounds. This assignment has been accomplished in the case of trimethylsilyl terminated poly(dimethyl-co-methylvinylsiloxanes) (Tables 12.12, 12.13 and 12.14, Figure 12.25) (60). The

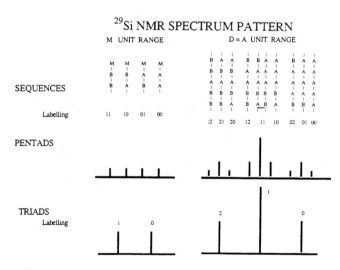

Figure 12.22. ^{29}Si NMR pattern calculated for $M(A_x)(B_x)M$ random copolymer (59). Reprinted with permission from *Plaste und Kautschuk*.

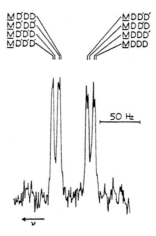

Figure 12.23. ^{29}Si NMR pattern observed for M units as influenced by the D and D' units in a (Me$_2$SiO)$_x$(MeHSiO)$_y$ copolymer (54). Reproduced from *J. Organometallic Chem.*, **70**, 43, © 1974 Elsevier Sequoia SA, Lausanne.

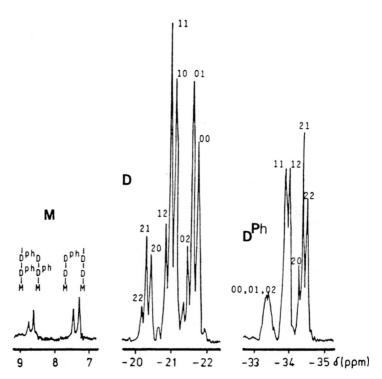

Figure 12.24. ^{29}Si NMR spectrum of a Me(Me$_2$SiO)$_x$(MePhSiO)$_y$SiMe$_3$ copolymer (58). Courtesy Springer-Verlag, © 1981.

388

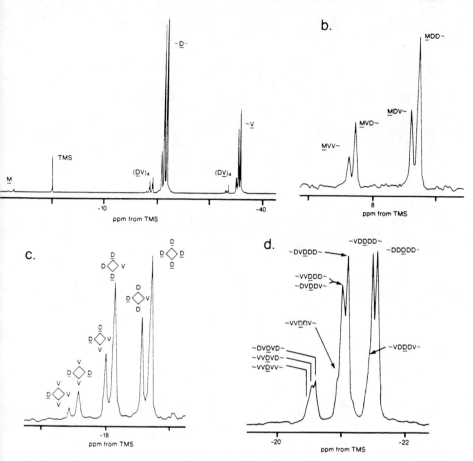

Figure 12.25. Typical ^{29}Si NMR spectra of a 30 mol% V unit, D_xV_y, copolymer where D = Me$_2$SiO and V = MeViSiO, Me$_3$SiO end blocked (60). Reprinted with permission from *Macromolecules*, **22**, 2111 © 1989 American Chemical Society.

approach is rigorous but time consuming. Alternatively, by recording spectra of respective homopolymers and assuming additivity of the chemical shift by replacing one D unit by a D′ unit, one can reasonably assign all the signals observed for a random copolymer.

The microstructure of the copolymer may be described quantitatively by calculating several parameters derived from statistical considerations and from measurement of the intensities of triad signals (56, 58, 59, 61). Intensities are directly related to the probabilities of occurrence of each signal corresponding to a given triad sequence centered on either the A or B unit.

Table 12.12. Trimethylsilyl End-Stopped Cooligomers Synthesized for ^{29}Si NMR Peak Assignments[a]

Compound	Unit[b]	δ(ppm) from TMS	Relative Intensity
MD_3VD_3M	M	7.28	2.0
	D(1, 2, 3)	-21.35 to -21.96	6.1
	V	-35.87	0.95
$MD_3V_2D_3M$	M	7.22	2.0
	D(1, 3)	-21.42	3.9
	D(2)	-21.99	2.1
	V	-35.47	2.0
$MD_3V_3D_3M$	M	7.23	2.0
	D(1, 3)	-21.41	3.9
	D(2)	-21.99	2.2
	V(1)	-35.41	2.0
	V(2)	-35.07	1.1
$MD_3V_4D_3M$	M	7.23	2.0
	D(1, 3)	-21.41	3.9
	D(2)	-21.99	2.1
	V(1)	-35.40	2.0
	V(2)	-35.02	2.0
$MD_3V_5D_3M$	M	7.24	2.0
	D(1, 3)	-21.40	4.0
	D(2)	-21.99	2.3
	V(1)	-35.40	2.1
	V(2, 3)	-34.99	3.4
$MD_3V_6D_3M$	M	7.24	2.0
	D(1, 3)	-21.41	4.1
	D(2)	-21.99	2.0
	V(1)	-35.41	2.1
	V(2, 3)	$-34.95, -35.00$	4.2
MD_6VD_6M	M	7.25	2.0
	D(1, 6)	-21.42	3.8
	D(2)	-22.06	2.0
	D(3, 4, 5)	$-21.91, -21.84$	5.7
	V	-35.84	1.0
$MD_6V_4D_6M$	M	7.25	2.0
	D(1, 6)	$-21.42, -21.35$	4.0
	D(2)	-22.05	2.2
	D(3, 4, 5)	$-21.92, -21.81$	5.7
	V(1)	-35.38	2.0
	V(2)	-34.99	2.0

Table 12.12. (*Contd.*)

Compound	Unit[b]	δ(ppm) from TMS	Relative Intensity
MVD$_n$VM	M	8.26	2.0
	D$_n$	-21.92	147
	V	-35.25	2.0
MDVD$_n$VDM[c]	M	7.34	2.0
	D(1)	-21.00	2.0
	V	-36.09	
	D$_n$	-21.93	

[a] Reprinted with permission from *Macromolecules*, **22**, 2111 © 1989 American Chemical Society.
[b] Numbering for D begins at D adjacent to M unit. Numbering for V units begins at V adjacent to D unit.
[c] Sample contaminated with about 15% MDVDM.

Table 12.13. ^{29}Si NMR Chemical Shifts of Comonomer Sequences in the Chains[a]

Sequence	δ(ppm) from TMS[b]	Sequence	δ(ppm) from TMS[b]
VVDVV	(-20.82)	DDVDD	-35.84
VVDVD	(-20.89)	DDVDV	(-35.74)
DVDVD	(-20.96)	VDVDV	(-35.64)
VVDDV	(-21.28)	DVVDD	-35.47
DVDDV	(-21.35)	DVVDV	(-35.38)
VVDDD	-21.35	VVVDD	-35.38
DVDDD	-21.42	VVVDV	(-35.28)
VDDDV	(-21.77)	DVVVD	-35.07
VDDDD	-21.84	VVVVD	-34.99
DDDDD	-21.91	VVVVV	-34.89

[a] Reprinted with permission from *Macromolecules*, **22**, 2111 © 1989 American Chemical Society.
[b] Values in parentheses are estimated.

The run number R, defined as the average number of comonomer sequences in 100 repeating units, is a single numeric parameter that quantifies the random character of the copolymer (62). Its value can be calculated for a copolymer with all the units statistically distributed within the chains by

$$R_{random} = 200 X_A X_B \qquad (12.10)$$

where X_A and X_B are the mole fraction of A and B siloxy units, respectively.

Table 12.14. ^{29}Si NMR Chemical Shifts
for Chain Ends^a

Sequence	δ(ppm) from TMS
MDD	7.25
MDV	7.34
MVD	8.26
MVV	8.40

^{a} Reprinted with permission from *Macromolecules*, **22**, 2111 © 1989 American Chemical Society.

Experimentally, the R number is related to the ratios of intensities I_i of specific signals of the triads [1], [2], and [3] by the following equation:

$$R = 100q_A X_A = 100q_B X_B \qquad (12.11)$$

The constants q_A and q_B are calculated from the following equations:

$$q_i = 4I_2/(2I_2 + I_1) \qquad q_i = 2I_1/(2I_0 + I_1)$$
$$q_i = 2 - 2(I_0/I_q)^{1/2} \qquad q_i = 2(I_2/I_q)^{1/2} \qquad (12.12)$$

By comparing the experimental (ex) and theoretical (th) run numbers R, one can assign the average microstructure of the copolymer in terms of blocky, random, or alternating character. When $R_{ex} > R_{th}$ there are more alternating sequences than predicted by statistical theory, and conversely, when $R_{ex} < R_{th}$, the monomeric units tend to agglomerate in blocks. Finally, when $R_{ex} = R_{th}$, the two kinds of siloxy units are randomly distributed along the copolymeric chains.

Another important quantitative parameter is the number average sequence length l_A and l_B of the A and B runs, respectively. These are calculated from the run number and the mole fraction of A and B units using the following expressions:

$$l_A = 200X_A/R \qquad l_B = 200X_B/R \qquad (12.13)$$

Adding to the macrostructure parameters described above, namely, the nature of the chain and terminating units, the molecular weight distribution and related average molecular weights $\bar{M}_n$ and $\bar{M}_w$; a set of microstructure parameters, defined as X_A, X_B, l_A, l_B, R_{ex} and R_{th}, completes the description of the copolymer.

Although the theory and its application to siloxane copolymers is elegant, one must be cautious about using a single numeric parameter R to describe such complex polymeric systems. Errors made in the measurement of integrated intensities propagate in the computation of the microstructural parameters; or put conversely, a deviation of the experimental values from the theoretical values does not necessarily mean a significant departure from the statistical distribution of the comonomeric units within the copolymer chains. Here, a comparison of the experimental and theoretical spectral patterns remains the best approach to this type of characterization. Furthermore, the run number is not a linear function of the degree of randomness between the pure alternating and pure blocky copolymer and its interpretation should be made with care.

5.5.3.2. Cocyclics. The equilibrium nature of siloxane polymerization always results in a mixture of cyclosiloxanes of different ring sizes and of linear chains of different lengths. The distribution of the cocyclic species can be observed in ^{29}Si NMR for the three, four, and five-membered rings (2). The signals from higher membered rings overlap with the main linear chain units. Furthermore, different isomers of a given cyclic can be resolved (1, 2, 4, 53,

Table 12.15. Cocyclotetrasiloxanes Synthesized for ^{29}Si NMR Peak Assignments[a]

Compound	Unit	δ(ppm) from TMS	Relative Intensity
D_4	D	− 19.10	
$\overline{D_{(1)}D_{(2)}D_{(1)}V}$	D(1)	− 18.58	2.0
	D(2)	− 18.99	1.0
	V	− 33.50	0.9
$\overline{DDVV}$	D	− 18.44	2.0
	V	− 33.02	2.0
$\overline{DVDV}$	D	− 18.01	2.0
	V	− 33.39	2.0
$\overline{DV_{(1)}V_{(2)}V_{(1)}}$	D	− 17.91	1.0
	V(1)	− 32.95	2.1
	V(2)	− 32.55	1.1

[a] Reprinted with permission from *Macromolecules*, **22**, 2111 © 1989 American Chemical Society.

Table 12.16. ^{29}Si NMR Chemical Shifts^a for Comonomer Sequences in Cocyclic Tetramers^a

Sequence	D Units	V Units
⌐DDDD⌐	− 19.10	
⌐VDDD⌐	− 18.99	− 33.50
⌐VDDD⌐	− 18.58	− 33.50
⌐VDDV⌐	− 18.44	− 33.02
⌐VDVD⌐	− 18.01	− 33.39
⌐DVVV⌐	− 17.91	− 32.95
⌐DVVV⌐	− 17.91	− 32.55
⌐VVVV⌐		− 32.44

^{a} In ppm from TMS.
^{b} Reprinted with permission from *Macromolecules*, **22**, 2111
© 1989 American Chemical Society.

58). A typical spectrum of cyclodimethyl-*co*-methylvinyltetrasiloxane is displayed in Figure 12.25. The same remarks stated for the assignment of the signals of linear *n*-ads are valid here. Model compounds can be synthesized and analyzed (Table 12.15). All cocyclic tetramers can then be properly assigned (Table 12.16) (60).

Similarly to the linear chains, a run number can be independently determined for cocylics and interpreted in terms of distribution of the comonomer units within the cyclic chains. The distribution of individual cyclic species for a given copolymer at equilibrium can be followed by GC to complement the ^{29}Si NMR results. Finally, this distribution can be calculated by a statistical binomial distribution under the assumption that the reactivity of siloxy units is independent of the nature of the substituents attached to the silicon atoms.

5.5.4. *Siloxane Alternating Copolymers*

In pure alternating siloxane copolymers, the two siloxy units A and B alternate regularly giving a single copolymeric sequence (59).

$$-A–B–A–B–A–B–A–B–$$

Only one type of heptad sequence exists in which only two types of silicon atoms, A and B, can be distinguished. The ^{29}Si NMR spectrum shows two single peaks arising from these A and B siloxy units. The ^{29}Si spectrum of a high molecular weight alternating poly(dimethyl-co-diphenylsiloxane) has been reported (63). A more detailed analysis of the spectrum obtained by increasing the S/N ratio reveals other sequences, leading to the conclusion that some randomization did occur and that it can be quantified.

5.5.5. *Intermediate Situations*

Quite often the polymerization process is not driven to thermodynamic equilibrium and/or the copolymers are prepared via nonrearranging synthetic routes. The microstructure of the copolymers then ranges between random and alternating or between random and blocky character. Rather than the quantitative parameters defined above, a close examination of the spectral pattern yields a better description of the system.

Poly(dimethyl-co-methyltrifluoropropylsiloxane) illustrates these situations (Figure 12.26). All three copolymers represent a 1:1 mole ratio of dimethyl- to methyltrifluoropropylsiloxy comonomeric units and are trimethylsilyl end blocked. Copolymer a is a totally random copolymer that can be represented as

$$M-(D_{146}-F_{146})_{random}-M$$

Copolymer b does not display the characteristic fine structure attributed to the 10 different pentad structures. Instead, equimolar quantities of pentads of the type –F–D–D–D–D– and –D–F–F–F–F– are distributed within the chains. Each D or F unit in these sequences has a different chemical shift and results in the observed spectrum. An idealized average structure of this polymer can probably be best described as a random multiblock copolymer:

$$M-[(D-F-F-F-F-)_{30}(F-D-D-D-D)_{30}]_{random}-M$$

Copolymer c has an average degree of polymerization of 134. It displays a fine structure typical of a random copolymer. However, a single signal dominates the F unit pattern and is attributed to an –F–F–F–F–F–F–F– heptad sequence. In the M region, only a single signal is observed and assigned to the M–F–F–F pseudoheptad. One then may conclude that the –F– block is also the terminating block of the copolymer. The relative abundance of F units forming this particular sequence leads directly to the average length of the –F– block. The average structure may, therefore, be

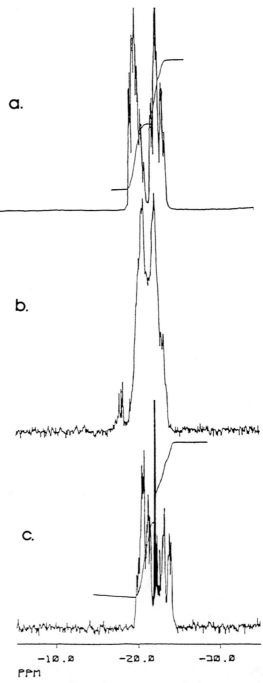

a.

b.

c.

−10.0 −20.0 −30.0
PPM

Figure 12.26. Examples of ^{29}Si NMR spectra resulting from different microstructures found in copolymers of PDMS and poly(methyl-1,1,1-trifluoropropylsiloxane).

396

written as

$$M-F_{13}-(D_{64}-F_{42}-)_{random}F_{13}-M$$

(C)

However, there is an alternative way of interpreting the same spectrum. The F units found in the pure F heptad may arise from a low molecular weight trimethylsilyl end-blocked homopolymer and the remaining F and D units arise from a higher molecular weight random copolymer:

$$M-F_{26}-M \qquad M-(D_{64}-F_{42})_{random}-M$$

(B)

Because of the higher degree of polymerization of the random copolymer, the S/N ratio of the pseudoheptads terminating the chains is low and the M end groups cannot be observed. The sample may, therefore, consist of a blend of a low molecular weight homopolymer with a high molecular weight copolymer. Analysis by gel permeation chromatography is necessary to distinguish between the blend (B) and the pure copolymer (C).

5.5.6. Tacticity in Siloxane Polymers

Homopolymers made of monosubstituted D units can be considered as configurational copolymers as a result of stereoisomerism arising from the pseudoasymmetry of the silicon atoms (4, 54). The D siloxane units adopt two different stereochemical configurations that have different chemical shifts and, as shown for the copolymers, triad and pentad sequences can be observed. Poly(methylhydrogensiloxane) reveals these stereochemical configurations. Figure 12.27 shows the D^H signals of $M-D_5^H-M$ oligomer. The three main signals are assigned to D_1, D_2, and D_3 units. The splitting of D_1 into a doublet is because D_2 has two stereoconfigurations, d or l, thus giving two microenvironments. The resonances of D_2 and D_3 consist of three lines each with a $1:2:1$ relative intensity ratio, each unit having two possible asymmetric neighbors. These features have also been observed in a poly(methylhydrogensiloxane) with a degree of polymerization of 50 (54). The relative intensities of the triplet peaks suggest that the polymer is atactic.

5.5.7. Siloxane–Organic Block Copolymers

Both ^{13}C and ^{29}Si NMR can be used to study block copolymers made of organic and siloxane polymers. The structure, the chemical composition, and the different block lengths of the copolymer can be obtained.

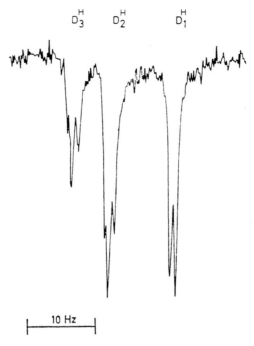

D_3^H D_2^H D_1^H

10 Hz

Figure 12.27. Tacticity effects in $Me_3SiO(MeHSiO)_5SiMe_3$ (54). Courtesy Royal Society of Chemistry.

A polybisphenol carbonate (BPAC)–block–PDMS copolymer illustrates this class of macromolecules (40).

a   b

The ^{13}C NMR experiment distinguishes between the bisphenol A carbonate (**a**) and the dimethylsiloxy units and, in addition, distinguishes single or isolated units such as the bisphenol A (BPA) (**b**) (of Section 4.4.3) (40). The ^{29}Si NMR spectrum of a typical copolymer with an average siloxane block length of 10 units is shown in Figure 12.28. Peaks A and A′ correspond to silicon atoms adjacent to polycarbonate blocks. The peak B corresponds to the second siloxy unit in the siloxane block and the rest of the internal units overlap with the main signals B′. The siloxane block length can be calculated directly from the following expression:

$$n_{PDMS} = 2\left[(I_B + I_{B'})/(I_A + I_{A'}) + 1\right] \tag{12.14}$$

where I_A, $I_{A'}$, I_B, and $I_{B'}$ are the integrated areas of the respective peaks.

Although a considerable amount of information is available from ^{13}C NMR, the use of ^{29}Si NMR reveals that apart from the –BPAC–b– PDMS–block copolymers, BPA–b–PDMS block copolymers are also produced (53).

The block length has also been determined for a poly[(tetramethyl-p-silphenylene)siloxane-co-poly(dimethylsiloxane)] block copolymer (53):

51

$$\delta_{Si}^A = -1.11 \text{ ppm}$$
$$\delta_{Si}^B = -2.45 \text{ ppm}$$
$$\delta_{Si}^C = -20.65 \text{ ppm}$$
$$\delta_{Si}^D = -21.86 \text{ ppm}$$

5.6. Analytical Applications

Starting in approximately 1980, ^{13}C and ^{29}Si NMR microstructural analysis has been applied and exploited in various ways. Analysis of different organo-functionalized siloxy units allows the simultaneous quantitative determination of different chemical groups and of a given group in different environments. The accuracy and precision of ^{29}Si NMR can exceed those of wet chemical methods of analysis. The specificity of the analysis is unique for ^{29}Si NMR. Furthermore, the determination of chemical functions is relative to the polysiloxane species and not to the bulk. Therefore, non-Si containing solvent, impurities, or additives do not interfere with the analysis.

Better characterization of copolymers allows the monitoring of polymerization process parameters and, consequently, their control. The sequence distribution in a copolymer can be correlated with physical properties only when the polysiloxane has reached the thermodynamic equilibrium. The

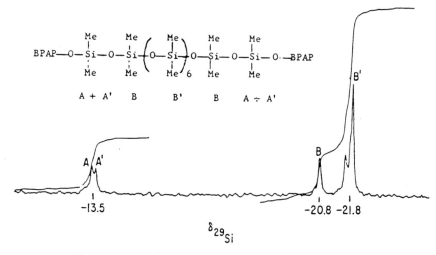

Figure 12.28. ^{29}Si NMR spectrum of a bisphenol A polycarbonate–PDMS block copolymer with average silicone block length of 10 (40). Reprinted with permission from *Macromolecules*, **10**, 782, © 1977 American Chemical Society.

glass transition of poly(dimethyl-*co*-diphenylsiloxane) has been related to the structure of the copolymer and shown to be linearly dependent on the composition of the copolymer (64).

The mechanism of the polymerization process can also be determined as one may follow the fate of specific silicon atoms as a function of time, temperature, conversion of reactants, and catalyst. One such study has led to the proposal of a thermodynamic model of siloxane copolymerization and to the evidence that the copolymerization process is entropy driven (60).

6. TWO-DIMENSIONAL FT–NMR

Two-dimensional (2-D) FT–NMR has become an important and powerful tool widely used for the analysis of organic compounds because of the increased resolution and the ability it offers to correlate interacting nuclei directly (65). These techniques can be applied to ^{29}Si, analogously to ^{13}C. To date, however, 2-D NMR has seldom been used in the analysis of organosilicon compounds, and then generally has been limited to the analysis of the organic portion of the molecule, where all of the standard 2-D methods for analysis apply. The utility of 2-D NMR methods for the characterization of the silicon backbone, however, has not been demonstrated and suffers from several limitations. First, all organosilicon compounds are synthetic (as contrasted to, e.g., natural products in organic chemistry). Because their basic building blocks are well known, silicone structures can usually be adequately characterized by other methods. Second, the most popular 2-D NMR techniques rely on scalar coupling to correlate bonded or neighboring nuclei as in 2-D homonuclear and heteronuclear correlation spectroscopies. Homonuclear correlation spectroscopy for silicon would require significant Si–Si coupling between chemically and magnetically distinct silicon nuclei and the use of a time-consuming low-sensitivity INADEQUATE-2-D-type experiment. Heteronuclear correlation is most often focused on directly bound protons or long-range protons removed by two or three bonds. For silicon, the SiH functionality is generally limited to well-known cases and the utility of long-range Si–H coupling is limited by the common condition that each Si is coupled to five or more protons by similar coupling constants. Perhaps the most limiting feature of ^{29}Si 2-D NMR is that the spectral resolution available from 1-D NMR directly answers many of the questions for which 2-D NMR is commonly required in organic compounds.

Most of the published applications of 2-D NMR to organosilicon compounds directly involving ^{29}Si have been for the analysis of oligosaccharides (66–69). Henge et al. (70) have applied INADEQUATE-2-D to Si frameworks in cyclosilanes, but mention that all the required information could be obtained from a 1-D experiment. It is possible that 2-D NMR will find increased utility for the analysis of organosilicon compounds in the future as new classes of compounds emerge, but it is unlikely that it will ever gain the widespread popularity or application that it has for the analysis of organic compounds.

7. SOLID-STATE NMR

For many years chemists were unable to apply NMR to the study of chemistry in the solid state because spin–lattice interactions broadened resonances to the point that little chemical information was available. X-ray diffraction patterns were useful to determine structure in solids that were regularly arrayed, but this approach was limited to a relatively small number of crystalline solids and extrapolation to solution structure was questionable. EXAFS gives nearest neighbor information but is a difficult and expensive technique. Infrared spectroscopy is useful but somewhat limited in detailed structural information without other supporting spectroscopies. With the advent of new NMR techniques to study directly molecular structure and dynamics in the solid state, interest in this area is great, and a rich, rapidly growing literature is available. In this section, we offer a brief introduction to the technique and some selected examples of applications to the solid state NMR analysis of organosilicon compounds. The discussion of the solid state NMR experiment presented here is primarily taken from Fyfe's text (71).

7.1. Practical Considerations

The principal difference between solution and solid-state NMR lies in the anisotropic or orientation-dependent interactions present in the solid state where the molecules are effectively fixed in space. These interactions include dipole–dipole interactions with other nuclei, magnetic shielding by surrounding electrons resulting in chemical shifts, scalar or spin–spin coupling to other nuclei, and quadrupolar interactions. Anisotropic interactions are not normally observed in solution because rapid molecular motions average dipolar and quadrupolar couplings to zero and chemical shifts and scalar couplings to discrete isotropic or average values.

Anisotropic interactions in solid state NMR typically result in broad featureless spectra. The ranges of these interactions are given in Table 12.17 and are dependent on the rates and types of motion present in the solid-state sample. In this discussion we deal only with ^{13}C and ^{29}Si nuclei in the solid state. Both nuclei are magnetically dilute because of their low isotopic natural abundance. This dilution effectively removes any homonuclear dipolar or scalar interactions. Both are spin $\frac{1}{2}$ nuclei so no quadrupolar interactions are present. However, the remaining heteronuclear dipolar, heteronuclear scalar, and anisotropic chemical shift interactions can result in resonances that are 10^4–10^5 Hz broad.

The complete cross-polarization–magic angle spinning (CP–MAS) NMR experiment, first demonstrated by Schaefer and Stejskal in 1976 (72), involves the combination of (a) cross-polarization of magnetization from abundant ^1H nuclei to the dilute ^{29}Si or ^{13}C nuclei to increase S/N ratios and to avoid the typically longer T_1 relaxation times of the dilute nuclei; (b) high power proton decoupling during acquisition to remove the heteronuclear dipolar and scalar

Table 12.17. Approximate Ranges of the Different Spin
Interactions (Hz)

Zeeman	10^6–10^9
Dipolar	0–10^5
Chemical shift	0–10^5
Scalar coupling	0–10^4
Quadrupolar	0–10^9

interactions; and (c) MAS at kilohertz frequencies to remove the chemical shift anisotropy interactions.

Taken together, the CP–MAS techniques can dramatically reduce or remove the anisotropic broadening interactions in solid samples to yield high resolution solution-like spectra as shown in Figure 12.29 (73). Cross-polarization produces a maximum sensitivity enhancement equal to the ratio of magnetogyric ratios of the abundant and dilute spins. For ^{1}H–^{13}C cross-polarization, this enhancement is $\gamma_H/\gamma_C = 4$, which translates to a reduction in experimental acquisition times by a factor of 16, plus the added benefit of a repetition rate dependent on the ^{1}H relaxation time. In compounds with no protons, such as ceramic materials, zeolites, or silica, cross-polarization is not possible and high power proton decoupling is not required. Thus, MAS with the standard one-pulse solution state NMR experiment may be used.

For the study of surface interactions with proton-bearing adsorbents or modified surfaces, the CP–MAS approach can be advantageous. It is now possible to look at bulk zeolite or silica using MAS NMR or to selectively study the surface by CP–MAS techniques.

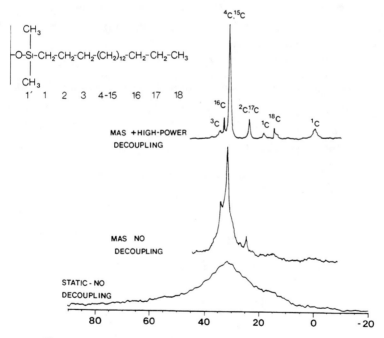

Figure 12.29. ^{13}C solid state NMR spectra showing effects of MAS and high power decoupling (73). Reproduced by permission.

7.2. Quantitation

Quantitation in the solid state by MAS follows the same rules as for solutions, requiring a relaxation delay time of 5X the longest T_1 in the sample between pulses. This sequence can be time consuming because ^{29}Si and ^{13}C T_1 values may be quite long in organosilicon compounds. Relaxation reagents like Cr(acac)$_3$ are not generally useful because the effect of the electron-nuclear relaxation is inversely dependent on the sixth power of the electron-nuclear distance. It is difficult to grind a sample fine enough to get significant or consistent results. One approach that has been suggested for sealed rotors is to saturate the sample with oxygen gas before sealing. This is appropriate for O_2-permeable samples like silica but would not be useful for glasses. However, significant reduction in T_1 relaxation times may be observed for appropriate samples with long T_1 values.

When available, the CP–MAS technique, diagrammed in Figure 12.30 (73), is generally preferred to the MAS technique because of the sensitivity enhancement afforded and the dependence of the repetition rate on the proton relaxation times. In many cases, with homogeneous solids of a single chemical phase or domain, the dependence on repetition rate for quantitation is entirely removed. This improvement results from the strength of dipolar interactions of the high abundance proton spins and is referred to as the establishment of spin temperature (71). In this case, all the protons will experience a common T_1 so that, independent of repetition rate, saturation effects will be equal for all nuclei and quantitative conditions are maintained.

A greater concern is that different ^{13}C or ^{29}Si nuclei can have different cross-polarization efficiencies, resulting in different degrees of enhancement

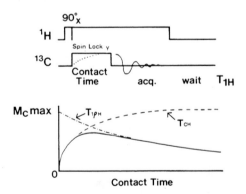

Figure 12.30. Diagram of ^{13}C CP/MAS NMR experiment. The observed magnetization of the ^{13}C spins, M_C, is dependent on both the proton relaxation time, $T_{1\rho h}$, and the cross-polarization contact time, T_{CH} (73). Reproduced by permission.

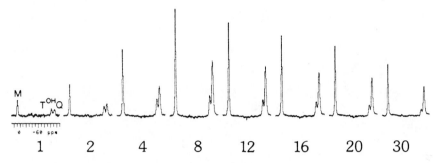

Figure 12.31. ^{29}Si NMR spectra of an MTQ resin at various contact times: 1, 2, 4, 8, 12, 16, 20, and 30 ms. Note the reversal of relative signal intensity for T and Q signals at 1- and 8-ms contact times.

for nuclei in different environments. Cross-polarization efficiency is dependent on molecular dynamics as well as on the distance from each ^{13}C or ^{29}Si nucleus to the proton. Calibrations for contact time, T_{SiH}, show different intensities for different nuclei at different times (Figure 12.31). A T_{SiH} may be chosen such that all responses are the same, or a calibration curve can be established giving correction factors at the T_{SiH} selected. Optimum conditions for quantitative CP–MAS (74) require

$$T_{\text{SiH}}/5 \approx T_1^{\text{Si}} \ll T_{1\rho}^{\text{H}} \qquad (12.15)$$

where T_1^{Si} and $T_{1\rho}^{\text{H}}$ are the laboratory and rotating frame spin-lattice relaxation times, respectively. Tetrakis(trimethylsilyl)silane has been proposed as a chemical shift standard for solid-state NMR spectroscopy (75).

7.3. Applications

The ^{29}Si and ^{13}C MAS or CP–MAS NMR have been applied to the study and characterization of a wide variety of organosilicon compounds including surface modified silicas, homogeneous polymers and copolymers, resins, and elastomers. Examples from three of these areas are presented here.

7.3.1. Organosilane Modified Silica

The ^{29}Si and ^{13}C CP–MAS NMR have been used to study silane–silica surface bonding interactions under systematically varied pre- and posttreatment conditions. DeHaan et al. (76) reported on the reactions of (3-aminopropyl)triethoxysilane (APTS) and (3-methacryloxypropyl)trimethoxysilane (MPTS) with porous (silica gel) and nonporous (Cab–O–Sil) silica

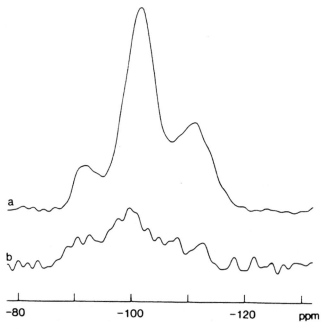

Figure 12.32. ^{29}Si CP–MAS NMR spectra: (*a*) silica gel, (*b*) Cabosil. Reprinted by permission from (76). Courtesy Academic Press.

powders. The influence of heat treatment and the addition of water at different stages was studied. The ^{29}Si CP–MAS NMR spectra of the two powders obtained under similar conditions are shown in Figure 12.32. Figure 12.33 shows ^{29}Si CP–MAS NMR spectra of the products of the reaction of APTS with silica gel in toluene under different conditions. The ^{29}Si spectra give information about the type and number of Si–O–Si bonds in the silane moiety. Mono-, bi-, and tridentate surface-bonded silane structures and chemical shift assignments are shown in Figure 12.34. The ^{13}C CP–MAS NMR spectra of the APTS modified silica gel are presented in Figure 12.35. Although these spectra are not quantitative, they clearly offer direct information on the effects of heat and water treatment on ethoxy content. The authors concluded that the reaction of APTS with silica gel in dry toluene resulted in the formation of mainly mono- and bidentate linkages. Heating to 200 °C produced the tridentate linkage while monodentate linkages decreased. Water treatment led to the formation of cross-linked structures and the complete loss of monodentate structures. Reaction of APTS with Cabosil

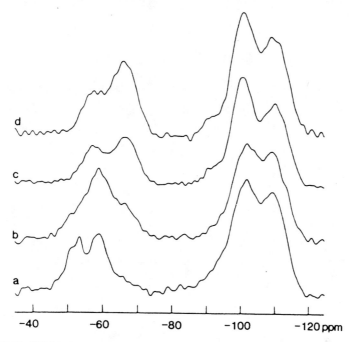

-40   -60   -80   -100   -120 ppm

Figure 12.33. ^{29}Si CP–MAS NMR spectra: (a) derivatization product of silica gel with APTS in toluene, (b) product after heating at 200 °C in an Ar atmosphere, (c) product after water treatment, and (d) product after treatment as in (c) and heating at 200 °C in an Ar atmosphere. Reprinted by permission from (76). Courtesy Academic Press.

under the same conditions gave qualitatively analogous results except that heat treatment resulted in conversion of **I** to **I**′ (Figure 12.34), which was attributed to the release of H_2O from the Cabosil surface.

Caravajal et al. (77) reported investigations on the derivatization of silica gel with APTS at about the same time. They reported similar trends for the reaction as DeHaan but went further to measure the relevant NMR relaxation parameters (T_{HSi}, T_{HC}, the ^1H–^{13}C cross-polarization time constants, and $T_{1\rho}^H$) that are necessary for quantitative comparison of results. The number average of residual ethoxy groups per attached silane moiety was calculated from both ^{29}Si and ^{13}C NMR data. Additionally, evidence for the protonation of the terminal amino group was gained from the ^{13}C CP–MAS spectra in which a shift toward greater shielding from 27 to 21 ppm for C_2 of the n-propyl group was observed on addition of H_2O (Figure 12.36). They observed the highest silane loading levels to occur with silica gel that had

```
        R                              R
        |                              |
EtO — Si — OEt                 HO — Si — OH
        |                              |
        O    -53                       O    -48
        |                              |
        Si                             Si
       /|\                            /|\
        I                             I'


    OEt   R                        OH    R
      \  /                           \  /
       Si    -58                      Si    -58
      / \                            / \
     O   O                          O   O
    /     \                        /     \
  Si       Si                     Si      Si
  /|\      /|\                    /|\      /|\
       II                             II'


        R                         R         R
        |                         |         |
        Si     -66                Si — O — Si
       /|\                       /| |      |\    -66
      O O O                     O  O  O   O  O
     /  |  \                   /    |   |     \
   Si  Si  Si                 Si   Si  Si     Si
   /|\ /|\ /|\                /|\  /|\  /|\    /|\
        III                             IV


        R          R          R
        |          |          |
HO — Si — O —[— Si — O —]— Si — OH
        |          |          |
        O          O   -66    O    -58
        |          |          |
        Si         Si        Si
       /|\        /|\ n     /|\
                    V
```

Figure 12.34. Correlation of mono-, bi-, and tri-dentate attachments of APTS to the silica surface with ^{29}Si NMR chemical shifts. Reprinted by permission from (76). Courtesy Academic Press.

been dried at 25 °C and 10^{-2} torr and silylated in dry toluene at 25 °C. This result confirmed the importance of the role of surface water in the silylation process but also showed that too much water can interfere with the reaction.

7.3.2. Characterization of Poly(di-n-alkylsilane)

Poly(di-n-alkylsilanes) comprise a class of polymers that exhibit a wide range of physical properties depending on the type of alkyl groups employed.

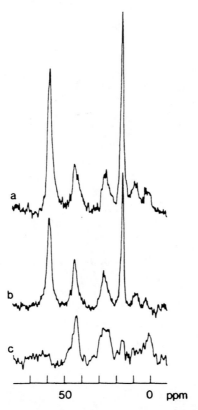

Figure 12.35. ^{13}C CP–MAS NMR spectra: (*a*) derivatization product of silica gel with APTS in toluene, (*b*) product after heating at 200 °C in an Ar atmosphere, and (*c*) product after water treatment. Reprinted by permission from (76). Courtesy Academic Press.

Several publications have dealt with solid-state characterization of poly(di-*n*-hexylsilane) (PDHS) by variable temperature ^{13}C and ^{29}Si CP–MAS NMR (78–80). Two phases coexist in solid PDHS at room temperature and below: a crystalline phase that has an all-trans silicon backbone and fully extended side chains (Phase I); and a disordered phase that has gauche conformations in the side chains, and nontrans conformations introduced into the silicon backbone (Phase II). Figure 12.37 compares solution and solid state ^{13}C NMR spectra of PDHS (78). At 50 °C, (Figure 12.37*c*), six resonances are seen for the hexyl side chains, which are all in the Phase II conformation. The increasing width of the peaks for carbon atoms closer to the silicon backbone results from reduced freedom of motion. At 23 °C, an additional set of peaks

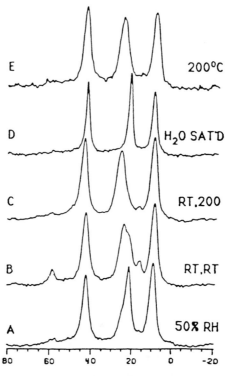

Figure 12.36. ^{13}C CP–MAS spectra of APTS—derivatized silica gel samples for which the silica was dried for 24 h at 25 °C and 10^{-2} torr, derivatized with APTS and heated 24 h at 25 °C and 10^{-2} torr (A, B, C) or 200 °C and 10^{-2} torr (D, E). A, D, and E were additionally treated: A at 50% relative humidity; D saturated w/H_2O; E, sample D heated at 200 °C and 10^{-2} torr for 24 h. Reprinted by permission from *Anal. Chem.*, **60**, 1776 © 1988 American Chemical Society.

is seen and assigned to the Phase I conformation, indicated by underlined numbers. The phase transition is better monitored by ^{29}Si CP–MAS NMR (Figure 12.38), which shows two distinct peaks for the two phases and the progression from predominantly Phase I at 21 °C to completely Phase II at 45 °C (78). A comparison of ^{29}Si CP–MAS spectra of poly(di-*n*-pentylsilane) (PDHS), and poly(di-*n*-butylsilane) taken near the midpoint of the Phase I–Phase II transitions is presented in Figure 12.39 (78). The upfield shift of Phase I silicon beyond the position of Phase II silicon in PDPS and PDBS was attributed to a 7:3 helical conformation of the silicon backbone, as determined by X-ray diffraction studies, resulting from a shortening of the alkyl side chain.

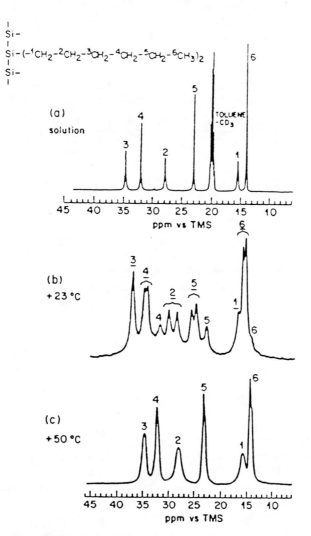

Figure 12.37. The 50.31 MHz ^{13}C spectra of poly(di-n-hexylsilane) (PDHS); (a) in toluene-d_8 solution at $\sim 25\,°C$; (b) in the solid state with CP–MAS at 23 $°C$; and (c) in the solid state at 50 $°C$ with MAS (no CP). Reprinted by permission from Ref. 78. © 1988, Division of Polymer Chemistry, Inc., American Chemical Society.

7.3.3. Topology of PDMS Elastomer Networks

Elastomers are widely used in industrial applications. Correlation of the molecular structure of the elastomeric network with the macroscopic proper-

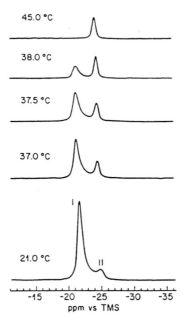

Figure 12.38. The 39.75 MHz ^{29}Si CP–MAS spectra of solid-state crystallized PDHS at five temperatures. Reprinted with permission from (78). © 1988, Division of Polymer Chemistry, Inc., American Chemical Society.

ties of the elastomer is necessary to be able to design elastomers with predictable properties. Beshah et al. (81) have applied solid state ^{29}Si CP–MAS NMR to investigate the molecular details of PDMS model networks. They were able to assess microscopic network properties such as junctionality, number density and dangling ends, of which the first two are major determinants of modulus. They found that cross-polarization was useful and necessary to give adequate sensitivity in these systems with long ^{29}Si relaxation times. Room temperature CP–MAS spectra of low $\bar{M}_n$ model networks provided excellent resolution on the order of 0.4 ppm FWHH (full width at half-height). Figure 12.40 shows ^{29}Si CP–MAS spectra for $\bar{M}_n = 520$, DP = 7 vinyl end-linked PDMS networks. Assignments are given. These networks were prepared by reacting Me_2ViSi- terminated oligomers with tetrafunctional $Si(OSiMe_2H)_4$ using chloroplatinic acid. The sample shown in spectrum A was extracted to remove all unreacted and unbound species. The Q resonance in spectrum B appearing at -105.2 ppm that does not appear in spectrum A was assigned to a double loop "butterfly" structure:

Single loop "bowtie" and zero loop junctions of four different chains are also shown.

The separation of ^{29}Si chemical shifts for different functionalities enabled identification of both gross and subtle network chemical and topographical features. Imperfections in the networks such as dangling ends and unreacted sites of the cross-linking reagent could be identified.

Static studies of networks below their glass transition temperatures were also presented. The ^{13}C CP–MAS spectra were obtained but found to be less informative than those of ^{29}Si because the silicon atoms occupy the key positions in the PDMS network structure. Further work is necessary to define conditions for quantitative acquisitions.

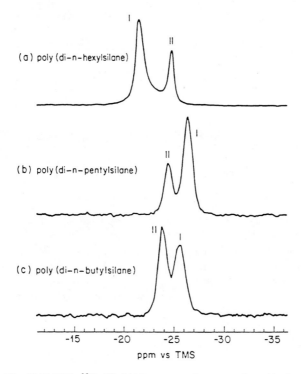

Figure 12.39. The 39.75 MHz ^{29}Si CP–MAS spectra taken near the midpoint of the Phase I–Phase II disordering transition for (a) PDHS, (b) PDPS, and (c) PDBS. Reprinted with permission from (78). © 1988, Division of Polymer Chemistry, Inc., American Chemical Society.

VINYL ENDLINKED (DP 7)
CP/MAS

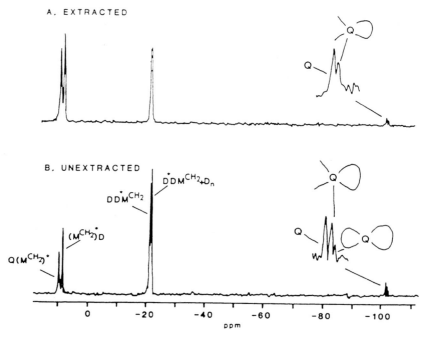

Figure 12.40. Room temperature CP–MAS spectra of vinyl-end linked PDMS networks, $M_n = 520$. Note the multiplet in the Q region. (*A*) Network extracted to remove unreacted and unbound species. One of the Q resonances of (*B*) is removed. (*B*) Unextracted network (81). Reprinted with permission from *J. Polym. Sci. Part B: Polym. Phys.*, **24**, 1207 © 1986 John Wiley & Sons, Inc.

In a related paper Beshah et al. (82), reported evidence from variable temperature ^{29}Si CP–MAS NMR experiments on model end-linked PDMS networks at 300 and 150 K that support the assignments for the Q structures given above. Different chemical shift dependencies on temperature were observed for different Q structures. These differences were attributed to different degrees of motional freedom that correlated with the number of loops present at each type of junction.

REFERENCES

1. E. A. Williams and J. D. Cargioli, "Silicon-29 NMR spectroscopy," in *Annual Reports on NMR Spectros.*, Vol. 9, G. A. Webb, Ed., Academic, London, 1979, p. 221.

2. R. K. Harris, J. D. Kennedy, and W. McFarlane, "Group IV—silicon, germanium, tin and lead," in *NMR and the Periodic Table*, Chapter 10, R. K. Harris and B. E. Mann, Eds., Academic, New York, 1978.

3. G. C. Levy and J. D. Cargioli, "^{29}Si Fourier transform NMR," in *Nuclear Magnetic Resonance Spectroscopy of Nuclei Other than Protons*, T. Axenrod, Ed., Wiley, New York, 1974, p. 251.

4. R. K. Harris and B. J. Kimber, "Silicon-29 NMR as a tool for studying silicones," *Appl. Spectrosc. Rev.*, **10**, 117 (1975).

5. J. Schraml and J. M. Bellama, in *Determination of Organic Structures by Physical Methods*, Chapter 4, Vol. 6, F. C. Nechal, J. J. Zuckerman, and E. W. Randall, Eds., Academic, New York, 1976.

6. R. K. Harris, "Introduction," in *NMR and the Periodic Table*, R. K. Harris and B. E. Mann, Eds., Academic, New York, 1978, Chap. 1.

7. G. Engelghardt, H. Jancke, M. Magi, T. Pehk, and E. Lippmaa, "Uber die ^{1}H–, ^{13}C–, and ^{29}Si–NMR Chemischem Verschiebungen Einiger Linearer, Verzweigter und Cyclischer Methylsiloxaneverbindungen," *J. Organomet. Chem.*, **28**, 293 (1971).

8. S. Macomber, *NMR Spectroscopy: Basic Principles and Applications*, Harcourt Brace Javonovich, San Diego, CA, 1988.

9. M. L. Martin, J-J. Delpuech, and G. J. Martin, *Practical NMR Spectroscopy*, Heyden, London, 1980.

10. D. E. Williams, "Nuclear magnetic resonance spectroscopy," in *Analysis of Silicones*, Chapter 11, A. L. Smith, Ed., New York, 1974.

11. P. Laszlo, in *Progress in NMR Spectroscopy*, Chapter 6, Vol. 3, J. W. Emsley, J. Feeney, and L. H. Sutcliffe, Eds., Pergamon Press, New York, 1967.

12. D. E. Williams, Dow Corning Corp., unpublished work.

13. J. Homer, A. W. Jarvie, A. Holt, and H. J. Hickton, "Organosilicon compounds. II. The methyl proton resonance spectra of methylphenyldisiloxanes," *J. Chem. Soc. B*, **1967**, 67.

14. T. L. Brown and K. Stark, "Effect of aromatic solvents on proton magnetic resonance spectra," *J. Phys. Chem.*, **69**, 2679 (1965).

15. E. A. V. Ebsworth and S. G. Frankiss, "Proton magnetic resonance spectra of di- and trisubstituted derivatives of methylsilane," *Trans. Faraday Soc.*, **63**, 1574 (1967).

16. J. M. Bellama and A. G. MacDiarmid, "Synthesis and proton nuclear magnetic resonance spectra of silylmethyl halides," *J. Organometal. Chem.*, **18**, 275 (1969).

17. C. Dybowski and R. L. Lichter, Eds., *NMR Spectroscopy Techniques*, Marcel Dekker, New York, 1987.

18. E. D. Becker, J. A. Ferretti, and P. N. Gambhir, "Selection of optimum parameters for pulse Fourier transform nuclear magnetic resonance," *Anal. Chem.*, **51**, 1413 (1979).

19. F. Kassler, *Quantitative Analysis by NMR Spectroscopy*, Academic, New York, 1973.

20. K. M. MacKay, A. E. Watt, and R. Watt, "Proton magnetic resonance spectra of the ethylhalosilanes," *J. Organometal. Chem.,* **12**, 49 (1968).

21. B. K. Hunter and L. W. Reeves, "Chemical shifts for compounds of the Group IV elements silicon and tin," *Can. J. Chem.,* **46**, 1399 (1968).

22. C. H. VanDyke and A. G. MacDiarmid, "The proton nuclear magnetic resonance spectra of silicon analogs of simple ethyl compounds," *Inorg. Chem.,* **3**, 1071 (1964).

23. E. A. V. Ebsworth and J. J. Turner, "Nuclear magnetic resonance spectra of silicon hydrides and derivatives. I. Coupling constants involving hydrogen," *J. Chem. Phys.,* **36**, 2628 (1962).

24. H. Elser and H. Dreeskamp, "Indirect nuclear spin coupling in organic phosphorus and silicon compounds," *Ber. Bunsenges. Phys. Chem.,* **73**, 619 (1969).

25. A. N. Egorochkin, M. L. Khudekel, V. A. Ponomarenko, and N. A. Zadorozhnyi, "Some regularities in proton magnetic spectra of trisubstituted silanes," *Izv. Akad. Nauk SSSR Ser. Khim.,* **1963**, 1868.

26. J. Schraml and V. Chvalovsky, "Organosilicon compounds. XLII. P.M.R. spectra of certain vinylsilanes," *Coll. Czech. Chem. Commun.,* **31**, 503 (1966)

27. R. T. Hobgood, J. H. Goldstein, and G. S. Reddy, "Proton magnetic resonance spectra of vinylmethylsilanes," *J. Chem. Phys.,* **35**, 2038 (1961).

28. D. E. Williams and G. M. Ronk, unpublished work, Dow Corning Corp., Midland, MI.

29. J. F. Hampton, C. W. Lacefield, and J. F. Hyde, "Identification of organosilanols by nuclear magnetic resonance spectroscopy," *Inorg. Chem.,* **4**, 1659 (1965).

30. B. D. Kay and R. A. Assink, "High field ^1H NMR studies of sol–gel kinetics," *Mater. Res. Soc. Symp. Proc.,* **73** (Better Ceram. Chem. 2), 157 (1986).

31. F. C. Shilling, F. A. Bovey, and J. M. Zeigler, "The characterization of polysilanes by carbon-13, silicon-29, and proton NMR," *Macromolecules,* **19**, 2309 (1986).

32. A. Garcia-Raso, P. Ballester, R. Bergueiro, I. Martinez, J. Sanz, and M. L. Jimeno, "Estimation of the polarity of stationary phases by proton nuclear magnetic resonance spectroscopy. Application to phenyl and methyl silicones (OV and SE series)," *J. Chromatogr.,* **402**, 323 (1987).

33. J. R. Parsonage and E. A. Vidgeon, "Determination of molecular weight of polyorganosiloxanes by ^1H NMR using anisotropic solvent-induced shifts (ASIS)," *Chem. Ind. (London),* **14**, 488 (1987), through *Chem. Abstr.,* **107**: 141748q (1987).

34. G. C. Levy, L. Lichter, and L. Nelson, *Carbon-13 Nuclear Magnetic Resonance Spectroscopy,* 2nd ed., Wiley, New York, 1980.

35. F. W. Wehrli and T. Wirthlin, *Interpretation of Carbon-13 NMR Spectra,* Heyden and Son, London, 1976.

36. E. A. Williams, "NMR spectroscopy of organosilicon compounds," in *The Chemistry of Organic Silicon Compounds,* Chapter 8, S. Patai and Z. Rappoport, Eds., Wiley, London, 1989.

37. M. Jensen, "The prediction of ^{29}Si–H and ^{13}C–H coupling constants in substituted silane and methanes," *J. Organomet. Chem.*, **11**, 423 (1968).

38. P. J. Kanyha and S. Brey, "Multinuclear NMR of organosilanes," Congr. AMPERE Magn. Reson. Relat. Phenom., Proc., 22nd, 341–342 (1984).

39. E. A. Williams, P. E. Donahue, and J. D. Cargioli, "Nuclear magnetic resonance determination of the effect of end-capping ratio on the structure of bisphenol A polycarbonate—poly(dimethylsiloxane) block copolymers," *Macromolecules*, **14**, 1016 (1981).

40. E. A. Williams, J. D. Cargioli, and S. Y. Hobbs, "The ^{13}C and ^{29}Si nuclear magnetic resonance analysis of bisphenol A polycarbonate-polydimethyl siloxane block copolymers," *Macromolecules*, **10**, 782 (1977).

41. M. E. Gangoda and R. K. Gilpin, "NMR investigations of ^{13}C labeled alkyl modified silica," *J. Magn. Resonance*, **53**, 140 (1983).

42. G. R. Holzman, P. C. Lauterbur, J. H. Anderson, and W. Koth, "Nuclear magnetic resonance field shifts of ^{29}Si in various materials," *J. Chem. Phys.*, **25**, 172 (1956).

43. P. C. Lauterbur, in *Determination of Organic Structure by Physical Methods*, Vol. 2, F. C. Nachod and D. W. Phillips, Eds., Academic, New York, 1962, p. 465.

44. E. Liepins, I. Zicmane, and E. Lukevics, "A multinuclear NMR spectroscopy study of alkoxysilanes," *J. Organometal. Chem.*, **306**, 167 (1986).

45. R. Z. Radeglia, "Theoretical interpretation of NMR chemical shifts of tetracoordinated central atoms. I. Principles and application to silicon compounds," *Phys. Chem. (Leipzig)*, **256**, 453 (1975).

46. J. M. Bellama, J. D. Nies, and N. Ben-Zvi, "Nuclear magnetic resonance study of selected derivatives of 2,8,9-trioxa-5-aza-1-silatricyclo[3.3.3.0$^{1.5}$]undecane-(silatrane)," *Magn. Reson. Chem.*, **24**, 748 (1986).

47. E. Petrakova, J. Schraml, J. Hirsch, M. Kvicalova, J. Zeleny, and V. Chvalovsky, "Analysis of oligosaccharides. Silicon-29 and carbon-13 NMR spectra of pertrimethylsilylated oligosaccharides derived from xylopyranose," *Collect. Czech. Chem. Commun.*, **52**, 1501 (1987).

48. T. C. Carr, personal communication, Dow Corning Corp., Midland, MI.

49. J. Schraml, V. Blechta, M. Kvicalcva, L. Nondek, and V. Chvalovsky, "Polar functional group analysis of mixtures by silicon-29 nuclear magnetic resonance," *Anal. Chem.*, **58**, 1892 (1986).

50. C. Brevard and P. Granger, *Handbook of High Resolution Multinuclear NMR*, Wiley, New York, 1981.

51. J. M. Dereppe and B. Parbhoo, "Quantitative protonated heteroatom determination by silicon-29 nuclear magnetic resonance spectrometry and polarization transfer pulses. Application to asphaltene," *Anal. Chem.*, **58**, 2641 (1986).

52. D. A. Laude, Jr., and C. L. Wilkins, "Applications of a recycled-flow Fourier transform nuclear magnetic resonance spectrometer system: molecular weight determination of siloxane polymers by silicon-29 NMR," *Macromolecules*, **19**, 2295 (1986).

53. E. A. Williams, "Recent advances in silicon-29 NMR spectroscopy," in *Annual Reports on NMR Spectroscopy*, Vol. 15, G. A. Webb, Ed., Academic, London, 1983, p. 235.

54. R. K. Harris and B. J. Kimber, "End-groups and tacticity in polymeric silicones detected by ^{29}Si nuclear magnetic resonance," *J. Chem. Soc. Chem. Commun.*, **1975**, 559.

55. E. A. Williams, J. D. Cargioli, and R. W. Larochelle, "Silicon-29 NMR. Solvent effects on chemical shifts of silanols and silylamines," *J. Organomet. Chem.*, **108**, 153 (1976).

56. T. C. Kendrick, B. Parbhoo, and J. W. White, "Polymerization of cyclosiloxanes," in *Comprehensive Polymer Chemistry*, Vol. 4, Chapter 25, G. C. Eastmond, A. Ledwith, S. Russo, and P. Sigwalt, Eds., Pergamon Press, 1989, p. 459.

57. Von D. Hoebbel, G. Garzo, G. Engelhardt, H. Jancke, P. Franke, and W. Wieker, "Gaschromatographische und ^{29}Si-NMR Spektroskopisch Untersuchungen an Kieselsauretrimethylsilylestern," *Z. Anorg. Allg. Chem.*, **424**, 115 (1976).

58. G. Engelhardt and H. Jancke, "Structure investigation of organosilicon polymers by silicon-29 NMR," *Polym. Bull.*, **5**, 577 (1981).

59. H. Jancke, G. Engelhardt, H. Kriegmann, and F. Keller, "Quantitative Mikrostructureanalyse von Siloxakopolymeren mit Hilfe der ^{29}Si-NMR-Spectroscopie," *Plaste Kautsch.*, **26**, 612 (1979).

60. M. Ziemelis and J. C. Saam, "Sequence distribution in poly(dimethylsiloxane-*co*-methylvinylsiloxanes)," *Macromolecules*, **22**, 2111 (1989).

61. P. J. A. Brandt, R. Subramanian, P. M. Sormani, T. C. Ward, and J. McGrath, "^{29}Si NMR of functional polysiloxane oligomers," *Polymer Preprint*, **26**(2), 213 (1985).

62. H. J. Harwood and W. M. Ritchey, "The characterization of sequence distribution in copolymers," *J. Polym. Sci. Polym. Lett.*, **2**, 601 (1964).

63. G. N. Babu, S. S. Christofer, and R. A. Newmark, "Poly(dimethylsiloxanes-*co*-diphenylsiloxanes): Synthesis, characterization and sequence analysis," *Macromolecules*, **20**, 2654 (1987).

64. H. Jancke, G. Engelhardt, M. Magi, and E. Lippmaa, "Untersuchungen zur Structuraufklarung einiger Siloxanvervindungen mit Hilfe des ^{29}Si–NMR-Spektroskopie," *Z. Chem.*, **13**, 392 (1973).

65. W. R. Croasmum and R. M. K. Carlson, *Two-dimensional NMR spectroscopy. Applications for Chemists and Biochemists.*, VCH Publishers, New York, 1987.

66. J. Schraml, E. Petrakova, J. Pelnar, M. Kvicalova, and V. Chvalovsky, "Trimethylsilylation-Aid in NMR analysis of oligosaccharides. Assignment of ^{29}Si and ^{13}C NMR spectra of trimethylsilylated methyl β-D-xylobiosides by 2-D NMR," *J. Carbohydr. Chem.*, **4**, 393–403 (1985).

67. E. Liepinsh, I. Sekacis, and E. Lukevics, "Analysis of the NMR spectra of trimethylsilyl-substituted sugars by two-dimensional heteronuclear ^{29}Si–^{1}H shift correlation spectroscopy," *Magn. Reson. Chem.*, **23**, 10–11 (1985).

68. J. Schraml, E. Petrakova, and J. Hirsch, "^{29}Si and ^{13}C NMR spectra of all

possible pertrimethylsilylated β-D-xylopyranosyl-substituted methyl 4-O-β-D-xylopyranosyl-β-D-xylopyranosides assigned by 2D-heteronuclear ^{1}H–^{29}Si and ^{1}H–^{13}C chemical shift correlations," *Magn. Reson. Chem.*, **25**, 75–79 (1987).

69. J. Schraml, E. Petrakova, J. Hirsch, J. Cermak, V. Chvalovsky, R. Teeaar, and E. Lippmaa, "^{13}C MAS NMR and ^{1}H–^{29}Si and ^{1}H–^{13}C heteronuclear correlation study of model xylooligosaccharides," *Collect. Czech. Chem. Commun.*, **52**, 2460–2473 (1987).

70. E. Henge and F. Schrank, "^{29}Si-double-quantum coherence spectroscopy (INADEQUATE). An efficient method for the structure elucidation of silicon frameworks," *J. Organometal. Chem.*, **362**, 11 (1989).

71. Colin A. Fyfe, *Solid State NMR For Chemists*, C.F.C. Press, Guelph, Ontario, 1983.

72. J. Schaefer and E. O. Stejskal, "Carbon-13 nuclear magnetic resonance of polymers spinning at the magic angle," *J. Am. Chem. Soc.*, **98**, 1031 (1976).

73. J. Frye, Solid State NMR for Surface Studies Short Course, Chemically Modified Oxide Surfaces Symposium, Midland, MI, June 28–30, 1989.

74. G. R. Hays, "High-resolution carbon-13 solid-state nuclear magnetic resonance spectroscopy," *Analyst*, **107**, 241 (1982).

75. J. V. Munteau and L. M. Stock, "Tetrakis(trimethylsilyl)silane, a suitable chemical shift-standard for solid state NMR spectroscopy," *J. Magn. Reson.*, **76**, 540 (1988).

76. J. W. DeHaan, H. M. Van Den Bogart, J. J. Ponjee, and L. J. M. Van De Ven, "Characterization of modified silica powders by Fourier transform infrared spectroscopy and cross-polarization magic angle spinning NMR," *J. Colloid Interface Sci.*, **110**, 591 (1986).

77. G. S. Caravajal, D. E. Leydon, G. R. Quinting, and G. E. Maciel, "Structural characterization of (3-aminopropyl)triethoxysilane-modified silicas by silicon-29 and carbon-13 nuclear magnetic resonance," *Anal. Chem.*, **60**, 1776 (1988).

78. F. C. Schilling and F. A. Bovey, "The solid state ^{29}Si and ^{13}C NMR of poly(di-n-alkylsilanes)," *Polym. Prepr. (Am. Chem. Soc., Div. Polym. Chem.)*, **29**, 72–73 (1988).

79. G. C. Gobbi, W. W. Fleming, R. Sooriyakumaran, and R. D. Miller, "Variable-temperature ^{13}C and ^{29}Si CPMAS NMR studies of poly(di-n-hexylsilane)," *J. Am. Chem. Soc.*, **108**, 5624 (1988).

80. F. C. Schilling, F. A. Bovey, and A. J. Lovinger, J. M. Ziegler, "Characterization of poly(di-n-hexylsilane) in the solid state. 2. ^{13}C and ^{29}Si magic angle spinning NMR studies," *Macromolecules*, **19**, 2660 (1986).

81. K. Beshah, J. E. Mark, and J. L. Ackerman, "Characterization of PDMS model junctions and networks by solution and solid-state silicon-29 NMR spectroscopy," *J. Polym. Sci.: Part B: Polym. Phys.*, **24**, 1207–1225 (1986).

82. K. Beshah, J. E. Mark, and J. L. Ackerman, "Topology of poly(dimethylsiloxane) elastomeric networks studied by variable-temperature solid-state nuclear magnetic resonance," *Macromolecules*, **19**, 2194–2196 (1986).

CHAPTER

13

MASS SPECTROMETRY

JOHN A. MOORE

Dow Corning Corporation
Midland, Michigan

1. INTRODUCTION

The capabilities of mass spectrometry have grown enormously since the publication of the first edition of this book in 1974. Great advances in methods of sample introduction, production of useful ions, range and speed of mass analysis, and high speed data processing have occurred during the 15-year period. Many of the limitations facing the analyst at that time have been surmounted by the use of "soft" ionization techniques, secondary ion methods, and high performance "hyphenated" methods of interfacing the mass spectrometer to other analytical instruments such as gas and liquid chromatographs.

This chapter highlights the developments in mass spectrometry that provide new analytical capabilities to the field of organosilicon chemistry. It is meant to aid the experienced mass spectroscopist in selecting methods and interpreting data for analyses involving organosilicon compounds.

The literature contains numerous publications involving the mass spectra of silanes and trialkylsilyl derivatives of organic compounds, but very few deal with siloxanes, an area of great commercial importance. On the other hand, recent developments in mass spectrometry, such as extension of mass range to greater than 10,000 and ability to produce useful ions from nonvolatile samples, are more applicable to analysis of polysiloxanes than to the relatively small and volatile silanes.

1.1. Silicon versus Carbon

Both the mass deficiency of Si and its naturally occurring isotopic distribution result in some distinctive differences in the appearance of Si-containing

The Analytical Chemistry of Silicones, Edited by A. Lee Smith.
ISBN 0-471-51624-4 © 1991 John Wiley & Sons, Inc.

421

ions in the mass spectrum versus those containing C, H, N, and O only. Common siloxane repeat units such as Me_2SiO (74.0188) and PhMeSiO (136.0344) are only 250 ppm mass positive, as compared to the methylene unit (1100 ppm) or ethylene oxide unit (600 ppm). This smaller mass increment is a benefit in interpretation of high mass data since the exact mass remains within one mass unit of the nominal mass to nearly mass 4000

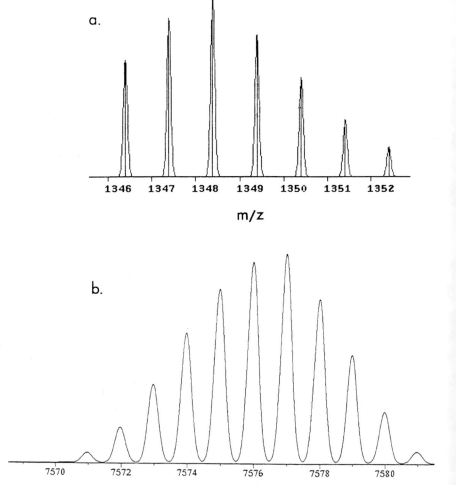

Figure 13.1. Isotopic distributions for molecular ion of (a) $Me_3SiO(Me_2SiO)_{16}SiMe_3$, MW 1346; and (b) $Me_3SiO(Me_2SiO)_{100}SiMe_3$, MW 7564.

($Me_2SiO \times 50 = 3700.94$). Exact masses of organic compounds shift three to four units from the nominal mass in that range. This small mass excess also makes poly(dimethylsiloxanes) good reference standards for exact mass measurements in chemical ionization (CI) work, as they are resolved from most organic ions with moderate resolution (1).

The natural isotopic distribution of silicon is 92.21% ^{28}Si, 4.70% ^{29}Si, and 3.09% ^{30}Si (2). The pattern displayed by this isotope combination is distinctive in the mass spectrum, particularly in the intensity of the M + 2 peak, for which there is little contribution from other common elements with the exception of halogens. Silicon's isotopic multiplicity has been exploited in a method of determining the number of trimethylsilyl groups in a silylated organic compound by comparing the experimental values to a table of theoretical combinations to find the best fit (3).

This distribution also results in a multiplicity of ion peaks in the high mass region of siloxane structures, however. Figure 13.1a is a computer generated pattern for peaks from the molecular ion of $Me_3SiO(Me_2SiO)_{16}SiMe_3$, MW 1346, illustrating the importance of ^{29}Si, ^{30}Si, and ^{13}C in the ions observed. This multiplicity can be a difficulty in determining the correct empirical formula of high molecular weight siloxanes as the abundance of the ion from the nominal isotopes ^{28}Si and ^{12}C becomes relatively low. Figure 13.1b shows the hypothetical distribution for $Me_3SiO(Me_2SiO)_{100}SiMe_3$, MW 7564, in which the peak representing an ion with all ^{28}Si and ^{12}C is less than 1% the most abundant ion. The first ion to exceed 5% appears 7 amu above the nominal mass; the most abundant ion is 13 amu higher.

1.2. Silicon-Containing Ions as Background Contaminants

Mass spectroscopists may encounter silicones at times and places they do not want or expect them (Chapter 4, Section 1.2). The most likely sources of background silicone are diffusion pump fluids, silicone rubber seals and septa, and chromatographic stationary phases based on polysiloxanes. Figure 13.2 shows the mass spectra of Dow Corning® 704 and 705 Diffusion Pump Fluids. Table 13.1 lists common ions that arise as the result of thermal degradation of silicone rubber and liquid phases used for chromatography.

2. SAMPLE INTRODUCTION

In addition to conventional sample introduction methods by heated sample inlets or direct probe, high performance separations of samples may be obtained by using the various chromatographies as sample introduction systems. Direct introduction methods are also available in which the sample

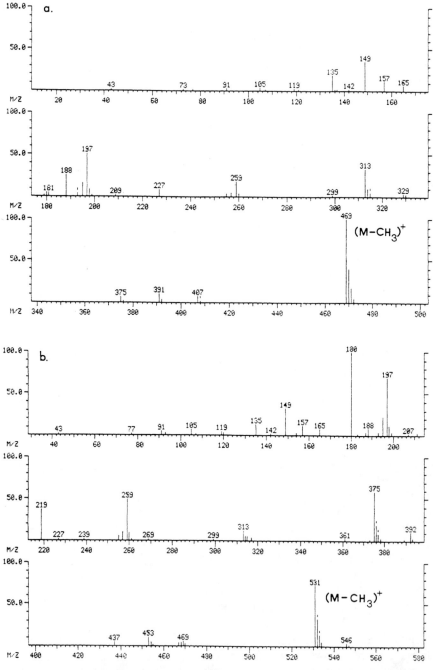

Figure 13.2. Mass spectra of (a) Ph$_2$MeSiOSiMe$_2$OSiPh$_2$Me (Dow Corning® 704 diffusion pump fluid), and (b) Ph$_2$MeSiOSiPhMeOSiPh$_2$Me (Dow Corning® 705 diffusion pump fluid).

Table 13.1. Common Ions from Silicones Used in GC–MS

m/z	Ion	Compounds	Source
73	Me_3Si^+	Methyl-containing silicones > 400 MW	GC liquid phases and septa
147	$Me_5Si_2O^+$	Same as for m/z 73	GC liquid phases
149	$C_8H_5O_3^+$	Phthlates	Plasticizers in silicone rubber
207	$Me_5Si_3O_3^+$	$(Me_2SiO)_3$	Thermal degradation of silicones
221	$Me_7Si_3O_2^+$	Same as for m/z 73	GC liquid phases
281	$Me_7Si_4O_4^+$	$(Me_2SiO)_4$	Thermal degradation of silicones
295	$Me_9Si_4O_3^+$	Same as for m/z 73	GC liquid phases
355	$Me_9Si_5O_5^+$	$(Me_2SiO)_5$	Thermal degradation of silicones

is admitted in the condensed phase and subsequently vaporized, pyrolyzed, or bombarded so as to emit useful ions. Each of these methods can be useful in the mass spectroscopic analysis of silicones.

2.1. Gas Chromatography–Mass Spectroscopy

The speed, resolution, and inertness of GC columns have reached a high level of performance with the widespread availability of fused silica open tubular columns containing bonded stationary phases. Their low volumetric gas flows and the use of efficient pumping systems on modern mass spectrometers allow the highly optimal arrangement of connecting the end of the column within a few millimeters of the ionization region of the mass spectrometer with no interface device, providing a high performance sample introduction system. This combination results in a productive analytical technique, as mass spectra can be obtained on nearly 100 compounds per hour when analyzing complex mixtures of oligomeric siloxanes (Figure 13.3). Speeds of 1 scan/s or better provide the best data from narrow chromatographic peaks. For the analysis of mixtures of very low molecular weight silanes and chlorosilanes, packed or PLOT (porous layer open tubular) or other wide bore open tubular columns may be required. Their higher gas flows prohibit the use of a direct interface, but either the open split interface or the jet separator can be used effectively for the analysis of even the most reactive silanes (such as $MeHSiCl_2$). Experience has shown that if an organosilicon

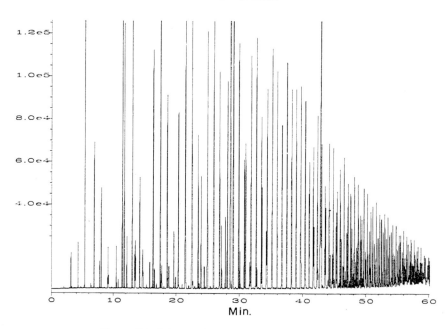

Figure 13.3. GC–MS of Me$_2$–MeH siloxane copolymer.

compound can be eluted from a GC column, a useful mass spectrum can be obtained.

2.2. Supercritical Fluid Chromatography–Mass Spectroscopy

The range of oligomeric siloxanes that can be separated and introduced to the mass spectrometer can be significantly increased through use of supercritical fluid chromatography (SFC) (4). With capillary columns and carbon dioxide mobile phase, a direct interface similar to GC–MS is feasible. Additional heating may be added if desired. Alternatively, an interface can be inserted through the solids probe inlet (4). The limitations to date in use of SFC–MS have been poor sensitivity and the necessity of using the chemical ionization mode.

2.3. Liquid Chromatography–Mass Spectroscopy

The dearth of published examples of the use of LC–MS for the analysis of silicones may be attributed to the nonpolar nature of most silicones, which thus lack a basis for LC separation and for ion formation by thermospray or continuous flow FAB. Where the siloxane backbone has been substituted with functional groups such as amines, amides, glycols or carboxylates, LC–MS may prove to be a useful technique.

2.4. Pyrolysis–Mass Spectroscopy

Pyrolysis, either directly into the mass spectrometer or prior to GC–MS, is a relatively simple and inexpensive means of sample introduction. This technique has been used by several authors to identify or characterize polysiloxanes (5–8).

Characteristic cyclosiloxanes are typically obtained from the pyrolysis of polysiloxanes, although some Si–C cleavage is also observed, for example, benzene in the case of phenylmethyl polysiloxanes. Pyrolysis experiments progress with an initial evolution of cyclic impurities in a simple evaporation step. At higher temperatures (400–1000 °C) thermal decomposition occurs with further formation of cyclic structures. This process can be made to occur at lower temperatures by addition of a catalyst such as NaOH.

2.5. Direct Probes

Introduction of a sample directly into the ion source can be a rapid and effective way to obtain mass spectra in cases where chromatographic introduction is not feasible or not necessary. The probe's temperature programability allows fractionation of mixtures as well as study of the thermal stability of organosilicon polymers. Where the amount of sample is limited, or the sample is thermally or hydrolytically unstable, or the sample cannot be chromatographed unless derivatized, or the sample is insoluble, direct probe MS may be a successful approach. Modern probes can be heated to 350–400 °C, and with high vacuum, can produce vapor from quite nonvolatile compounds. In silicon chemistry, analysis of materials containing functional groups such as SiOH, amine, or carboxy, as well as of higher molecular structure materials with nonpolar substituents benefit most from the use of a direct probe. The use of other more specialized probes, which involve "soft" ionization modes, are discussed in the next section.

3. IONIZATION METHODS

3.1. Electron Impact

Since publication of the first edition of this book in 1974, many spectroscopists have studied the electron impact (EI) mass fragmentation behavior of a variety of silanes and silyl derivatives of organic compounds. Much less work has occurred for the siloxanes, for which publications have mainly appeared from the Soviet Union. Since it is by far the most widely available technique and the mode for which library spectra are available, EI is the first technique to try for any sample that can be vaporized into the mass spectrometer. For silanes, it provides rich structural information because of the distinctive fragmentations and rearrangements that occur, which with experience can be very informative in interpretation of unknowns. For siloxanes containing four or fewer silicon atoms and simple alkyl or aryl substituents, EI is also informative. It becomes less and less useful as the chain lengths increase, eventually providing no information on the complete molecule, although still revealing clues to the repeat unit from characteristic R_3Si^+ and tri and tetracyclo rearrangement ions.

An EI spectrum should always be obtained, even if chemical ionization is also required. The combination of the structural information provided by EI and the molecular weight information from CI is often the key to identification of unknown organosilicon compounds. As a generalization, EI alone is usually sufficient for the industrially significant silanes and chlorosilanes, low molecular weight (< 500) siloxanes containing short alkyl or aryl groups, mono- and dimeric silanol compounds, and polysilanes and silahydrocarbons of modest chain length. Chemical ionization is necessary for all the siloxanes containing the 3,3,3-trifluoropropyl group, silanols longer than two siloxane units, many of the organofunctional silanes (which typically contain two or more hydrolyzable groups and a functional organic group), and alkyl substituted siloxanes with six or more repeat units (this rule varies with the R group). These generalizations are summarized in Table 13.2.

Table 13.2. Classes of Silicon Compounds for which CI is Recommended

Class	CI	Reagent Gas
R_xSiCl_z	No	
R_4Si, R < 4 carbon atoms or R aromatic	No	
R_4Si, R > 3 carbon atoms	Yes	Methane
$R_xSi(OR)_z$	Yes	Isobutane
Methyl siloxanes MW < 500	No	
Methyl siloxanes MW > 500	Yes	Ammonia
Phenyl siloxanes MW < 1000	No	
Phenyl siloxanes MW > 1000	Yes	Ammonia
Alkyl siloxanes R > 2 carbon atoms	Yes	Ammonia
$CF_3CH_2CH_2$ substituted siloxanes	Yes	Isobutane
Silanols, except monomer and dimer	Yes	Ammonia
Polysilanes	No	

3.2. Chemical Ionization

In the absence of information other than the fact that EI provided no ions from which the molecular weight can be deduced, the first CI experiment is usually with methane as the reagent gas. If no molecular or quasimolecular ions can be obtained from this experiment, a compound that produces a weaker Brønsted acid as reagent ion, such as isobutane or ammonia, is chosen (9). For many siloxanes, a mixture of methane and ammonia can be an effective reagent gas.

Alkyl and aryl polysiloxanes and siloxanols have been found to exhibit the best CI spectra when ammonia is the reagent gas. Exceptions are the 3,3,3-trifluoropropyl substituted siloxanes, which produce intense protonated molecular ions with isobutane as the reagent, while ammonia is ineffective in obtaining useful spectra.

Alkoxy silanes of the type $R_xSi(OR')_y$ can be protonated readily using methane or isobutane, but the subsequent ejection of the corresponding alcohol is a common process. This behavior illustrates the preference for protonation of the alkoxy oxygen even when the molecule contains other sites readily protonated under CI conditions. This conclusion is not based on labeling experiments but is a logical assumption based on the observed alcohol losses. Figure 13.4 compares the EI and the isobutane CI spectra of methacryloxypropyl trimethoxy silane, MW 248, which illustrates this point. Recommended reagent gases for various classes of organosilicon compounds are listed in Table 13.2.

Although EI is perfectly adequate for poly(dimethylsiloxanes) with two to five repeat units, the CI spectra for this series are also interesting. With isobutane, protonated molecular ions are obtained, although some fragmentation also occurs. With ammonia, three competing processes are apparent, and their relative importance depends on the length of the siloxane chain. Quasimolecular ions are formed either by hydrogen or ammonium ion addition, and subsequent loss of methane from these ions is common. Therefore ions can be observed at $M + 18$, $M + 2(18 - 16)$, $M + 1$, and $M - 15$. For the cyclic tetramer and pentamer, each of these processes is observed (Figure 13.5). The net $M + 2$ ion was shown not to be due to H_2 addition by analysis of perdeutero cyclic tetramer, MW 320. No ion is present at m/z 322 (other than the m/z 321 isotope peak) but the expected $M + NH_4 - CD_3H$ process is observed at m/z 319 (Figure 13.6).

Direct chemical ionization (DCI) shows promise for analysis of siloxanes and siloxanols that have insufficient volatility for introduction through the GC. With ammonia as reagent gas, PDMS with molecular weights over 3000 have been detected (Figure 13.7). There is also evidence that silanol ended linears produce $(M + NH_4)^+$ ions in the CI mode. This fact is of great utility

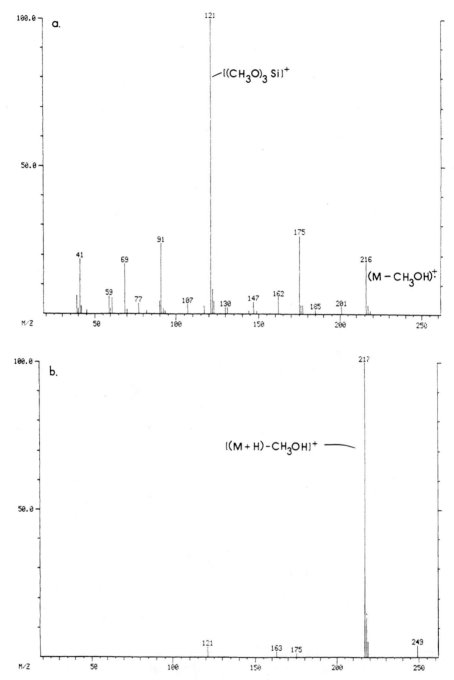

Figure 13.4. (*a*) EI mass spectrum of (MeO)₃SiCH₂CH₂CH₂OC(O)C(CH₃)=CH₂ and (*b*) iso-butane CI mass spectrum of (MeO)₃SiCH₂CH₂CH₂OC(O)C(CH₃)=CH₂.

430

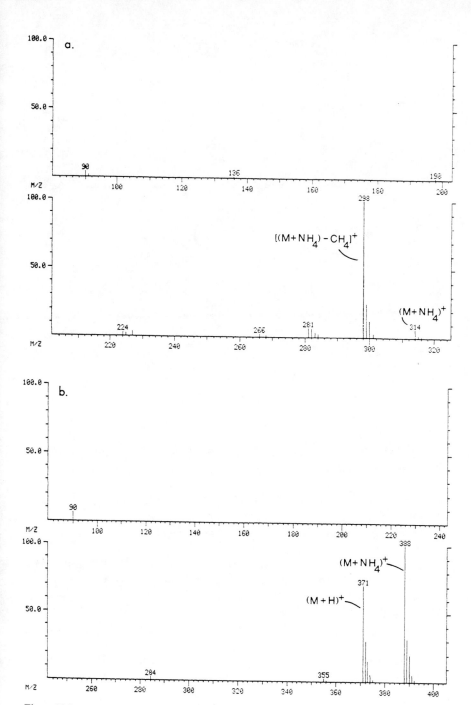

Figure 13.5. (*a*) Ammonia CI mass spectrum of (Me$_2$SiO)$_4$ and (*b*) ammonia CI mass spectrum of (Me$_2$SiO)$_5$.

431

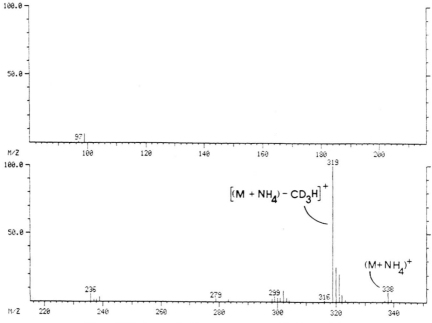

Figure 13.6. Ammonia CI mass spectrum of $[(CD_3)_2SiO]_4$.

in the analysis of these compounds, whose EI spectra are indistinguishable from the cyclosiloxanes of the same chain length. This result is attributed to a facile methyl loss followed by water ejection from the linear diols, which produces the same ion as the methyl loss ion for the cyclic.

3.3. Fast Atom Bombardment

This secondary ion technique, commonly referred to as FAB, liquid SIMS (secondary ion mass spectrometry), or fast ion bombardment with the use of Cs ion guns rather than xenon or argon guns, has not proven to be nearly the success in the field of silicones as it has in many other areas. Only a single reference to the use of FABMS for characterization of organosilicon compounds was found (10).

The FAB spectrum obtained for a silicone oil (neat) was in fact only a series of small fragments of the polymer ($m/z < 400$) and resembled the electron impact spectra of PDMS. The utility of FAB is apparently limited to the identification of methyl-substituted silicone, and no information on the distribution or molecular weight is obtained.

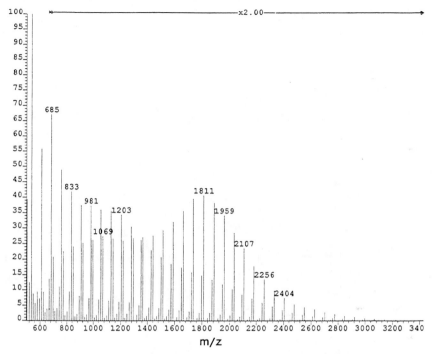

Figure 13.7. Ammonia DCI mass spectrum of $Me_3SiO(Me_2SiO)_nSiMe_3$.

Recently, cationization using sodium salts has shown some promise in the use of FAB–MS for characterization of mixtures of linear siloxanediols. The M + Na ions were obtained by mixture of the sample with NaI in nitrobenzyl alcohol. With the exception of this class of silicones, it is not expected that the generally nonpolar silicones of commercial significance will benefit from the use of FAB–MS.

3.4. Field Desorption

The application of FDMS (field desorption mass spectrometry) to polymer characterization is seeing a resurgence in recent years, and major MS manufacturers are responding to an increased demand for the technique with improved FD capabilities. While the technique is still relatively insensitive, as well as incompletely understood, it may prove to be very useful for polysiloxane analysis. Preliminary data indicate that $(M - Me)^+$ fragments are the only ions formed for PDMS under FD conditions. Analysis of polymer distributions have shown an ion every 74 mass units (Me_2SiO) to over mass 5000 (Figure 13.8). Design of activated emitters and control of

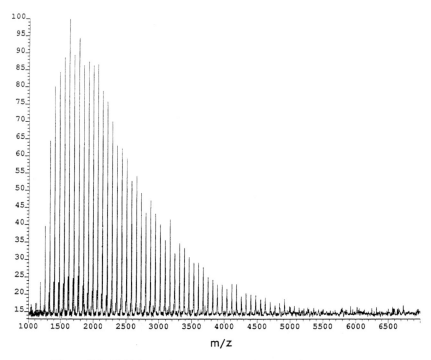

Figure 13.8. Field desorption mass spectrum of $Me_3SiO(Me_2SiO)_nSiMe_3$.

emitter current and field gradient are all areas that require further study for more routine use of FDMS in silicone analysis. No other technique offers as much promise, however, for the analysis of high molecular weight silicones by mass spectrometry.

3.5. Other Secondary Ion Techniques

Laser desorption (LD), static SIMS, and Cf-252 plasma desorption are other techniques used for analysis of condensed phase samples that may also be useful for silicones. Laser desorption with Fourier transform MS has been a useful combination. Similarly the use of time-of-flight MS with these techniques has some advantages. A few publications have appeared that apply desorption techniques to silicones (11–13).

3.6. Negative Ions

The detection of negative ions is possible on most modern mass spectrometers. While this capability has not been a benefit to the analysis of silicones in the past, the use of different ionization modes may make it more useful. Although negative ion CI is not expected to be useful unless the silicon compound contains electron capturing groups, techniques such as FAB and other desorption modes may prove to be worth pursuing for the formation of analytically useful anions.

4. INTERPRETATION

A discussion of the mechanisms for the unimolecular decomposition processes that occur for the various classes of organosilicon compounds provides a basis for interpretation of unknown mass spectra. For that purpose the text will be divided into the mass spectral behavior of silanes, siloxanes, silazanes, polysilanes, and silyl derivatives of organic compounds. The latter is distinguished from the section on silanes in that these compounds are generally (but not exclusively) comprised of three identical simple alkyl groups on silicon and one more complicated organic group.

Silicon is more electropositive than the elements to which it is normally bound (carbon, oxygen, hydrogen, nitrogen, and halogen). Therefore, the initial ion formed by electron ejection, $M^{+\cdot}$, will be more likely to dissociate to give $R\cdot + R_3Si^+$ than $R^+ + R_3Si\cdot$, that is, fragments containing silicon will tend to carry the positive charge. The decomposition of the molecular ion is facile since the unstable odd-electron ion can give an even-electron silicon-containing ion by elimination of an odd-electron radical fragment. Consequently, many organosilicon compounds exhibit a low or nonexistent parent molecular ion abundance. Ternary R_3Si^+ (siliconium) ions are therefore particularly stable, while radical ions $R_2Si^{+\cdot}$ are less likely to be observed.

The mass spectra of organosilicon compounds are thus dominated by ions containing silicon. They are a result of cleavage of radicals to give siliconium ions, and of skeletal rearrangements with preservation of charged silicon species. *This process is very common and must be recognized if fragment ions are to provide useful structural data.* The understanding of these rearrangement processes is in fact the key to successful interpretation of organosilicon mass spectral data.

4.1. Silanes

4.1.1. Compounds with Si–C or Si–H Bonds

The R_4Si compounds, where R is alkyl, aryl, hydride, and others (symmetrical and unsymmetrical), have been studied to determine their mechanisms of fragmentation. A comprehensive treatment of the mass spectrometry of organosilanes was given by Spalding in 1973 and is a good source for additional references (14).

The series Me_xSiH_{4-x} was reported by van der Kelen in 1965 (15). For each of the compounds the molecular ion is very small and the most intense peaks are due to siliconium ions R_3Si^+. The presence of the m/z 45 ion H_2SiMe^+ in the spectrum of Me_4Si is an example of the rearrangement process common to silicon compounds, and has been attributed to loss of ethylene from the trimethylsilyl ion m/z 73 (16, 17). The rearrangement process is even more facile for tetraethyl silane, as the R_3Si^+ ion is no longer the base peak, but the rearrangement ion formed by ethylene loss from Et_3Si^+ m/z 115 produces the base peak at m/z 87 Et_2SiH^+.

The presence of vinyl groups on R_4Si molecules has been shown to alter the course of fragmentation (18–20). As the number of vinyls increases, the appearance of odd-electron fragments becomes more probable. The compound $MeSiVi_3$ eliminates ethylene to give the base peak at m/z 96 and also eliminates butadiene to give a peak at m/z 70. Processes analogous to the behavior of alkanes on silicon, fragmentation of vinyl or methyl to the siliconium ion, and subsequent elimination of acetylene to rearranged siliconium ions, also occur for Me_3SiVi, Me_2SiVi_2, and $MeSiVi_3$. The compound Vi_4Si undergoes each of the processes mentioned above, and also exhibits a

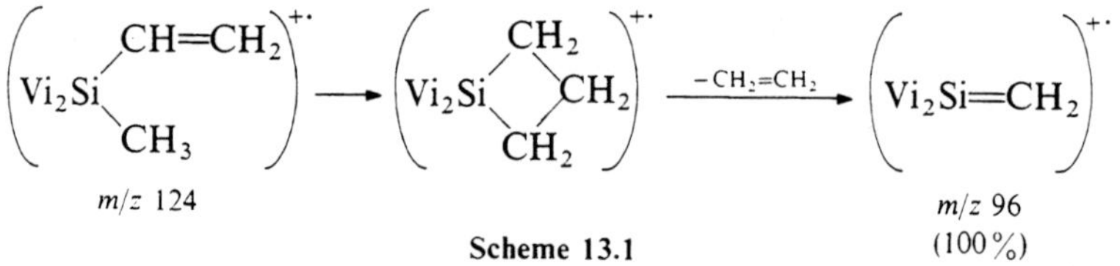

Scheme 13.1

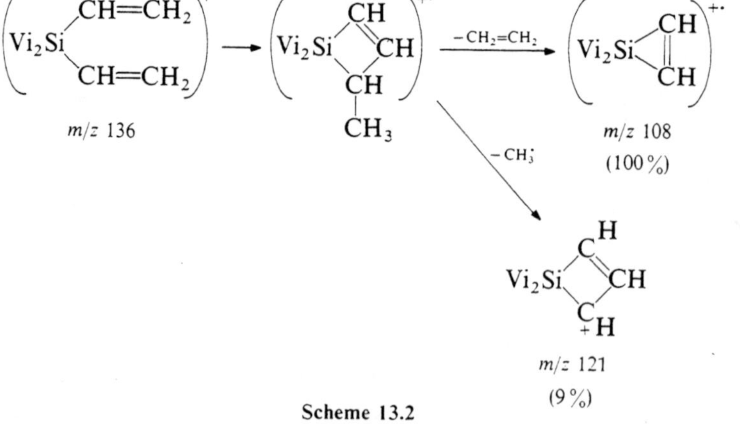

Scheme 13.2

peak for M − Me. Schemes 13.1 and 13.2 have been proposed (19) to account for the facile elimination of ethylene from odd-electron vinyl-containing ions and for methyl loss from the tetravinyl compound.

Soderquist and Hassner (20) studied further the acetylenic elimination process from vinyl siliconium ions $(Me_2SiCR=CHR')^+$. They concluded that transfer of the substituent on the alpha carbon to silicon is an important process as well as the earlier reported beta transfer. For compounds where the alpha R group was deuterium, methyl, bromine, methoxy, trimethyl-siloxy, or phenyl, the corresponding Me_2SiR^+ ion was prominent.

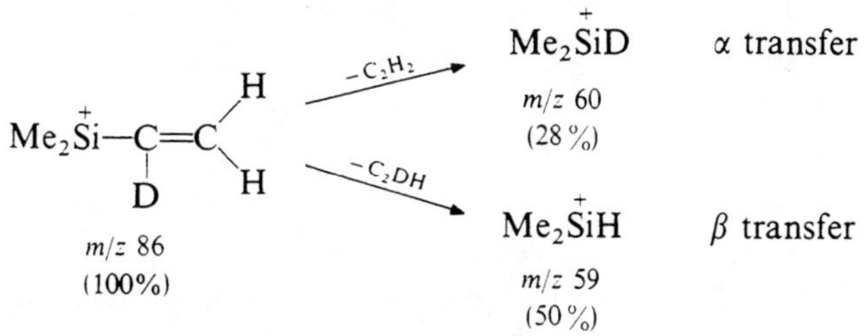

$$Me_2\overset{+}{Si}D \qquad \alpha \text{ transfer}$$
$$m/z\ 60$$
$$(28\%)$$

$$Me_2\overset{+}{Si}H \qquad \beta \text{ transfer}$$
$$m/z\ 59$$
$$(50\%)$$

Scheme 13.3

The effect of longer alkyl groups on silicon has been studied for Me_3SiR compounds (21) and for $MeSiR_3$ compounds (22). Molecular ions are very weak and cleavage of the largest R group to form the Me_3Si^+ or $MeSiR_2^+$ ion is favored. In the later case the base peak is formed from olefin elimination, $MeSiR_2^+ \rightarrow MeSiHR^+$ with transfer of beta hydrogen to silicon (23). For $EtSiR_3$ compounds, ethyl cleavage competes with higher alkyl cleavage to a greater extent than for the corresponding methyl compounds. There is evidence that branched alkyl groups are more likely to cleave than straight chain groups, as Onopchenko et al. (22) report five times greater abundance for $(M - \text{isopropyl})^+$ than for $(M - \text{tetradecyl})^+$ with the compound $Me_2(i\text{-Pr})Si(C_{14}H_{29})$.

The stability of Si–C bonds in which the R group is unsaturated is greater than for simple alkyls. Thus methyl loss is highly favored compared to loss of the R group for Me_xSiVi_y and Me_xSiPh_y. The presence of molecular ions is also more likely for these types of compounds. Mass spectra of mixed phenyl–methyl silanes are very simple compared to similar compounds where the R group is alkyl. The increased stability of the aromatic group results in few rearrangement processes occurring. Almost all the ionization is contained in molecular ions or methyl loss siliconium ion. The even-electron ion $C_6H_5Si^+$ m/z 105 is typically observed in moderate abundance while the ion resulting from cleavage of phenyl from the molecular ion is unlikely to appear. Discussions of the spectra of Ph_4Si and Ph_3SiVi can be found in Ref. 14.

An important feature of the siliconium ions generated by radical cleavage of silanes is their propensity for remote attack of functional groups in order to form an even more stable ion. Processes in which the Me_2SiOH^+, Me_2SiCl^+, Me_2SiPh^+, Me_3Si^+, and Me_2SiOMe^+ ion form through intramolecular rearrangement have been reported (24–27). For the series Ph–X–SiMe$_3$, while $(M - Me)^+$ and Me_3Si^+ are the dominant ions, the rearrangement to Me_2SiPh^+ m/z 135 occurs where X is $(CH_2)_{2-4}$ or CH=CH, but not where X is C≡C (24).

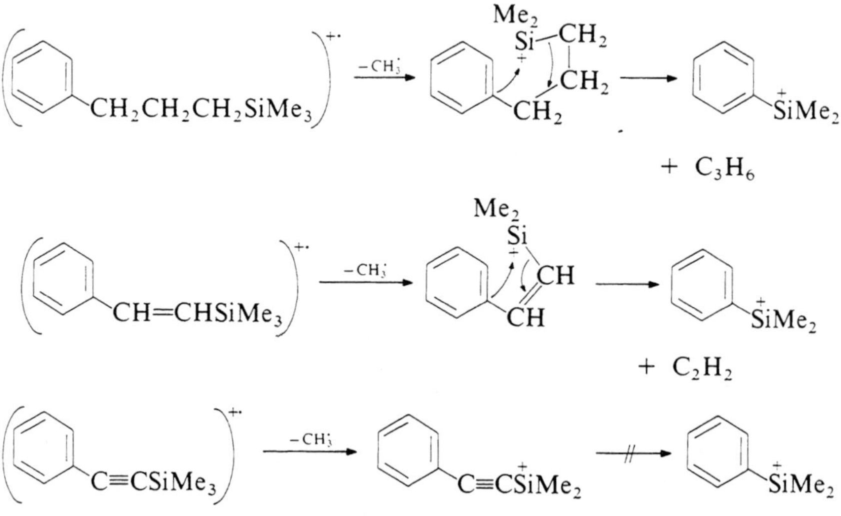

Scheme 13.4

Ho (25) has reported a similar effect for α-silyl phenyl ketones. The series $Me_xPh_{3-x}Si(CO)Ph$ was shown to form a siliconium ion with one more phenyl group than orignally on silicon, that is, the $MePh_2Si^+$ m/z 197 ion is present for the Me_2PhSi compound.

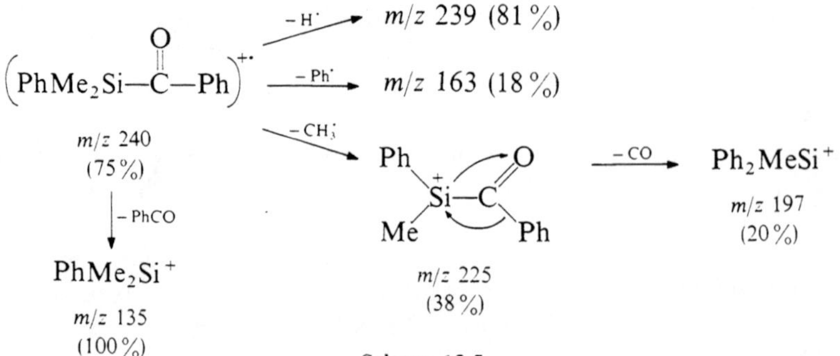

Scheme 13.5

Similarly the series $(MeO)CO-X-SiMe_3$ for $X = (CH_2)_{1-4}$ was shown to form the stable Me_2SiOMe^+ ion m/z 89 by intramolecular migration of the ester methoxy group to silicon after the initial siliconium ion formation from methyl cleavage (26).

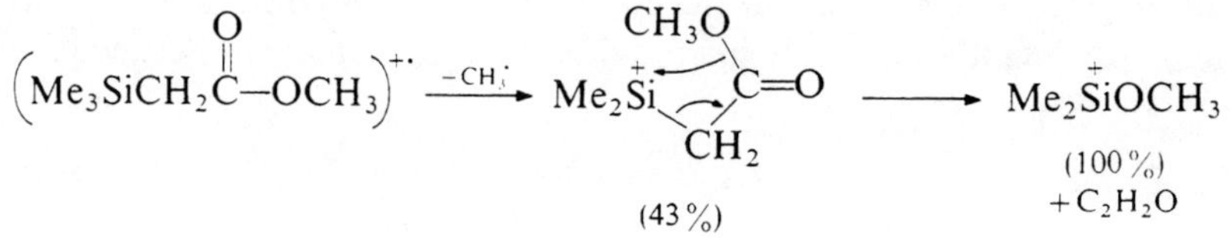

Scheme 13.6

Tsipis (27) has reported the importance of the Et_2SiOH^+ ion m/z 103 in the mass spectra of compounds of the type $Et_3SiCH=CHCR_2(OH)$. This fragment can be explained only by silyl ion attack of the remote hydroxyl. The observations by Dimmel (28) of the $MeSiCl_2^+$ m/z 113 ion in the spectra of $Me_3SiCHCl_2$ and Me_3SiCCl_3, and of Me_2SiCl^+ m/z 93 as well as $MeHSiCl^+$ m/z 79 in the spectrum of Me_3SiCH_2Cl is another illustration of this theme. A series of chlorine and methyl shifts is proposed to lead to formation of these stable siliconium ions.

The silacyclohydrocarbons can be viewed as silanes, but their mass spectral behavior bears little resemblance to that of the silanes previously

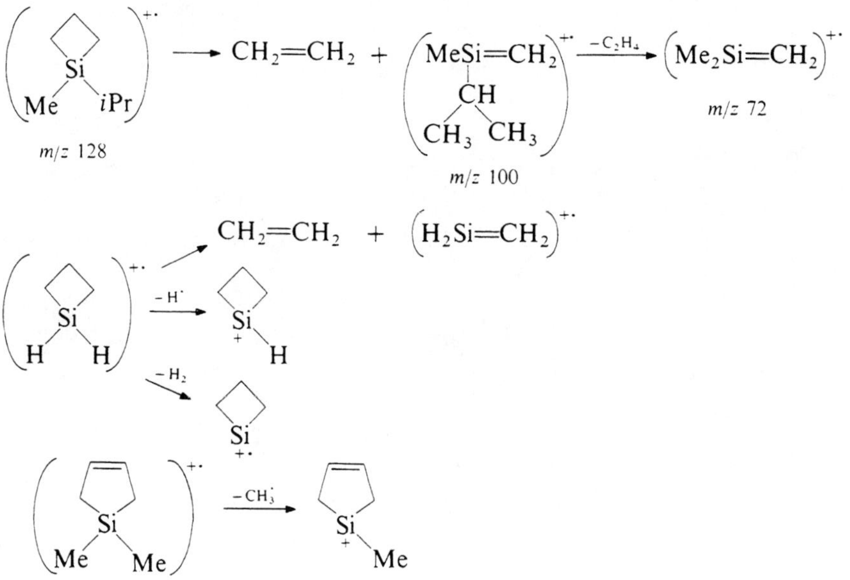

Scheme 13.7

discussed. Literature reports (29–31) are available for the silacyclobutanes and silacyclopentanes, and the data show the presence of stable molecular ions. The elimination of ethylene from both ring sizes is observed, followed by another ethylene loss if the R groups are alkyl. If R is H, $(M - H)^+$ and $(M - H_2)^+$ are also evident. If the cyclohydrocarbon is unsaturated (32), the ring is left intact and mainly intense molecular ions and Si–X cleavages are observed.

4.1.2. *Halosilanes*

Because of the commercial importance of the chlorosilanes as starting materials for polysiloxanes, this category of silanes is discussed separately, although the fragmentation behavior follows the trends for silanes in general. An important consideration is the stability of the Si–X bond, whose bond energies are greater than Si–C, Si–H, and some C–C bonds. Consequently, the most intense ions in chloro or fluoro silane mass spectra are almost invariably siliconium ions containing one or more Si–X bonds. The spectra are quite straightforward and consist almost entirely of simple radical cleavage products from the molecular ion. The mass spectrum of ethyl-methylchlorosilane is shown in Figure 13.9 and cleavage of each of the four

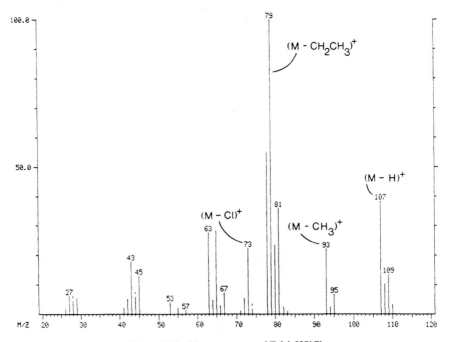

Figure 13.9. Mass spectrum of EtMeHSiCl.

substituents is observed. The simple chlorosilanes containing small alkyl groups produce molecular ions of moderate intensity, while phenyl-containing chlorosilanes produce intense molecular ions. The presence of the molecular ion, predictable radical cleavages to ternary ion, and an isotopic distribution pattern that allows the determination of the number of chlorines, make interpretation easy for this class of compounds.

The fluorosilanes do not offer a diagnostic isotope pattern, but do exhibit a considerable preference for Si–C cleavage products compared to the stable Si–F bond. An additional clue that the compound is a halosilane is usually available from the presence of the $SiCl^+$ m/z 63 or SiF^+ m/z 47 ions in the spectrum.

Alkyl groups on a halosilane behave as discussed previously; the larger R groups cleave preferentially, and unsaturates such as vinyl or phenyl have additional stability. As reported by Polivanov et al. (19) the presence of vinyl groups also increases the probability of odd-electron ions such as $(M - ethylene)^{+\cdot}$ or $(M - HCl)^{+\cdot}$. For silanes containing only hydride, methyl, and chlorine, Si–H cleavage typically produces the base peak, but for silanes with ethyl or higher groups the base peak generally arises from cleavage of the R group. Two other processes are operative for alkyl chlorosilanes containing ethyl or higher groups. As the alkyl chain gets longer, C–C bond breakage occurs, even in preference to Si–Cl cleavage. This point is illustrated in the mass spectrum of isobutyl trichlorosilane (Figure 13.10) in which an intense methyl loss peak is observed. For compounds in which the breaking of C–C bonds is not as facile, a second process competes, in which HCl is lost from the molecular ion, leaving another odd-electron ion apparently through action of the silyl center with a remote carbon. The mass spectrum of n-butyl trichlorosilane (Figure 13.10) contains evidence for both this process (m/z 154) and alkyl chain rupture $(M - CH_3CH_2)^+$ m/z 161. The loss of the HCl neutral is most pronounced for the n-propyl and n-butyl compounds, and apparently also occurs for ethyltrichlorosilane (Figure 13.11), but becomes less important for the higher alkyls as alkyl fragmentations predominate.

An interesting rearrangement process for Ph_2SiCl_2 is noteworthy (33). An intense molecular ion and a siliconium ion from phenyl cleavage are present, but the base peak is a rearrangement ion that has been attributed to the biphenyl radical m/z 154. This radical likely arises by loss of the $SiCl_2$ neutral from the molecular ion, since the m/z 154 ion is not important in $Ph_2MeSiCl$. It is reported (34) that the biphenyl radical ion is prominent in the mass spectrum of Ph_3SiCl and the neutral loss of PhSiCl from the molecular ion is suggested. The fragmentation of Ph_2SiF_2 proceeds similarly to that of the chlorosilane with the exception that hydride loss from the base molecular ion peak occurs for the former but is not evident for the chlorosilane. The

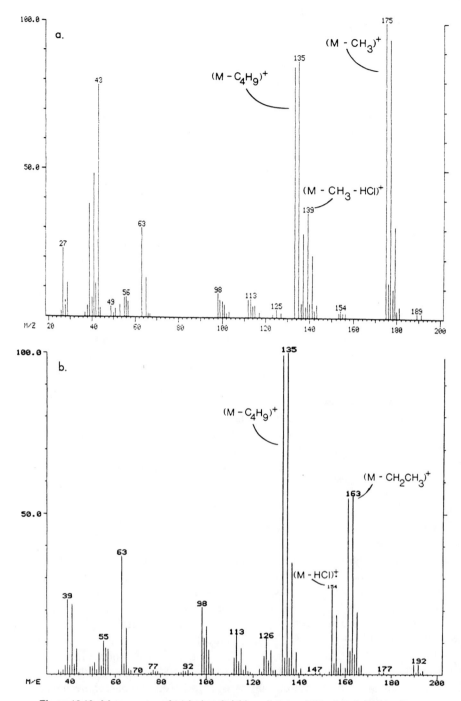

Figure 13.10. Mass spectra of (*a*) isobutyltrichlorosilane and (*b*) *n*-butyltrichlorosilane.

442

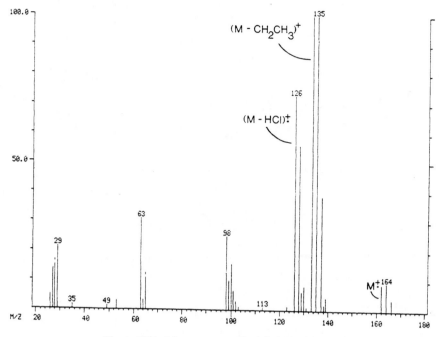

Figure 13.11. Mass spectrum of ethyltrichlorosilane.

fragmentation characteristics of $(Me_3SiCH_2)_xSiCl_{4-x}$ are somewhat more complex (35).

4.1.3. Silanes with Si–OR Bonds

The inclusion of compounds in this section rather than in the section on silyl ethers is somewhat arbitrary, but is meant to cover silanes with (usually more than one) SiOR bond, which are sometimes referred to as "organofunctional" silanes. It also includes the organic silicates, $Si(OR)_4$.

The simplest examples are the methoxy silanes, which are characterized by a rearrangement process that results in expulsion of formaldehyde from a siliconium ion. These 30 mass unit losses can be counted to determine the number of methoxy groups, as illustrated in the spectrum of $MeSi(OMe)_3$ (Figure 13.12). Three successive CH_2O losses are observed from the $(M - Me)^+$ ion, m/z 121, 91, 61, and 31, as well as two successive CH_2O losses from the $(M-OMe)^+$ ion, m/z 105, 75, and 45. Silicon–carbon cleavage is greatly favored over that of silicon–oxygen, so that the "inorganic" end of the molecule [e.g., $Si(OR)_3$] can usually be identified, while the *organic* end is cleaved as an uncharged radical and cannot be determined. For this reason

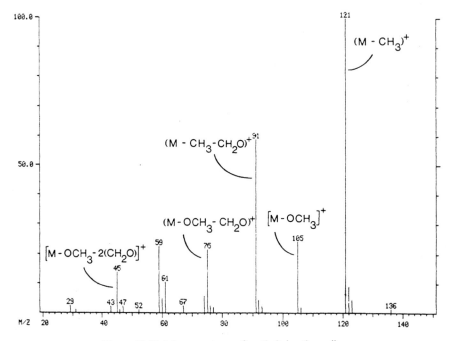

Figure 13.12. Mass spectrum of methyltrimethoxysilane.

chemical ionization is frequently needed for identification of compounds in this class. Figure 13.13 illustrates the EI versus isobutane CI mass spectra for *n*-hexyl trimethoxy silane. The EI data shows that the compound is of the type $RSi(OMe)_3$ but gives no definitive data on the R group, while the CI run provides a protonated molecular ion, indicating a probable R group of 85 mass units. The ion at m/z 224 is postulated to be $(M + NH_4)^+$, arising from traces of ammonia in the reagent gas line.

If the R group is phenyl, a significant molecular ion as well as other evidence allows interpretation of the EI spectrum. The compound $Ph_2Si(OMe)_2$ exhibits the biphenyl radical ion discussed earlier for Ph_2SiCl_2 (33), as well as the other processes that normally occur for methoxy silanes. The compound $PhSi(OMe)_3$ has a base peak of m/z 120, which can be postulated as the product of hydride loss from the $PhSi(OMe)_2^+$ siliconium ion.

The presence of certain hetero atoms in the organic ligand can have interesting effects on the fragmentation of methoxy silanes. The aminofunctional compound $NH_2CH_2CH_2NHCH_2CH_2CH_2Si(OMe)_3$ has an abundant ion from cleavage at the beta carbon (loss of $-CH_2NH_2$), and then forms

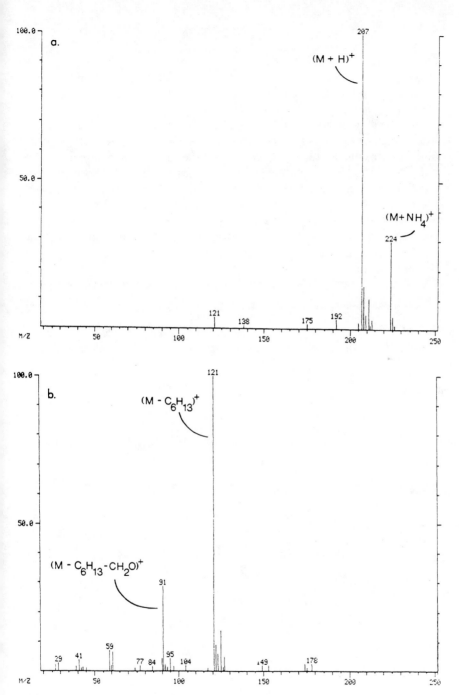

Figure 13.13. (a) Isobutane CI mass spectrum of n-hexyltrimethoxysilane and (b) EI mass spectrum of n-hexyltrimethoxysilane.

445

the base peak by elimination of methanol. Chloropropyl trimethoxy silane ejects C_3 groups to give ions with chlorine on silicon as shown in Figure 13.14. Mercaptopropyl trimethoxy silane has a base peak at m/z 164, from methanol ejection from the molecular ion (36).

For the ethoxy silanes and higher, both olefin and aldehyde elimination from the alkoxy groups are operative. As observed in Figure 13.15, losses of both ethylene and acetaldehyde occur from the various siliconium ions. Thus both Si(OH) and SiH containing ions are ultimately formed. The former type [$MeSi(OH)_2$ m/z 77, Me_2SiOH m/z 75, etc.] are more prevalent, especially for higher alkoxy groups and are good indicators that the compound contains SiOR groups. As the R group on oxygen gets larger, the likelihood of C–C bond cleavage within the R group is also increased. Branching in the R group also gives increased C–C bond cleavage. More examples of this pattern will be discussed in the section on silyl ethers.

The mass spectra of the organosilicates $Si(OR)_4$ become more difficult to interpret as the R group becomes larger. The spectrum of tetramethoxy silane is simple, consisting of an abundant molecular ion and a base peak of $Si(OMe)_3^+$. The rearrangement ion resulting from formaldehyde loss is also present. Tetraethoxy silane has a weak molecular ion and its base peak is

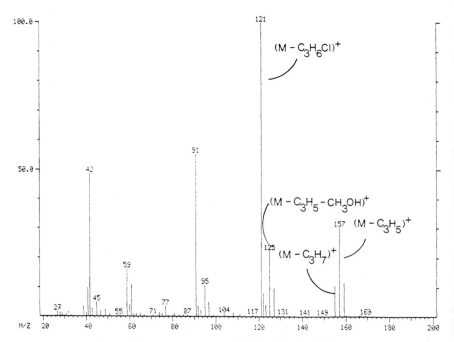

Figure 13.14. Mass spectrum of $ClCH_2CH_2CH_2Si(OCH_3)_3$.

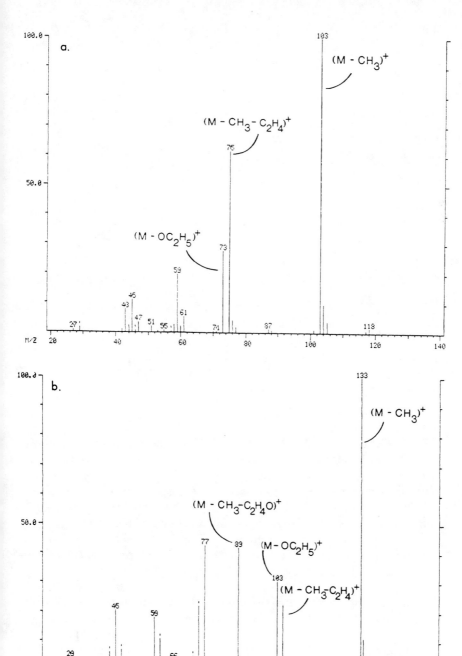

Figure 13.15. Mass spectra of methyl ethoxysilanes: (*a*) trimethylethoxysilane; (*b*) dimethyl-diethoxysilane, and (*c*) methyltriethoxysilane.

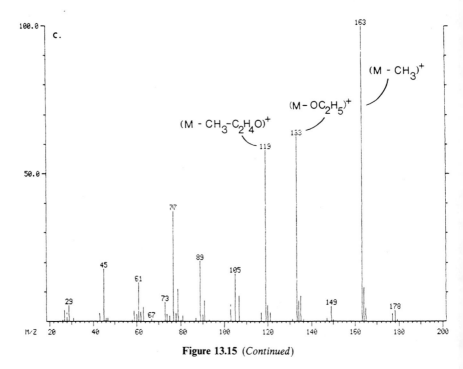

Figure 13.15 (*Continued*)

from C–C cleavage, $(M - Me)$, m/z 193. Also intense are the ion due to acetaldehyde loss from the base peak, $(M - Me - C_2H_4O)^+$, and the siliconium ion formed by ethoxy cleavage, $(EtO)_3Si^+$. Because of the possibility of both acetaldehyde and ethylene losses from each of the groups, the spectrum is rather complex (Figure 13.16). The $Si(OH)_3^+$ ion m/z 79 is prevalent for compounds of this type. Tetrapropoxy silane behaves similarly to the ethoxy; however, the molecular ion is very weak. The C–C bond cleavage produces the base peak at $(M - CH_2CH_3)$, and a series of propylene and propionaldehyde eliminations provide all the pieces of the puzzle.

The mass spectra of the methyl acetoxy silanes have been reported (37). Behavior is similar to the alkoxys in that a rearrangement from a siliconium ion to leave SiOH, in this case by elimination of a ketene, is a predominant process. Acetoxy cleavage from the molecular ion is more likely to be observed than for the alkoxys. Regardless of which group cleaves, successive $CH_2=C=O$ losses for each of the acetoxy groups is characteristic. Interestingly, for the series Me, Et, and vinyl triacetoxy silanes, the m/z 205 ion of $Si(OAc)_3$ has abundances of < 1, 6, and 27%, respectively. The spectra of $MeSi(OAc)_3$ and $EtSi(OAc)_3$ are shown in Figure 13.17.

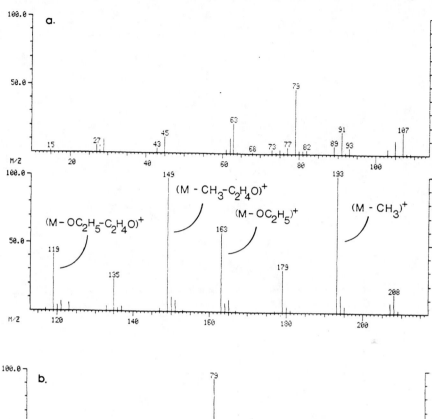

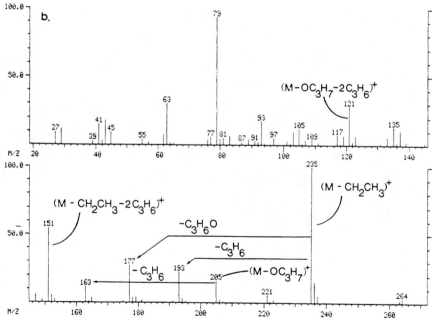

Figure 13.16. Mass spectra of (*a*) tetraethoxysilane and (*b*) tetra(*n*-propoxy)silane.

449

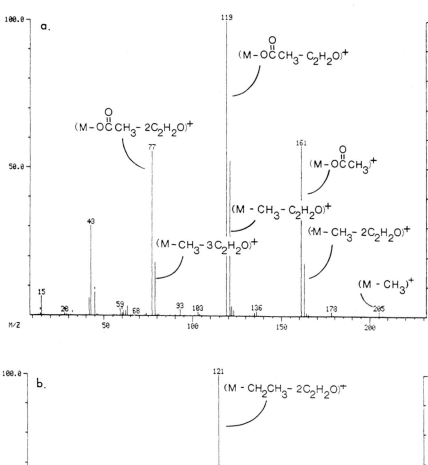

Figure 13.17. Mass spectra of (*a*) methyltriacetoxysilane and (*b*) ethyltriacetoxysilane.

450

4.2. Siloxanes

Electron impact induced fragmentation of siloxanes proceeds by several distinct pathways that can be generalized for most of the cyclic and linear polysiloxanes normally encountered. There are a few notable exceptions, which are discussed separately. Only when there are aromatic substituents is an appreciable molecular ion observed, otherwise the spectra are characterized by rapid dissociation to the even-electron ion formed from Si–C cleavage of the organic radical with the lowest bond energy. For linear di- and trisiloxanes and cyclic tri- and tetra siloxanes, this ion is often the base peak and nearly all the ionization is contained in it and other ions formed on the same siloxane skeleton. For siloxanes larger than these, however, much of the ionization becomes concentrated in ions with fewer silicon atoms than the starting molecule because of skeletal rearrangements with ejection of neutrals containing Si. As a result, identification using EI becomes less probable as the number of siloxane repeat units increases. The base peak for the larger siloxanes is often the one-silicon species, SiR_3^+.

4.2.1. Poly(dimethylsiloxanes)

The mechanisms for fragmentation and rearrangement of a dimethyl cyclic trimer through hexamer were published in 1972 (38). Each of the compounds has a sufficiently abundant $(M - Me)^+$ ion to allow easy identification of the molecule. A major difference between the two smaller cyclic compounds compared to the two larger ones is that the smaller compounds show a significant doubly charged ion from loss of two methyl groups, whereas the larger ones exhibit the rearrangement process that results in a net loss of 103 mass units from the molecular ion. They also have a Me_3Si^+ base peak. The mechanism for formation of these ions was proposed to be the transannular process depicted in Scheme 13.8.

The spectra of larger cyclics (D_6–D_{15}) have also been studied (39). It was shown that extensive skeletal rearrangement occurs, leading to ions for smaller cyclics minus a methyl group, and linear ions of the series $Me_3Si(OSiMe_2)_n^+$ for $n = 0$–4, depending on the size of the starting molecule. This process, coupled with the fact that the $(M - Me)^+$ ion becomes $< 1\%$ of the total ionization product, results in the spectra of cyclics of 10 silicons and higher appearing to be identical to one another. The processes leading to the formation of cyclic and linear siliconium ions with fewer silicons than the original molecule were proposed to occur analogously to the mechanism of VandenHeuval (Scheme 13.8) (38). For the larger cyclics the $(M - 103)^+$ ion from Me_4Si elimination is no longer observed, but is replaced by processes involving elimination of neutral cyclic molecules.

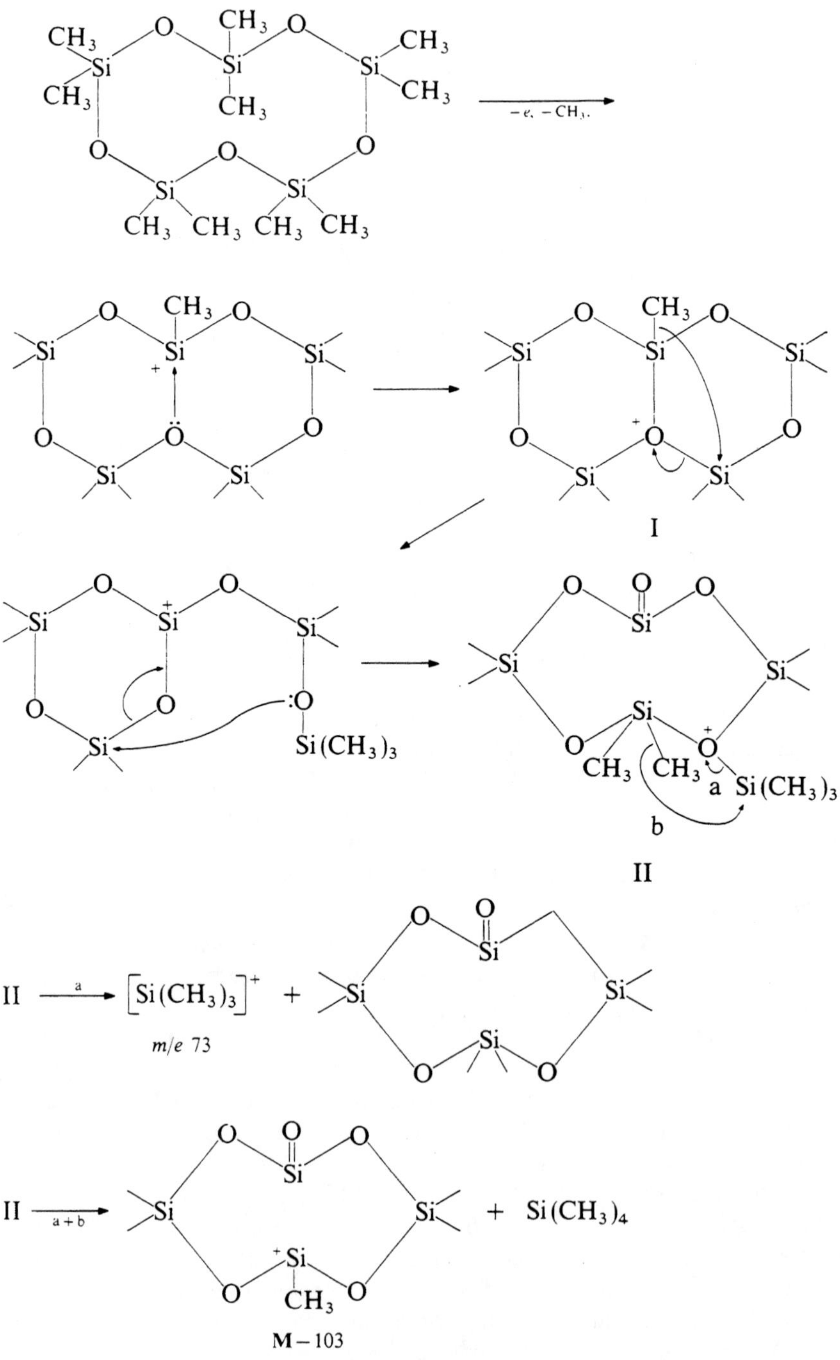

Scheme 13.8 (38)

452

The mass spectra of linears 2–6 were published in 1967 (40). As in the case for the cyclics, the larger the starting molecule, the less likely it is that ions containing the original number of silicons are observed. The $(M - Me)^+$ ion is the base peak for the linear dimer and trimer, but the rearrangement process becomes facile for longer chains and produces the base peak for the linear tetramer and higher. Also analogous to the cyclics, Me_4Si elimination from the siliconium ion occurs for the intermediate chain lengths, but is replaced by elimination of neutral cyclic molecules as the chain becomes longer. The spectra of linears are indistinguishable from one another for compounds above MD_5M. In contrast to the mass spectra of normal versus branched hydrocarbons, the branched siloxanes exhibit nearly identical spectra to the unbranched linears. This observation is consistent with the predominance of rearrangement ions in these spectra, resulting in the same ions regardless of the configuration of the original molecule. Figure 13.18 contains two examples of this point, a comparison of the spectra of MD_2M versus M_3T and MD_3M versus M_4Q.

Another manifestation of the virtual loss of identity of the starting molecule because of rearrangement processes is that the spectra of two compounds as dissimilar as D_6 and MD_3D^HM look very similar. These compounds have the same molecular weight to within 85 ppm and exhibit only small differences based on involvement of the SiH moiety in the fragmentation processes. Similarly, a linear methyl siloxane containing two SiH groups has many of the same ions in the low resolution mass spectrum as the cyclic with one SiH group as well as the cyclic with all methyls but containing a bridging oxygen between nonadjacent silicon atoms.

$(Me_2SiO)_6$ versus $Me_3SiO(Me_2SiO)_3(MeHSiO)SiMe_3$

$(M - Me)^+ = 429.0893$ versus 429.1257

$(Me_2SiO)_5(MeHSiO)$ vs. $Me_2HSiO(Me_2SiO)_4SiMe_2H$ vs. $\begin{matrix} & Me & \\ Me_2SiOSiOSiMe_2 \\ O\quad O\quad O \\ Me_2SiOSiOSiMe_2 \\ & Me & \end{matrix}$

$(M-Me)^+ = 415.0736$ versus 415.1100 versus 415.0372

Scheme 13.9

With care, assignable differences can be observed for these compounds; their GC retention times and a knowledge of the chemistry involved in the sample being studied are also useful clues.

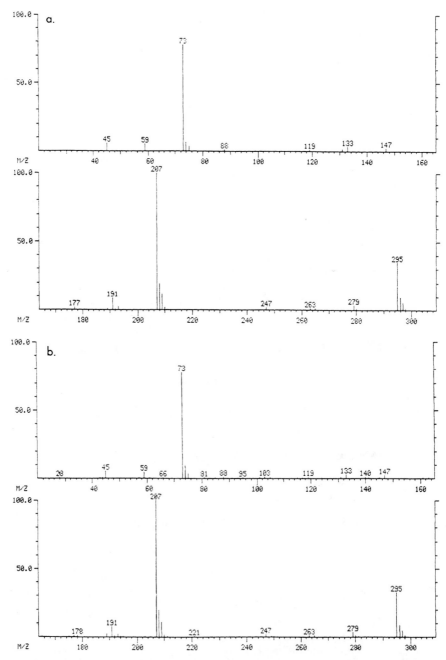

Figure 13.18. Mass spectra of (*a*) tris(trimethylsiloxy)methylsilane, (*b*) decamethyltetrasiloxane, (*c*) dodecamethylpentasiloxane, and (*d*) tetrakis(trimethylsiloxy)silane.

454

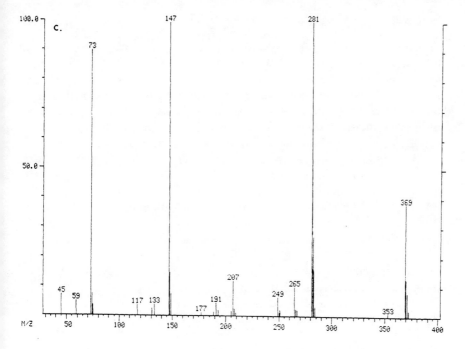

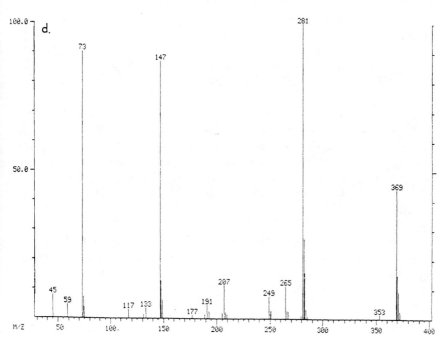

Figure 13.18 (*Continued*)

455

Another area of possible confusion can be described with the example of $(Me_2SiO)_4(EtMeSiO)$ (MW 384) versus MD_2D^HM (MW 370). In the former compound, cleavages of methyl and ethyl provide ions at m/z 369 and m/z 355. For the second compound, cleavages of hydrogen and methyl result in ions at m/z 369 and m/z 355. Again care is necessary and all the clues available from the rest of the spectrum should be examined. Interpretation of copolymer spectra will be discussed further in a later section.

As mentioned earlier, for identification of PDMS oligomers $> n = 6$, a combination of both EI and ammonia CI mass spectra is recommended. For those that cannot be eluted from a GC column, both ammonia DCI $(M + NH_4)^+$ and field desorption $(M - Me)^+$ sampling are good choices.

4.2.2. Siloxane Copolymers

Because the mass spectra of $(Me_2SiO)_x(MeRSiO)_y$ and $(MeRSiO)_z$ compounds are quite similar in their mechanisms of fragmentation, the polysiloxanes containing a MeRSiO repeat unit are discussed with those in which that unit is also part of a copolymer with Me_2SiO.

4.2.2.1. Me_2=MeH. The copolymers behave much like the permethylated compounds but can show additional fragments from $(M - H)^+$ and rearrangements involving SiH such as Me_3SiH elimination from the $(M - Me)^+$ or $(M - H)^+$ ions. There is little difference expected in the spectra for random versus block copolymers, for SiH located on the end groups rather than in the middle of the chain, or for cis–trans configurations on a cyclic backbone.

The mass spectra of MeHSiO cyclic tetramer–hexamer have been published (41) and are shown in Figure 13.19. Following the same pattern as the dimethyl system, for the tetramer most of the ionization is carried in ions having the same number of silicons as the original molecule, but rearrangement ions having fewer silicons are important for the larger compounds. The apparent ability to eliminate $MeSiH_3$ from the $(M - H)^+$ ion of the pentamer m/z 299 − m/z 253 suggests a facile transfer of hydrogen atoms between silicon atoms in these molecules. In contrast, no ion for Me_3SiH elimination from the $(M - Me)^+$ ion is observed m/z 285 → m/z 211. Elimination of Me_2SiH_2 is also expected to contribute to the observed pattern. By high resolution measurements it was concluded (41) that the series of ions 60 mass units below the above-mentioned rearrangement ions is from siloxane group expulsion, that is, MeHSiO and SiO_2.

4.2.2.2. Me_2–MeVi. A combination of the principles discussed for methyl siloxanes and vinyl silanes apply to the behavior of this class of compounds.

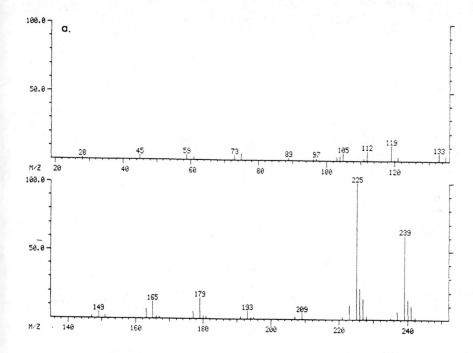

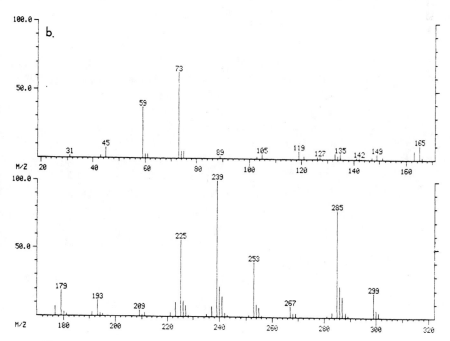

Figure 13.19. Mass spectra of (a) $(MeHSiO)_4$, (b) $(MeHSiO)_5$, and (c) $(MeHSiO)_6$.

457

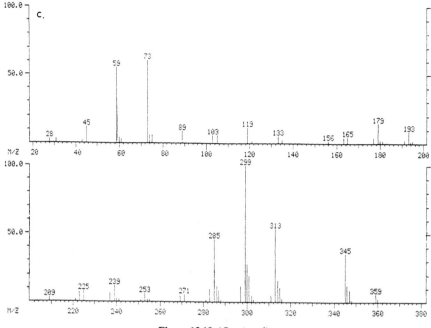

Figure 13.19 (*Continued*)

The size of the molecule determines whether the ionization will be concentrated in ions of the same silicon number or in single silicon ions. The difference between the spectra of the MeViSiO cyclic tetramer and pentamer can be seen in Figure 13.20. The pentamer has a base peak of m/z 97, Vi_2MeSi^+, and has significant ions due to $ViMe_2Si^+$, m/z 85, and Vi_3Si^+, m/z 109. These ions also rearrange to form ions at m/z 83, 71, and 59 by expulsion of acetylene from m/z 109, 97, and 85 (42). The authors explain the "vigorous" randomization of substituents that form Me_3Si^+ and Vi_3Si^+ from MeViSiO repeat units as resulting from a bridging process. By formation of a cyclic

$$AlMe_3 + AlEt_3 \rightleftharpoons Me_2Al \overset{Me}{\underset{Et}{\diamond}} AlEt_2 \rightleftharpoons AlMe_2Et + AlEt_2Me$$

$$-\overset{+}{\underset{R}{Si}}O\underset{}{Si}- \rightleftharpoons \left(>Si \overset{O}{\diamond} Si< \right)^+ \rightleftharpoons -\underset{R}{Si}O\overset{+}{\underset{R}{Si}}-$$

Scheme 13.10

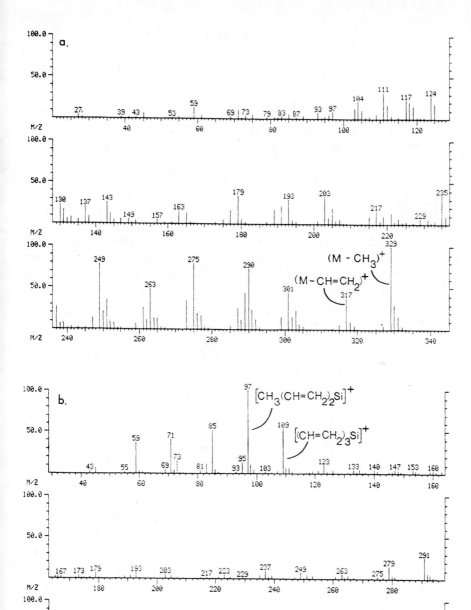

Figure 13.20. Mass spectra of (a) (MeViSiO)$_4$ and (b) (MeViSiO)$_5$.

459

transition state, the exchange process is postulated to occur in the same manner that trialkyl aluminum compounds exchange substituents, since aluminum and the silyl cation are isoelectronic.

As in the vinyl silanes, acetylene and ethylene losses from even-electron ions are abundant, resulting in a mass spectrum with many more ions than observed in the dimethyl system.

4.2.2.3. Phenyls. For siloxanes containing methyl and phenyl groups there is little difference in the spectra of compounds with the same number of phenyls, regardless of the arrangement of substituents (43, 44). Processes related to chain size hold the same as discussed for the permethyl system. For example, cyclic trimers and tetramers containing phenyl groups retain nearly all the ionization in species with the original number of silicons, whereas pentamers and higher oligomers rearrange to single silicon species (45). The elimination of benzene from the $(M - Me)^+$ ion is common for the small rings (with at least two phenyls) but not important for the larger ones. The larger ones are characterized by intense peaks at m/z 73, 135, 197, or 259 Me_xSiPh_{3-x} and by peaks from Me_xSiPh_{4-x} elimination from the $(M - Me)^+$ peak, depending on the number of phenyls in the molecule. A molecular ion is often observed for phenyl-containing siloxanes. A process unique to phenyl siloxanes is phenyl methyl or dimethyl silanone elimination from the base siliconium ion in the spectra of phenyl–methyl disiloxanes (46).

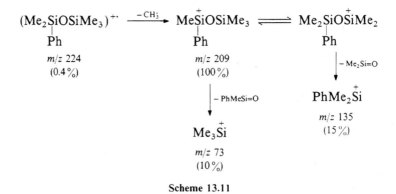

Scheme 13.11

4.2.2.4. Alkyls. The general rules previously discussed for siloxanes can be applied to compounds with higher alkyl groups, but the rearrangement behavior of the alkyl as discussed in the section on silanes must also be kept in mind. Loss of the largest alkyl from the molecular ion is usually observed, but exceptions occur, especially with a mixture of large R groups (47). The spectra become quite complex if a variety of different R groups is present

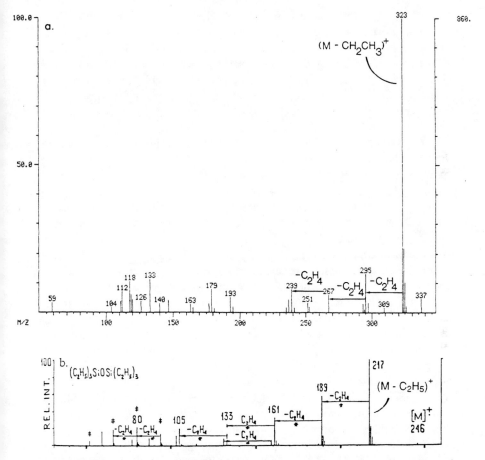

Figure 13.21. Mass spectra of (a) (EtMeSiO)$_4$ and (b) (Et$_3$Si)$_2$O; (48). Reprinted with permission from E. Mysov, et al., *Org. Mass Spectrom.*, **11**, 358 © 1976, John Wiley & Sons, Inc.

since (M − alkyl) and (M − alkyl − olefin) occurs for all possible combinations. One clue that the compound contains alkyl substituents other than methyl is the presence of the monosilicon ions such as m/z 87 Me$_2$EtSi$^+$, m/z 101 Me$_2$PrSi$^+$, and the like, for siloxanes of sufficient size to produce such ions. Figure 13.21 shows two examples of ethyl-containing siloxanes, hexaethyl disiloxane (48) and (EtMeSiO)$_4$.

4.2.2.5.   SiOH. Siloxanes that contain silanol groups are a problem in the EI mode because of their elimination of methane and methanol from

$(M - Me)^+$ for monosilanols, or rapid elimination of H_2O from $(M - Me)^+$ if two or more SiOH groups are present. Except for short molecules (< 4 Si atoms), this process does not allow observation of the methyl loss peak and in the case of the linear diols the spectra appear identical to those of the corresponding cyclosiloxanes. A good solution is to run the sample by ammonia CI, which provides abundant $M + NH_4$ peaks for these types of compounds.

4.2.2.6. Other groups. Chlorine-containing methyl siloxanes should behave very much like the methyl compounds, but the chlorine is observed to participate in the rearrangements, forming the $(M - Me - Me_3SiCl)^+$ ion and the Me_2SiCl^+ ion, analogous to Me_3Si^+.

Alkoxy-containing siloxanes share the rules for methyl siloxanes and alkoxy silanes. Thus, methoxy groups exhibit formaldeyde elimination, ethoxy groups lose C_2H_4O, and C_2H_4, and so on (49).

Siloxanes containing the 3,3,3-trifluoropropyl substituent provide a case where the rules discussed for earlier examples do not apply (50). The large driving force for migration of the electronegative fluorine to electropositive silicon results in no ions being detected in which the trifluoropropyl group is left intact on the original siloxane backbone. A common process is elimination of difluorocyclopropane to leave $=SiF^+$. Even short chain compounds show skeletal contraction, so m/z 211 ($Me_4FSi_3O_3$) and 215 ($Me_3F_2Si_3O_3$) are intense peaks for tetramer molecules. One- and two-silicon species are also prominent in the spectra: Me_xSiF_{3-x} (m/z 73, 77, and 81); the proposed $Me(OH)SiCH_2CH=CF_2$ m/z 137, $CF_3CH_2CH_2SiMe_xF_{2-x}$ (m/z 151, 155, and 159); and the disiloxanes of those three with Me_2SiO, 74 mass units higher (m/z 229, 233, and 237). This behavior under EI means that different molecules that contain this substituent cannot be distinguished.

Isobutane CI has been successful in forming abundant protonated molecular ions for cyclosiloxanes containing these substituents.

4.3. Silazanes

The mechanisms of EI fragmentation of hexamethyldisilazane were studied in detail using isotope labels and N-substituted derivatives (51). Silazanes are more likely than the corresponding siloxanes to exhibit a molecular ion, and the $(M - Me)^+$ fragment is also prominent. Both methane and ammonia elimination from this ion are often observed. Cleavage of the Si–N bond is possible (in contrast to Si–O in analogous siloxanes), so the Me_3Si^+ ion is observed for hexamethyldisilazane. An indication the spectrum might be that of a silazane is the series of even-numbered ions, resulting from the odd-numbered molecular weight if the compound has an odd number of nitrogen

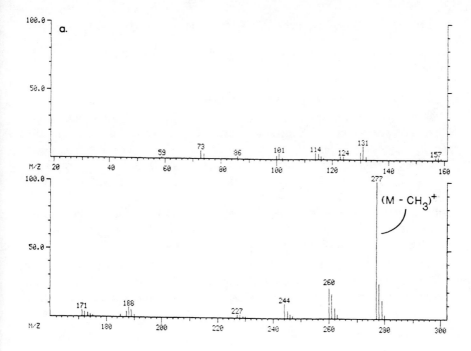

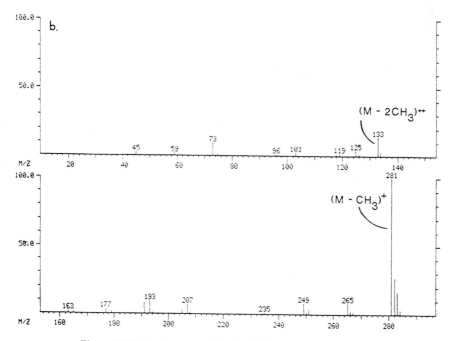

Figure 13.22. Mass spectra of (a) $(Me_2SiNH)_4$ and (b) $(Me_2SiO)_4$.

463

atoms. Figure 13.22 compares cyclic tetramers with siloxane versus silazane bonds.

4.4. Polysilanes

The mass spectrometry of polysilanes bears little resemblance to that of the polysiloxanes containing the same organic substituents. The Si–Si cleavages are generally more important than Si–C, thus $R_3Si^{\cdot}$ as well as neutral R_4Si losses are observed from the molecular ions of linear polysilanes, while formation of $(M - R)^+$ is a minor process. Molecular ions are more likely to be observed for the polysilanes, although they are weak for the linears but more intense for the cyclic polysilanes in the permethylated system (52, 53). Successive losses of R_2Si after the initial cleavage of a radical are also

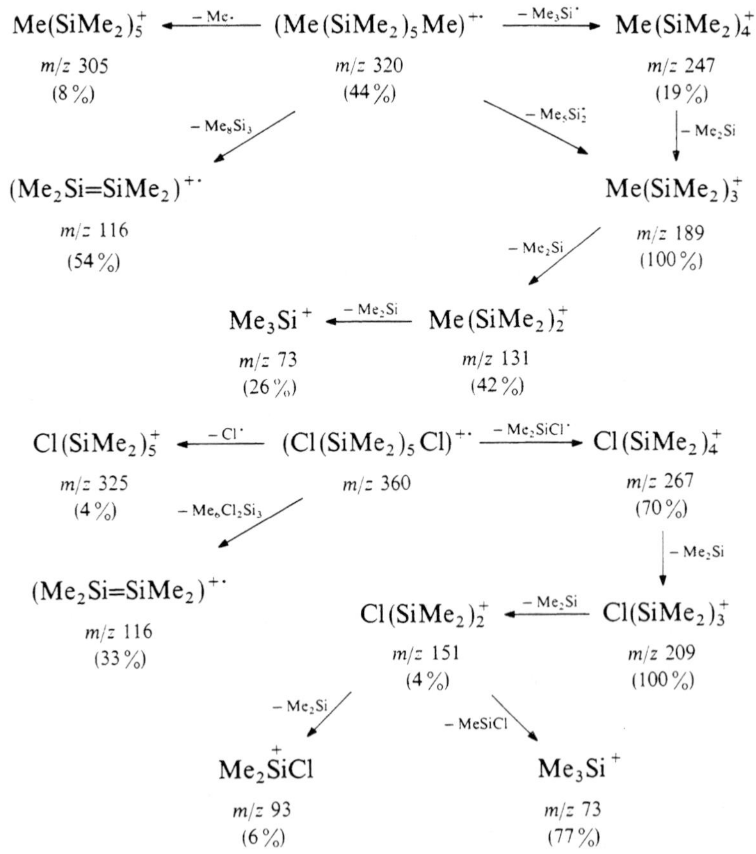

Scheme 13.12

common. The R_3Si^+ ion is abundant, even for the cyclics where methyl transfer would be required. For larger R groups, olefin eliminations are also observed. Halogen terminated linears have been shown to involve the halogen in many of the important ions, thus giving unusually stable high mass ions containing halogen, in competition with Me_3Si^+. Scheme 13.12 contrasts the fragmentation pattern of dichlorodecamethylpentasilane with that of the permethylated compound (54).

4.5. Silyl Derivatives of Organic Compounds

Alcohols are silylated for GC and mass spectroscopic analysis to provide thermal stability, inertness to the GC column, greater volatility, and more informative mass spectra. A review covering trialkylsilyl derivatization has been published (55).

The mass spectra of the trimethylsilyl ethers of aliphatic alcohols have been studied (56, 57). The molecular ion is usually weak, but the $(M - Me)^+$ ion is useful for molecular weight determination. Cleavage of the group alpha to the ether oxygen can proceed by Si–Me cleavage, or C–C cleavage, to expel the aliphatic radical. The preferred path is dependent on branching at the carbon site. For the derivative of 1-pentanol, loss of methyl from silicon is greatly preferred over loss of butyl. For the 2-pentanol derivative, loss of the propyl radical is much more abundant than methyl loss. Also in the branched compounds, loss of the larger R group is favored over the smaller. It was reported (58) that Me_2HSi derivatives exhibit less Si–Me cleavage relative to alpha cleavage in the R group. Use of both Me_2HSi and Me_3Si derivatives can be useful in structure elucidation.

The appearance of the ions m/z 73 and 75 is common for these derivatives. The Me_2SiOH^+ ion is a result of olefin elimination from the $(M - Me)^+$ ion, while the Me_3Si^+ ion is formed by formaldehyde elimination from the other alpha fission product $(M - R)^+$. Benzylic derivatives undergo a skeletal rearrangement resulting in formation of the $PhMe_2Si^+$ ion m/z 135. This rearrangement has been shown to occur by formaldehyde ejection from the $(M - Me)^+$ ion with migration of Si from oxygen to the aromatic ring.

The bis(trimethylsilyl) ethers of diols form the disiloxane ion m/z 147 by elimination of $(CH_2)_nO$ from the $(M - Me)^+$ ion. This has also been reported to be the mechanism by which the Me_2Si^+ ion is formed from the t-butyl dimethylsilyl derivatives of diols (59). The use of silylation reagents with larger R groups (C_2–C_4) can result in considerable differences in the mass spectra compared to Me_3Si– (60). Ions from $(M - R)^+$ were seen to be of greater abundance than for the Me_3Si compounds. Additionally, ions from alpha C–C cleavage were less significant for the R_3Si derivatives, but olefin elimination processes involving the R group on primary ions such as R_3Si^+

and R_2SiOH^+ were common. Similarly, the disiloxane ions formed in the spectra of derivatized diols exhibited successive olefin eliminations from alternate silicon atoms.

Some other interesting phenomena have been reported involving certain trimethylsilyl ethers. With ortho substituted aromatics, a cyclic ion has been proposed to explain the process by which Me_4Si is ejected (61).

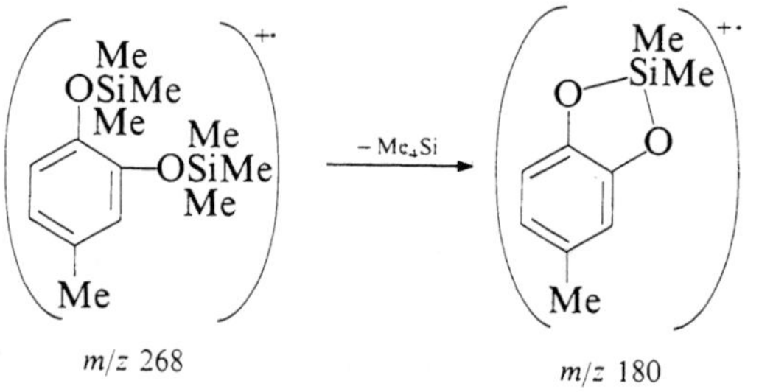

m/z 268 m/z 180

Scheme 13.13

A chlorine migration has been reported in the TMS derivatives of chloro alcohols, producing the base peak of Me_2SiCl^+, m/z 93 (62).

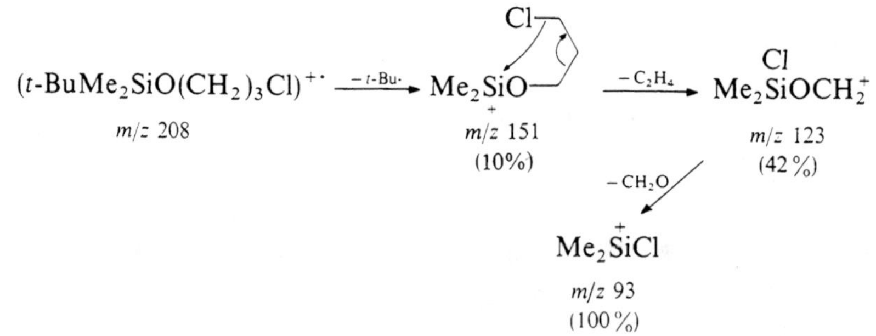

$(t\text{-}BuMe_2SiO(CH_2)_3Cl)^{+\cdot}$ $\xrightarrow{-t\text{-}Bu\cdot}$ Me_2SiO- $\xrightarrow{-C_2H_4}$ $Me_2SiOCH_2^+$

m/z 208 m/z 151 m/z 123
                   (10%)   (42%)

$\xrightarrow{-CH_2O}$

$Me_2\overset{+}{SiCl}$

m/z 93
(100%)

Scheme 13.14

Finally, certain TMS derivatives exhibit the loss of the neutral Me_3SiOH, depending on the availability of an active hydrogen that can be abstracted by the trimethylsiloxy group (63).

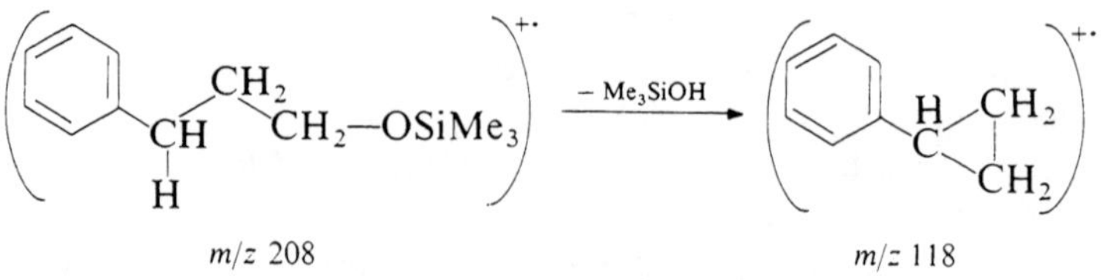

m/z 208 m/z 118

Scheme 13.15

REFERENCES

1. M. Bertrand, L. Maltais, and M. Evans, "Poly(dimethylsiloxane) as a reference standard for exact mass measurement in chemical ionization mass spectrometry," *Anal. Chem.*, **59**, 194 (1987).

2. R. Weast, Ed., *Handbook of Chemistry and Physics*, The CRC Press, Cleveland, Ohio, 55th ed., 1975.

3. R. Anderegg, A. Brajter-Toth, and J. Toth, "Determination of extent of silylation by isotope cluster analysis of a single mass spectrum," *Anal. Chem.*, **56**, 1351 (1984).

4. J. D. Pinkston, G. Owens, L. Burkes, T. Delaney, D. Millington, and D. Maltby, "Capillary supercritical fluid chromatography–mass spectrometry using a 'high mass' quadrupole and splitless injection," *Anal. Chem.*, **60**, 962 (1988).

5. J. Kleinert and C. Weschler, "Pyrolysis gas chromatographic–mass spectrometric identification of polydimethylsiloxanes," *Anal. Chem.*, **52**, 1245 (1980).

6. J. Chih-An Hu, "Pyrolysis mass spectrometric method for polymer characterization," *Anal. Chem.*, **53**, 942 (1981).

7. A. Ballistreri, D. Garozzo, and G. Montaudo, "Mass spectral characterization and thermal decomposition mechanism of poly(dimethylsiloxane)," *Macromolecules*, **17**, 1312 (1984).

8. S. Fujimoto, H. Ohtani, and S. Tsuge, "Characterization of polysiloxanes by high-resolution pyrolysis–gas chromatography–mass spectrometry," *Z. Anal. Chem.*, **331**, 342 (1988).

9. A. Harrison, *Chemical Ionization Mass Spectrometry*, CRC Press, Boca Raton, FL, 1983.

10. R. Freas and J. Campana, "Fast-atom bombardment and chemical ionization/fast-atom bombardment mass spectrometry of lubricants," *Anal. Chem.*, **58**, 2434 (1986).

11. I. Bletsos, D. Hercules, D. vanLeyen, and A. Benninghoven, "Time-of-flight secondary ion mass spectrometry of polymers in the mass range 500–10,000," *Macromolecules*, **20**, 407 (1987).

12. I. Bletsos, D. Hercules, J. Magill, D. VanLeyen, E. Niehuis, and A. Benninghoven, "Time-of-flight secondary ion mass spectrometry: detection of fragments from thick polymer films in the range $m/z < 4500$," *Anal. Chem.*, **60**, 938 (1988).

13. S. Graham and D. Hercules, "Laser desorption mass spectra of biomedical polymers: Biomer and Avcothane," *Spectrosc. Lett.*, **15**, 1 (1982).

14. M. Litzow and T. Spalding, "Mass spectrometry of inorganic and organometallic compounds," in *Physical Inorganic Chemistry*, (Monograph 2) Chapter 7, M. Lappert Ed., Elsevier Scientific, Amsterdam-London-New York, 1973.

15. G. van der Kelen, O. Volders, H. Van Onckelen, and Z. Eeckhaut, "Mass spectra of silanes and the methylsilanes," *Z. Anorg. Allg. Chem.*, **338**, 106 (1965).

16. J. J. de Ridder and G. Dijkstra, "Mass spectrum of the tetramethyl and tetraethyl compounds of carbon, silicon, germanium, tin, and lead," *Rec. Trav. Chim.*, **86**, 737 (1967), *Chem. Abstr.*, **67**: 103569e.

17. G. Groenewold, M. Gross, M. Bursey, and P. Jones, "Unimolecular and collision induced decompositions of gas-phase group IVA enium ions $(CH_3)_3M^+$," *J. Organometal. Chem.*, **235**, 165 (1982).

18. G. Smolinsky and M. Vasile, "Mass spectra of vinyltrimethylsilane and vinyltri(methyl-d_3)silane," *Org. Mass Spectrom.*, **7**, 1069 (1973).

19. A. Polivanov, T. Slyusarenko, V. Zhun, V. Bochkarev, and V. Sheludyakov, "Mass-spectrometric study of vinylsilanes," *Zh. Obshch. Khim.*, **49**, 1311 (1979).

20. J. Soderquist and A. Hassner, "Vinylmetalloids IV. Intramolecular bond- forming reactions of vinylsilanes upon electron impact," *J. Organometal. Chem.*, **217**, 151 (1981).

21. N. Chernyak, R. Khmel'nitskii, T. D'Yakova, and V. Vdovin, "Mass spectral study of alkylsilanes," *Zh. Obshch. Khim.*, **36**, 89 (1966).

22. A. Onopchenko, E. Sabourin, and D. Danner, "Mass spectra of some high molecular weight tetra(alkyl)silanes," *Org. Mass Spectrom.*, **20**, 505 (1985).

23. T. Kinstle, P. Ihrig, and E. Goettert, "Site-specific hydrogen transfer in even-electron ions," *J. Am. Chem. Soc.*, **92**, 1780 (1970).

24. W. Weber, A. Willard, and H. Boettger, "Mass spectroscopy of organosilicon compounds. Examples of interaction of the silyl center with remote phenyl groups," *J. Org. Chem.*, **36**, 1620 (1971).

25. B. Ho, E. Dexheimer, L. Spialter, and L. Smithson, "Alpha-silyl ketones: A high resolution mass spectromeric study," *Org. Mass Spectrom.*, **14**, 185 (1979).

26. W. Weber and H. Boettger, "Mass spectrometry of organosilicon compounds-some examples of interaction of the silyl center with remote functional groups," *Intra-Sci. Chem. Rep.*, **7**, 109 (1973).

27. C. Tsipis, "Mass spectral study of organosilanes: A novel 1,4-hydroxyl migration from carbon to silicon in hydroxy-silyl-alkenes," *Chim. Chron.*, **9**, 111 (1980).

28. D. Dimmel, C. Wilkie, and F. Ramon, "Mass spectra of silanes. Multiple rearrangements and bonding to silicon," *J. Org. Chem.*, **37**, 2665 (1972).

29. V. Zaikin, A. Mikaya, V. Vdovin, E. Babich, V. Traven, and B. Stepanov, "Mass-spectrometeric investigation of alkyl-substituted silacyclobutanes," *Zh. Obshch. Khim.*, **49**, 1307 (1979).

30. J. Laane, "Synthesis of silacyclobutane and some related compounds," *J. Am. Chem. Soc.*, **89**, 1144 (1967).

31. A. Duffield, H. Budzikiewicz, and C. Djerassi, "Mass spectrometry in structural and stereochemical problems. LXXI. A study of the influence of different heteroatoms on the mass spectrometric fragmentation of five-membered heterocycles," *J. Am. Chem. Soc.*, **87**, 2920 (1965).

32. V. Bochkarev, A. Polivanov, N. Komalenkova, S. Bashkirova, and E. Chernyshev, "Silicon-containing heterocyclic compounds XIV. Mass-spectrometric study of some silacyclopent-3-ene derivatives," *Zh. Obshch. Khim.*, **43**, 785 (1973).

33. B. Ho, L. Spialter, and L. Smithson, "Fragmentation and rearrangement processes in the high resolution mass spectra of diphenylsilyl compounds," *Org. Mass Spectrom.*, **10**, 361 (1975).

34. J. Bowie and B. Nussy, "Skeletal rearrangement and hydrogen scrambling processes in the positive and negative-ion mass spectra of phenyl derivatives of elements of groups IV and V," *Org. Mass Spectrom.*, **3**, 933 (1970).

35. K. Pope and P. Jones, "Silenes and silenoids. 7. Electron-impact-induced decomposition of (trimethylsilyl)methyl-substituted chlorosilanes," *Organometallics*, **3**, 354 (1984).

36. J. E. Coutant and R. J. Robinson, "Mass spectrometry," in *Analysis of Silicones*, Chapter 12, A. L. Smith, Ed., Wiley, New York, 1974.

37. G. Dube, U. Pape, and V. Chvalovsky, "The effect of substituents on mass spectra of acetoxy(methyl)silanes," *Coll. Czech. Chem. Commun.*, **42**, 2898 (1977).

38. W. VandenHeuvel, J. Smith, R. Firestone, and J. Beck, "Mass spectrometry of methylcyclosiloxanes, sources of anomalous peaks in gas–liquid chromatography," *Anal. Lett.*, **5**, 285 (1972).

39. G. Pickering, C. Olliff, and K. Rutt, "The mass spectrometeric behaviour of dimethylcyclosiloxanes," *Org. Mass Spectrom.*, **10**, 1035 (1975).

40. V. Orlov, "Dissociation of methylsiloxanes under electron bombardment," *Zh. Obshch. Khim.*, **37**, 2300 (1967).

41. E. Pelletier and J. Harrod, "Mass spectra of some substituted methylcyclosiloxanes," *Can. J. Chem.*, **61**, 762 (1983).

42. A. Polivanov, A. Bernadskii, V. Bochkarev, V. Zhun, and V. Sheludyakov, "Massspectrometric study of vinylsiloxanes," *Zh. Obshch. Khim.*, **50**, 614 (1980).

43. A. Polivanov, A. Bernadskii, V. Fal'ko, N. Telegina, V. Khvostenko, and V. Bochkarev, "Mass spectra of the positive and negative ions of cyclotri- and cyclotetra-siloxanes containing dimethyl-, methylphenyl-, and diphenylsilyleneoxy units," *Zh. Obshch. Khim.*, **48**, 399 (1978).

44. V. Bochkarev, A. Polivanov, V. Fal'ko, L. Blekh, V. Khvostenko, and E. Chernyshev, "Mass spectra of the positive and negative ions of linear methylphenylsiloxanes," *Zh. Obshch. Khim.*, **48**, 858 (1978).

45. V. Bochkarev, A. Polivanov, N. Telegina, and E. Chernyshev, "Mass-spectrometric study of methylphenylcyclopenta- and methylphenylcyclohexasiloxanes," *Zh. Obshch. Khim.*, **44**, 955 (1974).

46. R. Swaim, W. Weber, H. Boettger, M. Evans, and F. Bockhoff, "Mass spectrometry of aryl substituted di- and trisiloxanes," *Org. Mass Spectrom.*, **15**, 304 (1980).

47. R. Pfefferkorn, H. Grutzmacher, and D. Kuck, "The mass spectra of some hexaalkyl dilsiloxanes," *Int. J. Mass Spectrom. Ion Phys.*, **47**, 515 (1983).

48. E. Mysov, I. Akhrem, and D. Avetisyan, "Mass spectrometric investigation of ethyl-containing siloxanes and aromatic silyl ethers," *Org. Mass Spectrom.*, **11**, 358 (1976).

49. A. Polivanov, A. Bernadskii, N. Silkina, B. Klimentov, and V. Bochkarev, "Massspectrometric study of alkoxy-and chloro-siloxanes," *Zh. Obshch. Khim.*, **50**, 1780 (1980).

50. V. Bochkarev, A. Polivanov, A. Bernadskii, and N. Rodzevich, "Anomalous mass-

spectrometric fragmentation of cyclosiloxanes containing 3,3,3-trifluoropropyl substituents," *Zh. Obshch. Khim.*, **50**, 1074 (1980).

51. J. Tamas and P. Miklos, "Mass spectrometric study of hexamethyldisilazane and some of its N-substituted derivatives," *Org. Mass Spectrom.*, **10**, 859 (1975).

52. H. Sakurai, Y. Nakadaira, and Y. Kobayashi, "Mass spectroscopy of organosilicon compounds. Fragmentation patterns of linear organopolysilanes," *J. Organomet. Chem.*, **63**, 79 (1973).

53. T. Kinstle, I. Haiduc, and H. Gilman, "Mass spectrometry of cyclopolysilanes," *Inorg. Chim. Acta*, **3**, 373 (1969).

54. C. Middlecamp, W. Wojnowski, and R. West, "The preparation of dibromopermethylpolysilanes and characterization of these and related chloropermethylpolysilanes by mass spectrometry," *J. Organomet. Chem.*, **140**, 133 (1977).

55. C. Poole and A. Zlatkis, "Trialkylsilyl ether derivatives for gas chromatography and mass spectrometry," *J. Chrom. Sci.*, **17**, 115 (1979).

56. A. Sharkey, R. Friedel, and S. Langer, "Mass spectra of trimethylsilyl derivatives," *Anal. Chem.*, **29**, 770 (1957).

57. J. Diekman, J. Thomson, and C. Djerassi, "Mass spectrometry in structural and stereochemical problems. CXLI. The electron impact induced fragmentations and rearrangements of trimethylsilyl ethers, amines, and sulfides," *J. Org. Chem.*, **32**, 3904 (1967).

58. W. Richter and D. Hunnemann, "Use of dimethylsilyl ethers for characterizing primary aliphatic alcohols: A comparison of mass spectrometive fragmentation of di- and trimethylsilyl derivatives," *Helv. Chim. Acta* **57**, 1131 (1974).

59. G. Phillipou, "Genesis of the trimethylsilyl cation in the electron impact mass spectra of t-butyldimethylsilyl derivatives," *Org. Mass Spectrom.*, **12**, 261 (1977).

60. D. Harvey, "Mass spectrometry of the triethylsilyl, tri-n-propylsilyl and tri-n-butylsilyl derivatives of some alcohols, steroids and cannabinoids," *Org. Mass Spectrom.*, **12**, 473 (1977).

61. R. Horvat and S. Senter, "Cyclic ions in the mass spectra of trimethylsilyl derivatives of substituted o-dihydroxybenzenes," *Org. Mass Spectrom.*, **18**, 413 (1983).

62. S. Hill and M. Mokotoff, "Chlorine migration in the mass spectra of tert-butyldimethylsilyl derivatives of chloro alcohols," *J. Org. Chem.*, **49**, 1441 (1984).

63. D. Kingston, B. Hobrock, M. Bursey, and J. Bursey, "Intramolecular hydrogen transfer in mass spectra. III. Rearrangements involving the loss of small neutral molecules," *Chem. Rev.*, **75**, 714 (1975).

CHAPTER

14

ATOMIC SPECTROSCOPY

NELSON W. LYTLE

Dow Corning Corporation
Midland, Michigan

1. INTRODUCTION

Since 1859, when Kirchhoff and Bunsen developed the first practical spectroscope, atomic spectroscopy has contributed immeasurably to our knowledge and understanding of matter. Among the techniques that employ UV–vis radiation for elemental analysis the methods used to produce the signal vary considerably. They include flame spectrometric methods based on atomic emission, absorption, and fluorescence; DC arc excitation; high voltage spark spectroscopy; and plasma-atomic emission sources. Flameless sources for atomic absorption (AA) spectroscopy such as electrothermal atomization, cold vapor, and hydride generation are also used.

Atomic spectroscopy techniques have been employed for elemental analysis for many years, and a wealth of information is found in the literature. Several recent texts describe the theory and application of atomic spectroscopy techniques to elemental analysis (1–4). General literature reviews in journals (e.g., *Anal. Chem., J. Anal. At. Spectrom.,* and *At. Spectrosc.*) may also be consulted for additional information and current awareness.

The past 20 years has seen many developments in instrumentation for atomic spectroscopy. Progress in instrument automation and the availability of computer assisted data collection have led to the incorporation of these techniques into most analytical laboratories. Because some of the newer sources presently require that samples be introduced as liquids or solutions, refinements in sample preparation, such as the microwave based digestion procedures, have simplified the conversion of solid materials to liquids suitable for analysis.

Flame atomic absorption (FAA) and flame emission (FE) remain as workhorses in the analytical laboratory because of their simple operation,

The Analytical Chemistry of Silicones, Edited by A. Lee Smith.
ISBN 0-471-51624-4 © 1991 John Wiley & Sons, Inc.

modest cost, and reasonable limits of detection. Electrothermal atomization (ETA), sometimes known as graphite furnace AA, continues to attract the majority of research efforts in fundamental studies and new methodologies for sample analysis. Routine determinations at the parts per billion level are commonplace with modern furnace technologies. Atomic fluorescence (AF), a relative newcomer to the repertoire of sources, represents one of the methods having potential for ultrahigh sensitivity in optical spectroscopy. The inductively coupled plasma (ICP) discharge is a relatively recent development and is perhaps the most promising emission spectroscopic source today. Because of the high temperatures available and the inert atmosphere of the ICP discharge, some of the difficulties found in flame, arc, and spark techniques are not present in the ICP discharge. The ICP is simple, safe, and inexpensive to operate and maintain. Its high sensitivity and multielement capability make it a strong contender to replace some of the less attractive, currently used atomic emission sources. In this chapter we discuss the application of atomic spectrometric techniques to elemental analysis of silicones. Theoretical aspects of atomic spectroscopy are not covered, but a more practical approach dealing with sample preparation and instrument considerations is presented. The choice of atomization source is left to the individual analyst (5).

2. ANALYTICAL CONSIDERATIONS

2.1. Applicability

Even though all elements give rise to characteristic emission spectra, care must be taken when choosing a technique for the analysis of silicones. Silicones present challenges not common to analytical procedures for other materials. At trace levels, contamination is a major problem (Chapter 4, Section 1.2). At macro levels, selection of appropriate standards is a necessity because the Si response for all techniques discussed in this chapter depends on the molecular form of the Si. This fact may require that the chemical history of a sample be removed through chemical decomposition to form a soluble silicate. If standards having a structure closely related to that of the sample can be obtained, solvent extraction, or direct dilution of a sample in a solvent usually gives good results.

Consideration must also be given to the sample state (solid, liquid, or gas), sample volatility, potential hazards (flammability, corrosiveness, and toxicity), type of analysis (qualitative or quantitative), and the accuracy and precision required for the analysis.

2.2. Accuracy and Precision

Quantitation of elements at the parts per million and parts per billion level requires excellent laboratory technique and careful attention to possible sources of contamination. Silicon, for example, is a ubiquitous contaminant, and its measurement at low levels may require extraordinary precaution against dust and other sources of contamination. The accuracy and precision required affect the approach used for analysis as well as the time involved. In routine laboratory work, relative precision of 2–10% is usually attained. In favorable cases, precision may range from 1 to 2% relative for samples that have concentration levels at least 100 times greater than the detection limit.

The evaluation of accuracy requires comparing analysis results against those obtained from standard materials or from data obtained by independent techniques. Optimum precision and accuracy are achieved when the total composition of standards most nearly duplicates that of the samples.

2.3. Detectability

More than 60 elements can be determined by AAS or FES, many at or below 1 ppm. Emission spectroscopy has high sensitivity for as many as 72 elements, and many instruments have simultaneous multielement detection capability, which make emission spectroscopy techniques attractive for rapid analysis at element levels of 100 ppm or less. It is possible to detect picogram quantities of many elements using ETA–AAS, but the sample size is generally in the microliter rather than milliliter range. On a relative concentration basis, furnace detection limits are 10–100 times better than flame AA or ICP.

Comparisons of experimentally determined detection limits for various optical spectroscopy techniques are available in the literature (6–9). The values reported in these tables generally represent a best case analysis. They are good for comparison, but were reached under optimum instrument conditions using carefully prepared, single element solution standards. Extrapolation to the real world may be risky.

2.4. Interferences

Interferences from chemical, ionization, and physical effects should be considered when evaluating a technique for application to silicone analysis. Physical interferences such as sample viscosity, surface tension, vapor pressure, and temperature may alter the aspiration rate, and affect the nebulization and atomization efficiency. Maintaining solvent temperature and sample solution compositions are critical for successful analysis.

Matching the matrix of both sample and standard solutions is critical as well. The signal suppressing effect of organosiloxanes in AAS is well known. Differing substituent groups on the siloxane chain also affect the AAS signal.

The attenuation of the absorption signal by organosiloxane interference presents a serious problem in ETA–AAS. The addition of a matrix modifier may be necessary to overcome this problem (Section 4.2.3). This attenuation is most severe when direct dilution analysis of a siloxane in an organic solvent is attempted.

3. MAJOR METALS IN SILICONES

The emphasis in atomic spectroscopy has generally been on trace elemental analysis. A brief scan of the literature makes it evident that much research and development work in the field is devoted to further lowering limits of detection. However, atomic spectroscopy can also be used to good advantage in determining major constituents. It is used routinely to determine macro amounts of Si in organosilicon materials.

3.1. Determination of Silicon in Organosilicon Compounds by Direct Dilution and Atomic Absorption Spectroscopy

A rapid and useful technique used to assay the silicon content of silicone polymers is direct dilution of the polymer in a suitable solvent followed by AAS. The solvent of choice is 4-methylpentane-2-one or methyl isobutyl ketone (MIBK). Methylisobutyl ketone has been found to have good solvating properties for a wide variety of silicone materials, along with excellent burning characteristics in the nitrous oxide–acetylene flame used for this analysis.

Reagents

Acetylene, atomic absorption grade.
Nitrous oxide, atomic absorption grade.
Methyl isobutyl ketone, reagent grade.

Apparatus

Burner, 5-cm, single slot laminar type.
Lamp, silicon hollow cathode.

Standards

Dow Corning® 200 Fluid, 350 cS viscosity (37.85% Si).

Dow Corning® 704 Fluid, (17.4% Si).

Dow Corning® 705 Fluid, (15.4% Si).

Dow Corning® FS-1265 Fluid, (18.0% Si), used for fluorine containing silicone polymers.

Procedure

Samples are prepared by accurately weighing them into the solvent chosen for the analysis. The concentration of the sample should fall in the 5–200 μg of silicon per milliliter of solution. Standards of similar chemical structure are prepared in the same solvent. The number of substituent groups, such as phenyl, methyl, amine, and so on should approximate those of the samples. The standards should cover the expected concentration range.

Adjust the spectrophotometer to peak on the silicon 2516-Å line, with the lamp current set to optimize the S/N ratio. When properly adjusted, the flame has a red "feather" approximately $1\frac{1}{2}$ in. in height as MIBK is aspirated. (The flow of acetylene must be greatly reduced from normal rates to achieve this condition.) The solvent must be continually aspirated while the flame is operating in this mode. Before flame shutdown, the acetylene flow must be returned to normal rates if burner flashback is to be avoided. All of these adjustments must be approached with caution. Adjust the burner height and alignment for a maximum signal while aspirating a standard solution. Use nitrous oxide at a pressure recommended by the instrument manufacturer to aspirate samples. Analyze each sample and standard solution and record the absorbance trace for 10–20 s after it has stabilized. Scrape the burner slot with a spatula frequently to prevent carbon buildup. Prepare an analytical curve by plotting micrograms of silicon per milliliter versus absorbance. Determine the concentration of silicon in the sample solutions and calculate the concentration of silicon in the original sample. Prepare duplicate samples and average the results to obtain the concentration of silicon in the samples. The accuracy is dependent on the closeness of the match between the sample structure and the standard that is used to prepare the calibration curve.

3.2. Determination of Silicon by Matrix Decomposition–Atomic Absorption

In cases where the sample is not soluble in an appropriate solvent, or where standards approximating the sample composition are not available, the sample can be prepared for analysis by the Parr bomb sodium peroxide fusion technique. The peroxide fusion method for the preparation of organo-silicone materials for silicon analysis was described by McHard et al. in 1948 (10). Subsequent modifications have improved the reliability of the procedure. This procedure converts organic and most inorganic materials into water-soluble silicate and carbonate salts. The fusionate solutions were acidified before use with hydrochloric acid prior to 1978, when it was discovered that the AA signal quality was improved if organic acids were substituted in the acidification step (11). Acidification with glacial acetic acid increased the S/N ratio during measurement. A signal intensity increase for silicon of 16% was also observed. The enhancement is not specific for silicon but is observed for most metals.

The sodium peroxide fusion sample preparation method is not confined to the analysis for silicon but can be extended to other elements including P, B, As, Al, Ca, Mg, Ti, Y, Cu, Fe, Zn, Ba, S, and Cr. Samples in the form of solid polymers, ceramic fibers, or fluids are decomposed by the Parr bomb procedure outlined in Chapter 8, Section 2.1.2.

4.   TRACE METALS IN SILICONES

Numerous descriptions of spectrographic methods for trace analysis are found in the literature. In one 2-year period, more than 2000 articles were written on AA, AF, and FE spectrometry alone (12). Although relatively few of these dealt with silicones, atomic spectroscopic techniques are extremely valuable for the analysis of trace elements in organosilicone materials.

The importance of elements present in trace quantities in silicones is well known. Residual polymerization catalyst in a PDMS fluid may have detrimental effects on its thermal stability. Conversely, the presence of certain other metallic compounds may increase thermal stability. Platinum or rhodium are often used as catalysts in cure systems. Trace analysis is performed to determine if the catalyst is present in sufficient quantities to give the desired reaction rates. Stringent requirements are placed on the metal content of silicone resins and fluids used in the electronics industry. The semiconductor industry requires trace metals analysis of silicone materials used for encapsulation of devices. There are also mandatory requirements on the levels of trace element contaminants in food and cosmetics additives.

4.1.   Chlorosilanes

The requirements for chlorosilanes of extremely high purity established by the semiconductor industry have stimulated much research on trace metals analysis. Techniques developed for these analyses are applicable to some of the chlorosilane precursors of silicone polymers. Nonvolatile metallic contaminants in chlorosilanes or other volatile monomers are readily isolated by distilling the organosilicon compound away from the residue. Krylov et al. (13) used a graphite container to distill Me_2SiCl_2 under reduced pressure. Detection limits of around $10^{-10}-10^{-7}\%$ were claimed. Martynov et al. (14) has described a simple method for determining 16 trace elements in silicon tetrachloride by evaporating the sample in the presence of carbon and analyzing the residue by emission spectroscopy. Gooch (15) has determined trace B and Al in trichlorosilane by adsorbing the metallic impurities onto a column of amorphous silica. The silica was dissolved in an aqueous NaOH solution and analyzed by ICP–atomic emission spectroscopy (AES). Lin et al. (16) has also used ICP–AES to determine trace boron in trichlorosilane. Pchelintseva (17) extracted boron from silicon tetrachloride with triphenylchloromethane, and Vecsernyes (18) extended the procedure to include nine other elements. Kawasaki and Higo (19) further lowered the detection limits for boron in trichlorosilane and silicon tetrachloride to 0.06 ppb by using sodium hydroxide to eliminate excess carbinol resulting from the triphenylchloromethane extraction.

4.2. Siloxanes

4.2.1. Sample Preparation Methods

Most atomic spectroscopy techniques can be adapted to determining trace metal impurities in silicone polymers. Generally, some sample preparation is involved, either to remove the silicone matrix or dilute it in a suitable solvent. If solvent dilution is used, detection limits are somewhat higher than are obtained from the matrix destruction method. On the other hand, opportunities for contamination are significantly reduced. An ICP emission method has been described (20) for the determination of metallic impurities in the $10^{-6}-10^{-8}\%$ range in high purity siloxanes. Samples are decomposed with HNO_3–HF vapors in an autoclave to give a dry residue, which is taken up in HCl–HNO_3 and analyzed on a multichannel spectrometer.

Traditionally, trace element analysis by atomic spectroscopy has often required a tedious wet ashing procedure to eliminate the sample matrix and solubilize the sample (Chapter 8, Section 2.1.2). Alternatives are available, however.

Microwave digestion offers the analyst a simple and rapid method for eliminating the effects of structure. Time required for ashing and subsequent analysis may be reduced by a factor of 4–5 when compared to more traditional wet ashing methods. Closed Teflon® vessels sharply reduce the chances of airborne contamination that may be experienced during long periods of atmospheric exposure in open vessels.

Microwave Digestion Procedure

Either liquid or solid samples may be digested using this procedure. Solid samples should be ground or pulverized to increase surface area and aid in the digestion.

Reagents

Hydrofluoric acid (48%).
Nitric acid (70%).
Hydrochloric acid (5%).

Procedure

Add a 1.0-g sample to the sample vessel and to this add 5.0 mL each of nitric and hydrofluoric acids. After any effervescence and foaming subsides, place a safety valve and cap on the vessel and tighten the cap. Cycle at 100% power for 1 min and 50% power for 5 min. Cool to room temperature, uncap, add 2.0 mL each of nitric and hydrofluoric acids, and reseal. Repeat microwaving cycle. Cool to room temperature, uncap, and replace Teflon® seals. Dry in the microwave at 100% power for 1 min, 50% for 10 min. Repeat cycle. Add 10.0 mL of hydrochloric acid and recap with seals in place. Cycle at 100% power for 1 min and 30% for 3 min. Open the vessels and wash the contents into appropriate containers.

Caution. Manual venting of closed vessels should only be performed when the vessel contents are at or below room temperature to avoid the potential for chemical burns. When venting vessels, it is recommended that hand, eye, and body protection be worn.

4.2.2. Direct Determination by ICP–AES

Poly(dimethylsiloxane) samples of 100–12500-cS viscosity may be analyzed for trace elements with concentrations at levels of 1–100 ppm by dilution in xylene, followed by elemental analysis by ICP–AES. The accuracy and relative standard deviation for elements are comparable to solvent dilution AAS techniques. Additionally, the probability of elemental loss during wet ashing procedures is eliminated and the possibility of contamination during long atmospheric exposure is sharply reduced. Standards consisting of a solution of the element of interest in a 10% w/v solution of PDMS in xylene should be prepared that cover the concentration range. A 10% w/v solution of the sample in xylene is directly aspirated into the ICP and element levels are determined. A caveat should be mentioned: molecular emission bands are present in the spectrum from organosilicon compounds in the sample. These bands increase the noise in the spectrum and may interfere with element emission lines. Diluting the sample to minimize these interferences reduces sensitivity, and a compromise between sensitivity and background noise reduction must be reached. Weaker lines in a spectrally clean region can also be used. Complex mathematical background corrections or instrument optimization procedures (21) are also available to improve the S/N ratio and to moderate interferences when this situation is encountered.

4.2.3. Electrothermal Atomization AAS

One technique that shows promise for direct determination of trace impurities at the parts per billion level is ETA–AAS. Improvements in platform technology, high quality graphite materials, fast photometric detectors, and Zeeman background correction have greatly alleviated problems commonly associated with this technique. A matrix modifier that is soluble in organic solvents is needed to control the attenuation of absorbances due to the silicone matrix (15). Matrix modifiers (22) are used to retain the analyte at higher temperatures in the furnace, and allow volatilization of interfering matrix compounds. For silicones, some alkali metal salts meet this requirement by catalyzing decomposition of the siloxanes while retaining other elements. Trace levels of platinum or rhodium catalyst can be determined in a silicone by dilution in MIBK and direct injection into the ETA–AAS furnace.

Although a few workers have apparently used ETA–AAS successfully for trace Si analysis, it is often difficult to obtain credible results. First, the extreme sensitivity of the method means that dust particles and siloxane vapors (the furnace tubing is often made of a silicone elastomer) can cause major interferences. Second, when silicone polymers are heated in the graphite furnace, they tend to form cyclic and other volatile oligomers that are swept away by the gas stream before being sampled. Thus, graphite furnace excitation methods must overcome these difficulties and demonstrate consistent recoveries on known standards at an acceptable level of precision.

4.2.4. Flame Photometric Determination of Sodium and Potassium

Analysis for trace levels of sodium and potassium is important in both process and product control in the silicone industry. The method described below is unique in that very small samples can be used, thus eliminating the more laborious task of ashing large samples. Both elements can be simultaneously determined in the range 0.1–1000 ppm with precision and accuracy of +/− 10% in the range 0.1–1.0 ppm and +/− 2% at higher levels.

Reagents

Hydrofluoric acid, ultrapure.
Lithium solution, 1500 meq/L.
Nitric acid, ultrapure.
Sulfuric acid, ultrapure.
Propane gas.

Apparatus

Crucible, platinum, 5 mL.
Flame photometer.

Procedure

Weigh a 50–60-mg sample into a platinum crucible. Prepare duplicate
reagent blanks for each set of samples. Add 1 drop of H_2SO_4 to each sample.
Heat on a hot plate to incipient SO_3 fumes. Remove from the hot plate, cool
to room temperature, and add 1 drop of HNO_3 and 1 drop of HF. Place on
the hot plate and continue heating to SO_3 fumes. Repeat the addition of
HNO_3 and HF and heating until SiO_2 is no longer present. Transfer the
crucible to a muffle furnace at 600 °C and burn off the remaining carbonaceous matter. Inspect the crucible for SiO_2. If necessary, add 1 drop of
H_2SO_4 and 1 drop of HF and heat. Heat again in the muffle furnace for
5–10 min. Remove and cool to room temperature. Add an appropriate
amount of lithium solution. If anticipated values for sodium and potassium
are greater than 50 ppm, dilute the sample in a volumetric flask to bring the
concentration into the range 10–50 ppm. Optimize the instrument operating
conditions to obtain the maximum response. Place a sheet of plastic film over
each crucible, hold between thumb and forefinger and shake vigorously.
Aspirate each sample and record the digital readout. Prepare analytical
curves from standard solutions, plotting sodium and potassium concentration versus digital readout. Read the concentration directly from the analytical curves and make the appropriate correction for dilution.

5.   TRACE SILICONES IN OTHER MATERIALS

The increasing use of silicones in products and processes as additives or
process aids has necessitated the development of analysis techniques capable
of detecting these materials at trace levels in a variety of matrices. Silicones
are added in small quantities to products for a desired effect, for example,

antifoams in the food processing industry, or release additives to molded plastic parts. There is also a need to detect trace levels of silicones in biological and environmental samples. Because of the complexity and diversity of matrices in which silicones are found, it is sometimes impossible to analyze specifically for the silicone. An elemental analysis for silicon may be the only means to determine the silicone content of a sample. Atomic spectroscopy techniques, of course, are unable to distinguish between inorganic silicon present as silicates and the organic silicone additive. Considerable effort has gone into developing sample preparation techniques that extract the silicone from the sample matrix into an organic solvent, leaving behind any inorganic silicates. Procedures for extraction of silicones from bioenvironmental matrices have been developed for waste water, sediments, foods and juices, and animal tissues and are discussed in detail in Chapter 4. Once extracted, the silicone may be quantified by using AA or ICP–AES as discussed earlier. In some cases, extraction is difficult or impossible. Then direct determination of silicon is necessary. An example of such a procedure follows.

5.1. Poly(dimethylsiloxane) in Vegetable Cooking Oil

Poly(dimethylsiloxane) is recognized as an effective defoamer for cooking oils. Various researchers have stated that PDMS is effective at concentrations in the range 1–3 ppm. The silicone not only acts as a defoamer, but oil oxidation is reduced and the smoke point temperature is increased.

Atomic absorption spectroscopy is the most commonly used method for determining trace concentrations of PDMS in cooking oils. However, its limited sensitivity makes it only marginally useful for detecting silicone concentrations in the range 1–5 ppm. Inductively coupled plasma spectrometry is several times more sensitive to silicon than AAS, and for this reason is the preferred analytical method (15, 23). The sensitivity of ICP is on the order of 1 ppm PDMS in cooking oil.

Reagents

Vegetable oil (with no defoamer added, similar to oil to be analyzed).
Xylene (of known metals content).
Argon gas (ICP grade).
PDMS (viscosity similar to defoamer).

Procedure

Standards Preparation. Vegetable oil–PDMS standards may be prepared by saturating the oil with an excess of PDMS. Add 5 g of PDMS and 15 g of oil to a plastic sample bottle and agitate for 2 h. Centrifuge this mixture to separate the PDMS phase from the cooking oil. Determine the concentration of silicone in the concentrated PDMS–oil solution by diluting the PDMS–oil concentrate in MIBK and analyzing by AAS. A calibration curve of the PDMS defoamer in MIBK will provide the concentration of PDMS in the concentrate. Portions of the PDMS–oil concentrate are diluted with vegetable oil to prepare standards covering the 1–10 ppm PDMS in cooking oil.

Analysis. To 25 mL of xylene add 10 g of the cooking oil samples or silicone standards. It is necessary to keep the oil–solvent ratio of the calibrating solution and unknowns constant to control solution viscosity and matrix effects from the oil. Aspirate the standards and sample solutions directly into the ICP monitoring the 288.1-nm silicon line. Prepare a calibration curve using the standards and a blank. Calculate the percent silicon concentration from the calibration curve and convert to PDMS concentration in the sample.

REFERENCES

1. B. Welz, *Atomic Absorption Spectrometry*, 2nd ed., Verlagsgellschaft mbH, Weinheim, FRG, 1985.

2. A. Montaser and D. W. Golightly, Eds., *Inductively Coupled Plasmas in Analytical Spectrometry*, VCH Publishers, New York, 1987.

3. P. W. J. M. Boumans, Ed., *Inductively Coupled Plasma Emission Spectroscopy. Part 1: Methodology, Instrumentation and Performance*, Wiley, New York, 1987.

4. P. W. J. M. Boumans, Ed., *Inductively Coupled Plasma Emission Spectroscopy. Part 2: Applications and Fundamentals*, Wiley, New York, 1987.

5. W. Slavin, "Flames, furnaces, plasmas. How do we choose?" *Anal. Chem.*, **58**, 589A (1986).

6. V. A. Fassel and R. N. Kniseley, "Inductively coupled plasma. Optical emission spectroscopy," *Anal. Chem.*, **46**, 1110A (1974).

7. V. Svoboda and I. Kleinmann, "High-current impulse argon arc method for the spectrographic analysis of microsamples," *Anal. Chem.*, **40**, 1534 (1968).

8. J. P. Faris, *Proc. 6th Conf. Anal. Chem., Nucl. Reactor Tech.*, TID-76655, Gatlingburg, TN, 1967.

9. G. D. Christian and F. J. Feldman, "Comparison study of detection limits using flame-emission spectroscopy with nitrous oxide–acetylene flame and atomic-absorption spectroscopy," *Appl. Spectrosc.*, **25**, 660 (1971).

10. J. A. McHard, P. C. Servais, and H. A. Clark, "Determination of silicon in organosilicon compounds," *Anal. Chem.*, **20**, 325 (1948).

11. E. G. Gooch and P. R. Roupe, "Use of organic acids for improved signal stability and sensitivity for the determination of metals in high salt solutions by atomic absorption spectroscopy," *Anal. Chem.*, **51**, 2410 (1979).

12. J. A. Holcombe and D. A. Bass, "Atomic absorption, atomic fluorescence, and flame emission spectrometry," *Anal. Chem.*, **60**, 226R (1988).

13. V. A. Krylov, A. V. Loginov, V. N. Shishov, G. L. Murskii, I. V. Filimonov, I. Yu. Durinov, and Yu. V. Gorshkov, "Analysis of high-purity dimethyldichlorosilane," *Vysokochist. Veshchestva, 1989*, 217, through *Chem. Abstr.*, **112**: 15799x (1989).

14. Yu. M. Martynov, I. I. Kornblit, N. P. Smirnova, and R. V. Dzhagatspanya, "Spectrographic determination of impurities in silicon tetrachloride and silicon dioxide," *Zavod. Lab.*, **27**, 839 (1961).

15. E. G. Gooch, private communication, Dow Corning Corp., Midland, MI.

16. C. Y. Lin, F. C. Chang, Y. C. Yeh, and S. C. Wu, "Determination of trace boron in trichlorosilane," *Hua Hsueh*, **40**, 133 (1982).

17. A. F. Pchelintseva, N. A. Rakov, and L. P. Slyusareva, "Spectral determination of traces of boron in high purity silicon tetrachloride," *Zavod. Lab.*, **28**, 677 (1962).

18. L. Vecsernyes, "Spectrochemical determination of trace elements in silicon tetrachloride and in trichlorosilane," *Mag. Kem. Foly.*, **72**, 377 (1966).

19. K. Kawasaki and M. Higo, "Spectrographic determination of ultramicroamounts of boron in high-purity silicon tetrachloride and trichlorosilane," *Anal. Chim. Acta*, **33**, 497 (1965).

20. V. Z. Krasil'scchik, M. F. Tikhonov, V. S. Tsar'kova, E. I. Voropaev, S. R. Nanush'yau, E. I. Alekseeva, and G. N. Konareva, "Spectrochemical determination of trace impurities in extrapure organosilicon compounds," *Vysokochist. Veshchestva* **1989**, 211, through *Chem. Abstr.*, **112**: 36994x (1989).

21. R. J. Thomas and J. B. Collins, "The benefits of a multiparameter optimization algorithm for the analysis of difficult samples using inductively coupled plasma-optical emission spectrometry," *Spectroscopy*, **5**, 38 (1990).

22. W. Slaven, *Graphite Furnace AAS. A Source Book.* Perkin-Elmer Corp., Norwalk, CT, 1984.

23. A. Bocca, A. Massucotelli, and S. Baragli, "Spectrochemical determination of silicon trace amounts in vegetable oil," *Riv. Ital. Sostanze Grasse*, **61**, 559 (1984).

CHAPTER

15

X-RAY METHODS

D. R. PETERSEN and M. J. OWEN

Dow Corning Corporation
Midland, Michigan

and

R. D. PARKER

Dow Corning Corporation
Glamorgan, South Wales
United Kingdom

INTRODUCTION

X-radiation lies in that part of the electromagnetic spectrum that falls between, and partly overlaps, UV and γ radiation. It occupies the spectral region between ~ 0.01 and 100 keV (0.1–1200 Å). Several physical processes lead to the production of X-rays: radioactive decay, the transition of energetically excited atoms from a higher to a lower state, and the deceleration of rapidly moving particles. Data that describe aspects of the X-ray spectrum and its uses are gathered in the several volumes of the *International Tables* (1).

The most common method of generating X-rays in the laboratory involves inducing electrons ejected from a heated-filament source cathode to move in vacuum across an electric field to be stopped by a cold-metal target anode. With a potential difference V_s, the electron acquires kinetic energy eV_s; if it gives up all this energy at once when it strikes the target, it produces a photon with energy $hv_s = hc/\lambda_s$. Thus $eV_s = hc/\lambda_s$ or $V_s = hc/e\lambda_s = 12.3984/\lambda_s$, with V_s in units of kiloelectron volts (keV) and λ_s in units of Ångstroms Å (2). The term V_s defines the high-energy end of a *continuous spectrum* of emitted radiation, since the decelerating electron usually is involved in not one but many collisions before it comes to rest. The wavelength λ_s associated with V_s is called the *short wavelength limit*; the generated spectrum continues on

The Analytical Chemistry of Silicones, Edited by A. Lee Smith.
ISBN 0-471-51624-4 © 1991 John Wiley & Sons, Inc.

through the UV into the vis and IR regions. The distribution of intensity of scattered radiation in the continuous spectrum is a function of the potential difference V_s, the electron current, and the target material, as well as such geometrical parameters as the configuration of the filament and the angles of observation with respect to the surface of the target and to the incident beam of electrons (3).

X-radiation passes readily through thin layers of matter, suffering only partial absorption. The fraction absorbed by the layer is a smooth function of the incident X-ray photon energy, except that at specific energy thresholds characteristic of the atoms of a given species, much more of the incident X-ray energy is absorbed on the high side of the threshold than on the low side. This phenomenon is called an *absorption edge*, and arises from the formation of inner-shell ions; in the most common instance, the incoming photon dislodges and ejects a $1s$ orbital electron from the innermost (K) shell. The excited atom that results then spontaneously drops to a lower energy state by moving an electron from one of its outer shells into the vacancy in the K shell, at the same time emitting a quantum of energy. The respective energy levels of the before-and-after states are such that the energy difference appears as a precisely defined photon in the X-ray region. For example, when an electron from the L_{III} level of an excited Cu atom drops into the K-level vacancy (the $K\alpha_1$ transition), radiation of energy 8.04782 keV (1.54058 Å) is emitted. Should the L_{II} level be involved ($K\alpha_2$), the radiation is at 8.02780 keV (1.54441 Å). The transition probabilities are such that $K\alpha_1$ has twice the intensity of $K\alpha_2$. The collective radiation produced by these two events gives rise to a close energy doublet with mean location 8.04121 keV (1.54186 Å). This doublet is superimposed on the continuous X-ray spectrum from a Cu target as a very sharp intense spike. Other electronic transitions, such as from the M shell to the K shell, produce other spikes. Collectively, these are known as the *characteristic spectrum* of an atomic species and are dependent primarily on the species and to a lesser extent on the exciting energy. Of course, a minimum energy threshold—the absorption edge—must be reached before the characteristic spectrum appears (8.980 keV for Cu Kα). Characteristic lines are used to generate X-ray diffraction patterns, as shown in Section 2.

Because of the precise location of the absorption edges and the peaks in the characteristic spectrum for a given atomic species, both the absorption spectrum and the emission spectrum can be used as bases for elemental analyses. This topic is further discussed in Section 3.

When an excited atom drops to a lower energy state, the emitted X-ray photon may interact with an electron in the same atom and cause ejection. This ejected electron has energy characteristic of the atom; it is known as an Auger electron. Auger electron spectroscopy (AES) is described in Section 4.

Energetic electrons interacting with matter can produce X-rays; in similar fashion, X-rays interacting with matter can produce photoelectrons, which reflect the energy levels present in the atoms from which they are emitted. Examination of these characteristic energies forms the basis of X-ray photoelectron spectroscopy (XPS), also described in Section 4.

2. X-RAY DIFFRACTION

2.1. Nature of Diffraction

Diffraction is a physical phenomenon associated with the wave nature of radiation. An electron in the path of an X-ray beam oscillates in response to the sinusoidally varying field it sees, and reemits radiation of the same frequency as that of the driving radiation. Another way to say this is that a small fraction of the energy of the incident beam is scattered (in random directions) by the interaction with the electron. The set of electrons bound around an atom nucleus collectively dissipates energy in this manner; the more electrons, the greater the scattering. This effect is not the same as the disruptive interaction mentioned in the section above in which an electron is ejected from the atom. Should the electrons in a pair of bonded atoms be so driven to oscillate and emit radiation, in certain directions the intensity of the emitted radiation is reduced because of destructive interference, while in other directions it is reinforced (constructive interference).

Atoms in a crystalline solid are arranged in a fixed array that exhibits manifold translational periodicity in space. Because the spacing of the atoms is of the same magnitude as X-ray wavelengths, this array acts as a three-dimensional diffraction grating for X-radiation. The directions in which the scattered radiation is maximized are a function of the incident wavelength and of the various interatomic distances in the solid, and the symmetrical arrangement of atoms within the array. Equations developed by von Laue express this relation in terms of the wavelength and of distances between identical atoms in three noncolinear lattice *rows* in the structure. The three corresponding vectors identify the smallest entity of the crystal lattice, called the *unit cell*. Bragg simplified the von Laue equations by using the distance d between equivalent lattice *planes*, and by introducing the concept of "reflection" from these planes, to give

$$n\lambda = 2d \sin \theta \qquad (15.1)$$

where θ is the Bragg angle at which the diffraction maximum associated with d occurs (4, 5).

The entities that make up the structure are related not only by the translational elements, but also by symmetry elements such as mirror planes or rotation axes. The collection of symmetry elements for a given structure is called a *space group* (1, 4, 5); there are 230 unique space groups. The unit cell is described by the lengths of its three edges a, b, c and by its three interedge angles α, β, γ.

The three-dimensional crystal lattice and its associated reciprocal lattice are mathematical constructions unique to the material represented.

2.2. Experimental Methods

X-radiation is generated in the laboratory by applying 30–50 kV between a heated-filament cathode and a water-cooled grounded metal target anode separated in an evacuated tube. The target may be a stationary block or a spinning cylinder. A selected fraction of the emitted X-rays is allowed to escape the shielded tube enclosure through thin beryllium foil windows. Depending on the viewing aspect relative to the position of the cylindrical filament coil, the X-ray source on the surface of the anode may look like a spot or a line. The former is collimated through a pair of pinholes and the fine parallel beam that results is used for camera work; the latter is allowed to pass through narrow slits, and the wide divergent beam is used for diffractometer work (5).

In diffraction experiments it is important to have an essentially monochromatic beam of X-radiation. The selection of a narrow wavelength range from the composite continuous–characteristic spectrum of an operating X-ray tube can be accomplished in several ways (absorption filters, diffractive monochromators, and detected-energy discriminators) (6). An alternate source of X-radiation, of extremely high intensity, is the synchrotron, which has found increasing application in diffraction experiments (7).

Diffracted X-radiation is detected either by capture on photographic film or by electronic counting. The X-ray diffraction pattern is scattered in a full solid angle in space around the specimen material. Cameras may be used to capture appropriate parts of this pattern. Since a spherical camera is impractical, one deals commonly with either a flat film or a cylindrical film arrangement (8).

The diffractometer uses an electronic counting device to measure the intensity of the scattered radiation. The consistency, efficiency, sensitivity, linearity, and response over a wide range of counting rates of these devices are important considerations. Most diffractometers now use dedicated computers to make angular settings and to gather and analyze intensity data. Automated diffractometers for both single crystal and for powder work are commercially available.

2.3. Application to Single Crystals

Crystalline materials are arrangements of molecules, atoms, or ions ordered in three dimensions. When a single crystal is used as a diffraction grating for monochromatic X-rays, the resulting diffraction pattern can be interpreted to give information about the molecular arrangement in the crystal.

2.3.1. Sample Preparation

A small (0.1 mm) single crystal free of such flaws as twinning or inclusions is mounted on the end of a short glass fiber or inside a glass capillary so that it may be completely bathed in a beam of parallel monochromatic X-radiation. It is held in a goniometer head with translation slides and rocking arcs, which allow centering and adjustment to make one of the crystal axes coincident with the rotation axis of the head.

2.3.2. Determination of Crystal Parameters

The crystal is rotated inside a cylindrical camera to capture a segment of the diffraction pattern. From Bragg's law and simple geometry one can find the repeating distance (length of the cell edge) along the crystal axis. Different camera orientations together with the linear dimensions and angles measured from the film, allow unit cell parameters to be calculated (8, 9). In most cases, the space group can be determined from the symmetry and the extinctions observed; the number of molecules in the cell can be found from the cell volume and the measured density of the crystal.

2.3.3. Identification

The unit cell symmetry and dimensions are unique, and so can serve as the basis for an identification scheme. Such has been developed, and tabulated unit cell data for many thousands of materials are available for manual searching (10) or for computer scanning (11–13). A significant number of organosilicon compounds is included in these collections.

2.3.4. Crystal Structure Determination

Rather than to use camera data, it is most common now to gather the full diffraction pattern from a single crystal with an automatic (computer-driven) diffractometer that positions the crystal and counter successively to the proper places to collect the full pattern, typically several thousand reflections. Each is associated with a point in the reciprocal lattice. Symmetry is used to minimize the total number of reflections gathered. The three-dimensional intensity distribution in the reciprocal lattice may be treated in such a way that its Fourier transformation gives the electron distribution in the crystal lattice (14–16). Since the electrons are bound at atoms, this process locates the atoms within the unit cell in terms of x, y, z coordinates (fractions of the respective edges). These data in turn may be used to calculate the internal molecular configuration and the external molecular architecture within the crystal. Inter- and intramolecular distances and angles are interesting for individual molecules, but are most important for broader studies that involve many related molecules (17–19). Both manual searching (20) and computer scanning (21) of tabulated crystal structure data are possible. The Cambridge Structural Data Base now holds data on 74,000 organic crystals, of which a thousand-odd are of organosilicon compounds (22, 23). Crystal structure determination of organosilicon compounds has been applied widely as an analytical procedure (24–27).

2.4. Application to Polycrystalline Aggregates

Perhaps the most common form in which a solid sample is presented to an analyst is as a polycrystalline aggregate, that is, a powder. A camera or a diffractometer may be used to gather a diffraction pattern from such a material. The powder pattern can then be interpreted to give d spacings with associated relative intensities, a set characteristic of the material that makes up the sample.

2.4.1. Sample Preparation

In one sense, sample preparation for powder diffraction is very simple, since any crystalline material, regardless of form, placed in an X-ray beam can be made to give a diffraction pattern. The resulting pattern, however, may not be useful for analytical purposes. The preparation of a sample for powder diffraction analysis is based on the need to minimize the preferred orientation of components of the sample. This is done through reduction of particle size and proper loading to avoid the introduction of orientation. Both cameras and diffractometers commonly incorporate mechanisms to move the sample appropriately during exposure—sliding, rocking, or spinning—to further reduce the effect of preferential orientation.

Many crystalline organosilicon compounds are inherently difficult to grind because of their relative softness, and hard to load randomly because of their morphology, often needlelike or platy. Also, most are electrical insulators, which may introduce the complicating effects of static charge. Some of these problems may be overcome by grinding at low temperatures, or by

incorporating a noncrystalline dilutant such as acacia gum or finely powdered silica glass. Specific procedures to prepare samples have been described
in detail (28–30).

An internal standard may be mixed with the sample to allow the determination of accurate d-spacing values for the sample (31). The standard is
chosen on the basis of its known parameters, lack of peak interference with
the sample, and good peak spread over the recorded angular range. Com-
ommonly used is NIST Standard Reference Material 640b, Si powder, with
$a = 5.430825\,\text{Å}$. For spacings at low angles, important for many organosilicon crystals, SRM 675, fluorophlogopite, is used, with $d(001) = 9.98104\,\text{Å}$.

2.4.2. Identification

In practical experience, X-ray powder diffraction is not usually the analytical
method of choice for the *identification* of unknown organosilicon solids,
because reference data (previously recorded diffraction patterns of substances
of known composition and purity) may not be available. The description of a
new crystalline organic compound in the literature rarely includes a powder
pattern, in part because other methods—NMR, MS, and IR—have been used
to characterize the compound, and in part because molecular crystals
typically have complex structures with low symmetry and large nonortho-
gonal unit cells, leading to patterns with many overlapping diffraction
maxima at low angles 2θ, in turn making interpretation difficult and
ambiguous.

A large collection of X-ray powder reference data, now including repres-
entations of about 51,000 phases in 39 sets, does exist in convenient form for
comparison matching; it is called the Powder Data File, or PDF (32). About
2000 patterns are added annually. The data for a given phase are tabulated as
d, I pairs, and the tables are available alternately on individual cards,
gathered into books, printed on microfiche, held on magnetic tape, or on CD-
ROM. Printed indexes arranged in alphabetic order by compound name and
in search order by principal line positions are part of the file. Unfortunately,
in this wealth of data only ~ 100 reference patterns of organosilicon
compounds are included. Where a reference pattern does exist, however,
matching it with the pattern of the unknown phase does provide positive
identification.

Another approach is to compute a *calculated* powder pattern from atomic
position data taken from a single-crystal structure determination. A calculated pattern, moreover, may be more reliable than an existing reference
pattern, particularly if the latter is based on an early photographic record.
The published single-crystal structures give a rich source for "new" powder
patterns. Upwards of a thousand complete structures for organosilicon

compounds are waiting in the literature as a base for calculating analogous powder patterns (21, 33).

Analysts concerned with using X-ray powder diffraction in the identification of organosilicon compounds will find it necessary to build their own reference files of such materials to augment the PDF, either by gathering experimental powder data from known compounds, or by calculating from single-crystal data, or both.

The procedure for identification based on matching the pattern of the unknown material with that of a known phase appears to be simple and straightforward, a fingerprint comparison. So it is, should the unknown be made up of a single pure component. The reality of the situation is that the unknown material is usually a composite of several phases, and the interpretation of a diffraction pattern with contributions from multiple phases is anything but trivial. Some preliminary information about the unknown, such as a qualitative indication of elements present, is virtually a requirement. Strategies for resolving the puzzle of multiple phases have been proposed (34) and even programmed for computer solution (35), but these are found to be not universally applicable. Interpretation of multiphase patterns remains the principal difficulty in X-ray powder diffraction analysis.

2.4.3. Quantitative Estimation

Most analytical procedures cannot be used to identify specific phases present in a mixture or to estimate the amount of each constituent. Since the peaks in a diffraction pattern are characteristic of the *phases* and not of the *elements* present, powder diffraction provides a proper approach to the quantitative analysis of solid mixtures. Also, since the intensity of the contribution of each component to the total pattern is proportional to the concentration of that component, except for an adjustment due to absorption, analysis is relatively staightforward (28).

2.4.4. Other Characterization Procedures

Other structure-related properties of a material may be exhibited in its powder diffraction pattern. Most commonly, these are incorporated in the shape of the diffraction maxima, and profile analysis is used to provide information on these properties. A simple example is the effect of the size of the diffracting entities, or crystallites. The peaks in a diffraction pattern become noticeably narrower as the dimensions of the crystallites increase with respect to the wavelength of the diffracted radiation, up to perhaps 1000-fold, after which no further sharpening is apparent. The departure of the breadth of a peak from this minimum can be used to estimate the linear dimension of the diffracting region in the direction within the crystallite that is associated with the peak (averaged over all the irradiated crystallites, of course) (28).

2.4.5. *Crystal Structure Determination*

Full structure determinations are possible from X-ray powder data alone. An example, written as a student exercise, is the elucidation of the structure of K_2SiF_6 from its powder diffraction pattern (36). A more general approach is through a procedure known as the Rietvelt method (37), which uses a full-pattern fitting procedure. While valid, crystal structure determination from powder data is used much less frequently than from single-crystal data because of the ambiguous nature of the powder pattern resulting from overlapping maxima.

2.5. Application to Partly Crystalline Materials (Polymers)

Polymeric substances generally are found to be only partly crystalline in nature; the molecules that make up the material are in part ordered into crystalline regions and in part not. The aligned regions tend to be small, with dimensions suggesting that any given molecule may participate in a sequence of alternate ordered and disordered regions. The diffraction patterns that result from polymer samples usually show composite features characteristic of both ordered and disordered phases, that is, both sharp and diffuse maxima. While it is not strictly correct to consider a polymer to be an intimate mixture of a crystalline phase and a noncrystalline phase, its diffraction pattern may be so interpreted. This allows the analyst to estimate from the pattern the extent of crystallization of the polymer. Procedures for doing this have been described (38).

Identification of polymers based on their diffraction patterns is less satisfactory than for fully crystalline materials, because the pattern changes with extent of crystallization and because the number of peaks contributed by the crystalline and the noncrystalline components is small. In the case of the latter, one or two diffuse maxima are typical, and these tend to fall in about the same regions for most polymers. Very few polymers are included in the PDF (only *three* organosilicon polymers), and the analyst who wants to use diffractometry for polymer identification will find that building one's own file of reference patterns is necessary (39, 40).

2.6. Application to Noncrystalline Materials

In noncrystalline materials such as completely disordered polymers, glasses, oils or solutions, enough short-range periodicity may occur to provide detectable diffraction effects. The sharply defined peaks in the diffraction pattern, which are characteristic of long-range periodicity in crystals, are replaced by diffuse undulations. These can be interpreted through a mathematical treatment called radial-distribution analysis to give information about interatomic distances, which in turn allows a description of the short-range structure (28, 38).

2.7. Related and Complementary Procedures

A collimated electron beam can be used to give a diffraction pattern of a crystalline solid in much the same way as an X-ray beam. Some differences: the electron beam requires a vacuum path; the electron beam is much less penetrating than the X-ray beam, so that the sample must be much thinner; the interaction of the charged electrons with the sample changes the relative intensities in an complex manner, making interpretation of the pattern difficult. An electron diffraction experiment is usually carried out within an electron microscope, and under normal operating conditions of 100–200 kV, the electrons have an effective wavelength of a few hundredths of an angstrom. An example of the application of single-crystal electron diffraction in conjunction with X-ray powder diffraction to establish unit-cell data is given for poly(dipropylsiloxane) (40).

Electron beam diffraction also has been applied to the determination of the structure of molecules in the gas phase. It is most appropriately carried out with relatively small molecules. Many simple organosilicon compounds have been studied. Helpful tabulations of completed structures established by gas-phase electron diffraction are available (41), and other collections also include such data (20, 21, 33).

A beam of thermal neutrons provides another way to secure a diffraction pattern. The wavelengths used are comparable to those used in X-ray diffraction. A significant difference lies in scattering power of the elements for X-rays and for neutrons. With X-rays, hydrogen is the weakest scatterer in the periodic table, and is thus difficult to locate in structure experiments, while with neutrons hydrogen is a relatively strong scatterer, making location simpler.

Small angle X-ray scattering (7, 42) is used to examine that part of a diffraction pattern within a few angular degrees of the direct beam. It gives structural information on large periodicities within the sample substance. It is described separately from powder diffraction (wide-angle scattering) because the equipment for gathering the pattern and the procedures for interpreting it differ markedly. Small angle neutron scattering is also applied in similar studies (43).

Structural information over very small distances from selected atoms of interest is derived from the interpretation of displacements and fluctuations in the absorption close to an X-ray absorption edge. The procedures used are referred to as XANES (X-ray absorption near-edge structure) or NEXAFS (near-edge X-ray absorption fine structure), EXAFS (extended X-ray absorption fine structure), and SEXAFS (surface EXAFS) (7, 44).

3. X-RAY FLUORESCENCE

X-ray fluorescence spectrometry utilizes one or more of the characteristic X-ray lines emitted by an element when it is bombarded with high energy radiation, such as X-rays from a heavy element (Section 1). It affords a rapid, precise, nondestructive technique for determining the presence and/or concentration of elements present in a sample. It may be used for all elements in the periodic table from uranium to boron, and is applicable to some gases, and most liquids, solids, and powders.

Unlike many other analytical techniques, these analyses are often made on materials with minimal sample preparation, so one avoids reagent impurity effects and the like.

Sensitivities of a few parts per million are achieved for many elements and depending on sample composition, concentrations of a single element often may be measured from the limit of detection to 100%. Calibrations within a single matrix are normally linear with intensities of radiations at specific wavelengths being proportional to the concentrations of the corresponding element. Deviations from this rule may arise from interelement absorption or enhancement effects. Tables of absorption coefficients for most elements have been published (45) and mathematical adjustments can be made for these effects.

Other difficulties may arise from line overlap, when the major analytical line of one element occurs at the same wavelength as a secondary line of another element. Such coincidences severely restrict the sensitivity of the major line element if the concentration of the secondary line element is high relative to the major line element. Other blind spots may occur from the overlap of second or third order lines from the target element in the X-ray tube, but these can usually be removed by the detector pulse height processing electronics. Apart from these overlap problems, which affect a relatively small number of elements, the technique of XRF has wide application in analyses of organosilicon materials.

3.1. Instrumentation

X-ray fluorescence analysis of silicones requires a vacuum spectrograph with either a helium or vacuum radiation path, X-ray target tube, X-ray generator of 3 kW capacity, a scintillation detector and a proportional gas flow detector with electronic counting circuits, and a set of analyzing crystals. A lithium fluoride crystal is normally used for elements with atomic numbers between 92 and 19, and a variety of crystals is used to cover elements down to atomic number 5; for example, germanium for P, S, and Cl, indium antimonide for Si, pentaerythritol for Al, tungsten–silicon multilayer for elements Si to F and nickel–carbon multilayer for elements O to B. Detection limits for these latter elements are rather high with a lower detection limit of 0.4% claimed for boron. The use of an energy dispersive spectrometer system expedites qualitative analysis but with a sacrifice in light element sensitivity and a decrease in resolution for midatomic number elements.

Choice of an X-ray target tube may depend on the main elements of interest, for example, a rhodium target is used for general applications, chromium or scandium for light elements, and tungsten for high atomic number elements. Dual anode tubes such as scandium–molybdenum are also available to combine the advantages of light and heavy element excitation within a single target tube. The use of a computer function is also recommended for data handling.

Some analyses for relatively high levels of a limited number of elements can be made on "benchtop" instruments, which use radioactive source excitation and pulse height wavelength selection. Applications for these latter instruments include measurement of silicone coating thickness on papers and sulfur in fluids.

Accessory equipment required for elemental analysis includes an analytical balance, vibratory grinder, and a press for compressing powders into disks. Cutter mills or liquid N_2-cooled mills can be used to reduce textiles and other fibrous materials to a fine powder.

3.2. Sample Handling

Liquid samples may be analyzed directly in cells with a polymer (e.g., Mylar or polypropylene) film window that is compatible with the sample and the spectral region scanned. Plastic disposable cells are available that eliminate cell clean-up. Low volume cells are also available for the analysis of materials in limited supply.

Solid samples with homogeneous smooth surfaces may also be analyzed directly by cutting a disk of material to fit the sample cell (e.g., paper, rubber sheet, and metals). Other solid materials can be ground and homogenized in a mill to a fine particle size (ideally 1 μm), after which they are compressed to a pellet with a suitable binding agent. Techniques for the analysis of powders and solids are discussed fully in the literature (46). Solid samples that are soluble in acids or water may be dissolved and analyzed in aqueous solution.

3.3. Detection Limits

Typical detection limits for various elements in a PDMS matrix are listed in Table 15.1. The exact values depend on the type of instrumentation available. Detection limits are matrix dependent and lower limits may be achieved for many elements in organic or aqueous media.

3.4. Qualitative Analysis

Elements can be identified in a sample from a sequential scan on a wavelength dispersive instrument, in which measurement is made of the 2θ angles corresponding to peaks in the spectra. Identification can then be made by consulting reference tables relating to the relevant analyzing crystal (47),

Table 15.1. Detection Limits for Various Elements in a PDMS Matrix

Element	Lower Limit of Detection mg Element/kg (ppm)
Al	20
As	3
Ba	2
Bi	5
Br	5
Ca	3
Cl	10
Co	2
Cr	10
Cu	2
Fe	1
K	0.3
Mn	3
Ni	2
P	3
Pb	3
Pt	10
S	2
Sn	3
Ti	0.2
V	1
Zn	2
Zr	2

which list the 2θ angles of the elements. A full spectrum can be obtained in a much shorter time using an energy dispersive detector. In this case elements are identified from their characteristics energy levels, and many modern instruments are programmed for automatic element identification and semi-quantitative analysis.

3.5. Quantitative Analysis

The intensities of fluorescent X-rays emitted from a sample are strongly influenced by the matrix, and several analytical methods have been devised to overcome this problem. The following basic techniques have proved useful in silicone applications.

The External Standard Method, which uses standards for calibration prepared in a matrix closely matching the samples. This method is mainly used for the analysis of minor elements in systems where the major matrix elements remain reasonably constant.

The Internal Standard Method is applicable to variable matrices. A constant amount of a selected element of similar properties to the element sought is added to the sample. Both elements are influenced to the same extent by absorption or enhancement effects, and hence the intensity ratio of the element sought to the reference element is matrix independent and is proportional to the concentration of the element sought.

The Standard Addition Method utilizes comparison of the intensities before and after the addition of known amounts of the element sought, for compensation of matrix effects. Two or more additions are required to verify linearity of the concentrate–concentration relationship.

The Dilution Method utilizes dilution with an excess of diluent to minimize matrix variations and render the fluorescent intensities nearly proportional to concentration. This approach is not recommended for samples of unknown composition but can be usefully applied for assay purposes on known materials.

3.6. Applications

3.6.1. Trace Metals in Silicone Fluids

Analyses for trace elements such as wear metals, catalyst residues, and additive elements in silicone fluids can be made directly by the external standard method. Calibration standards for several elements can be prepared by dissolving suitable metal compounds in a "clean" fluid of the same functional type as the sample and a direct comparison can be made between the intensities from the standard and the sample. A wide range of "Certified Analysis" oil soluble standards is available, which are compatible with silicones. Generally these are salts of organic acids—cyclohexane butyrates, and decanoates. Other compounds such as triphenyl arsine, diphenyl sulfide, triphenyl phosphate, titanium isopropoxide, and zirconium tetraisopropoxide have been used to prepare standards for these elements in silicone fluids. These silicone-soluble compounds can also be used for standard addition analyses on unknown fluids. Calibrations for many elements are linear over the range from the limit of detection to ~ 300 ppm or over. Analyses have been made by this technique for As, Ba, Br, Ca, Cl, Cu, Pb, Pt, Sn, Sr, Ti, Zn, and Zr.

On exposure to X-rays, platinum compounds undergo reduction to platinum black. This material settles to the bottom of the liquid cell during analysis to give high results, if counting times in excess of 20 s are used. The short counting time limits the precision and sensitivity of the platinum analysis. Count rate instability in the analysis for C and P in liquid samples has also been noted in WDXRF but not with EDXRF.

Wear metals may also be present in samples as suspended solid particles, which may settle during analysis. An approximate analysis can be obtained on these suspensions by mixing the sample with molten petroleum jelly to form a stable grease within the sample cell. However, if precise data are required on both dissolved and solid contaminants, the latter can be removed by either filtration or centrifugation, and separate analyses made on the fluid and solid residue. Wear metal residues are usually soluble in mineral acids and can be quantified in aqueous media.

Problems from settling of the analyte can also arise in analyses of polymers for potassium catalyst residues that may be present as a mixture of soluble potassium and insoluble particulate potassium carbonate. This element is best extracted into dilute acid and determined in an aqueous media. Quantitative recovery of parts per million concentrations of potassium can be obtained by extraction of a 1 : 5 dilution of sample in dichloromethane with $0.1 M$ HCl.

Line overlaps occur, the most notable being with As $K\alpha$ and Pb $L\alpha$, and corrections must be made when these two elements occur in the same sample. The Pb $L\beta$ line can be used to measure lead, and if the lead concentration is low, a correction can be made to the As $K\alpha$ intensity. Otherwise if the lead concentration is high the As $K\beta$ line must be used at the expense of arsenic sensitivity.

Some catalysts may be used in a series of similar products of differing silicon content and a single calibration can be applied to these provided a mathematical correction is computed to correct the interelement effects. An example of this is zinc catalyst in silicone resin where there is significant absorption of the zinc radiation by silicon.

Analyses are made directly on these resins without sample preparation, and measurement is made of both the Zn Kα and Si Kα radiation. The concentration of zinc is computed from the Lucas = Tooth Pyne model (48) or other mathematical correction system, for example,

$$C_i = (D_i + E_i R_i)(1 + \sum \alpha_{ij} R_j) \qquad (15.2)$$

where C_i is the concentration of element i; R_i, R_j are the count rates for elements i and j; D and E are constants; and α_{ij} is the interelement factor.

The interelement factors (α_{ij}) may be determined by solving the algorithm from a calibration involving a range of samples covering a random distribution of both elements. If only a few samples are available, theoretical α values can be calculated.

3.6.2. Silicone Fluids—Percent Silicon

Assays for percent silicon in silicone fluids can conveniently be made by direct comparison of the intensity of a 1:5 dilution of the sample in xylene against a standard of the same material. The technique is applicable to many organosilicon compounds of low vapor pressure. However, for fluids of unknown composition, the silicone assay is best performed by incorporating an internal standard to compensate for interelement absorption–enhancement effects (49, 50).

A suitable internal standard for silicon is phosphorus, which can be added to silicone material as tributyl phosphate. The analysis can conveniently be made by diluting a known weight of sample with a mixture of 100 parts xylene, 55 parts perchlorethylene, and 45 parts tributyl phosphate.

Calibration is made against a series of standards prepared by adding varying weights of silicone in a constant volume of dilution–spiking mixture when measurement is made on the Si Kα and P Kα radiation. An intensity ratio is established for each standard according to the following formulas:

$$R(\text{Si/P}) = \frac{\text{net intensity of Si}}{\text{net intensity of P}} \qquad (15.3)$$

The intensity ratio for each standard is plotted against silicon concentration to provide a calibration curve.

3.6.3. Emulsions and Particulate Suspensions

Analyses for silicon and trace metals can be made on aqueous emulsions by the standard addition method using, respectively, aqueous standards of soluble silicates or soluble metal standards for calibration. Some emulsions may segregate during measurement and these can be stabilized by gelatine, added as a 15% w/w aqueous solution at a ratio of 1:9 to the sample. The mixture is melted in a water bath and homogenized by shaking. The mixture is allowed to cool in a sample cell where a stable gel is formed. Emulsions or suspensions in oil can be stabilized similarly by the addition of 10% w/w of petroleum jelly to form a grease. Silicone fluid can be used as a standard addition calibrant for silicon, and oil-soluble metal salts are available for metals calibration.

3.6.4. Silicone on Surfaces

Quality control assays of silicone coatings on greaseproof papers or release papers can conveniently be made by XRF spectrometry. Direct measurements can be made on paper by cutting a disk to fit the sample cell and measuring the intensity of the Si Kα radiation. Blank analyses on uncoated paper should be made to correct for any natural inorganic silicate within the paper. A comparison of XRF, chemical analysis, IR spectroscopy, and neutron activation methods for quantitative analysis of silicone paper coatings (51) led to the conclusion that the X-ray method was preferred.

Calibration standards for this analysis can be made by coating a suitable substrate with varying weights of silicone coating preparation, preferably of the same type as that being analyzed. Several options, none of them ideal, are available for standard preparation.

A gravimetric approach may be used in which the silicone is sprayed on aluminum disks. The coating weight after cure is easily determined. With energy dispersive systems, however, emission from the Al substrate limits the power that can be applied to the X-ray tube because of detector dead time, and thus interferes with the Si measurement. A polyethylene film substrate can be coated at predetermined thickness using one of the standard adjustable roller or blade coater instruments available within the paper industry for producing test samples of coated paper. A simple hand operated coater is also available that uses standard rods known as "Mayer Rods" to draw coating fluid over a surface. The coated sheets are then dried and cured under conditions specified for the coating material (usually heating in an air oven for 30 s at 120°C). A series of sheets is prepared to cover the working range 0.2–2.0 g of silicone/m², and intensity measurements are made on disks cut from each standard sheet. A calibration plot of count rate versus concentration can then be prepared. A second-order fit to the calibration data usually gives a better calibration curve over a wide coating weight range, because of self-absorption effects. Coating concentrations may be measured on the standards by a chemical analysis for silicon (Chapter 8).

Some types of paper contain a very high concentration of inorganic silicate from clays used as fillers, and these often give variable blanks, which prevent direct analysis. A correction can be applied for the inorganic silicate by measuring the ratio of the intensity of the Si Kα radiation against the Al Kα radiations from the clay surface on a piece of uncoated blank paper, for example,

$$R(Si/Al) = I_{Si}/I_{Al} \qquad (15.4)$$

Then the X-ray intensities from these two elements on the coated paper is measured, and the intensity from the silicone coating is given by

$$I_{silicone} = I_{total\ Si} - I_{Al}R(Si/Al) \qquad (15.5)$$

Some materials such as greaseproof paper allow penetration of the silicone into the core of the paper, and measurement of the total amount of silicone present may be required. These samples are ground to a powder in a cutter mill, mixed with a binding agent such a polyvinyl alcohol, and pressed into a pellet to fit the sample cell. Calibration standards for total silicon on paper can be prepared by absorbing known amounts of silicone fluid in acetone solvent onto cellulose powder. After evaporation of the solvent and addition of binding agent the powders can be pressed into standard disks.

Since cotton, cloth, and leather are also similar in nature to cellulose, these materials may be analyzed by the same methods and using the same cellulose calibration standards. Some synthetic fibers such as polyester may melt during the cutting–grinding process, and techniques involving cooling the sample and cutting mill with liquid nitrogen have been used to prevent this problem.

The same approach has also been used with silicone coatings on metal surfaces such as "nonstick" cooking utensils. In this case, silicone resins were used to coat metal disks for calibration standards.

3.6.5. *Other Solid Samples*

The pressed pellet technique has also been applied to trace element analysis in metallurgical silicon, natural silicas, diatomite, and clays. Elements determined have included, Al, As, Bi, Ca, Cr, Fe, Mn, Ni, P, Pb, S, Ti, and V. Calibration normally relies on chemically analyzed standards or on synthetic standards, prepared by milling stable metal oxides of 400 mesh or finer particle size into high purity sample material. The standards cover the expected concentration ranges of the elements sought. "Certified Analysis" standards, when available, can be used to check calibration.

Elemental analyses can also conveniently be made directly on cured sheets of silicone elastomers from disks cut to fit the sample cell. Calibration is made using synthetic standards prepared by milling known weights of pure metal oxides into the composition before curing. All surfaces presented for measurement must be perfectly smooth and level.

4. X-RAY PHOTOELECTRON AND AUGER ELECTRON SPECTROSCOPY

4.1. Introduction

X-ray photoelectron spectroscopy (XPS) and Auger electron spectroscopy (AES) are widely accepted as key techniques for the characterization and analysis of material surfaces. X-ray photoelectron spectroscopy is also known as electron spectroscopy for chemical analysis (ESCA). Because electron beams can be easily focused, a scanned image is possible with AES. An AES instrument used in this mode is called a SAM, scanning Auger microprobe.

These methods, particularly XPS, have expectedly been applied to a variety of organosilicon surfaces. The fundamentals of these techniques and their relationship to the numerous other surface analytical probes are found elsewhere (52, 53). A general organosilicon surface analysis perspective is offered in Chapter 5. One of our major interests is the application of XPS to polymer surfaces. Andrade's review (54) is one of the most helpful and readable on this subject. Other useful sources are the biennial reviews of XPS and AES in *Analytical Chemistry* (55).

The fundamental processes underlying these techniques are not new. Einstein described the photoelectron effect in 1905, and the Auger effect was discovered in 1923. However, it was not until reliable ultrahigh vacuum (UHV) conditions could be provided that commercial instruments became possible around 1970. The first reports on organosilicon materials appeared soon after; XPS in 1972 (56) and AES in 1974 (57). Ultrahigh vacuum (pressure of 10^{-7} Pa or better) is needed for studies of clean surfaces as there is time at these pressures to make measurements while the surface is still clean. For most "as-received" samples, UHV is not required and the upper pressure limit is set by the operating characteristics of each particular instrument. This UHV condition means samples must be stable in vacuum. Surfaces with volatile components are usually better studied by other techniques such as IR.

4.2. Capability and Limitations

The advantages and disadvantages of XPS and AES are summarized in a review by Brundle (58). Because XPS has much broader applicability to organosilicon surfaces, it is the primary focus of the following sections.

For organosilicon materials under normal conditions the sampling depth is of the order of 50 Å down to 20 Å with angle resolved capability. An unpublished Dow Corning study by Williams showed the escape depth of Si-2p photoelectrons using a Mg anode X-ray source to be 51 Å for $(Me_2SiO)_n$, 47 Å for $(MeSiO_{1.5})_n$, and 45–36 Å for a variety of other organo-siloxane polymers.

The $\pi–\pi^*$ shakeup transition in aromatic groups and the wide separation (~ 8 eV) of hydrocarbon and fluorocarbon make the C-1s binding energies the most useful for obtaining chemical information in organosilicon XPS. Clark et al. (59) first provided such data on silicone polymers. Gardella et al. (60) have shown that aromatic shakeup features are also present in the O-1s and Si-2p XPS peaks of aromatic silicone polymers.

X-ray beam damage is rarely a problem with organosilicon materials, although there is a report that long X-ray exposure is damaging to the organofunctional part of phenylmethylcyanopropylsiloxane (61). Electron beam damage is severe in many cases, particularly of organic groups, but despite this disadvantage there are several AES studies of silane coupling agents; and a few studies of silicone fluids, elastomers, resins, and copolymers.

Ultrahigh vacuum (10^{-7}–10^{-9} Pa) can be a concern with silicone materials containing cyclics and other low molecular weight constituents, and charging and energy referencing can be a problem as almost all organosilicon samples are nonconducting. Referencing to the ubiquitous C-1s peak is usually satisfactory.

4.3.   Sample Preparation

Sample preparation is usually not a critical problem so long as the sample is vacuum stable and is small enough or can be cut to fit the sample probe. Analysis is possible on a wide variety of sample types including films, foils, other solids, powders, and fiber meshes or bundles. The best form of attachment is by a metal clip ensuring electrical contact but double-sided adhesive tape is also used. The tape is useful for securing powders, which can also be studied by pressing them into indium foil. A low volatility tape should be selected and checked in the instrument to make sure contamination by plasticizers, silicone release coatings, and the like is not occurring. This point is particularly critical when low levels of silicone are being investigated. Dilks (62) has shown that not only will silicone release agents used in such tapes migrate across samples with time but in the case of low density polyethylene, can even diffuse through the film. When one has the choice, samples prepared without processing aids are preferred.

Generally, the as-received surface should be analyzed for reference, and sample preparation minimized as almost any procedure will likely modify the surface. The main problem lies with volatile material. If such is of particular interest, liquid nitrogen cooled probes can be used. For instance, Ratner et al. (63) have used this approach to study hydrated and dehydrated hydrogel coatings on silicone elastomers. Otherwise ultrasonic extraction should be tried. Argon ion sputter etching is commonly used to obtain depth profiles in XPS and AES and will obviously remove surface contaminants but can also change the surface. For very clean surface preparation with suitable equipment, the material can be fractured or scraped within the introduction or test chamber under vacuum conditions. Williams and Davis (64) successfully used a vacuum fracture accessory with silicone elastomer and resin samples. Silicone elastomers adsorb and absorb atmospheric gases and vapors readily. Relatively long outgassing times in the sample introduction chamber, sometimes a few hours, are not unknown. A system for preliminary outgassing helps if many such samples need to be routinely analyzed. Cryogenically pumped instruments also help speed analysis of polymeric materials.

4.4. X-Ray Photoelectron Spectroscopy Elemental Analysis

General concepts of instrument operation and analysis procedures are well covered in the many reviews, and specific details are dealt with in the instrument manufacturers' handbooks. Elemental identification is by the binding energy of the emitted photoelectrons, which is characteristic of the atomic orbital from which they came, and quantitation is by the number of such photoelectrons. Essentially XPS thus consists of determining the number of photoelectrons emitted as a function of binding energy. Usually a broad scan (0–1000 eV) spectrum is first obtained for elemental identification. Figure 15.1 is an example of such a spectrum for a poly(trifluoropropyl-methylsiloxane) (PTFPMS) elastomer (the silica filler was also treated with PTFPMS).

The groups most commonly found on silicon can be categorized as stable, organofunctional, and reactive. The stable groups include methyl, higher alkyl, phenyl, and trifluoropropyl. The organofunctional groups, which are usually present in small amounts in organofunctional silicone polymers but are present on every silicon atom in silane coupling agents, include amino, epoxy, methacryloxy, and mercapto groups, as well as others. The reactive groups, usually present in low levels in most samples encountered, include SiOH, SiOR, SiCl, and SiH groups. Thus Si, C, and H (not detected by XPS) are always present, O almost always so, and F, N, Cl, and S sometimes present. The photoelectron binding energies and photoionization cross-sections for these elements are given in Table 15.2.

They are presented in increasing binding energy, the most useful form for elemental identification. Only the significant cases of photoionization cross-

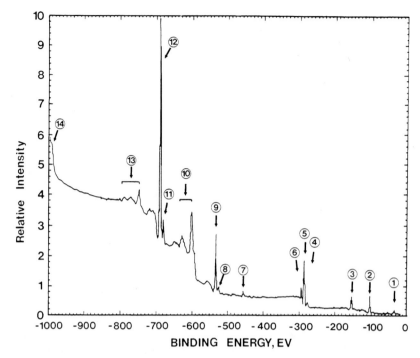

Figure 15.1. XPS survey spectrum of poly(trifluoropropylmethylsiloxane) taken on a Perkin-Elmer Physical Electronics model 550 ESCA–AES instrument, 100 eV pass energy, Mg anode X-ray source (1253.6 eV).

section > 0.1 are included and the most intense line for each element is noted. This useful form of the table is taken from Andrade (54) who gives a complete version for all the elements.

Elemental quantification is achieved by high resolution scans across the regions of the identified elements and the use of sensitivity factors based on the photoionization cross section, electron mean free path (sample depth), and the instrument response function. Examples of such atomic composition determination are given in Table 15.3 for some silicone polymers including the example shown in Figure 15.1. The sensitivity factors used can both be calculated and empirically derived from standard samples.

In addition to the photoelectron lines in an XPS spectrum, numerous other kinds of lines are present including

- Auger line clusters, independent of primary beam energy.
- X-ray satellites, characteristically placed pattern of lower binding energies than main peak.

Table 15.2. Characteristics of Significant Photoelectron Lines for Key Organo-silicon Elements

Binding Energy (eV)	Element	Orbital	Photoionization Cross Section (Relative to C-1s = 1.00)
7	Cl	$3p's$	0.143
16	S	$3s^{1/2}$	0.147
18	Cl	$3s^{1/2}$	0.185
24	O	$2s^{1/2}$	0.141
31	F	$2s^{1/2}$	0.210
99	Si	$2p^{3/2}$	0.541
100	Si	$2p^{1/2}$	0.276
149[a]	Si	$2s^{1/2}$	0.955
164	S	$2p^{3/2}$	1.11
165	S	$2p^{1/2}$	0.567
200	Cl	$2p^{3/2}$	1.51
202	Cl	$2p^{1/2}$	0.775
229[a]	S	$2s^{1/2}$	1.43
270[a]	Cl	$2s^{1/2}$	1.69
284[a]	C	$1s^{1/2}$	1.00
399[a]	N	$1s^{1/2}$	1.80
532[a]	O	$1s^{1/2}$	2.93
686[a]	F	$1s^{1/2}$	4.43

[a] Indicates most intense line for each element.

- X-ray "ghosts," impurities in anode, accessories, and so on.
- Shakeup lines, peaks of slightly higher binding energy (such as the characteristic π-bond shakeup line for carbon in aromatic compounds) that arise from transitions between electronic levels occurring during the primary photoionization event.
- Multiple splitting, most noticeable in the ionization of p levels.
- Energy loss lines.
- Valence lines and bands.

Some of these features are evident in Figure 15.1. The numbers in parentheses in the element identification column of Table 15.3 correspond to the numbered spectral features in Figure 15.1. The other numbered features in this figure are

1	F-2s/O-2s
3	Si-2s
4, 8, 11	X-ray satellites
6	C-1s fluorocarbon peak
7	Ti-2p from sample support clip
10	F Auger bands
13	O Auger bands
14	C Auger bands

Table 15.3. XPS Analysis of Organosiloxane Polymers

Binding Energy (Referred to C-1s at 285.0 eV)	Element Identification	Atomic Composition (%)	
		Actual	Theory
Poly(dimethylsiloxane)[a]			
102.4	Si-2p	24.7	25
285.0	C-1s	50.0	50
532.6	O-1s	25.2	25
Poly(methylphenylcyanopropylsiloxane)[b]			
102.2	Si-2p	11.6	11.8
285.0	C-1s	70.7	70.6
400.2	N-1s	5.3	5.9
532.7	O-1s	12.3	11.8
Poly(trifluoropropylmethylsiloxane)[c]			
102.2	Si-2p (2)[d]	11.2	11.1
285.0	C-1s (5)	39.7	44.4
532.6	O-1s (9)	12.8	11.1
688.7	F-1s (12)	36.4	33.3

[a] Dow Corning Syl-off® 292 release coating, our work (65).
[b] Alltech Associates OV-225 fluid, Toth et al. (61).
[c] Dow Corning fluorosilicone elastomer, our work (66).
[d] Numbers in parentheses refer to spectral features in Figure 15.1.

4.5. Other Aspects of X-Ray Photoelectron Spectroscopy Analysis

4.5.1. Chemical Group Information

Binding energy assignments useful for molecular information (chemical shifts) result from a combination of theoretical and experimental studies with model compounds. For example, as first discovered by Clark et al. (59), the $\pi-\pi^*$ shakeup is useful for identifying aromatic silicones. Wagner's compilation of Auger and photoelectron energies (67) contains useful Si-2p data illustrating the magnitude of this chemical shift. For example the Si-2p binding energy for vapor deposited Si metal is 99.7 eV, for PDMS is 102.4 eV, and for silica gel is 103.6 eV. Unfortunately the XPS chemical information useful for identifying functional groups on surfaces is often insufficient for a clear choice to be made between plausible possibilities in organosilicon systems. A common example is oxidized polymer surfaces. Hydroxyl groups, be they carbinol or silanol, produced by exposure to such environments as UV, plasma, corona discharge, flame, and so on, cannot readily be distinguished from backbone siloxane oxygen by binding energy chemical shifts (68).

One approach to this problem is to derivatize the surface with a reactant containing a readily detected labeling atom that is not already present in the

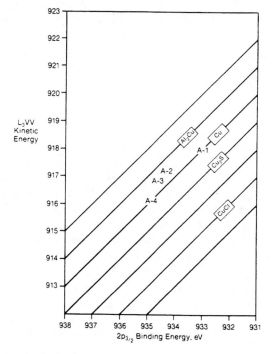

Figure 15.2. The data for the four contact masses, A-1, A-2, A-3, and A-4, are plotted on this 2-D scatter plot. The Y axis is the kinetic energy of the Cu L3VV peak, the X axis is the binding energy of Cu 2$p^{3/2}$ peak, and the diagonal lines are the Auger parameter. All four contact masses fall near the diagonal for the zero oxidation state of copper. [From (70).] Courtesy of *J. Catalysis.*

surface region of interest. Such studies should be entered into cautiously, as determining suitable conditions can be very time consuming. Used properly this approach can provide important molecular information. For example, Williams and Tangney (69) have shown that 3,3,3-trifluoropropyldimethyl-N-methylsilylacetamide (TFSA) is an excellent silylating agent for this purpose. It is one of the few silylating agents that reacts from the vapor phase at room temperature with surface hydroxyl groups; it provides a useful F elemental tag; and the products of reaction are easily removed, volatile species. A gradual loss of fluorine as HF is observed during prolonged XPS measurement of TFSA-tagged silica, but the rate is slow enough that reliable initial surface compositions can be obtained before appreciable X-ray damage has occurred. A tabular summary of such reactions use for derivatizing polymer surfaces has been provided by Andrade (54). Another way of getting molecular information such as elemental oxidation state is to make use of both the XPS and Auger data inherently present in the XPS spectrum in a 2-D scatter plot. This approach is possible only with certain metals and is not generally useful for common organosilicon systems, although it can be helpful in characterizing silicone–metal interfaces. An example of its use in analyzing silicone intermediate manufacturing "direct process" contact masses is given in Figure 15.2 (70). The key catalytic element in this process, copper, is shown to be in the zero oxidation state rather than $+1$ by this approach (the lack of shakeup satellites showed it not to be in the paramagnetic $+2$ state either).

4.5.2. Depth Profiling

There are a number of ways of obtaining information on the depth of an element in a sample by XPS. The methods utilizing spectral characteristics such as energy loss peaks and intensity ratioing of peaks widely spaced in kinetic energy are rather limited and not suitable for the elements usually encountered in organosilicon surface studies. Some information is also available when a dual X-ray source is available as the energy of emitted photoelectrons depends on the incident X-ray energy. More widely applicable than any of these, and also applied to surfaces of current interest, are (a) angularly resolved XPS providing information concerning depths less than the usual sampling, and (b) argon ion sputtering providing depth profiles to any chosen depth greater than the original analysis.

Angularly resolved XPS alters the angle between the plane of the sample surface and the angle of acceptance by the analyzer, providing a more surface sensitive analysis for smaller angles. Table 15.4 provides an example of such data from the comprehensive study of silicone-containing commercial segmented polyurethane biomaterials by Grobe et al. (71).

Table 15.4. Angle Resolved XPS Analysis of Silicone-Containing Biomaterials

Material	Angle (°)	Atomic Composition Ratios				
		C/O	Si/O	Si/C	N/O	N/C
Avcothane®	90	0.790	0.308	0.391	0.005	0.006
	45	0.730	0.286	0.392	0.009	0.013
Biomer®	90	0.991	0.095	0.081	0.131	0.108
	45	0.857	0.157	0.184	0.006	0.005
Avcomat®	90	1.452	0.010	0.006	0.086	0.059
	45	1.478	0.008	0.008	0.082	0.056
PDMS	90	0.630	0.289	0.464		
	45	0.661	0.287	0.436		

This study was also supported by IR (ATR and PA) and considered XPS binding energy changes not listed here. The XPS results show that the silicone layer of Biomer® changes significantly in the top 20–50 Å, whereas the silicone component of Avcothane® is thicker than 50 Å (of the order of 1 μm, according to IR). Avcomat® is a supposedly silicone-free segmented polyurethane but silicone is present in very small amounts (2 × background). Interestingly, the PDMS comparison sample also showed compositional depth variation, possibly attributable by IR to alkoxysilane or hydrocarbon contamination.

Argon ion sputtering gives depth profile information limited only by the analyst's patience (typical sputter rates are of the order of 50 Å/min). The technique can be used to demonstrate differences in composition throughout a film or coating, particularly in the surface (such as oxidized layers), and can detect buried interfaces and thereby give a measure of the thickness of the overlying layer.

4.6. Selected X-Ray Photoelectron Spectroscopy and Auger Electron Spectroscopy Applications

To illustrate the diversity of the applications of XPS and AES, including the techniques described in the preceding section such as angle resolved XPS and depth profiling, and to suggest further reading, we offer the summary Table 15.5 of selected (far from comprehensive) organosilicon studies. Note that in many of these studies, particularly the more recent ones, other

Table 15.5. Summary of Selected Organosilicon XPS–AES Studies

Sample Type	Technique	Key Information Obtained	References
Silane coupling agent metal oxide–polymer interface	XPS depth profile	Evidence of IPN formation	Sung et al. (68), 1982 Chaudhury et al. (72), 1987
	XPS	Locus of failure in adhesive joints	Gettings et al. (73), 1977
Silane coupling agents on mica	XPS	Adsorption isotherms effect of pH, temperature, electrolyte concentration	Herder et al. (74), 1988
Silane coupling agent review	XPS	Adhesion mechanisms	Anderson and Sachdev (75), 1989
Alkylamine silanized metal oxide electrodes	XPS depth profile	Different forms of surface nitrogen	Moses et al. (76), 1978
Silane coupling agents on silicon wafers	AES	Film thicknesses of deposited silane layers	Cain and Sacher (77), 1978
Trimethylsilylated silica	XPS	$SiO_x(OH)_y$ surface stoichiometry	Linton et al. (78), 1985
Plasma-treated silicone elastomer	XPS	Chemistry of plasma treatment	Triolo and Andrade (79), 1983
Plasma-polymerized silanes	XPS	Different states of C and Si identified	Inagaki et al. (80), 1985

512

Sample	Technique	Information	Reference
Plasma-polymerized hexamethyl-disiloxane	XPS	Surface elemental composition	Tajima and Yamamoto (81), 1985
Contaminated, failed electrical relays	AES	Detection of silicone films	Haque and Spiegler (82), 1980
Siloxane fluid	XPS depth profile	Effect of X-ray and argon ion exposure	Toth et al. (61), 1985
Silicone elastomers and resins	XPS depth profile	Effect of argon ion exposure	Williams and Davis (64), 1976
PDMS adhesive/organic copolymer interface	XPS	Identification of cure-inhibiting species	Kardos and Fountain (83), 1974
Silicone homopolymers	XPS	Limitations of XPS compared to ISS	Hook et al. (84), 1986
PDMS–polyimide copolymers	XPS angle resolved	Surface concentration of PDMS component	Bott et al. (85), 1987
PDMS–polystyrene block copolymers	XPS	Thickness of surface PDMS layer, first copolymer study	Clark et al. (86), 1976
PDMS–polycarbonate blends	XPS angle resolved	Surface composition and morphology	Schmitt et al. (87), 1986
Hydrogel grafted silicone elastomers	XPS "cold probe" study	Marked difference in composition of dry and hydrated surfaces	Ratner et al. (63), 1978
PDMS–PVA graft copolymers	XPS angle resolved	Surface concentration of PDMS	Tezuka et al. (88), 1986

Table 15.5. (*Contd.*)

Sample Type	Technique	Key Information Obtained	References
Silicone-containing biomaterials	{ AES XPS XPS angle resolved	Differences in air and substrate formed surface composition Surface domain segregation	Sung and Hu (89), 1979 Graham and Hercules (90), 1981 Grobe et al. (71), 1987
Silicone-based release coatings	XPS	Characterization of release surfaces	Duel and Owen (65), 1983 Hsu et al. (91), 1986
MeCl/Si	AES	Mechanism of "direct process" for silicone intermediate manufacture	Frank and Falconer (92), 1985
Polyacrylic acid	Surface derivatized XPS	Surface acid groups derivatized using organosilyl reactant	Davies and Munro (93), 1988

complementary techniques such as SEM, ATR–FTIR and $^{29}Si/^{13}C$ solid-state NMR are used to augment the study and provide a more comprehensive understanding of the surface than is possible with any single technique.

REFERENCES

1. N. F. M. Henry and K. Lonsdale, Eds., *International Tables for X-ray Crystallography*, Vol. I, Symmetry Groups, Kynoch Press, Birmingham, England, 1952. J. S. Kaspar and K. Lonsdale, Eds., *International Tables for X-ray Crystallography*, Vol. II, Mathematical Tables, Kynoch Press, Birmingham, England, 1959. C. H. Macgillavry and G. D. Rieck, Eds., *International Tables for X-ray Crystallography*, Vol. III, Physical and Chemical Tables, Kynoch Press, Birmingham, England, 1962. J. A. Ibers and W. C. Hamilton, Eds., *International Tables for X-ray Crystallography*, Vol. IV, Revised and Supplementary Tables, Kynoch Press, Birmingham, England, 1974. T. Hahn, Ed., *International Tables for Crystallography*, Vol. A, Space Group Symmetry (replaces Vol. I), Riedel, Dordrecht, The Netherlands, 1983.

2. E. R. Cohen and B. N. Taylor, *The 1986 Adjustment of the Fundamental Physical Constants, a Report of the CODATA Task Group on Fundamental Constants*, CODATA Bulletin 63, Pergamon Press, New York, 1986. Also by the same authors, "The 1986 CODATA recommended values of the fundamental physical constants," *J. Res. Natl. Bur. Stand.* **92**(2), 85 (1987) and *J. Phys. Chem. Ref. Data*, **17**(4), 1795 (1988).

3. J. A. Small, D. E. Newbury, and R. L. Myklebust, *X-ray Bremsstrahlung Intensities from Elemental Targets*, NBS Technical Note 1245, National Bureau of Standards, Gaithersburg, MD, 1988.

4. M. J. Buerger, *Elementary Crystallography*, Wiley, New York, 1956.

5. B. D. Cullity, *Elements of X-ray Diffraction*, 2nd ed., Addison-Wesley, Reading, PA, 1978.

6. F. H. Herbstein, Ed., *Methods of Obtaining Monochromatic X-rays and Neutrons*, Bibliography 3, Commission on Crystallographic Apparatus, International Union of Crystallography, A. Oosthoek's Uitgevers Mij N.V., Utrecht, The Netherlands, 1967.

7. H. Winnick and S. Doniach, Eds., *Synchrotron Radiation Research*, Plenum Press, New York, 1980.

8. N. F. M. Henry, H. Lipson, and W. A. Wooster, *Interpretation of X-ray Diffraction Photographs*, 2nd ed., Macmillan, New York, 1961.

9. M. J. Buerger, *Crystal Structure Analysis*, Wiley, New York, 1960.

10. Crystal Data Determinative Tables, 3rd ed., Vol. 1, 1972; Vol. 2, 1973; Vols. 3–4, 1978; Vols. 5–6, 1983; National Bureau of Standards and JCPDS—International Centre for Diffraction Data, Swarthmore. Holds bibliographic information, unit cell, composition, formula, and symmetry data for 70,500 organic and inorganic compounds.

11. NBS CRYSTAL DATA, a crystallographic data base compiled by the NIST Crystal Data Center, National Institute of Standards and Technology, Gaithersburg. Currently holds bibliographic information, unit cell, composition, formula, symmetry, and reduced cell data for approximately 115,000 organic and inorganic compounds. CIS (Chemical Information System), an online integrated collection of data bases suitable for search and analysis, including NBS CRYSTAL DATA (as component XTAL). Fein-Marquart Associates, Baltimore, MD.

12. CRYSTDAT (CRISTALDON), an online search and analysis system for NBS CRYSTAL DATA, developed jointly by the NIST Crystal Data Center and the Canada Institute for Scientific and Technical Information, National Research Council of Canada, Ottawa. Available through the Canadian Scientific Numeric Database Service, CAN/SND. CAN/SND maintains an online collection of bibliographic and numeric data bases for search and analysis, including the NBS Crystal Data Identification File (as component CRYSTDAT), the Inorganic Crystal Structure Database (as component CRYSTIN), the NRC Metals Crystallographic Data File (as component CRYSTMET), and the Cambridge Structural Database (as component CRYSTOR). Canada Institute for Scientific and Technical Information, National Research Council of Canada, Ottawa.

13. T. Siegrist, "Applications of the crystallographic search and analysis system CRYSTDAT in materials science," *J. Res. Natl. Inst. Stand. Technol.*, **94**, 49 (1989).

14. J. D. Dunitz, *X-ray Analysis and the Structure of Organic Molecules*, Cornell University Press, Ithaca, NY, 1979.

15. M. F. C. Ladd and R. A. Palmer, *Structure Determination by X-ray Crystallography*, 2nd ed., Plenum Press, New York, 1985.

16. G. H. Stout and L. H. Jensen, *X-ray Structure Determination: a Practical Guide*, 2nd ed., Wiley, New York, 1989.

17. S. R. Wilson and J. C. Huffman, "Cambridge Data File in organic chemistry. Applications to transition-state structure, conformational analysis and structure/ activity studies," *J. Org. Chem.*, **45**, 560 (1980).

18. F. H. Allen, O. Kennard, and R. Taylor, "Systematic analysis of structural data as a research technique in organic chemistry," *Acc. Chem. Res.*, **16**, 146 (1983).

19. H. B. Buergi and J. D. Dunitz, "From crystal statics to chemical dynamics," *Acc. Chem. Res.*, **16**, 153 (1983).

20. O. Kennard and D. G. Watson, *Molecular Structures and Dimensions. Bibliography*, Reidel, Dortrecht, The Netherlands. Organosilicon compounds are in Class 63: Vol. 1 (covering 1935–1969), Classes 1–59, 1970. Vol. 2 (covering 1935–1969), Classes 60–86, 1970. Vol. 3 (1969–1971), 1971, and subsequent annual volumes through Vol. 15 (1982–1983), 1984.

21. Cambridge Structural Database (CSD), Crystallographic Data Centre, Cambridge, England. Currently holds bibliographic information, unit cell, composition, symmetry, connection tables and three-dimensional structural data for 74,000 organic and organometallic compounds. Available online from CAN/SND (as component CRYSTOR).

22. F. H. Allen, et al., "The Cambridge Crystallographic Data Centre: Computer-based search, retrieval, analysis and display of information," *Acta Cryst.*, **B35**, 2331 (1979).

23. F. A. Allen and M. F. Lynch, "The storage and retrieval of chemical structures," *Chem. Br.*, **1989**, 1101.

24. V. E. Shklover and Yu. T. Struchkov, "Crystal and molecular structure of organosilicon compounds," in *Advances in Organosilicon Chemistry*, M. G. Voronkov, Ed., Mir Publishers, Moscow, 1985.

25. V. E. Shklover, Yu. T. Struchkov, and M. G. Voronkov, "Structure of organic derivatives of non-tetrahedral silicon," *Main Group Metal Chem.*, **11**, 109 (1988).

26. E. Lukevics, O. A. Pudova, and R. Y. Sturkovich, *The Molecular Structure of Organosilicon Compounds*, Zinatne Publishers, Riga, USSR, 1988.

27. F. J. Feher and T. A. Budzichowski, "New polyhedral oligosilsesquioxanes via the catalytic hydrogenation of aryl-containing silsesquioxanes," *J. Organometallic Chem.*, **373**, 153 (1989).

28. H. P. Klug and L. E. Alexander, *X-ray Diffraction Procedures for Polycrystalline and Amorphous Materials*, 2nd ed., Wiley, New York, 1974.

29. H. F. McMurdie, M. C. Morris, E. H. Evans, B. Paretzkin, W. Wong-Ng, and C. R. Hubbard, "Methods of producing standard X-ray diffraction powder patterns," *Powder Diffr.*, **1**, 40 (1986).

30. R. Jenkins, T. G. Fawcett, D. K. Smith, J. W. Visser, M. C. Morris, and L. K. Frevel, "Sample preparation methods in X-ray powder diffraction," *Powder Diffr.*, **1**, 51 (1986).

31. W. Wong-Ng and C. R. Hubbard, "Standard reference materials for X-ray diffraction. Part II. Calibration using d-spacing standards," *Powder Diffr.*, **2**, 242 (1988).

32. JCPDS Powder Diffraction File, PDF-2, JCPDS—International Centre for Diffraction Data, Swarthmore. Currently holds bibliographic information, single-phase X-ray powder diffraction patterns, and related data for 51,000 compounds. Printed indexes and computer-centered search/match tools are available.

33. E. W. Abel and F. G. A. Stone, *Organometallic Chemistry*, Vol. 1, Royal Society of Chemistry, London, 1972, and following annual volumes.

34. L. K. Frevel, "Computational aids for identifying crystalline phases by powder diffraction," *Anal. Chem.*, **37**, 471 (1965).

35. Lin Tian-Hui, Zhang Sai-Zhu, Chen Li-Jun, and Cai Xin-Xing, "An improved program for searching and matching X-ray powder diffraction patterns," *J. Appl. Cryst.*, **16**, 150 (1983).

36. J. H. Loehlin and A. P. Norton, "Crystallographic determination of molecular parameters for K_2SiF_6," *J. Chem. Educ.*, **65**, 480 (1988).

37. J. Plevert, M. Louer, and D. Louer, "The *ab initio* structure determination of $Cd_3(OH)_5(NO)_3$ from X-ray powder diffraction data," *J. Appl. Crystallogr.*, **22**, 470 (1989).

38. L. E. Alexander, *X-ray Diffraction Methods in Polymer Science*, Krieger, Huntington, NY, 1979.

39. H. Tadokoro, *Structure of Crystalline Polymers*, Wiley, New York, 1979.

40. D. R. Petersen, D. R. Carter, and C. L. Lee, "Analysis of X-ray and electron beam diffraction patterns from poly(dipropylsiloxane)," *J. Macromol. Sci.—Phys.*, **B3**, 519 (1969).

41. L. E. Sutton, *Tables of Interatomic Distances and Configuration in Molecules and Ions*, Special Publication No. 11, 1958 and *Supplement 1956–1959*, Special Publication No. 18, 1965, The Chemical Society, London.

42. O. Glatter and O. Kratky, Eds., *Small Angle X-ray Scattering*, Academic Press, New York, 1982.

43. S. Mallam, A.-M. Hecht, and E. Geissler, "Structure of swollen poly(dimethylsiloxane) gels," *J. Chem. Phys.*, **91**, 6447 (1989).

44. B. K. Teo and D. C. Joy, Eds., *EXAFS Spectroscopy Techniques and Applications*, Plenum Press, New York, 1981.

45. E. P. Bertin, *Principles and Practice of X-Ray Spectrometric Analysis*, Plenum Press, New York, 1970, p. 638.

46. R. Tertian and F. Claisse, *Principles of Quantitative X-Ray Fluorescence Analysis*, Heyden & Son Ltd., London, 1982, p. 319.

47. E. W. White, G. V. Gibbs, G. G. Johnson, and G. R. Zechman, *X-Ray Emission Line Wavelength and Two-Theta Tables*, ASTM Data Series DS 37, ASTM Philadelphia, 1965.

48. J. Lucas-Tooth and C. Pyne, "The accurate determination of major constituents by X-ray fluorescent analysis in the presence of large interelement effects," in *Advances in X-ray Analysis*, Vol. 7, W. M. Mueller, G. Mallett, and M. Fay, Eds., Plenum Press, New York, 1964, p. 523.

49. R. Tertian and F. Claisse, *Principles of Quantitative X-Ray Fluorescence Analysis*, Heyden & Son Ltd., London, 1982, p. 218.

50. R. Jenkins and J. L. DeVries, *Practical X-Ray Spectrometry*, Springer Verlag, New York, 1970, p. 129.

51. E. D. Lipp, P. S. Rzyrkowski, and R. F. Geiger, "Comparison of X-ray fluorescence with other techniques for the quantitative analysis of silicone paper coatings," *Tappi J.*, **70**(5), 95 (1987).

52. J. M. Walls, *Methods of Surface Analysis*, Cambridge University Press, Cambridge, England, 1988.

53. D. P. Woodruff and T. A. Delchar, *Modern Techniques of Surface Science*, Cambridge University Press, Cambridge, England, 1986.

54. J. D. Andrade, "X-ray photoelectron spectroscopy (XPS)," in *Surface and Interfacial Aspects of Biomedical Polymers*, Vol. 1, J. D. Andrade, Ed., Plenum Press, New York, 1985, p. 105.

55. N. H. Turner, "Surface analysis: X-ray photoelectron spectroscopy and Auger electron spectroscopy," *Anal. Chem.*, **60**, 377R (1988).

56. M. Millard and A. Pavlath, "Surface analysis of wool fibers and fiber coatings by X-ray photoelectron spectroscopy," *Text. Res. J.*, **42**, 460 (1972).

57. G. Stupian, "Auger spectroscopy of silicones," *J. Appl. Phys.*, **45**, 5278 (1974).

58. C. R. Brundle, "Ultra-high vacuum techniques of surface characterization," in *Industrial Applications of Surface Analysis* (ACS Symp. Ser.) **199**, 13 (1982).

59. D. T. Clark, A. Dilks, J. Peeling, and H. R. Thomas, "Application of ESCA to studies of structure and bonding in polymers," *Faraday Discuss. Chem. Soc.*, **60**, 183 (1976).

60. J. A. Gardella, Jr., S. A. Ferguson, and R. L. Chin, "$\pi^* \leftarrow \pi$ Shakeup satellites for the analysis of structure and bonding in aromatic polymers by X-ray photoelectron spectroscopy," *Appl. Spectrosc.*, **40**, 224 (1986).

61. A. Toth, I. Bertoti, T. Szekely, and M. Mohai, "XPS study of ion-induced changes on the surface of an organosilicon model polymer," *Surf. Interface Anal.*, **7**, 282 (1985).

62. A. Dilks, "Polymer surfaces," *Anal. Chem.*, **53**, 802A (1981).

63. B. D. Ratner, P. K. Weathersby, A. S. Hoffman, M. A. Kelly, and L. H. Scharpen, "Radiation-grafted hydrogels for biomaterial applications as studied by the ESCA technique," *J. Appl. Polym. Sci.*, **22**, 643 (1978).

64. D. E. Williams and L. E. Davis, "Sputter induced compositional changes during ESCA/sputtering of polymers," *ACS Div. Org. Coat. Plast. Chem. Preprints*, **36**, 249 (1976).

65. L. A. Duel and M. J. Owen, "ESCA studies of silicone release coatings," *J. Adhesion*, **16**, 49 (1983).

66. M. J. Owen, "Surface tension of polytrifluoropropylmethylsiloxane," *J. Appl. Polym. Sci.*, **35**, 895 (1988).

67. C. D. Wagner, "Auger and photoelectron energies and the Auger parameter: A data set," in *Practical Surface Analysis by Auger and X-ray Photoelectron Spectroscopy*, D. Briggs and M. P. Seah, Eds., Wiley, New York, 1983, p. 477.

68. N. H. Sung, A. Kaul, I. Chin, and C. S. P. Sung, "Mechanistic studies of adhesion promotion by γ-aminopropyl triethoxy silane in α-Al_2O_3/polyethylene joint," *Polym. Eng. Sci.*, **22**, 637 (1982).

69. D. E. Williams and T. J. Tangney, "Surface silylation mechanism studies," in *Silanes, Surfaces and Interfaces*, D. E. Leyden, Ed., Gordon & Breach, New York, 1986, p. 471.

70. T. M. Gentle and M. J. Owen, "X-ray photoelectron spectroscopic investigation of direct process contact masses," *J. Catalysis*, **103**, 232 (1987).

71. G. L. Grobe, III, J. A. Gardella, Jr., W. L. Hopson, W. P. McKenna, and E. M. Eyring, "Angular dependent ESCA and infrared studies of segmented polyurethanes," *J. Biomed. Mater. Res.*, **21**, 211 (1987).

72. M. K. Chaudhury, T. M. Gentle, and E. P. Plueddemann, "Adhesion mechanism of polyvinyl chloride to silane primed metal surfaces," *J. Adhesion Sci. Technol.*, **1**, 29 (1987).

73. M. Gettings, F. S. Baker, and A. J. Kinloch, "Use of Auger and X-ray photo-electron spectroscopy to study the locus of failure of structural adhesive joints," *J. Appl. Polym. Sci.*, **21**, 2375 (1977).

74. P. Herder, L. Vagberg, and P. Stenius, "ESCA and contact angle studies of the adsorption of aminosilanes on mica," *Colloids Surf.*, **34**, 117 (1988/1989).

75. H. R. Anderson, Jr., and K. G. Sachdev, "X-ray photoelectron spectroscopy in understanding adhesion phenomena," in *Treatise on Adhesion and Adhesives*, Vol. 6, R. L. Patrick, Ed., Marcel Dekker, New York, 1989, p. 213.

76. P. R. Moses, L. M. Wier, J. C. Lennox, H. O. Finklea, J. R. Lenhard, and R. W. Murray, "X-ray photoelectron spectroscopy of alkylamine–silanes bound to metal oxide electrodes," *Anal. Chem.*, **50**, 576 (1978).

77. J. F. Cain and E. Sacher, "Auger electron spectroscopy of deposited silane layers," *J. Colloid Interface Sci.*, **67**, 538 (1978).

78. R. W. Linton, M. L. Miller, G. E. Maciel, and B. L. Hawkins, "Surface character-ization of chemically modified (trimethylsilyl) silicas by ^{29}Si solid state NMR, XPS, and IR photoacoustic spectroscopy," *Surf. Interface Anal.*, **7**, 196 (1985).

79. P. M. Triolo and J. D. Andrade, "Surface modification and evaluation of some commonly used catheter materials. I. Surface properties," *J. Biomed. Mater. Res.*, **17**, 129 (1983).

80. N. Inagaki, S. Kondo, M. Hirata, and H. Urushibata, "Plasma polymerization of organosilicon compounds," *J. Appl. Polym. Sci.*, **30**, 3385 (1985).

81. I. Tajima and M. Yamamoto, "Spectroscopic study on chemical structure of plasma-polymerized hexamethyldisiloxane," *J. Polym. Sci. Polym. Chem. Ed.*, **23**, 615 (1985).

82. C. A. Haque and A. K. Spiegler, "Study of silicone oil contamination on relay contacts using Auger electron spectroscopy," *Appl. Surf. Sci.*, **4**, 214 (1980).

83. J. L. Kardos and R. Fountain, "The use of X-ray photoelectron spectroscopy (ESCA) to characterize cure inhibition in a polysiloxane adhesive," *J. Polym. Sci. Polym. Lett. Ed.*, **12**, 161 (1974).

84. T. J. Hook, R. L. Schmitt, J. A. Gardella, Jr., L. Salvati, Jr., and R. L. Chin, "Analysis of polymer surface structure by low-energy ion scattering spectro-scopy," *Anal. Chem.*, **58**, 1285 (1986).

85. R. H. Bott, J. D. Summers, C. A. Arnold, L. T. Taylor, T. C. Ward, and J. E. McGrath, "Synthesis and characteristics of novel poly(imide siloxane) segmented copolymers," *J. Adhesion*, **23**, 67 (1987).

86. D. T. Clark, J. Peeling, and J. M. O'Malley, "Application of ESCA to polymer chemistry. VIII. Surface structures of AB block copolymers of polydimethylsilox-ane and polystyrene," *J. Polym. Sci. Polym. Chem. Ed.*, **14**, 543 (1976).

87. R. L. Schmitt, J. A. Gardella, Jr., and L. Salvati, Jr., "Studies of surface composi-tion and morphology in polymers. 2. Bisphenol A polycarbonate and poly-(dimethylsiloxane) blends," *Macromolecules*, **19**, 648 (1986).

88. Y. Tezuka, A. Fukushima, S. Matsui, and K. Imai, "Surface studies on poly(vinyl alcohol)-poly(dimethylsiloxane) graft copolymers," *J. Colloid Interface Sci.*, **114**, 16 (1986).

89. C. S. P. Sung and C. B. Hu, "Application of Auger electron spectroscopy for surface chemical analysis of Avcothane," *J. Biomed. Mater. Res.*, **13**, 45 (1979).

90. S. W. Graham and D. M. Hercules, "Surface spectroscopic studies of Avcothane," *J. Biomed. Mater. Res.*, **15**, 349 (1981).

91. T. Hsu, S. S. Kantner, and M. Mazurek, "Surface effects in siloxane-containing methacrylates," *Polym. Mater. Sci. Eng.*, **55**, 562 (1986).

92. T. C. Frank and J. L. Falconer, "Silane formation on silicon: Reaction kinetics and surface analysis," *Langmuir*, **1**, 104 (1985).

93. C. Davies and H. S. Munro, "Surface derivatization of poly(acrylic acid)," *Polym. Commun.*, **29**, 47 (1988).

APPENDIX

PHYSICAL PROPERTIES OF ORGANOSILICON COMPOUNDS

ORA L. FLANINGAM

Dow Corning Corporation
Midland, Michigan

Physical properties of organosilicon compounds are tabulated for convenience. This listing is not intended to be comprehensive.

Physical property values for linear and cyclic poly(dimethylsiloxanes) are given in Tables 7.2 and 7.3.

Boiling points are at 760 torr unless otherwise indicated.

Densities (grams per milliliter, g/mL), refractive indexes, and viscosities (centipoise, cP) are given at 20°C unless another temperature is specified.

Reference numbers refer to the references at the end of this section.

An empirical formula index is given at the end of this table.

The Analytical Chemistry of Silicones, Edited by A. Lee Smith.
ISBN 0-471-51624-4 © 1991 John Wiley & Sons, Inc.

Table A. Physical Properties of Some Chlorosilanes and Related Compounds

Number	Compound	Melting Point (°C)	Boiling Point (°C) at P (torr)	d_4^{20}	n_D^{20}	Viscosity. (cP at 20°C)	References
1	H₃SiCl	−118	−30.5	1.007 at −30°C		0.160 at −19.9°C	14, 95
2	H₂SiCl₂	−122	8.3	1.180	1.436	0.234	19, 95
3	HSiCl₃	−127	31.9	1.3417	1.4020	0.341	19, 58, 76
4	SiCl₄	−68.4	57.1	1.48	1.411 at 25°C	0.473	19, 23, 76
5	MeSiCl₃	−75.8	66.1	1.273	1.412	0.488	23, 35, 78
6	Me₂SiCl₂	−74.2	70.1	1.034	1.4025 at 25°C	0.490	19, 23, 35, 78
7	Me₃SiCl	−55.2	57.6	0.861	1.389	0.395	23, 33, 35
8	MeHSiCl₂	−92.5	41.2	1.1118	1.3992	0.344	9, 23, 33, 81
9	MeH₂SiCl	−134	9.8	0.935 at −80°C			95
10	Me₂HSiCl	−102	35	0.869	1.381	0.293	23, 26, 96
11	MeEtSiCl₂		101.6	1.057	1.419	0.640	23
12	MeEtHSiCl		69.1	0.882		0.387	23
13	Me₂EtSiCl		90				23, 94
14	MeEt₂SiCl		120	0.898	1.419		88, 94
15	EtSiCl₃	−105	98.9	1.2342	1.4256	0.606	8, 17, 23, 105
16	Et₂SiCl₂	−99.1	130.4	1.052	1.4232	0.707	23, 39, 62
17	Et₃SiCl	−76	146.8	0.897	1.4315		4, 17, 23, 62
18	EtH₂SiCl		43	0.90	1.397		23, 85
19	EtHSiCl₂	−102.5	75.6	1.0926	1.4148	0.474	23, 58, 62
20	Et₂HSiCl	−99	100	0.8895	1.415	0.49	27, 44, 66
21	n-PrSiCl₃		124.5	1.195	1.4290	0.703	23,74
22	i-PrSiCl₃		120 at 748	1.1934	1.4319		74
23	PrMeSiCl₂		125	1.036	1.425		23
24	ClPrSiCl₃		182.8	1.358		1.847	23
25	ClPrMeSiCl₂		186	1.205	1.461		23
26	n-BuSiCl₃		148	1.1603	1.4352		23, 74

No.	Compound	m.p. (°C)	b.p. (°C)	d	n_D		Ref.
27	i-BuSiCl₃	−78	142	1.162	1.4381		23
28	s-BuSiCl₃		147	1.176	1.4410		23
29	t-BuSiCl₃	99	134				23, 102
30	ViHSiCl₂		66.5	1.122	1.4254		23
31	ViH₂SiCl		26 (est.)	0.925 (est.)			23
32	Vi₂HSiCl		86 (est.)	0.933 (est.)			23
33	ViSiCl₃	−95	90.7	1.270	1.4295		23, 73
34	Vi₂SiCl₂		118	1.0813	1.451		23, 36, 87
35	MeViSiCl₂	−95	93.8	1.084	1.451		23, 36
36	Me₂ViSiCl		83	0.892	1.4162	0.368	18, 23
37	MeVi₂SiCl		106	0.914	1.4409		23
38	Vi₃SiCl		130	0.9342	1.4602		23
39	cyclo-HexSiCl₃		204	1.223	1.477		23
40	PhH₂SiCl		162	1.076	1.5257		23
41	Ph₂HSiCl		286	1.118	1.581		23
42	PhSiCl₃	−39.8	201.8	1.318	1.5240 at 25°C	1.438	23, 77
43	Ph₂SiCl₂	−22	304	1.214	1.5814	5.816	23, 68, 77
44	Ph₃SiCl	96	378	1.038 at 100°C		2.27 at 100°C	77, 106
45	PhHSiCl₂		186	1.211	1.5246		23, 38, 55
46	PhMeSiCl₂	−43.5	204	1.173	1.518	1.619	25, 23
47	PhMe₂SiCl		195	1.025	1.5082	1.468	20, 23
48	Ph₂MeSiCl		299	1.089	1.5747	7.01	23
49	PhMeHSiCl		178	1.055	1.5171		23
50	PhEtSiCl₂		230	1.156	1.519		23
51	PhViSiCl₂		225	1.170	1.5307		23
52	Cl₃SiSiCl₃	−1	149	1.551	1.4748	1.666	59, 66, 91
53	Cl₃SiOSiCl₃	−28	135	1.560		1.04	23, 46, 82, 83
54	SiF₄	−86.8	Sublimes				19
55	(Prf₃)SiCl₃		114	1.4076	1.3870		23
56	(Prf₃)MeSiCl₂		122	1.214	1.3840	0.991	23
57	(Prf₃)Me₂SiCl		118	1.112	1.3745		23

Table B. Physical Properties of Some Silicon-Functional Species

Number	Compound	Melting Point (°C)	Boiling Point (°C) at P (torr)	d_4^{20}	n_D^{20}	Viscosity (cP at 20°C)	References
58	Me_3SiOH	9.5	99	0.8139	1.3889	6.569	5, 23, 31, 34, 93
59	$Me_2Si(OH)_2$	96–98	Decomposes	1.097	1.454 at 25°C		37
60	$HSi(OMe)_3$	−114	84	0.958	1.446 1.3561		23
61	$HOMe_2SiOSiMe_2OH$	68	231(?)	1.10 Solid	1.46 Solid		23, 55
62	Et_3SiOH		63 at 12	0.8638	1.4329		34, 93
63	$Et_2Si(OH)_2$	96	140	1.134	1.413		30, 41
			Decomposes		1.493		
					1.517		
64	$(n\text{-}Pr)_3SiOH$		207				70
65	$(i\text{-}Pr)_3SiOH$		196 at 750		1.4542		25
66	Ph_3SiOH	155		1.1777	1.564		21, 40
67	$Ph_2Si(OH)_2$	137–141			1.648 1.656		65

526

No.	Compound	m.p.	b.p.	d	n_D		Ref.
68	PhMeSi(OH)$_2$	84		0.8961	1.388		23
69	Me$_3$SiOAc	−32	104	1.0485	1.403		7, 28, 84
70	Me$_2$Si(OAc)$_2$		45 at 3	1.1697	1.407		7, 84
71	MeSi(OAc)$_3$	41	95 at 9				7, 9, 84
72	Si(OAc)$_4$	110					84
73	ViSi(OAc)$_3$		220(?)	1.1702	1.423		23
74	EtSi(OAc)$_3$	5	227(?)	1.1437	1.412		23
75	Me$_3$SiNHMe		71	0.7295	1.3905		80
76	Me$_2$Si(NMe$_2$)$_2$	−98	128	0.809	1.4169 at 22°C		2
77	MeSi(NMe$_2$)$_3$	−11	161	0.850	1.4324 at 22°C		2
78	Si(NMe$_2$)$_4$	−2	180	0.973 at 22°C			2
79	Ph$_3$SiNH$_2$	56					48
80	Ph$_3$SiNMe$_2$	81					32
81	Ph$_2$Si(NMe$_2$)$_2$		156 at 3				51
82	(Me$_3$Si)$_2$NH		126	0.7741	1.4078	0.624	23, 80
83	(Me$_2$ViSi)$_2$NH		164	0.819	1.4405		23
84	(Me$_3$Si)$_2$NMe		148 at 740	0.794 at 25°C	1.4190 at 25°C		23, 67
85	(Me$_2$SiNH)$_3$	−10	188	0.9196	1.4448	1.523	11, 23
86	(Me$_2$SiNH)$_4$	97	236	0.818 at 102°C		1.48 at 102°C	11, 23

Table C. Physical Properties of Some Alkoxy and Aryloxy Silanes and Siloxanes

Number	Compound	Melting Point (°C)	Boiling Point (°C) at P (torr)	d_4^{20}	n_D^{20}	Viscosity (cP at 20°C)	References
87	Si(OMe)$_4$	5	122	1.033	1.3688	0.643	54, 97, 104
88	H$_2$Si(CH$_2$)$_3$		43	0.783	1.4422	0.387	23
89	MeSi(OMe)$_3$	−78	103	0.955	1.3689	0.54	23, 42, 103
90	Me$_2$Si(OMe)$_2$	−80	82	0.853	1.3708	0.378	23, 69
91	Me$_3$SiOMe		57	0.7592	1.3675		23, 79
92	Si(OEt)$_4$	−77	168	0.9346	1.3831	0.721	24, 54, 53, 104
93	MeSi(OEt)$_3$		143	0.899 at 25 °C	1.3844 at 25 °C		98
94	Me$_2$Si(OEt)$_2$		112	0.827 at 25 °C	1.3840 at 25 °C		98
95	Me$_3$SiOEt		76	0.755 at 25 °C	1.3737 at 25 °C		98
96	(MeO)$_3$SiOSi(OMe)$_3$		58 at 1	1.1222	1.3806		103
97	(MeO)$_3$SiSi(OMe)$_3$		210	1.095	1.4087		23
98	(EtO)$_3$SiOSi(OEt)$_3$		235	0.9979	1.3915		82, 103

No.	Compound	mp	bp	d	n		Ref.
99	Me₃SiOPh		182 at 742	0.9256	1.4782		50
100	MeViSi(OMe)₂		104.5	0.8889	1.392	0.452	23
101	Me₂ViSiOMe		82.3	0.797	1.392	0.344	23
102	ViSi(OMe)₃		125	0.971	1.3915	1.21 at 25°C	23
103	EtSi(OMe)₃		125	0.9494	1.3838 at 25°C	0.562	23
104	PrSi(OMe)₃		144	0.938	1.3883	0.682	23
105	(ClPr)MeSi(OMe)₂		185	1.024	1.4242 at 25°C		23
106	(ClPr)Si(OMe)₃		199	1.079	1.4189 at 25°C	0.600	23
107	PhSi(OMe)₃	− 25	218	1.063 at 25°C	1.4710 at 25°C	2.10	15, 23, 90
108	Ph₂Si(OMe)₂		292	1.0771	1.5450	6.9 at 25°C	6
109	Ph₃SiOMe		235				23, 86
110	PhSi(OEt)₃		235	0.995	1.4580 at 25°C	1.65 at 25°C	45, 90
111	Ph₂Si(OEt)₂		299	1.0329	1.5270		63
112	Ph₃SiOEt	65	366				15, 23
113	PhMeSi(OMe)₂		205	0.9934	1.4694	1.645	13, 20
114	PhMeSi(OEt)₂		218	0.9627	1.4701		58
115	HSi(OEt)₃		132	0.8903 at 25°C	1.3767		

Table D. Physical Properties of Some Alkyl and Aryl Silanes and Siloxanes

Number	Compound[a]	Melting Point (°C)	Boiling Point (°C) at P (torr)	d_4^{20}	n_D^{20}	Viscosity (cP at 20°C)	References
116	SiH_4	−185	−112	0.709 at −185°C		0.20 at −140°C	19, 76
117	H_3SiSiH_3	−132	−14.6	0.788	1.5700		19
118	$MeSiH_3$	−157	−58	0.6277 at −58°C			99
119	Me_2SiH_2	−150	−20	0.6377 at −20°C			99
120	$MeEt_3Si$		128	0.7337	1.4160		23
121	Me_2Et_2Si		96	0.7129	1.4010		23
122	Me_3EtSi		62	0.6834	1.3820		23
123	Me_3SiH	−136	6.7	0.6375 at 6.7°C			99
124	$PhSiH_3$		120	0.8817	1.5125		6, 29
125	Ph_2SiH_2		260	0.9969	1.5795		6, 23
126	Me_4Si	−99	26	0.647	1.3582	0.242	3, 12, 99
127	Me_3ViSi	−132	54	0.692	1.3903	0.399	23
128	Me_2Vi_2Si		81	0.73	1.4172		23
129	$MeVi_3Si$		107	0.778	1.4405	0.40	23
130	$ViSiH_3$	−171	−23	0.675 at −30°C	1.418		23
131	Vi_2SiH_2		42	0.773	1.4475		23
132	Vi_3SiH		91	0.799	1.4620		23
133	Vi_4Si		130	0.799	1.4620		23
134	Et_4Si	−82	154	0.7662	1.4267	0.468	99
135	Et_3SiH	−157	109	0.7318	1.4119	0.412	58, 59
136	Et_2SiH_2	−132	56	0.683	1.3916	0.287 at 25°C	23
137	$EtSiH_3$	−180	−14	0.640 at −14°C			23
138	$(n\text{-}Pr)_4Si$	−46	213 at 742	0.787	1.4378		64, 71
139	$(n\text{-}Bu)_4Si$		156 at 22	0.8008	1.4465		32

530

No.	Compound	mp	bp	d	n_D	η	Ref.
140	Ph_3SiH	44	352	0.870 at 252°C			23
141	Ph_4Si	235	428	0.870			23, 75, 100
142	$PhMe_2SiH$	−124	159	0.9973	1.4990		23
143	Ph_2MeSiH	−73	266	0.869	1.5720	3.0 at 25°C	23
144	Me_3PhSi	−62	171	0.9867	1.4886	0.865 at 25°C	23, 57
145	Me_2Ph_2Si		277		1.5644	3.57 at 25°C	16, 23, 60
146	$MePh_3Si$	68	358				23, 52
147	$Me_3SiSiMe_3$	14	113	0.7265	1.4228	0.747	23
148	$Ph_3SiOSiMe_3$	51	349 at 750	1.032 at 25°C	1.5587 at 25°C		20
149	$Ph_3SiOSiPh_3$	226	494 at 750		1.59, 1.68, 1.71 at 25°C		20
150	$(MePh_2Si)_2O$	50	420	1.050 at 60°C	1.5866 at 25°C	11.3 at 70°C	23
151	$Me_3SiO(Ph_2SiO)SiMe_3$	−78	110 at 0.3	0.984	1.4981		89, 101
152	$Me_3SiO(PhMeSiO)_2SiMe_3$	−40	326	0.979	1.4774	1.22 at 25°C	23
153	$PhMe_2SiO(PhMeSiO)SiPhMe_2$		169 at 0.7	1.0227	1.528		56
154	$(Me_3SiO)_4Si$	−53	222	0.8657	1.3895	2.82	23, 92
155	$MeSi(OSiMe_3)_3$	−74	191	0.850	1.3880	1.44	23
156	$(HMe_2SiO)_4Si$		190	0.886	1.3869	0.982	23
157	$(HMe_2Si)_2O$	−103	71.5	0.757	1.3695	0.42	23
158	$(ViMe_2Si)_2O$		140	0.816	1.412		23
159	$(Ph_2SiO)_3$	192	600 +/− 30				23
160	$(Ph_2SiO)_4$	202	650 +/− 50				23
161	$(MeEtSiO)_3$	−2.9	198	0.9443	1.4110	2.06	23
162	$(MeEtSiO)_4$	−40	245	0.9513	1.4164	5.69	11
163	$(MeEtSiO)_5$	−122	286	0.9616	1.4208	9.37	23
164	$(MeHSiO)_3$		94	0.9677	1.377		10
165	$(MeHSiO)_4$	−69	135	0.9912	1.3870		81
166	$(MeHSiO)_5$	−108	169	0.9985	1.3912		81
167	$(MeHSiO)_6$	−79	93 at 21	1.006	1.3944		81
168	$(MeViSiO)_3$		188	0.9692	1.4245	1.452	23, 61
169	$(MeViSiO)_4$	−44	229	0.9875	1.4342	3.82	23, 43
170	$(MeViSiO)_5$	−140	277	0.9936	1.4373	6.8	23
171	$(MeViSiO)_6$	−119	311	1.002	1.4400	16.1	23

Table D. (Contd.)

Number	Compound[a]	Melting Point (°C)	Boiling Point (°C) at P (Torr)	d_4^{20}	n_D^{20}	Viscosity (cP at 20°C)	References
172	(Me₂SiCH₂)₂		119		1.4380		47
173	(Me₂SiCH₂)₃	80	200 at 745	0.846 at 25°C	1.4606 at 25°C		22
174	(Me₂SiCH₂)₄	35	104 at 2				49
175	((Prf₃)MeSiO)₃[b]	−15	241	1.247	1.3660		23
176	((Prf₃)MeSiO)₄[c]	77 41 −90	293	1.272	1.3720		19
177	(Prf₃)SiMe₃	64.5	95	0.945	1.3572		23
178	(PhMeSiO)₄[c]	59 100 74	427	1.12	1.5461		107

[a] Abbreviations: Bu = butyl, ClPr = 1-chloropropyl, Et = ethyl, Me = methyl, Ph = phenyl, Pr = propyl, Prf₃ = 3,3,3-trifluoropropyl Vi = vinyl.
[b] Two isomers.
[c] Four isomers.

Compound Number	Empirical Formula	Name
5	CH_3Cl_3Si	Methyltrichlorosilane
8	CH_4Cl_2Si	Methyldichlorosilane
9	CH_5ClSi	Methylchlorosilane
118	CH_6Si	Methylsilane
33	$C_2H_3Cl_3Si$	Vinyltrichlorosilane
30	$C_2H_4Cl_2Si$	Vinyldichlorosilane
31	C_2H_5ClSi	Vinylchlorosilane
15	$C_2H_5Cl_3Si$	Ethyltrichlorosilane
19	$C_2H_6Cl_2Si$	Ethyldichlorosilane
6	$C_2H_6Cl_2Si$	Dimethyldichlorosilane
130	C_2H_6Si	Vinylsilane
18	C_2H_7ClSi	Ethylchlorosilane
10	C_2H_7ClSi	Dimethylchlorosilane
59	$C_2H_8O_2Si$	$Me_2Si(OH)_2$
137	C_2H_8Si	Ethylsilane
119	C_2H_8Si	Dimethylsilane
55	$C_3H_4Cl_3F_3Si$	3,3,3-Trifluoropropyltrichlorosilane
35	$C_3H_6Cl_2Si$	Methylvinyldichlorosilane
24	$C_3H_6Cl_4Si$	3-Chloropropyltrichlorosilane
21	$C_3H_7Cl_3Si$	n-Propyltrichlorosilane
22	$C_3H_7Cl_3Si$	i-Propyltrichlorosilane
88	C_3H_7Si	Silacyclobutane
11	$C_3H_8Cl_2Si$	Ethylmethyldichlorosilane
12	C_3H_8ClSi	Ethylmethylchlorosilane
7	C_3H_9ClSi	Trimethylchlorosilane
58	$C_3H_{10}OSi$	Trimethylsilanol
60	$C_3H_{10}O_3Si$	Trimethoxysilane
123	$C_3H_{10}Si$	Trimethylsilane
164	$C_3H_{12}O_3Si_3$	$(MeHSiO)_3$
34	$C_4H_6Cl_2Si$	Divinyldichlorosilane
32	C_4H_7ClSi	Divinylchlorosilane
56	$C_4H_7Cl_2F_3Si$	3,3,3-Trifluoropropylmethyldichlorosilane
131	C_4H_8Si	Divinylsilane
36	C_4H_9ClSi	Dimethylvinylchlorosilane
25	$C_4H_9Cl_3Si$	3-Chloropropylmethyldichlorosilane
26	$C_4H_9Cl_3Si$	n-Butyltrichlorosilane

Compound Number	Empirical Formula	Name
27	$C_4H_9Cl_3Si$	*i*-Butyltrichlorosilane
28	$C_4H_9Cl_3Si$	*s*-Butyltrichlorosilane
29	$C_4H_9Cl_3Si$	*t*-Butyltrichlorosilane
16	$C_4H_{10}Cl_2Si$	Diethyldichlorosilane
23	$C_4H_{10}Cl_2Si$	*n*-Propylmethyldichlorosilane
13	$C_4H_{11}ClSi$	Ethyldimethylchlorosilane
20	$C_4H_{11}ClSi$	Diethylchlorosilane
91	$C_4H_{12}OSi$	Trimethylmethoxysilane
63	$C_4H_{12}O_2Si$	$Et_2Si(OH)_2$
90	$C_4H_{12}O_2Si$	Dimethyldimethoxysilane
89	$C_4H_{12}O_3Si$	Methyltrimethoxysilane
87	$C_4H_{12}O_4Si$	Tetramethoxysilane
126	$C_4H_{12}Si$	Tetramethylsilane
136	$C_4H_{12}Si$	Diethylsilane
75	$C_4H_{13}NSi$	$Me_3SiNHMe$
157	$C_4H_{14}OSi_2$	1,1,3,3-Tetramethyldisiloxane
61	$C_4H_{14}O_3Si_2$	Tetramethyldisiloxanediol
165	$C_4H_{16}O_4Si_4$	$(MeHSiO)_4$
37	C_5H_9ClSi	Methyldivinylchlorosilane
57	$C_5H_{10}ClF_3Si$	3,3,3-Trifluoropropyldimethylchlorosilane
68	$C_5H_{10}O_2Si$	$PhMeSi(OH)_2$
101	$C_5H_{12}OSi$	Dimethylvinylmethoxysilane
69	$C_5H_{12}O_2Si$	Me_3SiOAc
100	$C_5H_{12}O_2Si$	Methylvinyldimethoxysilane
102	$C_5H_{12}O_3Si$	Vinyltrimethoxysilane
127	$C_5H_{12}Si$	Trimethylvinylsilane
14	$C_5H_{13}ClSi$	Diethylmethylchlorosilane
95	$C_5H_{14}OSi$	Trimethylethoxysilane
103	$C_5H_{14}O_3Si$	Ethyltrimethoxysilane
122	$C_5H_{14}Si$	Ethyltrimethylsilane
166	$C_5H_{20}O_5Si_5$	$(MeHSiO)_5$
42	$C_6H_5Cl_3Si$	Phenyltrichlorosilane
45	$C_6H_6Cl_2Si$	Phenyldichlorosilane
40	C_6H_7ClSi	Phenylchlorosilane
124	C_6H_8Si	Phenylsilane
38	C_6H_9ClSi	Trivinylchlorosilane
132	$C_6H_{10}Si$	Trivinylsilane
39	$C_6H_{11}Cl_3Si$	Cyclohexyltrichlorosilane

Compound Number	Empirical Formula	Name
70	$C_6H_{12}O_4Si$	$Me_2Si(OAc)_2$
128	$C_6H_{12}Si$	Dimethyldivinylsilane
177	$C_6H_{13}F_3Si$	3,3,3-Trifluoropropyltrimethylsilane
105	$C_6H_{15}ClO_2Si$	3-Chloropropylmethyldimethoxysilane
106	$C_6H_{15}ClO_3Si$	3-Chloropropyltrimethoxysilane
17	$C_6H_{15}ClSi$	Triethylchlorosilane
172	$C_6H_{16}Si_2$	$(Me_2SiCH_2)_2$
62	$C_6H_{16}OSi$	Et_3SiOH
94	$C_6H_{16}O_2Si$	Dimethyldiethoxysilane
104	$C_6H_{16}O_3Si$	*n*-Propyltrimethoxysilane
115	$C_6H_{16}O_3Si$	$HSi(OEt)_3$
135	$C_6H_{16}Si$	Triethylsilane
121	$C_6H_{16}Si$	Diethyldimethylsilane
76	$C_6H_{18}N_2Si$	$Me_2Si(NMe)_2$
97	$C_6H_{18}O_6Si_2$	Hexamethoxydisilane
96	$C_6H_{18}O_7Si_2$	Hexamethoxydisiloxane
147	$C_6H_{18}Si_2$	Hexamethyldisilane
82	$C_6H_{19}NSi_2$	Hexamethyldisilazane
85	$C_6H_{21}N_3Si_3$	Hexamethylcyclotrisilazane
64	$C_6H_{22}OSi$	$(n\text{-}Pr)_3SiOH$
65	$C_6H_{22}OSi$	$(i\text{-}Pr)_3SiOH$
167	$C_6H_{24}O_6Si_6$	$(MeHSiO)_6$
46	$C_7H_8Cl_2Si$	Phenylmethyldichlorosilane
49	C_7H_9ClSi	Phenylmethylchlorosilane
71	$C_7H_{12}O_6Si$	Methyltriacetoxysilane
129	$C_7H_{12}Si$	Methyltrivinylsilane
93	$C_7H_{18}O_3Si$	Methyltriethoxysilane
120	$C_7H_{18}Si$	Triethylmethylsilane
77	$C_7H_{21}N_3Si$	$MeSi(NMe_2)_3$
89	$C_7H_{21}NSi_2$	$(Me_3Si)_2NMe$
51	$C_8H_8Cl_2Si$	Phenylvinyldichlorosilane
50	$C_8H_{10}Cl_2Si$	Ethylphenyldichlorosilane
47	$C_8H_{11}ClSi$	Dimethylphenylchlorosilane
73	$C_8H_{12}O_6Si$	Vinyltriacetoxysilane
72	$C_8H_{12}O_8Si_4$	$Si(OAc)_4$
142	$C_8H_{12}Si$	Dimethylphenylsilane
133	$C_8H_{12}Si$	Tetravinylsilane
74	$C_8H_{14}O_6Si$	Ethyltriacetoxysilane

Compound Number	Empirical Formula	Name
158	$C_8H_{18}OSi_2$	Tetramethyldivinyldisiloxane
83	$C_8H_{19}NSi_2$	Tetramethyldivinyldisilazane
92	$C_8H_{20}O_4Si$	Tetraethoxysilane
134	$C_8H_{20}Si$	Tetraethylsilane
78	$C_8H_{24}N_4Si$	$Si(NMe_2)_4$
86	$C_8H_{28}N_4Si_4$	Octamethylcyclotetrasilazane
156	$C_8H_{28}O_4Si_5$	Tetrakisdimethylsiloxysilane
99	$C_9H_{14}OSi$	Trimethylphenoxysilane
113	$C_9H_{14}O_2Si$	Methylphenyldimethoxysilane
107	$C_9H_{14}O_3Si$	Phenyltrimethoxysilane
144	$C_9H_{14}Si$	Trimethylphenylsilane
168	$C_9H_{18}O_3Si_3$	Trimethyltrivinylcyclotrisiloxane
173	$C_9H_{24}Si_3$	$(Me_2SiCH_2)_3$
161	$C_9H_{24}O_3Si_3$	Ethylmethylcyclotrisiloxane
155	$C_{10}H_{30}O_3Si_4$	(Tris)trimethylsiloxymethylsilane
114	$C_{11}H_{18}O_2Si$	$PhMeSi(OEt)_2$
43	$C_{12}H_{10}Cl_2Si$	Diphenyldichlorosilane
41	$C_{12}H_{11}ClSi$	Chlorodiphenylsilane
67	$C_{12}H_{12}O_2Si$	$Ph_2Si(OH)_2$
125	$C_{12}H_{12}Si$	Diphenylsilane
110	$C_{12}H_{20}O_3Si$	Phenyltriethoxysilane
175	$C_{12}H_{21}F_9O_3Si_3$	3,3,3-Trifluoropropylmethylcyclotrisiloxane
169	$C_{12}H_{24}O_4Si_4$	Tetramethyltetravinylcyclotetrasiloxane
138	$C_{12}H_{28}Si$	$(n\text{-}Pr)_4Si$
98	$C_{12}H_{30}O_7Si_2$	Hexaethoxydisiloxane
174	$C_{12}H_{32}Si_4$	$(Me_2SiCH_2)_4$
162	$C_{12}H_{32}O_4Si_4$	Ethylmethylcyclotetrasiloxane
154	$C_{12}H_{36}O_4Si_5$	Tetrakistrimethylsiloxysilane
48	$C_{13}H_{13}ClSi$	Methyldiphenylchlorosilane
143	$C_{13}H_{14}Si$	Methyldiphenylsilane
108	$C_{14}H_{16}O_2Si$	Diphenyldimethoxysilane
145	$C_{14}H_{16}Si$	Dimethyldiphenylsilane
170	$C_{15}H_{30}O_5Si_5$	Petamethylpentavinylcyclopentasiloxane
163	$C_{15}H_{40}O_5Si_5$	Ethylmethylcyclopentasiloxane
111	$C_{16}H_{20}O_2Si$	Diphenyldiethoxysilane
81	$C_{16}H_{22}N_2Si$	$Ph_2Si(NMe_2)_2$

Appendix—Empirical Formula Index (*Contd.*)

Compound Number	Empirical Formula	Name
176	$C_{16}H_{28}F_{12}O_4Si_4$	3,3,3-Trifluoropropylmethylcyclotetrasiloxane
139	$C_{16}H_{36}Si$	$(n\text{-}Bu)_4Si$
44	$C_{18}H_{15}ClSi$	Triphenylchlorosilane
66	$C_{18}H_{16}OSi$	Ph_3SiOH
140	$C_{18}H_{16}Si$	Triphenylsilane
79	$C_{18}H_{17}NSi$	Ph_3SiNH_2
151	$C_{18}H_{28}O_2Si_3$	$Me_3SiO(Ph_2SiO)SiMe_3$
171	$C_{18}H_{36}O_6Si_6$	Hexamethylhexavinylcyclohexasiloxane
109	$C_{19}H_{18}OSi$	Ph_3SiOMe
146	$C_{19}H_{18}Si$	Methyltriphenylsilane
112	$C_{20}H_{20}OSi$	Triphenylethoxysilane
80	$C_{20}H_{21}NSi$	Ph_3SiNMe_2
152	$C_{20}H_{34}O_3Si_4$	$Me_3SiO(PhMeSiO)_2SiMe_3$
148	$C_{21}H_{24}O_2Si$	$Ph_3SiOSiMe_3$
153	$C_{22}H_{27}O_2Si_3$	$PhMe_2SiO(PhMeSiO)SiPhMe_2$
141	$C_{24}H_{20}Si$	Tetraphenylsilane
150	$C_{26}H_{26}OSi_2$	1,3-Dimethyltetraphenyldisiloxane
178	$C_{28}H_{32}O_4Si_4$	$(PhMeSiO)_4$
149	$C_{36}H_{30}O_5Si$	$Ph_3SiOSiPh_3$
159	$C_{36}H_{30}O_3Si_3$	Hexaphenylcyclotrisiloxane
160	$C_{48}H_{40}O_4Si_4$	Octaphenylcyclotetrasiloxane
1	ClH_3Si	Chlorosilane
2	Cl_2H_2Si	Dichlorosilane
3	Cl_3HSi	Trichlorosilane
4	Cl_4Si	Silicon tetrachloride
53	Cl_6OSi_2	Hexachlorodisiloxane
52	Cl_6Si_2	Hexachlorodisilane
54	F_4Si	Silicon tetrafluoride
116	H_4Si	Silane
117	H_6Si_2	Disilane

REFERENCES

1. T. Alfrey, F. J. Honn, and H. Mark, *J. Polym. Sci.*, **1**, 102 (1946).
2. H. H. Anderson, *J. Am. Chem. Soc.*, **74**, 1421 (1952).

3. J. G. Aston, R. M. Kennedy, and G. H. Messerly, *J. Am. Chem. Soc.*, **63**, 2343 (1941).
4. D. L. Bailey, L. H. Sommer, and F. C. Whitmore, *J. Am. Chem. Soc.*, **70**, 435 (1948).
5. M. I. Batuev, M. F. Shostakovskii, V. I. Belyaev, A. D. Matveeva, and E. V. Dubrova, *Dokl. Akad. Nauk SSSR*, **95**, 531 (1954).
6. R. A. Benkeser, H. Landesman, and D. J. Foster, *J. Am. Chem. Soc.*, **74**, 648 (1952).
7. A. F. Bidaud and P. Dumont, US Patent 2,573,302 (1951).
8. H. S. Booth and P. H. Carnell, *J. Am. Chem. Soc.*, **68**, 2650 (1946).
9. H. S. Booth and R. L. Jarry, *J. Am. Chem. Soc.*, **71**, 971 (1949).
10. S. D. Brewer, *J. Am. Chem. Soc.*, **70**, 3962 (1948).
11. S. D. Brewer and C. P. Haber, *J. Am. Chem. Soc.*, **70**, 3888 (1948).
12. A. Bygden, Doctoral Dissertation, Uppsala University, Uppsala, Sweden (1916).
13. V. Capuccio and R. Pirani, *Chim. Ind. (Milan)*, **33**, 282 (1951).
14. J-S Cheng, C. L. Yaws, L. L. Dickens, and J. R. Hopper, *Ind. Eng. Chem. Process Des. Dev.*, **23**, 48 (1984).
15. S. Chrzczonowicz and Z. Lasocki, *Rocz. Chem.*, **34**, 1667 (1960).
16. V. Chvalovsky and V. Bazant, *Chem. Listy*, **46**, 158 (1952).
17. C. Curran, R. M. Witucki, and P. A. McCusker, *J. Am. Chem. Soc.*, **72**, 4471 (1950).
18. J. W. Curry, *J. Am. Chem. Soc.*, **78**, 1686 (1956).
19. T. E. Daubert, R. P. Danner, American Institute of Chemical Engineers, Design Institute for Physical Property Data, Data Compilation, American Institute of Chemical Engineers, New York (1989).
20. W. H. Daudt and J. F. Hyde, *J. Am. Chem. Soc.*, **74**, 386 (1952).
21. W. Dilthey and F. Eduardoff, *Ber. Deut. Chem. Ges.*, **37**, 1139 (1904).
22. Dow Corning Corporation, Br. Patent 667,435 (1952).
23. Dow Corning Corporation, unpublished data.
24. N. Dyachkova, E. Vigdorovich, and L. Ivanyutin, *Zh. Fiz. Khim.*, **47**, 456 (1973).
25. C. Eaborn, *J. Chem. Soc.*, **1952**, 2840.
26. H. J. Emeleus and M. Onyszchuk, *J. Chem. Soc.*, **1958**, 604.
27. H. J. Emeleus and S. R. Robinson, *J. Chem. Soc.*, **1947**, 1592.
28. Y. Etienne, *C.R. Acad. Sci. Paris*, **235**, 966 (1952).
29. A. E. Finholt, A. C. Bond, Jr., K. E. Wilzbach, and H. I. Schlesinger, *J. Am. Chem. Soc.*, **69**, 2692 (1947).
30. P. D. George, L. H. Sommer, and F. C. Whitmore, *J. Am. Chem. Soc.*, **75**, 1585 (1953).
31. H. Gilman and R. N. Clark, *J. Am. Chem. Soc.*, **69**, 967 (1947).
32. H. Gilman, B. Hofferth, H. W. Melvin, and G. E. Dunn, *J. Am. Chem. Soc.*, **72**, 5767 (1950).

33. V. P. Glushko, *Thermal Constants of Substances*, Vol. IV, Part 1. Academy of Sciences, Moscow, USSR, 1970.

34. W. T. Grubb and R. C. Osthoff, *J. Am. Chem. Soc.*, **75**, 2230 (1952).

35. R. L. Halm, private communication of evaluated data, Dow Corning Corp., Midland, MI.

36. D. T. Hurd, *J. Am. Chem. Soc.*, **67**, 1813 (1945).

37. J. F. Hyde, *J. Am. Chem. Soc.*, **75**, 2166 (1953).

38. J. F. Hyde and R. C. Delong, *J. Am. Chem. Soc.*, **63**, 1194 (1941).

39. A. C. Jenkins and G. F. Chambers, *Ind. Eng. Chem.*, **46**, 2367 (1954).

40. G. Jerusalem, *J. Chem. Soc.*, **97**, 2190 (1910).

41. M. Kakudo and T. Watase, *J. Chem. Phys.*, **21**, 167 (1953).

42. S. W. Kantor, *J. Am. Chem. Soc.*, **75**, 2712 (1953).

43. S. W. Kantor, R. C. Osthoff, and D. T. Hurd, *J. Am. Chem. Soc.*, **77**, 1685 (1955).

44. K. Karasharli, V. Kostryukov, O. Dzhafrov, and G. Geidarov, *Azerb. Khim. Kh.*, **1969**, (1), 32; CA., **71**, 74917s.

45. F. S. Kipping and A. G. Murray, *J. Chem. Soc.*, **130**, 2734 (1927).

46. O. Klejnot, *Inorg. Chem.*, **2**, 825 (1963).

47. W. H. Knoth, Jr., US Patent 2,850,514 (1958).

48. C. A. Kraus and R. Rosen, *J. Am. Chem. Soc.*, **47**, 2739 (1925).

49. W. A. Kriner, *J. Org. Chem.*, **29**, 1601 (1964).

50. S. H. Langer, S. Connell, and I. Wender, *J. Org. Chem.*, **23**, 50 (1958).

51. E. Larsson and L. Bjellerup, *J. Am. Chem. Soc.*, **75**, 995 (1953).

52. E. Larsson and E. van der Pals, *Sven. Kem. Tidskr.*, **63**, 177 (1951); *Chem. Abstr.*, **46**, 2516f (1952).

53. I. A. Lavygin, O. V. Leitan, M. G., Pomerantseva, L. A. Efimova, and A. M. Pribytko, Misc. Symp. Conf. Meet. USSR (1980), 11pp. (32), *Chem. Abstr.*, **97**, 91684t (1982).

54. A. Lazarev and M. Voronkov, *Opt. Spektrosk.*, **1958**, 4, 180, *Chem. Abstr.*, **52**, 7703e (1958).

55. R. N. Lewis, *J. Am. Chem. Soc.*, **69**, 717 (1947).

56. R. N. Lewis, *J. Am. Chem. Soc.*, **70**, 1115 (1948).

57. J. J. McBride, Jr., and H. C. Beachell, *J. Am. Chem. Soc.*, **74**, 5247 (1952).

58. C. A. MacKenzie, A. P. Mills, and J. M. Scott, *J. Am. Chem. Soc.*, **72**, 2032 (1952).

59. G. Martin, *J. Chem. Soc.*, **105**, 2836 (1914).

60. A. P. Mills and W. E. Becker, *J. Phys. Chem.*, **60**, 1644 (1956).

61. M. Momonoi and N. Suzuki, *J. Chem. Soc. Jpn. Pure Chem. Sect.*, **78**, 1324 (1957).

62. M. Nagiev, I. Dzafarov, K. Karasharli, V. Kostryukov, and K. Geidarov, *Dokl. Akad. Nauk SSSR*, **188**, 370 (1969).

63. N. S. Nametkin, A. V. Topchiev, and F. F. Machus, *Dokl. Akad. Nauk SSSR*, **83**, 705 (1952).

64. N. S. Nametkin, A. V. Topchiev, and L. I. Kartasheva, *Dokl. Akad. Nauk SSSR*, **93**, 667 (1953).
65. D. J. Neal and R. D. Blaumanis, *Anal. Chem.*, **32**, 139 (1960).
66. L. Niselson, L. Sokolova, and Y. Golubkov, *Teplofiz. Svoistva Veshchestv Mater.*, **1972**, 128.
67. J. E. Noll, J. L. Speier, and D. F. Daubert, *J. Am. Chem. Soc.*, **73**, 3867 (1951).
68. W. Noll, *Chemistry and Technology of Silicones*, Academic, New York, 1968.
69. R. Okawara, *J. Chem. Soc. Jpn. Ind. Chem. Sect.*, **61**, 690 (1958).
70. C. Pape, *Ber. Dtsch. Chem. Ges.*, **14**, 1872 (1881).
71. A. D. Petrov and E. A. Chernyshev, *Dokl. Akad. Nauk SSSR*, **86**, 737 (1952).
72. A. D. Petrov and G. I. Nikishin, and N. P. Smetankina, *Zh. Obshch. Khim.*, **28**, 2085 (1958).
73. A. D. Petrov, S. I. Sadykh-zade, and I. L. Tsetlin, *Dokl. Akad. Nauk SSSR*, **107**, 99 (1956).
74. A. D. Petrov, N. P. Smetankina, and G. I. Nikishin, *Zh. Obshch. Khim.*, **25**, 2332 (1955).
75. A. Polis, *Ber. Dtsch. Chem. Ges.*, **19**, 1012 (1886).
76. E. G. Rochow, *An Introduction to the Chemistry of the Silicones*, 2nd ed., Wiley, New York, 1951.
77. E. G. Rochow and W. F. Gilliam, *J. Am. Chem. Soc.*, **67**, 1772 (1945).
78. O. Samorukov and V. Kostryukov, *Zh. Fiz. Khim.*, **45**, 1311 (1971).
79. R. O. Sauer, *J. Am. Chem. Soc.*, **66**, 1707 (1944).
80. R. O. Sauer and R. H. Hasek, *J. Am. Chem. Soc.*, **68**, 241 (1946).
81. R. O. Sauer, W. J. Scheiber, and S. D. Brewer, *J. Am. Chem. Soc.*, **68**, 962 (1946).
82. W. C. Schumb and D. F. Holloway, *J. Am. Chem. Soc.*, **63**, 2753 (1941).
83. W. C. Schumb and C. M. Saffer, Jr., *J. Am. Chem. Soc.*, **61**, 363 (1939).
84. H. A. Schuyten, J. W. Weaver, and J. D. Reid, *J. Am. Chem. Soc.*, **69**, 2110 (1947).
85. Z. Sergeeva, S. Tszylan-lan, *Zh. Obshch. Khim.*, **33**, 1874 (1963).
86. L. M. Shorr, *J. Am. Chem. Soc.*, **76**, 1390 (1954).
87. M. F. Schostakovskii and D. A. Kochkin, *Izv. Akad. Nauk SSSR*, Otd. Khim. Nauk, **1954**, 174.
88. M. Shostakovskii, D. Kochkin, I. Shikiev, and V. Vlasov, *Zh. Obshch. Khim.*, **25**, 622 (1955).
89. W. Simmler, quoted in *Chemistry and Technology of Silicones*, W. Noll, Academic, New York, 1968, p. 252.
90. B. Smith, Doctoral Thesis, 1951. Chalmers. Tek. Hogskola, No. 6, 154 pp. (1951); *Chem. Abstr.*, **49**, 909g (1955).
91. T. Sokolova, Y. Golubkov, and L. Niselson, *Teplofiz. Svoistva Veshchestv Mater.*, **1972**, 133.
92. L. H. Sommer, L. Q. Green, and F. G. Whitmore, *J. Am. Chem. Soc.*, **71**, 3253 (1949).

93. L. H. Sommer, E. W. Pietrusza, and F. C. Whitmore, *J. Am. Chem. Soc.*, **68**, 2282 (1946).

94. O. Steward, W. Uhl, and B. Sands, *J. Organometal. Chem.*, **15**, 329 (1968).

95. A Stock and C. Somieski, *Ber. Dtsch. Chem. Ges.*, **52**, 695 (1919).

96. W. Sundermeyer, quoted in *Chemistry and Technology of Silicones*, W. Noll, Academic, New York, 1968, p. 122.

97. T. Takatani, *J. Chem. Soc. Jpn. Pure Chem. Sect.*, **75**, 948 (1953).

98. T. Takiguchi, *J. Chem. Soc. Jpn. Ind. Chem. Sect.*, **62**, 1262 (1959).

99. S. Tannenbaum, S. Kaye, and G. F. Lewenz, *J. Am. Chem. Soc.*, **75**, 3753 (1953).

100. A. V. Topchiev and N. S. Nametkin, *Dokl. Akad. Nauk SSSR*, **80**, 897 (1951).

101. L. J. Tyler, US Patent 2,611,744 (1950).

102. L. J. Tyler, L. H. Sommer, and F. C. Whitmore, *J. Am. Chem. Soc.*, **70**, 2876 (1948).

103. M. G. Voronkov, *Zh. Obshch. Khim.*, **29**, 907 (1959).

104. M. G. Voronkov and B. N. Dolgov, *J. Appl. Chem. USSR*, **24**, 103 (1951).

105. M. G. Voronkov and B. N. Dolgov, *Zh. Obshch. Khim.*, **24**, 1082 (1954).

106. U. Wannagat, H. Burger, and E. Ringel, *Monatsh. Chem.*, **93**, 1363 (1962).

107. C. Young, P. Servais, C. Currie, and M. J. Hunter, *J. Am. Chem. Soc.*, **70**, 1115 (1948).

INDEX